Microlocal Analysis and Spectral Theory

NATO ASI Series

Advanced Science Institutes Series

A Series presenting the results of activities sponsored by the NATO Science Committee, which aims at the dissemination of advanced scientific and technological knowledge, with a view to strengthening links between scientific communities.

The Series is published by an international board of publishers in conjunction with the NATO Scientific Affairs Division

A	**Life Sciences**	Plenum Publishing Corporation
B	**Physics**	London and New York
C	**Mathematical and Physical Sciences**	Kluwer Academic Publishers
D	**Behavioural and Social Sciences**	Dordrecht, Boston and London
E	**Applied Sciences**	
F	**Computer and Systems Sciences**	Springer-Verlag
G	**Ecological Sciences**	Berlin, Heidelberg, New York, London,
H	**Cell Biology**	Paris and Tokyo
I	**Global Environmental Change**	

PARTNERSHIP SUB-SERIES

1.	**Disarmament Technologies**	Kluwer Academic Publishers
2.	**Environment**	Springer-Verlag / Kluwer Academic Publishers
3.	**High Technology**	Kluwer Academic Publishers
4.	**Science and Technology Policy**	Kluwer Academic Publishers
5.	**Computer Networking**	Kluwer Academic Publishers

The Partnership Sub-Series incorporates activities undertaken in collaboration with NATO's Cooperation Partners, the countries of the CIS and Central and Eastern Europe, in Priority Areas of concern to those countries.

NATO-PCO-DATA BASE

The electronic index to the NATO ASI Series provides full bibliographical references (with keywords and/or abstracts) to more than 50000 contributions from international scientists published in all sections of the NATO ASI Series.
Access to the NATO-PCO-DATA BASE is possible in two ways:

– via online FILE 128 (NATO-PCO-DATA BASE) hosted by ESRIN, Via Galileo Galilei, I-00044 Frascati, Italy.

– via CD-ROM "NATO-PCO-DATA BASE" with user-friendly retrieval software in English, French and German (© WTV GmbH and DATAWARE Technologies Inc. 1989).

The CD-ROM can be ordered through any member of the Board of Publishers or through NATO-PCO, Overijse, Belgium.

Series C: Mathematical and Physical Sciences – Vol. 490

Microlocal Analysis and Spectral Theory

edited by

Luigi Rodino
Department of Mathematics,
University of Torino,
Torino, Italy

Springer Science+Business Media, B.V.

Proceedings of the NATO Advanced Study Institute on
Microlocal Analysis and Spectral Theory
Il Ciocco, Castelvecchio Pascoli (Lucca), Italy
23 September – 3 October 1996

A C.I.P. Catalogue record for this book is available from the Library of Congress.

ISBN 978-0-7923-4544-2 ISBN 978-94-011-5626-4 (eBook)
DOI 10.1007/978-94-011-5626-4

Printed on acid-free paper

Contents

Preface

The NATO Advanced Study Institute "Microlocal Analysis and Spectral Theory" was held in Tuscany (Italy) at Castelvecchio Pascoli, in the district of Lucca, hosted by the international vacation center "Il Ciocco", from September 23 to October 3, 1996.

The Institute recorded the considerable progress realized recently in the field of Microlocal Analysis. In a broad sense, Microlocal Analysis is the modern version of the classical Fourier technique in solving partial differential equations, where now the localization proceeding takes place with respect to the dual variables too. Precisely, through the tools of pseudo–differential operators, wave–front sets and Fourier integral operators, the general theory of the linear partial differential equations is now reaching a mature form, in the frame of Schwartz distributions or other generalized functions. At the same time, Microlocal Analysis has grown up into a definite and independent part of Mathematical Analysis, with other applications all around Mathematics and Physics, one major theme being Spectral Theory for Schrödinger equation in Quantum Mechanics.

Concerning general theory of linear PDE, contributions were presented in the following directions:

- discussion of new topics in the Gevrey–analytic category, as propagation of singularities and hypoellipticity in the case of multiple characteristics, higher analytic microlocalization and applications, diffractive boundary value problems, CR manifolds;

- advances in elliptic boundary value problems, in particular geometric invariants associated to them, the case of the manifolds with singularities and the corresponding edge–pseudo–differential calculus;

- new results on lower bounds for pseudo–differential operators with multiple characteristics.

Concerning Spectral Theory, different problems for the Scrödinger equation were discussed, in particular:

- asymptotic behaviour of the eigenvalues in the case of a polynomial potential;

- semiclassical analysis in large dimension and statistical mechanics;
- microlocal tunneling and adiabatic theory;
- asymptotic of resonances.

On the whole, the Institute was able to cover these topics in Microlocal Analysis, which have pre–eminence because of their novelty or importance in the applications, in the field of partial differential equations as well as in other areas of Mathematics and Theoretical Physics.

The Institute was attended by 82 participants: 72 from NATO countries (Belgium: 1, Denmark: 5, France: 17, Germany: 16, Italy: 22, Turkey: 2, U.K.: 1, U.S.A.: 8) and 10 from other countries (Armenia: 1, Bulgaria: 2, India: 1, Romania: 1, Russia: 5).

The lectures were held by 13 lecturers; moreover 24 advanced seminars were organized by the participants and devoted to the discussion of their contribution in the field.

These Proceedings, aiming at a state–of–the–art volume, present a selection of the aforesaid lectures. Lack of space does not allow publishing the texts of the seminars; an excuse to the editor is that they are addressed to more experienced readers, who would not find difficulty to find them in various journals.

We want to express our gratitude to NATO which was the main sponsor of this meeting. Our thanks go to the Scientific Affairs Division and specially to the NATO Science Committee, to Dr. L. Veiga da Cunha, Director of the ASI Programme, and to Barbara and Tilo Kester, of the NATO ASI series Publication Coordination Office.

It is our pleasure to mention other Institutions which supported financially the meeting: Dipartimento di Matematica, Universitá di Torino, covering in part the organization expenses; Politecnico di Torino, host of the home page of this NATO ASI; all the Institutions providing suppport to the travel expenses of the students, among them: National Science Foundation U.S.A., TUBITAK of Turkey, MURST of Italy, Max–Planck Institute of Germany.

On the behalf of all participants, we express our gratitute to Prof. M. Mascarello and Eng. B. Monastero, of the Politecnico di Torino, who performed smoothly and efficiently the work of secretary.

Finally, we wish to express our warmest thanks to Dr. S. Coriasco, Ph.D. student at the Universitá di Torino, who elaborated by computer this volume to its final form; his help has been invaluable for us.

Luigi Rodino
Director of the Institute

LINEAR PARTIAL DIFFERENTIAL EQUATIONS WITH MULTIPLE INVOLUTIVE CHARACTERISTICS

O. LIESS
Universitá di Bologna
Dipartimento di Matematica
Piazza di Porta S. Donato 5
I-40126 Bologna (ITALIA)

AND

L. RODINO
Universitá di Torino
Dipartimento di Matematica
V. C. Alberto 10
I-10129 Torino (ITALIA)

0. Introduction

1. Foreword. In this paper we consider linear partial differential operators with involutive characteristics of high multiplicity in the case when no assumptions of Levi-type is made on lower order terms; for such operators we shall prove results on propagation of singularities, local solvability and hypoellipticity in the frame of analytic, Gevrey and C^∞-classes.

The arguments in the proofs are based mainly on the machinery developed in [13], [15] and are given in the sections following this one. The present introduction, however, corresponds to part of the introductory lecture to the school, given by one of us (L.R.) and meant to help the non-specialized participant to become familiar with some of the basic notions in the general theory of linear partial differential operators.

The notations which we use are standard in the general theory of linear PDE; in particular, given a multi-index $\alpha = (\alpha_1, \ldots, \alpha_n) \in Z^n_+$ we write $D^\alpha = D_1^{\alpha_1} \ldots D_n^{\alpha_n}$, where $D_j = -i\partial/\partial x_j$.

L. Rodino (ed.), Microlocal Analysis and Spectral Theory, 1–38.

Moreover, we expect the reader is familiar with the spaces $C_0^\infty(\Omega)$, $C^\infty(\Omega)$, Ω being an open subset of R^n, $\mathcal{S}(R^n)$ and their topological duals $\mathcal{D}'(\Omega)$, $\mathcal{E}'(\Omega)$ and $\mathcal{S}'(R^n)$. The generic linear partial differential operator P can be written in the form

$$P = \sum_{|\alpha|\leq m} a_\alpha(x) D^\alpha, \tag{0.1}$$

where we assume initially the coefficients $a_\alpha(x)$ be given in $C^\infty(\Omega)$; then $P : C^\infty(\Omega) \to C^\infty(\Omega)$ can be extended to a linear map

$$P : \mathcal{D}'(\Omega) \to \mathcal{D}'(\Omega). \tag{0.2}$$

We call the symbol of P the function in $\Omega \times R^n$

$$p(x,\xi) = \sum_{|\alpha|\leq m} a_\alpha(x)\xi^\alpha. \tag{0.3}$$

The principal symbol of P is

$$p_m(x,\xi) = \sum_{|\alpha|=m} a_\alpha(x)\xi^\alpha, \tag{0.4}$$

homogeneous function of order m with respect to the dual variables ξ. The characteristic manifold of P is the subset of $\Omega \times R^n$ given by

$$\Sigma = \{(x,\xi) \mid p_m(x,\xi) = 0, \xi \neq 0\}. \tag{0.5}$$

An operator P is said to be elliptic in Ω if $\Sigma = \emptyset$, i.e. $p_m(x,\xi) \neq 0$ for all $x \in \Omega$ and all $\xi \neq 0$, whereas P is said to be of principal type if the characteristic manifold Σ is non-empty and

$$\forall (x,\xi) \in \Sigma : d_{x,\xi} p_m(x,\xi) \neq 0. \tag{0.6}$$

The theory of elliptic operators and operators of the principal type is nowadays well developed; in this paper we shall consider operators with multiple characteristics, i.e. satisfying

$$\exists (x,\xi) \in \Sigma : d_{x,\xi} p_m(x,\xi) = 0, \tag{0.7}$$

where we understand Σ is non-empty.

When dealing with multiple characteristics operators, it is convenient to enlarge the Schwartz frame (0.2). Precisely, let us first introduce the Gevrey space $G^s(\Omega), 1 \leq s < \infty$, consisting of all $f \in C^\infty(\Omega)$ satisfying in every $K \subset\subset \Omega$

$$\sup_{x\in K}|D^\alpha f(x)| \leq C^{|\alpha|+1}(\alpha!)^s, \tag{0.8}$$

with a constant C independent of α. Observe that $G^1(\Omega)$ is the class $\mathcal{A}(\Omega)$ of all the analytic functions in Ω. When $s > 1$, we write $G_0^s(\Omega)$ for $G^s(\Omega)\cap C_0^\infty(\Omega)$.

The spaces of s–ultradistributions $\mathcal{D}'_s(\Omega)$, $\mathcal{E}'_s(\Omega)$ are the duals of $G_0^s(\Omega)$, $G^s(\Omega)$. We have the inclusions $\mathcal{D}'(\Omega) \subset \mathcal{D}'_t(\Omega) \subset \mathcal{D}'_s(\Omega)$ for $s < t$. To have uniform notations, we shall also write $G^\infty(\Omega)), \mathcal{D}'_\infty(\Omega)$ for $C^\infty(\Omega), \mathcal{D}'(\Omega)$, and $\mathcal{D}_1(\Omega)$ for the space of generalized functions of Sato, including all the preceding classes.

From now on, we shall assume that the coefficients $a_\alpha(x)$ of P in (0.1) belong to $\mathcal{A}(\Omega)$; then for $1 \leq s \leq \infty$

$$P : G^s(\Omega) \to G^s(\Omega) \tag{0.9}$$

extends to a map

$$P : \mathcal{D}'_s(\Omega) \to \mathcal{D}'_s(\Omega) \tag{0.10}$$

that gives the expected more general frame to the study of the equation $Pu = f$.

2. Pseudo–differential operators. In the analysis of (0.10) one is led in a natural way to consider a larger class of operators, namely the pseudo–differential operators, which are defined in a broad sense by

$$Pu(x) = p(x, D)\, u(x) = (2\pi)^{-n} \int e^{ix\xi} p(x,\xi)\hat{u}(\xi)d\xi, \tag{0.11}$$

where $\hat{u} = \mathcal{F}(u)$ is the Fourier transform of u and $p(x,\xi)$ is the so–called symbol of P. Homogeneous classical symbols are functions in $C^\infty(\Omega \times R^n)$ admitting an asymptotic expansion

$$p(x,\xi) \sim \sum_{j=0}^{\infty} p_{m-j}(x,\xi), \tag{0.12}$$

where $m \in R$ and $p_{m-j}(x,\xi)$ is positively homogeneous of degree $m-j$ with respect to ξ. We may obviously refer the preceding definitions (0.5), (0.6), (0.7) to the leading term $p_m(x,\xi)$ in (0.12). Of course, a linear partial differential operator is a pseudo–differential operator, with symbol given by (0.3). We have from (0.11), (0.12)

$$P = p(x, D) : C_0^\infty(\Omega) \to C^\infty(\Omega) \tag{0.13}$$

with extension

$$P = p(x, D) : \mathcal{E}'(\Omega) \to \mathcal{D}'(\Omega). \tag{0.14}$$

As generalization of the linear partial differential operators with analytic coefficients, we may consider (homogeneous classical) analytic pseudo–differential operators. Their symbols can be expanded as in (0.12), with $p_{m-j}(x, \xi)$ admitting holomorphic extension in a complex neighborhood of $\Omega \times R^n$, satisfying there suitable uniform estimates. Beside (0.13), (0.14), we have in this case for $1 < s < \infty$

$$P = p(x, D) \quad : \quad G_0^s(\Omega) \to G^s(\Omega) \tag{0.15}$$

$$P = p(x, D) \quad : \quad \mathcal{E}_s'(\Omega) \to \mathcal{D}_s'(\Omega) \tag{0.16}$$

with continuous action also on the Sato space $\mathcal{D}_1'(\Omega)$.

3. Wave front sets and microfunctions. Basic ingredients of the microlocal analysis are the wave front sets, defined in the following way. Let us begin with the Gevrey case.

Definition 0.1 *Fix $(x^0, \xi^0) \in \Omega \times \dot{R}^n$. For $u \in \mathcal{D}_s'(\Omega), 1 < s < \infty$, we say that $(x^0, \xi^0) \notin WF^s u$ if there exist $\varphi \in G^s(\Omega)$ with $\varphi(x) = 1$ in a neighborhood of x_0, and positive constants C, ϵ such that*

$$|\mathcal{F}(\varphi u)(\xi)| \leq C \exp\left(-\epsilon |\xi|^{1/s}\right) \tag{0.17}$$

for all ξ in a conic neighborhood of ξ^0.

The projection on Ω of the s–wave front set $WF^s u$ is the s–singular support of u, defined as the complement of the largest open subset of Ω where u is of the class G^s.

The ∞–wave front set $WF^\infty u$ of a Schwartz distribution u is defined in the same way, by replacing (0.17) with

$$|\mathcal{F}(\varphi u)(\xi)| \leq C_M (1 + |\xi|)^{-M} \tag{0.18}$$

for arbitrary $M = 1, 2, \ldots$, where now we allow $\varphi \in C_0^\infty(\Omega)$. In the analytic case test–functions with compact support do not exist; we may however say that $(x^0, \xi^0) \notin WF^1 u$ if there exists a sequence $u_N, N = 1, 2, \ldots$, with compact support, and $u_N = u$ in a neighborhood of x^0, such that

$$|\hat{u}_N(\xi)| \leq C^{N+1} N^N (1 + |\xi|)^{-N}, \quad N = 1, 2, \ldots, \tag{0.19}$$

for all ξ in a neighborhood of ξ^0. Projections of $WF^\infty u$ and $WF^1 u$ in Ω are the respective singular supports.

Next step is to consider microfunctions. Precisely, let Λ be an open subset of $\Omega \times \dot{R}^n$ conic with respect to ξ, with relatively compact projection on Ω. For $u, v \in \mathcal{D}'_s(\Omega)$ we shall write $u \sim v$ to mean that $\Lambda \cap WF^s(u-v) = \emptyset$, and we shall denote by $\mathcal{M}^s(\Lambda)$ the factor space $\mathcal{D}'_s(\Omega)/\sim$, $1 \le s \le \infty$. The s–wave front set $WF^s u$ of a microfunction $u \in \mathcal{M}^s(\Lambda)$ is a well defined conic closed subset of Λ. If P is an analytic pseudo–differential operator, then

$$\forall u \in \mathcal{E}'_s(\Omega),\ 1 \le s \le \infty\ :\ WF^s Pu \subset WF^s u, \tag{0.20}$$

and therefore, by factorization

$$P : \mathcal{M}^s(\Lambda) \to \mathcal{M}^s(\Lambda),\ 1 \le s \le \infty, \tag{0.21}$$

the inclusion (0.20) keeping valid for all $u \in \mathcal{M}^s(\Lambda)$. As for (0.21), we may actually assume that $p(x,\xi)$, symbol of P, is defined only in a conic neighborhood of Λ. The theorems of the next sections will be stated in the frame of the micro–operators (0.21). Note that results in the more standard setting (0.10) are hence easily deduced, by covering $\Omega \times \dot{R}^n$ with conic sets Λ and projecting $WF^s u$ into s–singsupp$u$. So, for example, if P is s–micro–hypoelliptic in the sense that $WF^s u = WF^s Pu$ for all $u \in \mathcal{M}^s(\Lambda)$ in any Λ, then P is s-hypoelliptic in the standard sense, i.e. s–singsupp$u$ = s–singsuppPu for all $u \in \mathcal{E}'_s(\Omega)$.

4. Fourier integral operators. The setting (0.21) has at least two advantages, with respect to the standard local point of view. First, the classical theorem of regularity for the solutions of the elliptic equations can be refined by means of the formula

$$WF^s u \subset WF^s Pu \cup \Sigma, \tag{0.22}$$

where $u \in \mathcal{E}'_s(\Omega)$ or $u \in \mathcal{M}^s(\Lambda)$, $1 \le s \le \infty$, and Σ is the characteristic manifold in (0.5). So, if we are concerned with the singularities of the solutions u when $f = Pu$ is smooth, it will be actually sufficient to study (0.21) in an arbitrarily small neighborhood Λ of the points $(x^0, \xi^0) \in \Sigma$.

Moreover, the machinery of the Fourier integral operators may lead to relevant simplifications in the study of the micro–operator $P = p(x, D)$ in (0.21). Precisely, let χ be a homogeneous analytic canonical transformation acting from the conic neighborhood Λ of the point $\rho_0 = (x^0, \xi^0)$ to a conic neighborhood Γ of the point $\chi(\rho_0) = (y^0, \eta^0)$; that χ is canonical means that it preserves the symplectic two–form $\sigma = \sum_{j=1}^n d\xi_j \wedge dx_j$. Then we may consider the Fourier integral operator F with phase function corresponding to χ; this is a map $F : \mathcal{M}^s(\Lambda) \to \mathcal{M}^s(\Gamma)$, $1 \leq s \leq \infty$, with inverse $F^{-1} : \mathcal{M}^s(\Gamma) \to \mathcal{M}^s(\Lambda)$, such that

$$WF^s(Fu) = \chi(WF^s u), \;\; WF^s(F^{-1}v) = \chi^{-1}(WF^s v). \tag{0.23}$$

Moreover

$$\tilde{P} = FPF^{-1} : \mathcal{M}^s(\Gamma) \to \mathcal{M}^s(\Gamma) \tag{0.24}$$

is a (micro) pseudo–differential operator, with homogeneous classical analytic symbol $\tilde{p}(y,\eta)$ having principal part

$$\tilde{p}_m(y,\eta) = p_m(\chi^{-1}(y,\eta)). \tag{0.25}$$

In particular, if we assume $\rho_0 \in \Sigma$ and denote by $\tilde{\Sigma}$ the characteristic manifold of $\tilde{P}$, then $\chi(\rho_0) \in \tilde{\Sigma}$ and $\tilde{\Sigma} = \chi(\Sigma)$ in Γ.

In this way, by fixing a suitable canonical transformation χ, we may reduce ourselves to the study of operators $\tilde{P}$ of a truly elementary form. Additional simplifications in the expression of $\tilde{P}$ can be obtained by means of composition with elliptic pseudo–differential factors.

5. Other geometric invariants. In the following of the paper we shall suppose that Σ is an analytic regular manifold of codimension $n' \geq 1$ (more restrictively, Σ will be assumed involutive, cf. the next definition 1.1) and $p_m(x,\xi)$ vanishes exactly to the order $k \geq 2$ on Σ, i.e. there exists a constant $C > 0$ such that

$$C^{-1} d_\Sigma(x,\xi)^k \leq |p_m(x,\xi/|\xi|)| \leq C d_\Sigma(x,\xi)^k, \;\; (x,\xi) \in \Lambda, \tag{0.26}$$

where $d_\Sigma(x,\xi)$ is the distance from $(x,\xi/|\xi|)$ to Σ.

We have already observed that $p_m(x,\xi)$ has a geometric invariant meaning, i.e. (0.25) is valid after conjugation by Fourier integral operators. Consequently, the assumption (0.26) is also invariant. It is interesting, in the case of the multiple characteristics, to relate the second term in the asymptotic expansion

$$\tilde{p}(y,\eta) \sim \tilde{p}_m(y,\eta) + \tilde{p}_{m-1}(y,\eta) + \ldots \tag{0.27}$$

to the expression of $p_m(x,\xi)$ and $p_{m-1}(x,\xi)$. To this end, define

$$p'_{m-1}(x,\xi) = p_{m-1}(x,\xi) - \frac{1}{2i}\sum_{j=1}^{n} \frac{\partial^2}{\partial x_j \partial \xi_j} p_m(x,\xi), \tag{0.28}$$

which also has an invariant meaning, if we limit ourselves to the points (x,ξ) where $p_m(x,\xi)$ vanishes at least to the order 2. In fact, if we consider $\tilde{p}'_{m-1}(y,\eta)$ from (0.27), we may recognize that $\tilde{p}'_{m-1}(\chi(x,\xi)) - p'_{m-1}(x,\xi) = \mathcal{X}_0 p_m(x,\xi)$, where $\mathcal{X}_0$ is a vector field on Γ with homogeneous analytic coefficients depending on χ; therefore $\tilde{p}'_{m-1}$ and p'_{m-1} coincide at the double characteristics set. In the present case, under the assumption (0.26) with $k \geq 2$, we shall write

$$I_0(x,\xi) = p'_{m-1}(x,\xi)\big|_{\Sigma} \tag{0.29}$$

and call it sub–principal symbol of $P = p(x,D)$, as standard.

Consider now $X_1,\ldots,X_J \in T_\rho(\Lambda), \rho \in \Sigma$, and let $\mathcal{X}_j, j = 1,\ldots,J$, be analytic vector fields on Λ such that $\mathcal{X}_j(\rho) = X_j$. We define

$$\Phi(p_m(x,\xi),\rho,X_1,\ldots,X_J) = (\mathcal{X}_1\ldots\mathcal{X}_J p_m)(\rho) \tag{0.30}$$

which is also invariant, in the sense that

$$\Phi(p_m(x,\xi),\rho,X_1,\ldots,X_J) = \Phi(\tilde{p}_m(y,\eta),\chi(\rho),d_{\chi(\rho)}X_1,\ldots,d_{\chi(\rho)}X_J) \tag{0.31}$$

It is clear that such invariance fails if in (0.30) we replace p_m with p'_{m-1}. In fact, the derivatives of $\mathcal{X}_0 p_m$ do not vanish in general, even if evaluated at the double characteristics set.

However, if we assume $p_m(x,\xi)$ has a zero of order at least $k \geq 3$ at ρ, then if $J \leq k-2$

$$\begin{aligned} &\Phi(\tilde{p}'_{m-1}(y,\eta),\chi(\rho),d_{\chi(\rho)}X_1,\ldots,d_{\chi(\rho)}X_J) = \\ = \ &(d_\chi \circ \mathcal{X}_1\ldots d_\chi \circ \mathcal{X}_J \tilde{p}'_{m-1})(\chi(\rho)) \\ = \ &(\mathcal{X}_1\ldots\mathcal{X}_J p'_{m-1})(\rho) + (\mathcal{X}_1\ldots\mathcal{X}_J\mathcal{X}_0 p_m)(\rho) \\ = \ &\Phi(p'_{m-1}(x,\xi),\rho,X_1,\ldots,X_J) \end{aligned}$$

since $\mathcal{X}_1\ldots\mathcal{X}_J\mathcal{X}_0 p_m$ vanishes at ρ under our assumptions. Summing up, under our hypothesis (0.26) the following functions have an invariant meaning:

$$\Phi(p_m,\rho,X_1,\ldots,X_J) \ : \ (T_\rho(\Lambda))^J \to C, \quad \rho \in \Sigma; \tag{0.32}$$

$$\Phi(p'_{m-1},\rho,X_1,\ldots,X_J) \ : \ (T_\rho(\Lambda))^J \to C, \quad \rho \in \Sigma, J \leq k-2. \tag{0.33}$$

We may go further, considering $N(\Sigma) = T(\Lambda)/T(\Sigma)$, the normal bundle of Σ; for every $(\rho,X) \in N(\Sigma)$, with $\rho \in \Sigma, X \in T_\rho(\Lambda)/T_\rho(\Sigma)$ we take Y in the equivalence class of X and define

$$\Pi(\rho, X) = \frac{1}{k!}\Phi(p_m, \rho, \underbrace{Y, \ldots, Y}_{\text{k times}}) : N(\Sigma) \to C \tag{0.34}$$

$$I_j(\rho, X) = \frac{1}{j!}\Phi(p'_{m-1}, \rho, \underbrace{Y, \ldots, Y}_{\text{j times}}) : N(\Sigma) \to C, 1 \leq j \leq k-2 \tag{0.35}$$

In view of (0.26), we have $\Pi(\rho, X) \neq 0$ for $X \neq 0$. We reserve the notation $I_0(\rho)$ for the sub-principal symbol, according to (0.29).

Our results, in the next sections, will be formulated in terms of the invariants $p_m, \Pi, I_0, I_j, 1 \leq j \leq k-2$; the evident advantage will be that in the proofs we shall be free to change symplectic coordinates.

1. A result on propagation of singularities

1. Statements. We shall first give a result on propagation of singularities in the case when the characteristic manifold Σ is regular involutive of arbitrary codimension n', $1 \leq n' < n$. We recall the following definition:

Definition 1.1. *A homogeneous submanifold $\Sigma \subset T^*R^n$ is said to be regular involutive if:*

i) for every $\rho \in \Sigma$ the restriction of the two-form $\sigma = \sum_{j=1}^n d\xi_j \wedge dx_j$ to $T_\rho(\Sigma)^\perp$ is identically zero
and

ii) the radial vector field $\sum_{j=1}^n \xi_j \partial/\partial\xi_j$ does not belong to $T_\rho(\Sigma)^\perp$ for any $\rho \in \Sigma$.

($T_\rho(\Sigma)^\perp$ denotes the orthogonal complement of $T_\rho(\Sigma)$ with respect to σ.)

It is standard to observe that locally on conic sets a homogeneous analytic submanifold is regular involutive if and only it is of form $\Sigma = \{u_1(x,\xi) = \cdots = u_{n'}(x,\xi) = 0\}$ where the functions $u_j, j = 1, \ldots, n'$ are analytic real-valued in Λ, homogeneous of order 1 in ξ, the forms $du_1, \ldots, du_{n'}$ and $\sum_{j=1}^n \xi_j dx_j$ are linearly independent and

$$\{u_j, u_k\} = 0 \text{ on } \Sigma, j, k = 1, \ldots, n'. \tag{1.1}$$

(Here we have written, as is standard, $\{f, g\} = \sum_{j=1}^n (\partial_{x_j} f \partial_{x_j} g - \partial_{x_j} g \partial_{x_j} f) = H_f g$.)

This definition has the following important consequence: in view of (1.1), the vector fields H_{u_j} are tangent to Σ, and their restriction to Σ satisfies the integrability condition of Frobenius, i.e., we have $[H_{u_j}, H_{u_i}] = H_{\{u_j,u_i\}}$ and since $\{u_j, u_i\}$ is a linear combination with analytic coefficients of the u_k, then $[H_{u_j}, H_{u_i}]$ is, on Σ , a linear combination with analytic coefficients of the H_{u_k}. The manifold Σ can be therefore endowed with a canonical foliation $\mathcal{F}$, whose leaves L are the integral manifolds of the vector fields H_{u_j}, $j = 1, \dots, n'$. In the case of codimension 1, these leaves are of course just the bicharacteristic strips.

For $\rho \in \Sigma$, we shall denote by L_ρ the leaf through ρ.

Our condition

$$C^{-1} d_\Sigma(x,\xi)^k \le |p_m(x,\xi/|\xi|)| \le C d_\Sigma(x,\xi)^k \tag{1.2}$$

implies that we may write in Λ

$$p_m(x,\xi) = \sum_{|\gamma|=k} a_\gamma(x,\xi) u^\gamma(x,\xi), \tag{1.3}$$

for some analytic symbols $a_\gamma(x,\xi)$ of order $m-k$, where the symbols $u_j(x,\xi)$ are defined as above. Note that one may assume without loss of generality that (1.1) are satisfied in the whole Λ, what we shall suppose in the following.

Consider now the function $\Pi(\rho, X) : N(\Sigma) \to C$ defined in (0.34); in terms of (1.3), we have

$$\Pi(\rho, X) = \sum_{|\gamma|=k} a_\gamma(\rho) du^\gamma(X) \tag{1.4}$$

Note that the map $N(\Sigma) \to \left[T(\Sigma)^\perp\right]^\star$ defined by $(\rho, X) \to i_X\sigma$ is an isomorphism. Moreover we can identify $\left[T(\Sigma)^\perp\right]^\star$ with the cotangent space to the leaves of the foliation, $\cup_{L\in\mathcal{F}} T^\star(L)$. Therefore, $\Pi(\rho, X)$ defines for each leaf a function

$$\Pi_L : T^\star(L)/0 \to C \tag{1.5}$$

which in terms of (1.3) can be seen as the principal symbol of the operator on L:

$$P_L = (-i)^{|\gamma|} \sum_{|\gamma|=k} a_\gamma(x,\xi) H^\gamma_{u(x,\xi)} \tag{1.6}$$

We now begin by recalling some known results.

Theorem 1.2 *Let $p_m(x,\xi)$ satisfy (1.2), and Σ be regular involutive. Consider $\rho_0 \in \Sigma$ and write L_{ρ_0} for the leaf through ρ_0. For $1 \leq s < k/(k-1)$ we have: if $u \in \mathcal{M}^s(\Lambda)$ with $Pu = 0$, then $\rho_0 \in WF^s u$ implies $L_{\rho_0} \subset WF^s u$.*

The theorem was proved by Bony–Shapira [3] for $s = 1$, for $1 < s < k/(k-1)$ by Kessab [9]. The study of P for $k/(k-1) \leq s \leq \infty$ involves the sub–principal symbol I_0. The following result of hypoellipticity was proved in Liess–Rodino [14].

Theorem 1.3 *Let $p_m(x,\xi)$ satisfy (1.2), and Σ be regular involutive. Assume*

$$\Pi(\rho, X) + I_0(\rho) \neq 0 \tag{1.7}$$

for all $(\rho, X) \in N(\Sigma)$. Then for $k/(k-1) \leq s \leq \infty$:

i) P is s–micro–hypoelliptic, i.e. $WF^s Pu = WF^s u$ for all $u \in \mathcal{M}^s(\Lambda)$;

ii) For every $v \in \mathcal{M}^s(\Lambda)$ there exists $u \in \mathcal{M}^s(\Lambda)$ such that $Pu = v$.

We want to give results for $s \geq k/(k-1)$ in the case when the invariant in (1.7) vanishes. We shall limit ourselves to a special case, precisely we assume that, possibly after multiplying P by an elliptic factor:

$$p_m(x,\xi) \geq 0 \qquad \text{for } (x,\xi) \in \Lambda; \tag{1.8}$$

$$I_0(x,\xi) < 0 \qquad \text{for } (x,\xi) \in \Sigma; \tag{1.9}$$

$$I_j(\rho, X) \quad \text{is real valued on } N(\Sigma) \text{ for } 0 < j \leq k-2 \tag{1.10}$$

where I_j is defined as in (0.35). An equivalent expression of (1.10), whose invariant meaning is less transparent, is the following, in terms of $p'_{m-1}(x,\xi)$ from (0.28):

$$\partial_x^\alpha \partial_\xi^\beta p'_{m-1}(x,\xi) \text{ is real valued on } \Sigma, \text{ for } 0 < |\alpha+\beta| \leq k-2 \tag{1.11}$$

Of course the conditions (1.10), (1.11) are empty if $k = 2$.

Once (1.8), (1.9), (1.10) are assumed to be satisfied, we can multiply by a positive elliptic factor and suppose without further loss of generality

$$I_0(x,\xi) \text{ has constant value along each leaf } L \in \mathcal{F} \tag{1.12}$$

For example, multiplying by $q(x,D)$ with $q(x,\xi) > 0$, $q(x,\xi) = -1/I_0(x,\xi)$ on Σ, we get as new sub–principal symbol the constant -1. The setting (1.12) simplifies the following statement.

For $\rho \in \Sigma$, let us denote by $\mathcal{H}_{L_\rho}$ the Hamiltonian vector field of Π_{L_ρ} in (1.5). The related integral curves run in $T^\star(L_\rho)$. Let us call geodesics through ρ the projections on L_ρ of such curves; it will be easily seen from (1.2) that no one of the geodesics reduces to the point ρ.

Theorem 1.4 *Let $p_m(x,\xi)$ satisfy (1.2), and Σ be regular involutive. Assume further (1.8),(1.9),(1.10),(1.12). For $k/(k-1) \leq s \leq \infty, \rho_0 \in \Sigma$ we have: if $u \in \mathcal{M}^s(\Lambda)$ with $Pu = 0$ and $\rho_0 \in WF^s u$, then one at least of the geodesics through ρ_0 is included in $WF^s u$.*

For $k = 2$, the result was proved by Boutet de Monvel [4] in the case $s = \infty$, and stated by Lascar [10] in the case $2 < s < \infty$. As a model for theorem 1.4, consider in $R^n = R^{n'} \times R$, $x = (y,t), \xi = (\eta, \tau)$:

$$P = \sum_{|\gamma|=m} c_\gamma D_y^\gamma - D_t^{m-1} \tag{1.13}$$

where

$$\sum_{|\gamma|=m} c_\gamma \eta^\gamma \geq c|\eta|^m \tag{1.14}$$

for some $c > 0$. The geodesics are straight lines in the leaves $t = t_0, \eta = 0, \tau = \tau_0 > 0$.

The first step in the proof of theorem 1.4 is a standard application of Fourier integral operators with analytic phase and amplitude functions, cf. Liess–Rodino [14],[15], Rodino [24]. Precisely, we may define a canonical map on Λ, such that the first n' new coords are given by $u_1, \ldots, u_{n'}$. By conjugation with the corresponding Fourier integral operators, we are reduced to consider an operator with principal symbol

$$p_m(x,\xi) = \sum_{|\gamma|=k} a_\gamma(x,\xi)\xi'^\gamma \tag{1.15}$$

where $\xi = (\xi', \xi''), \xi' \in R^{n'}, \xi'' \in R^{n''}, n' + n'' = n$, and similarly we split $x = (x', x'')$. The symbols $a_\gamma(x,\xi)$ are analytic homogeneous of order $m-k$, defined in a conic neighborhood Γ of $\rho_0 = (x^0, \xi^0)$ with $\xi^{0'} = 0$. Multiplying by an elliptic factor, we may suppose $m = k$. The assumption (1.8) is now that all the a_γ are real valued, and for some $C > 0$ we have

$$C^{-1}|\xi'|^m \leq \sum_{|\gamma|=m} a_\gamma(x,\xi)\xi'^\gamma \leq C|\xi'|^m \text{ for } (x,\xi) \in \Gamma, \tag{1.16}$$

which implies m is even.

Writing

$$p_{m,0}(x,\xi) = \sum_{|\gamma|=m} a_\gamma(x,0,\xi'')\,\xi'^\gamma, \tag{1.17}$$

we also have

$$C^{-1}|\xi'|^m \le p_{m,0}(x,\xi) \le C|\xi'|^m \text{ for } (x,\xi) \in \Gamma. \tag{1.18}$$

The characteristic manifold Σ is now given by $\{\xi' = 0\}$ and the leaf of the foliation L through ρ_0 is $\{x'' = x^{0''}, \xi' = 0, \xi'' = \xi^{0''}\}$. From (0.28) we have $p'_{m-1} = p_{m-1}$; the sub-principal symbol I_0 reduces then to

$$p_{m-1,0}(x,\xi'') = p_{m-1}(x,0,\xi''), \tag{1.19}$$

and (1.9), (1.10) mean that

$$-C|\xi''|^{m-1} \le p_{m-1,0}(x,\xi'') \le -C^{-1}|\xi''|^{m-1}, \tag{1.20}$$

$$p_{m-1,j}(x,\xi'') = \sum_{|\gamma|=j} (\partial^\gamma_{\xi'} p_{m-1})(x,0,\xi'')\frac{\xi'^\gamma}{\gamma!} \text{ is real valued for } 0 < j \le k-2. \tag{1.21}$$

As for (1.12), here we shall not require it is satisfied. The function Π_L in (1.5) is given, in local coordinates, by

$$\Pi(x',\xi') = p_{m,0}(x',x^{0''},\xi',\xi^{0''}) = \sum_{|\gamma|=m} a_\gamma(x',x^{0''},0,\xi^{0''})\xi'^\gamma. \tag{1.22}$$

Consider the solutions $(x'(t),\xi'(t))$ of the system

$$\begin{cases} \dot{x}'_j &= \partial_{\xi'_j}\Pi(x',\xi') \\ \dot{\xi}'_j &= -\partial_{x'_j}\Pi(x',\xi') - \partial_{x'_j} p_{m-1,0}(x',x^{0''},\xi^{0''}) \\ x'(0) &= x^{0'}, \xi'(0) = \eta' \end{cases} \tag{1.23}$$

with $\eta' \in R^{n'}, \eta' \ne 0, p_{m,0}(x^0,\eta',\xi^{0''}) + p_{m-1,0}(x^0,\xi^{0''}) = 0$. In the case when $p_{m-1,0}$ does not depend on x', as we assume in (1.12), the projections $x' = x'(t)$ of the solutions are the geodesics through x'_0 on the leaf $\{x'' = x^{0''}, \xi' = 0, \xi'' = \xi^{0''}\}$. They never reduce to $x' = x^{0'}$, since $d_{\xi'}\Pi(x',\xi') \ne 0$ for $\xi' \ne 0$, in view of the ellipticity of $\Pi(x',\xi')$ from (1.18) and the Euler identity. The following result gives then theorem 1.4.

Theorem 1.5. *Let $p(x,\xi) = p_m(x,\xi)+p_{m-1}(x,\xi)+\cdots$ satisfy the preceding assumptions. In particular we assume that (1.18), (1.20), (1.21) hold in a conic neighborhood Γ of $\rho_0 = (x^0,\xi^0)$ with $\xi^{0\prime} = 0$. Let $m/(m-1) \le s \le \infty$. Shrinking Γ if necessary, we can then conclude that if $u \in \mathcal{M}^s(\Gamma)$ satisfies $Pu = 0$, and if on every curve $\gamma(t) = (x'(t), x^{0\prime\prime}; 0, \xi^{0\prime\prime})$ there exists $\tilde\rho \in \gamma \cap \Gamma$ with $\tilde\rho \notin WF^s u$, then $\rho_0 \notin WF^s u$. (By $x'(t)$ we mean the first n' components of a solution of (1.23), for η' running through $\dot R^{n'}$, and satisfying the condition $p_{m,0}(x^0,\eta',\xi^{0\prime\prime}) + p_{m-1}(x^0,\xi^{0\prime\prime}) = 0$.)*

Theorem 1.5 will be proved in the sections 2, 3, 4, 5. For brevity we shall limit ourselves to argue in the Gevrey case $s < \infty$, taking u a (Schwartz) distribution. The extension to the case $s = \infty$ and to Gevrey (ultra-) microfunctions do not present any additional difficulties. In the remaining part of this section, we fix some notations and terminology to be used in the proofs. In all what follows $k \ge 2$ is a fixed integer as in the statement of theorem 1.5. We assume henceforth that $k = m$ and denote $(k-1)/k$ by δ.

2. Further notations and preliminary properties.

Definition 1.6. *a) Let $R^n_\xi = R^{n'}_{\xi'} \times R^{n''}_{\xi''}$. $A \subset R^n_x \times R^n_\xi$, is called "quasihomogeneous" if $(x,\xi) \in A$ implies $(x, \nu\xi', \nu^{k/(k-1)}\xi'') \in A$ for any $\nu > 0$.*

b) $B \subset R^n_x \times R^n_\xi$ is called a quasihomogeneous neighborhood of A, if B contains an open quasihomogeneous set $\tilde B$ which contains A.

c) A function $(x,\xi) \to f(x,\xi)$ is called σ-quasihomogeneous (or quasihomogeneous of degree σ) if $f(x,\nu\xi',\nu^{k/(k-1)}\xi'') = \nu^\sigma f(x,\xi)$, $\forall \nu > 0$.

Actually, we should call functions as in c) of the preceding definition $(1, k/(k-1))$- quasihomogeneous of degree σ, but since k will be kept fixed throughout this paper, we omit explicit reference to k. An example of a quasihomogeneous function of degree 1 is for $k = 4$, $n = 3$, $n' = 2$, $f(x,\xi) = (\xi_3^3 - \xi_2^4)^{1/4}$ in the region $\xi_3^3 - \xi_2^4 \ge 0$. Interesting quasihomogeneous sets in the context of this paper are sets of form $U \times \{\xi \in R^n; \xi'' \in G, |\xi'| < c|\xi''|^\delta\}$, where $U \subset R^n$, G is a cone in $R^{n''}$ and $c > 0$. In principle we are interested to work here with c large. Since we will want to remain in a conic neighborhood of $R^{n'}_{\xi'}$, this will make sense only if we restrict our attention to sets of form $|\xi| \ge c'$.

There is another way to take advantage of the product structure of R^n_ξ, when we write it as $R^{n'}_{\xi'} \times R^{n''}_{\xi''}$. This is based on the following definition:

Definition 1.7. *Consider $x^0 \in R^n$, $\xi^0 \in \dot R^n$ with $\xi^{0\prime} = 0$, and $\xi'^1 \in \dot R^{n'}$. We say that V is a bineighborhood of (x^0, ξ'^1, ξ^0) if it contains a set of form $U \times \{\xi \in R^n; \xi \in G, \xi' \in G'\}$ where U is a neighborhood of x^0 in R^n, G is* *an open cone in $\dot{R}^n$ which contains ξ^0 and G' is an open cone in $\dot{R}^{n'}$ which contains ξ'^1.*

Bineighborhoods are objects of second microlocalization. It is perhaps illuminating to look at second microlocalization as an instance of general high order microlocalization: see e.g. [13]. It should then be said that definitions are chosen to be more intuitive here than are the corresponding definitions in [13]. The reason why we could stick to simpler definitions in this paper is of course that we do not need to microlocalize further. (In [13] the definitions were formulated in order to allow for rather high-order microlocalization.)

As is often the case in microlocalization, we need to distinguish for some given point (x, ξ) between the points $(\tilde{x}, \tilde{\xi})$ which are "close" to (x, ξ) in $R^n \times \dot{R}^n$ and points $(\tilde{x}, \tilde{\xi})$ which we consider "far" from (x, ξ). Sophisticated metrics to measure distances are sometimes considered in the literature. Here we will work with a rather simple metric which we denote by "$\mathrm{dist}_\sim$".

Definition 1.8. *We denote by $dist_\sim((x, \xi), (\tilde{x}, \tilde{\xi}))$ the expression*

$$dist_\sim((x, \xi), (\tilde{x}, \tilde{\xi})) = |x - \tilde{x}| + |\xi - \tilde{\xi}|/(|\xi|^\delta + |\tilde{\xi}|^\delta).$$

To explain the main idea of the proof of theorem 1.5, it is useful to introduce the following generalization of WF^s.

Definition 1.9. *Let $U \subset R^n$ be open, consider $u \in \mathcal{D}'(U)$ and $\Lambda \subset U \times \dot{R}^n$. Also consider $s > 1$ with $1/s \leq \delta < 1$. We write*

$$\Lambda \cap WF^s_\sim u = \emptyset \qquad (1.24)$$

if for any compact set $K \subset U$ there is $\varepsilon > 0$, so that if $\rho \in G^s_0(x \in R^n, |x - x^0| < \varepsilon)$, then there are $c > 0$, $c' > 0$, $c'' > 0$, for which

$$|\mathcal{F}_{x \to \xi}(\rho(x - x^0)u(x))(\xi)| \leq c \exp[-c'|\xi|^{1/s}] \text{ for } |\xi - \xi^0| < c''|\xi^0|^\delta,$$

provided $x^0 \in K$, $(x^0, \xi^0) \in \Lambda$.

One of the main features of this definition is that the "fibers" $A_x = \{\xi; (x, \xi) \in \Lambda\}$ are not necessarily conic, and indeed, a first idea which comes to mind in the situation from theorem 1.5 is to model them on the quasi-homogeneous structure of R^n. Let us also mention that $(x^0, \xi^0) \notin WF^s u$, in the sense of definition 0.1, if and only if there is $c > 0$ and an open cone G which contains ξ^0 so that $\{(x, \xi); |x - x^0| < c, \xi \in G\} \cap WF^s_\sim u = \emptyset$. Of course, (1.24) is interesting only if Λ is unbounded in the fiber-variables, but it may well happen that the fibers A_x considered above are non-trivial but bounded for some x and unbounded for some other x. In particular, our present definition is slightly more general than the corresponding definitions in [15], where we only considered the situation "$(x^0 \times G) \cap WF^s_{\sim}u = \emptyset$" for some set $G \subset R^n_\xi$. The reason for which we generalize the set-up in this paper is that with our present definition it is easier to see how wave front sets are transformed under the canonical transformations which we have to consider later on. It is easy (and in fact standard) to show that $[\Lambda_1 \cup \Lambda_2] \cap WF^s_{\sim}u = \emptyset$, if it is true that $\Lambda_i \cap WF^s u_{\sim} = \emptyset$ for $i = 1, 2$. Let p_m and $p_{m-1,0}$ be defined by (1.17) and (1.19) respectively; we denote by

$$W' = \{(x, \xi) \in \Gamma; p_{m,0}(x, \xi) + p_{m-1,0}(x, 0, \xi'') = 0\},$$

and, as before, by

$$\delta = 1 - 1/k = (k-1)/k.$$

Outside any quasihomogeneous neighborhood of W', p is essentially elliptic. The quantitative version of this is given in the following

Proposition 1.10. *Let $u \in \mathcal{D}'(U)$ be a solution of $p(x, D)u = 0$ on Γ, where Γ is a conic neighborhood of some point (x^0, ξ^0). Also consider some constant c and some compact set $K \subset U$. Denote $\Lambda_1 = \{(x, \xi); x \in K,\ dist_{\sim}((x, \xi), W') \geq c\}$. Then it follows that $\Lambda_1 \cap WF^s_{\sim}u = \emptyset$.*

The proof of this proposition is rather straightforward. We shall give it in section 2.

In view of proposition 1.10 it is now natural to distinguish essentially two regions, Λ_1 and Λ_2, in the co-tangent space of the set U. Λ_1 will be a small quasihomogeneous neighborhood of W', and we shall essentially regard $p(x, D)$ as an operator of principal type on Λ_1. In the region Λ_1 we shall apply one more microlocalization of a form which we shall explain in a moment. Arguing then essentially as for operators of principal type, we shall be able to show that in the assumptions from theorem 1.5, Λ_1 does not intersect $WF^s_{\sim}u$. On the other hand, Λ_2 will be the complement of some still smaller quasihomogeneous neighborhood of W'. On Λ_2 we shall thus stay away from W' and therefore p will be essentially elliptic on Λ_2 as a consequence of the proposition 1.10. The main thing is now that if Λ_1 and Λ_2 are chosen suitably, then their union will cover a set of form $\{x; |x - x^0| < c\} \times \Gamma'$, where Γ' is an open cone which contains ξ^0. We want to stress here again the fact that neither Λ_1 nor Λ_2 will be conic in the fiber variables.

We next describe how one has to microlocalize further within some fixed small quasihomogeneous neighborhood of W'. The idea is to take advantage of the bihomogeneous structure of R^n. The main step in the proof of theorem 1.5 will then be the following result:

Theorem 1.11. *Let the assumptions of theorem 1.5 be satisfied. Also consider* $\xi'^1 \in \dot{R}^{n'}$. *Then we can find a bineighborhood* V *of* (x^0, ξ'^1, ξ^0) *and* $c > 0$ *so that*

$$\{(x,\xi) \in V; |x| < c,\ dist_{\sim}((x,\xi), W') < c\} \cap WF^s_{\sim} u = \emptyset.$$

To see which is the relation between theorem 1.11 and theorem 1.5, we observe that it follows from theorem 1.11, if we also use a compacity argument, that there are constants $c' > 0, c'' > 0$ so that if $\Lambda' = \{(x,\xi); |x| < c',\ \text{dist}_{\sim}((x,\xi), W') < c''\}$, then $\Lambda' \cap WF^s_{\sim} u = \emptyset$. The proof of theorem 1.5 is then concluded as described above.

Remark 1.12. *When arguing on bineighborhoods, one will automatically stay away from the set* $\{(x,\xi); \xi' = 0\}$, *which is the characteristic variety of our operator. That this is still relevant for the Gevrey-s wave front set of the solutions of* $p(x,D)u = 0$ *comes from the fact that the whole region* $\{(x,\xi); |\xi'| < c|\xi''|^{\delta}\}$ *is taken care of by proposition 1.10.*

3. We conclude this section with some preliminary comments on the proof of theorem 1.11. Let us then fix some $\xi'^1 \in \dot{R}^{n'}$. Without loss of generality we may assume that ξ'^1 has the form $(1, 0, \ldots, 0)$. To simplify notations, we shall make a renotation for variables and n. We shall then in fact replace n', n, with $n'+1$, $n+1$, respectively ξ with λ and ξ' with λ', where $\lambda' \in R^{n'+1}$ is written as (τ, ξ'), $\tau \in R$, $\xi' \in R^{n'}$. Also denote $\lambda^0 \in R^{n+1}$, respectively $\lambda'^1 \in R^{n'+1}$, the points corresponding to ξ^0 and ξ'^1 in the new notations. In particular, we may assume that $\lambda^0 = (0, \ldots, 0, 1)$ and that $\lambda'^1 = (1, 0, \ldots, 0)$. Since p_m is transversally elliptic with respect to $\lambda' = 0$, $x_1 = 0$ will be noncharacteristic with respect to $p_{m,0}$. If V is a small bineighborhood of $(0, \lambda'^1, \lambda^0)$, and if $W'_{c,\delta} = \{(x,\lambda) \in U \times R^{n+1};\ \text{dist}_{\sim}((x,\lambda), W') \le c\}$, then we shall be able to write p on $V \cap W'_{c,\delta}$ in the form $p \sim \tilde{q} \circ (\tau - \sum b_j)$, where $\tilde{q}$ is a pseudodifferential operator which is elliptic on V in the Gevrey-s calculus and $\sum b_j$ is a formal sum of quasihomogeneous symbols. The precise meaning of this will be explained in section 3. In particular, the symbols b_j will not depend on τ. On $V \cap W'_{c,\delta}$ and with respect to the Gevrey-s wave front set, $p(x,D)u = 0$ will then be equivalent to $p'(x,D)u = 0$ where $p'(x,D)$ is the pseudodifferential operator associated with the symbol $\tau - \sum b_j$. The main thing is now that we can conjugate p' on a set of form $\{(x,\xi) \in V;\ \text{dist}_{\sim}((x,\xi), W') \le c\} \cap V$ with some Fourier integral operator so that after conjugation we are working with a pseudodifferential operator with principal symbol σ and which is defined on a set of form $V' \cap \{(y,\theta); |y| < \varepsilon, |(\sigma, \eta')| < c|\eta''|^{\delta}\}$, where V' is a bineighborhood of $(0, (1, 0, \ldots, 0), (0, \ldots, 0, 1))$. Here we have denoted the variables of the operator after conjugation by (y, θ), where $\theta = (\sigma, \eta)$ and $\eta = (\eta', \eta'')$. The characteristic variety of the new operator is then of course

$\sigma = 0$ and propagation of Gevrey-s wave front sets for the new operator is easy to establish. The idea of all this is of course well understood since quite some time. In the present situation we have a number of technical difficulties which come from the fact that we need to localize to bineighborhoods, while in addition we have to remain in a quasihomogeneous neighborhood of W'.

2. The geometry of $p_m + p_{m-1} = 0$

1. Let $(x^0, \xi^0) \in R^n \times \dot{R}^n$ with $\xi'^0 = 0$ and a conic neighborhood Γ of (x^0, ξ^0) be given. We assume that $(x, \xi) \in \Gamma$ implies $|\xi'| < d|\xi''|$ for some small constant $d > 0$. Also consider a classical analytic symbol $p \sim \sum_{j \geq m} p_j$ on Γ so that $p_m(x, \xi) = \sum_{|\alpha|=k} a_\alpha(x, \xi)\, \xi'^\alpha$ with a_α positively homogeneous of degree 0 for $(x, \xi) \in \Gamma$. In the present and in the next section we assume only (1.8), (1.9), i.e. that there are strictly positive constants c_i, $i = 1, 2, 3, 4$, so that

$$c_1|\xi'|^m \leq p_m(x, \xi) \leq c_2|\xi'|^m, \tag{2.1}$$

and denoting by

$$p_{m,0}(x, \xi) = \sum_{|\alpha|=m} a_\alpha(x, 0, \xi'')\, \xi'^\alpha,$$

$$p_{m-1,0}(x, \xi) = p_{m-1}(x, 0, \xi''),$$

that

$$-c_3|\xi''|^{m-1} \leq p_{m-1}(x, \xi) \leq -c_4|\xi''|^{m-1}. \tag{2.2}$$

Moreover, denote by

$$W' = \{(x, \xi) \in \Gamma; p_{m,0}(x, \xi) + p_{m-1,0}(x, 0, \xi'') = 0\}$$

and by

$$\delta = 1 - 1/k = (k-1)/k.$$

(It is again no loss of generality to assume that $k = m$, as we shall most often do henceforth.) In particular, W' is quasihomogeneous in an obvious sense and one can use quasihomogeneity to obtain part of the results which we need below. Since we cannot rely on quasihomogeneity alone, we shall prefer a more direct approach. The argument is very simple anyway.

2. Our first remark is that if d (in the condition $|\xi'| < d|\xi''|$) is small enough and if we shrink Γ (if necessary), then $p_{m,0}$ dominates p_m and $p_{m-1,0}$ dominates p_{m-1}. A more precise statement about this is

Remark 2.1. *Fix $\Gamma' \subset\subset \Gamma$. Then there are constants c_5, c_6, which do not depend on d, so that*

$$|p_m(x,\xi) - p_{m,0}(x,\xi)| \leq c_5 d\, |p_m(x,\xi)|, \tag{2.3}$$

and

$$|p_{m-1}(x,\xi) - p_{m-1,0}(x,\xi)| \leq c_5 d\, |p_{m-1}(x,\xi)|, \tag{2.4}$$

if $(x,\xi) \in \Gamma'$, $|\xi'| < d|\xi''|$. In particular, if $c_5 d < 1/2$, $c_6 d < 1/2$, then $p_m(x,\xi) \leq 2p_{m,0}(x,\xi)$, $|p_{m-1}(x,\xi)| \leq 2|p_{m-1,0}(x,\xi)|$, for such (x,ξ).
Indeed, we have e.g., $p_m(x,\xi) - p_{m,0}(x,\xi) = O(|\xi'|^{m+1}/|\xi|) \leq c_5' d\, |\xi'|^m \leq c_5' c_2 p_m(x,\xi)$, if $(x,\xi) \in \Gamma'$ and $|\xi'| < d|\xi|$.

Proposition 2.2. *Let $c > 0$ be given. Then we can find $c' > 0$, $c'' > 0$, $c''' > 0$, $\tilde{\Gamma}$, so that*

$$|p_{m,0}(x,\xi) + p_{m-1,0}(x,\xi)| \geq c'(|p_{m,0}(x,\xi)| + |p_{m-1,0}(x,\xi)|), \tag{2.5}$$

if $(x,\xi) \in \tilde{\Gamma}$, $|\xi'| < c''|\xi|$ and $dist_\sim((x,\xi), W') \geq c$.

Remark 2.3. *It is here important that we can fix c arbitrarily small. As a consequence, also c' might become in principle rather small.*

Proof of proposition 2.2. (Beginning.) We choose a conic neighborhood Γ'' of (x^0,ξ^0) in $(U \times R^{n''})$ and c_7 so that $(x,\xi'') \in \Gamma''$, $|\xi'| \leq c_7|\xi''|$ implies $(x,\xi',\xi'') \in \Gamma$. Let us in fact denote $\Gamma' = \{(x,\xi); (x,\xi'') \in \Gamma'', |\xi'| < c_7|\xi''|\}$. If we shrink Γ'', c_7, suitably, we will have $\Gamma' \subset\subset \Gamma$. We fix Γ'' with this property, but we will further shrink c_7 if necessary. In particular, we shall work from now on in the region $|\xi''| < c_7|\xi'|$ with c_7 as small as needed.

Let us next fix $(x,\xi'') \in \Gamma''$ and denote

$$W'_{(x,\xi'')} = \{\eta' \in R^{n'}; p_{m,0}(x,\eta',\xi'') + p_{m-1,0}(x,0,\xi'') = 0\}.$$

The greater part of our argument will be in the variables $\eta' \in R^{n'}$ and will refer to the set $W'_{(x,\xi'')}$. Let us in fact consider ξ' for which we assume

$$\text{dist}_\sim((x,\xi',\xi''), W') \geq c.$$

It follows of course that $|\xi' - \eta'| \geq c(1 + |\xi''|)^\delta$ for any $\eta' \in W'_{(x,\xi'')}$.

The following lemma is immediate:

Lemma 2.4. *If $\tilde{\Gamma} \subset\subset \Gamma$, we can find c_8, c_9, so that*

a) $|p_{m,0}(x,\eta',\xi'')| \leq (1/2)|p_{m-1,0}(x,0,\xi'')|$, *if* $|\eta'| < c_8|\xi''|^\delta$, $\eta' \in W'_{(x,\xi'')}$, *b)* $|p_{m-1,0}(x,\eta',\xi'')| \leq (1/2)|p_{m,0}(x,0,\xi'')|$, *if* $|\eta'| > c_9|\xi''|^\delta$, $\eta' \in W'_{(x,\xi'')}$. *In particular,* $\eta' \in W'_{(x,\xi'')}$ *implies*

$$c_8|\xi''|^\delta \leq |\eta'| \leq c_9|\xi''|^\delta$$

and $|\eta'| \leq c_8|\xi''|^\delta$, *respectively* $|\eta'| \geq c_9|\xi''|^\delta$, *both imply (2.5).*

Proof. We prove a), the other relations are similar. We have that $|p_{m,0}(x,\eta', \xi'')| \leq c_2|\eta'|^m < c_2 c_8^m|\xi'|^{m-1} \leq c_2c_8^m c_3^{-1}|p_{m-1,0}(x,0,\xi'')|$, so it suffices to shrink c_8 until $c_2c_8^m c_3^{-1} \leq 1/2$.

3. Proof of proposition 2.2 (End). We study the function $\theta \to F(x,\xi,\theta) = p_{m,0}(x,\theta\xi'/|\xi'|,\ \xi'') + p_{m-1,0}(x,0,\xi'')$ for $\theta \geq 0$. We want to show that $F(x,\xi,|\xi'|) \geq c|\xi''|^{m\delta}$. We recall that (x,ξ'') had been fixed and that ξ' was chosen so that $|\xi'-\eta'| > c(1+|\xi''|)^\delta$ for any $\eta' \in W'_{(x,\xi'')}$. We also observe that in view of lemma 2.4 we are left with the case $c_8|\xi''|^\delta \leq |\xi'| \leq c_9|\xi''|^\delta$, and in this region $|p_{m-1,0}(x,\xi)| \leq c_4|\xi''|^m \leq c_{13}|\xi'|^{m-1}$, so it suffices to show that $F(x,\xi,|\xi'|) \geq c|\xi'|^{m-1}$ for $c_8|\xi''|^\delta \leq |\xi'| \leq c_9|\xi''|^\delta$ if $|\xi'-\eta'| \geq c(1+|\xi''|)^\delta$ for any $\eta' \in W'_{(x,\xi'')}$. Here we note that $F(x,\xi,0) = p_{m-1,0}(x,0,\xi'') < 0$, whereas for $\theta > c_9|\xi''|^\delta$, $F(x,\xi,\theta) > 0$. It follows therefore that we can find θ^0 so that $F(x,\xi,\theta^0) = 0$. In particular, $|\theta^0\xi'/|\xi'| - \xi'| \geq c(1+|\xi''|)^\delta$. It follows that $F(x,\xi,|\xi'|) = (d/d\theta)F(x,\xi,\tilde\theta)|\xi' - \theta^0\xi'/|\xi'|| = m\tilde\theta^{m-1}p_{m,0}(x,\xi'/|\xi'|,\xi'')|\xi' - \theta^0/|\xi'|| \geq c_{14}\,\tilde\theta^{m-1}(1+|\xi''|)^\delta$ with $\tilde\theta$ on the segment $[\theta^0,|\xi'|]$. (Here we have used that $(d/d\theta)F(x,\xi,\theta) = (d/d\theta)(\theta^m p_{m,0}(x,\xi'/|\xi'|,\xi''))$.) Since $\theta^0 \geq c_8|\xi''|^\delta$, this gives $F(x,\xi,|\xi'|) \geq c_{15}|\xi''|^{(m-1)\delta}(1+|\xi''|)^\delta \geq c_{16}(1+|\xi''|)^{\delta m}$.

Corollary 2.5. *If* $d > 0$ *is sufficiently small,* $|p(x,\xi)| \leq c(|p_{m,0}(x,\xi) + p_{m-1,0}(x,\xi)|)$ *for* (x,ξ) *as in proposition 2.2.*

4. Before we now state our main result, we prove

Lemma 2.6. *For*

$$\text{dist}_\sim((x,\xi),W') \geq c,\ (x,\xi) \in \Gamma', |\xi''| \geq c_{10},$$

we have

$$|\partial_x^\alpha\partial_\xi^\beta p_m(x,\xi)| \leq c^{|\alpha|+|\beta|+1}\alpha!\beta!|\xi|^{-|\beta|\delta}|(p_{m,0}+p_{m-1,0})(x,\xi)| \tag{2.6}$$

Proof. We consider separately the cases $|\beta| \leq m$ and $|\beta| > m$.

I. In the first case we use that

$$|\partial_x^\alpha\partial_\xi^\beta p_m(x,\xi)| \leq c^{|\alpha|+|\beta|+1}\alpha!\beta!|\xi'|^{m-|\beta|}. \tag{2.7}$$

In the region $|\xi'| \geq |\xi|^\delta$ it follows that $|\xi'|^{-|\beta|} \leq |\xi|^{-\delta|\beta|}$. We also have $|\xi'|^m \leq p_{m,0}(x,\xi)$. Together with proposition 2.2 this gives (2.6) in this case. If, on the other hand, $|\xi'| \leq |\xi|^\delta$, then

$$|\xi'|^{m-|\beta|} \leq |\xi|^{\delta(m-|\beta|)} = |\xi|^{m-1-\delta|\beta|} \leq c|p_{m,0}(x,\xi'')||\xi|^{-\delta|\beta|},$$

since $\delta m = m-1$. We conclude the argument once more with the aid of proposition 2.2.

II. For $|\beta| > m$ we have instead:

$$|\partial_x^\alpha \partial_\xi^\beta p_m(x,\xi)| \leq c^{|\alpha|+|\beta|+1}\alpha!\beta!|\xi|^{m-|\beta|}, \tag{2.8}$$

since if we derivate β times a term $a_\alpha(x,\xi)\xi'^\alpha$, then at least $|\beta|-m$ derivatives will have to act on a_α. Also note that

$$m-|\beta| = m-(1-\delta)|\beta| - \delta|\beta| = m - |\beta|/m - \delta|\beta| \leq m-1-\delta|\beta|.$$

It follows that

$$|\xi|^{m-|\beta|} \leq c|\xi|^{-\delta|\beta|}|p_{m-1,0}(x,0,\xi'')| \leq c|\xi|^{-\delta|\beta|}|p_m(x,\xi)+p_{m-1}(x,\xi)|,$$

for the (x,ξ) as in the statement.

Theorem 2.7. *If $d>0$ (in the condition $|\xi'|<d|\xi|$) is sufficiently small and for the (x,ξ) from the statement of proposition 2.2, it follows that*

$$|\partial_x^\alpha \partial_\xi^\beta p(x,\xi)| \leq c^{|\alpha|+|\beta|+1}\alpha!\beta!|\xi|^{-|\beta|\delta}\,|p(x,\xi)|. \tag{2.9}$$

Proof. In view of corollary 2.5 and the preceding lemma it suffices to observe that

$$|\partial_x^\alpha \partial_\xi^\beta (p-p_m)(x,\xi)| \leq c^{|\alpha|+|\beta|+1}\alpha!\beta!\,|\xi|^{-|\beta|\delta}\,(1+|\xi|)^{m-1}.$$

Remark 2.8. *We have now proved theorem 2.7. The fact that proposition 1.10 is valid follows now if we combine theorem 2.7 with the results proved in [14].*

3. Factorization of the symbol

1. Notation for variables is as in nr. 3. from section 1 and assumptions are as in (1.15), with $k=m\geq 2$ an arbitrary integer, and (1.16), (1.20). If f is s–quasihomogeneous, then $(\partial/\partial\lambda_j)f$ is $(s-1)$-quasihomogeneous for $j\leq n'$ and $[s-k/(k-1)]$–quasihomogeneous if $j>n'$. The difference between the two degrees is $1/(k-1)$, which is one of the reasons why in this theory

it is natural to work with formal symbols of form $\sum_j q_j$ with q_j quasihomogeneous of degree $m - j/(k-1)$. To some extent, quasihomogeneity will be mixed up with bihomogeneity. Note that the variables λ' will here practically be given the weight 1 and the variables λ'' the (higher) weight $k/(k-1)$. A typical situation is when we regard terms of form $(\partial/\partial\lambda')^\beta a(x, 0, \lambda'')\lambda'^\gamma$, for some positively homogeneous function a of degree μ. It follows that such terms are $|\gamma|+(\mu-|\beta|)k/(k-1)$-quasihomogeneous. If we now consider some $f(x, \lambda)$ which is positively homogeneous of degree μ in a conic neighborhood of $(0, \lambda^0)$, then we can expand it as a formal sum of quasihomogeneous symbols:

$$f(x, \lambda) = \sum_{\alpha \geq 0} (\partial/\partial\lambda')^{\alpha'} f(x, 0, \lambda'')\lambda'^\alpha/\alpha!, \tag{3.1}$$

the sum being actually convergent if $\lambda'/|\lambda''|$ is small. In particular, $f(x, 0, \lambda'')$ is the quasihomogeneous principal part of f. In order to remain rooted in a more direct way in the quasihomogeneous theory, we will now prefer to work in situations when $f(x, \lambda)$ is a polynomial of order at most m in τ. This can be achieved with the aid of a well-established variant of the Weierstrass preparation theorem for symbols. Let us assume in fact at first that homogeneous analytical coordinates are chosen so that (after multiplication with an analytic symbol),

$$p_m(x, \lambda) = \tau^m + \sum_{|\alpha|=m, \alpha_1 < m} a_\alpha(x, \lambda)\lambda'^\alpha. \tag{3.2}$$

With the aid of the classical Weierstrass preparation theorem, we can rewrite p_m as a (pointwise) product

$$p_m(x, \lambda) = \psi(x, \lambda)[\tau^m + \sum_{j<m} \varphi_j(x, \xi)\tau^j], \tag{3.3}$$

where the φ_j and ψ are analytic and homogeneous of degree $m - j$ (the φ_j) and 0 (the ψ) respectively, and ψ is elliptic in a conic neighborhood of $(0, \lambda^0)$. Moreover, the Weierstrass preparation theorem gives that the φ_j vanish at $\lambda' = 0$, but more is true, as can be seen if we compare (3.2) and (3.3). Indeed, it follows that we must have that $(\partial/\partial\lambda')^\gamma(\tau^m + \sum_{j<m} \varphi_j(x, \xi)\tau^j) = 0$ for $|\gamma| < m$ and $\lambda' = 0$. It follows from this that $\partial_{\xi'}^\gamma \varphi_j(x, \xi) = 0$ if $\xi' = 0$ and $|\gamma| < m - j$. After composition with the symbol ψ^{-1} from the left, we can now assume henceforth that p_m has the form

$$p_m(x, \lambda) = \sum_{|\beta|+j=m} a_{\beta,j}(x, \xi)\tau^j\xi'^\beta, \tag{3.4}$$

with the coefficient of τ^m in (3.4) equal to 1. All this followed from the classical form of the Weierstrass preparation theorem and (3.4) is of course valid in a conic neighborhood of $(0, \lambda^0)$. To continue, we have now to apply the Weierstrass preparation theorem for symbols. It follows then that $p = \rho \circ [\sum_{j=0}^{m} g_j(x,\xi)\tau^j]$, with g_j classical analytic symbols, ρ elliptic. Again this is a relation in a conic neighborhood of $(0, \lambda^0)$. The coefficient of τ^m is here of form $(1+r)$ with r of order -1 and r is a symbol which does not depend on τ. We next compose from the left with $(1+r)^{-1} \circ \rho^{-1}$. It follows then that we can assume that p is of form $p = \sum_{j \geq 0} p_{m-j}$, with p_m as in (3.4) and the p_{m-j}, for $j \geq 1$, analytic symbols which are polynomials of degree at most $m-1$ in τ (and which are defined in a conic neighborhood of $(0, \xi^0)$). We next expand all the p_l as in (3.1). We conclude that p can be written as

$$p \sim \sum_{j} \sum_{\gamma \geq 0} (\partial/\partial\lambda')^\gamma p_{m-j}(x, 0, \xi'')\lambda'^\gamma/\gamma!,$$

where we have not yet ordered terms according to their degree of quasihomogeneity. The quasihomogeneous principal part of p is $p_{m,0}(x,\lambda) + p_{m-1,0}(x,\lambda)$ where, as above, $p_{m,0}(x,\lambda) = \sum_{|\beta|+j=m} a_{\beta,j}(x,0,\xi'')\tau^j\xi'^\beta$ and $p_{m-1,0}(x,\lambda) = p_{m-1}(x,0,\xi'')$. More generally, denote by

$$p_{m-j,r}(x,\lambda) = \sum_{|\gamma|=r} (\partial/\partial\lambda')^\gamma p_{m-j}(x,0,\xi'')\lambda'^\gamma/\gamma!.$$

As observed above it is quasihomogeneous of degree $|\gamma| + (m-j-|\gamma|)k/(k-1) = m - j - |\gamma|/(k-1)$ and is a polynomial of order at most $m-1$ in τ if $j \geq 1$. This gives then that

$$p \sim \sum_{r \geq 0} \sigma_{m-r/(k-1)},$$

where

$$\sigma_{m-r/(k-1)} = \sum_{j,\gamma} (\partial/\partial\lambda')^\gamma p_{m-j}(x,0,\xi'')\lambda'^\gamma/\gamma!, \tag{3.5}$$

where the sum is for $m - j - |\gamma|/(k-1) = m - r/(k-1)$ and, again, $\sigma_{m-r/(k-1)}$ is a polynomial of order at most $m-1$ in τ if $j \geq 1$.

2. Our main result is the following

Theorem 3.1. *Let $\lambda^0 \in \dot{R}^{n+1}$ be given with $\lambda^{0\prime} = 0$ and denote $\lambda'^1 = (1,0,\ldots,0)$. There are b_j, q_{ij}, $j \geq 0$, $i = 0,1,\ldots,m-1$, with the following properties:*

- *the b_j and the q_{ij} are defined and analytic on a set of form $\{(z,\zeta) \in U \times C_\zeta^n; |\zeta'| < c|\zeta''|^\delta, \zeta'' \in G''\}$, where U is a complex neighborhood of 0 in C_z^{n+1} and G'' is a complex conic neighborhood of $\lambda^{0\prime\prime}$,*

– *the b_j are quasihomogeneous of degree $1 - j/(k-1)$, the q_{ij} of degree $m-1-i-j/(k-1)$,*
– *the b_j and q_{ij} do not depend explicitly on τ, and we have that*

$$p(x,\xi) \sim [\tau^{m-1} + \sum_{i=1}^{m-1} \sum_{j\geq 0} q_{ij}(x,\xi)\tau^{m-1-i}] \circ (\tau - \sum_{j\geq 0} b_j(x,\xi)). \quad (3.6)$$

(3.6) is here valid for (x,λ) with (x,ξ) in the real part of the domain of definition provided that we also assume that $|\tau| < c_1|\xi''|^\delta$ for some constant $c_1 > 0$, which we have chosen suitably previously. In particular we can assume that it is satisfied on a set of form $V \cap W'_{c\delta}$ where V is a bineighborhood of $(0, \lambda'^1, \lambda^0)$ and $\lambda'^1 = (1,0,\ldots,0)$. Moreover, the b_j are real for $j \leq J$ if the $p_{m-1,j}$ are real for $j \leq J$ and b_0 is determined by the following two conditions:

$$b_0(x,0,\xi'') = [-p_{m-1}(x,0,\xi'')]^{1/m}, \quad (3.7)$$

$$p_{m,0}(x, b_0(x,\xi),\xi) + p_{m-1}(x,0,\xi'') = 0. \quad (3.8)$$

The m-th root $[\alpha]^{1/m}$ is considered real for real positive arguments α. When m is even, we can consider the positive $m-$th root, or also the negative determination of the $m-$th root. Recall that $-p_{m-1}(x,0,\xi'') > 0$ for real (x,ξ). Finally, there is a constant c' so that the b_j satisfy the inequality

$$|b_j(x,\xi)| \leq c'^{j+1} j!(|\xi'| + |\xi''|^\delta)^{1-j/(k-1)} \quad (3.9)$$

on their domain of definition.

3. The proof of the theorem is standard but technical. The first remark is that (3.7) and (3.8) determine an analytic function b_0 which is defined on a set of form $\{(z,\zeta) \in U \times C^n_\zeta; |\zeta'| < c|\zeta''|^\delta, \zeta'' \in G''\}$, which is 1–quasihomogeneous there and which is real-valued for real arguments. This is clear from the implicit function theorem. Once b_0 is found, we can find q_{i0} quasihomogeneous of degree $m-1-i$, so that

$$p_{m,0}(x,\lambda) + p_{m-1,0}(x,\lambda) = (\tau^{m-1} + \sum_{i=1}^{m-1} q_{i0}(x,\xi)\tau^{m-1-i})(\tau - b_0(x,\xi)), \quad (3.10)$$

the multiplication between brackets in (3.10) being pointwise.That this is possible, follows either by direct division or by an application of the Weierstrass preparation theorem. It is also interesting to note that $q_{i0}(x,0,\xi'') = [-p_{m-1,0}(x,\lambda)]^i = b_0^i(x,0,\xi'')$, as is clear from the formula $(\tau^{m-1} + \alpha\tau^{m-2} + \cdots + \alpha^{m-1})(\tau - \alpha) = \tau^m - \alpha^m$. In particular, we obtain that

$$b_0^i(x,\xi) + \sum_{i=1}^{m-1} q_{i0}(x,\xi) b_0^{m-1-i}(x,\xi) \neq 0 \quad (3.11)$$

in a suitable quasihomogeneous neighborhood of $(0,0,\xi^0)$.

Once we have found q_{i0} and b_0, we shall now construct the q_{ij} and b_j, for $j \geq 1$, by an iteration. We claim in fact that it is possible to find $b_0, \ldots, b_\nu$, $q_{i0}, \ldots, q_{i\nu}$, so that

$$(\tau^{m-1} + \sum_{i=1}^{m-1} \sum_{j=0}^{\nu} q_{ij}\tau^{m-1-i}) \circ (\tau - \sum_{j=0}^{\nu} b_j) = \sum_{r=0}^{r} \sigma_{m-r/(k-1)} + \sum_{j \geq 0} \rho_{\nu j},$$

where the $\sigma_{m-r/(k-1)}$ are from (3.5) and the $\rho_{\nu j}$ are quasihomogeneous of degree $m-(\nu+1)/(k-1) - j(k-1)$. For $\nu = 0$ this follows immediately from the conditions on the b_0, q_{i0}. To argue by induction, we shall now also show how one can find $b_{\nu+1}$, $q_{i,\nu+1}$, $i = 1, \ldots, m-1$, if one has already found $b_0, \ldots, b_\nu$, $q_{i,0}, \ldots, q_{i,\nu}$, $i = 1, \ldots, m-1$. Actually, we obtain for the $b_{\nu+1}$, $q_{i,\nu+1}$, among others, the following conditions, in which the b_j, q_{ij}, $j > \nu + 1$, are not yet involved:

$$\begin{aligned} -(\tau^{m-1} + \sum_{i=1}^{m-1} q_{i,0}\tau^{m-1-i})b_{\nu+1} \quad + & \\ (\sum_{i=1}^{m-1} q_{i,\nu+1}\tau^{m-1-i})(\tau - b_0) \quad = & \quad \sigma_{m-(\nu+1)/(k-1)} + \rho_{\nu 0}. \end{aligned} \quad (3.12)$$

Here we recall that the $\sigma_{m-(\nu+1)/(k-1)}$ and $\rho_{\nu 0}$ are polynomials of degree $m-1$ in τ. We can thus write

$$\sigma_{m-(\nu+1)/(k-1)} + \rho_{\nu 0} = \sum_{j=0}^{m-1} d_j \tau^j.$$

Identifying coefficients of τ^r in (3.12) now gives

$$-b_{\nu+1}q_{m-1,0} - b_0 q_{m-1,\nu+1} = d_0,$$

$$-b_{\nu+1}q_{m-2,0} - b_0 q_{m-2,\nu+1} + q_{m-1,\nu+1} = d_1,$$

$$\vdots$$

$$-b_{\nu+1}q_{1,0} - b_0 q_{1,\nu+1} + q_{2,\nu+1} = d_{m-2},$$

$$-b_{\nu+1} + q_{1,\nu+1} = d_{m-1}.$$

This is an $m\times m$ system for the m unknowns $b_{\nu+1}, q_{1,\nu+1}, \dots, q_{m-1,\nu+1}$ and the determinant is nonvanishing. Moreover, degrees of quasihomogeneity are correct. One can however also solve this $m \times m$-system in a more direct way. Indeed, it is possible to eliminate at first the $q_{i,\nu+1}$, calculate $b_{\nu+1}$ and then calculate successively $q_{1,\nu+1}, q_{2,\nu+1}, \dots, q_{m-1,\nu+1}$. The elimination of the $q_{i,\nu+1}$ is done multiplying the i-th equation by $(b_0)^{i-1}$ and summing. We thus get

$$-b_{\nu+1}(q_{m-1,0}+q_{m-2,0}b_0+q_{m-3,0}b_0^2+\dots+b_0^{m-1}) = d_0+d_1b_0+\dots+d_{m-1}b_0^{m-1}.$$

The coefficient of $b_{\nu+1}$ is here different from zero in view of (3.11). We have now seen how we can determine the b_j, q_{ij}, iteratively. The preceding arguments give also easily the extensions to the appropriate complex domains and the estimates required in (3.9).

Finally, let us comment on the fact that the b_j are real-valued for $j \leq J \leq m-2$ if the $p_{m-1,j}$ are real for $j \leq J$. Indeed, it is seen from the construction that in the calculation of the $\sigma_{m-r/(k-1)}$, $r \leq \nu$, ρ_ν, $\nu \leq m$, we do not consider but the values of $p_{m,j}$ and of the $p_{m-1,j}$ and that we do not have to consider terms in the formula of composition of symbols which have complex coefficients.

4. In the sequel we are mainly interested in the $b_0, \dots, b_J$. It is worthwhile to mention that they are related only to p_m and p_{m-1}. This is due to the assumption that $p_{m-1}(0,\xi^0) \neq 0$. We shall denote $b_0+\dots+b_J$ by b and $\sum_{j>J} b_j$ by b', quasihomogeneous of degree $1-(J+1)/(k-1)$. Also note that on a set of form $V\cap W'_{c,\delta}$, the equation $p(x,D)u = 0$ is equivalent with $\rho(x,D)u = 0$, where $\rho(x,D)$ is the pseudodifferential operator associated with $\tau - b(x,\xi) - b'(x,\xi)$. This shows then in particular that in some sense and in some bineighborhood of $(0,\lambda'^1,\lambda^0)$, $W = \{(x,\lambda); \tau = b(x,\xi)\}$ is the "true" characteristic variety associated with p. It is the main content of section 4 that we can choose canonical coordinates in which W becomes flat and in which the principal part of the operator ρ becomes $-i(\partial/\partial t)$, where we have written t for the variable x_1. In particular, ρ will transform after conjugation with some F.I.O.'s associated with this canonical map to the operator $-i(\partial/\partial t) + r$, where r is a pseudodifferential operator with symbol in $S^{1-(J+1)/(k-1)}$.

Proposition 3.2. *Any quasihomogeneous neighborhood of W is also a quasihomogeneous neighborhood of W' and viceversa.*

Proof. This is based on the fact that W' is parametrized by $(x, b_0(x,\xi),\xi)$, whereas W is parametrized by $(x,b(x,\xi),\xi)$. If $P = (x,b_0(x,\xi),\xi)$ and $Q = (x,b_0(x,\xi),\xi)$ correspond to the same parameters (x,ξ), then $|P-Q| =$

$|b(x,\xi) - b_0(x,\xi)| \le \sum_{j=1}^{s} |b_j(x,\xi)| < c(|\xi'| + |\xi''|^\delta)^{1-2/k}$. The proposition now follows easily.

4. The phase function

1. The notations for phase variables are as in the sections 2 and 3. Also assume again that $\lambda^0 = (0,\ldots,0,1) \in R^{n+1}$ and fix $\lambda'^1 \in \dot{R}^{n'+1}$. Without loss of generality, we shall assume that $\lambda'^1 = (1,0,\ldots,0)$ and we shall work in a bineighborhood of $(0,\lambda'^1,\lambda^0)$, while still remaining in a quasihomogeneous neighborhood $W'_{c,\delta}$ of W', where W' is the quasihomogeneous characteristic variety associated with p. Here p is of course an operator which satisfies the assumptions of the preceding section 3. What we want is to find a canonical transformation χ, so that χ^{-1} lives on $V \cap W'_{c,\delta}$ and so that in the new canonical variables, p has a very simple form: see (4.1) below. It will be possible to choose the canonical transformation in such a way that the x-variable associated with the phase-variable τ remains unchanged. We denote it by t, changing notation of the x-variables in R^{n+1} from x to (t,x). Thus $x \in R^n$ and $(t,x) \in R^{n+1}$. Let us also consider another set of canonical variables which we denote by (t,y,σ,η). Sometimes we shall also write θ for (σ,η). What we want is then to find a canonical map χ which maps $(t,y,\theta) \to (t,x,\lambda)$, and for which

$$(\tau - b(t,x,\xi)) \circ \chi(t,y,\theta) = \sigma, \tag{4.1}$$

where b is from section 3. We thus recall that $b = b_0 + b_1 + b_2 + \cdots + b_J$ is real-valued, and that we had

$$p \sim (\tau^{m-1} + \sum_{i,j} q_{ij}(t,x,\xi)\tau^{m-1-i}) \circ (\tau - \sum_j b_j(t,x,\xi)).$$

As for estimates, we have that the b_j satisfy for $j \ge 1$

$$|\partial_{t,x}^\alpha \partial_\xi^\beta b_j(t,x,\xi)| \le c^{|\alpha|+|\beta|+j+1}\alpha!\beta!j!(|\xi'| + |\xi''|^\delta)^{1-j/(k-1)}.$$

One of the main problems in the following, is to keep control of the domains of definition of our canonical transformations. Since we have some modest freedom in the choice of χ, we shall at first fix conditions in the η-variables. We shall in fact at first fix some $\tilde{\eta} \in R^n$ with $\tilde{\eta}' = 0$. We shall then start our construction for $|\eta - \tilde{\eta}| < c|\tilde{\eta}|^\delta$. As is standard, we shall also want to dispose of a generating function $\omega(t,x,\sigma,\eta)$ for χ. The requirements for ω are thus:

$$\omega_t(t,x,\sigma,\eta) - b(t,x,\nabla_x\omega(t,x,\sigma,\eta)) = \sigma,$$

to which we add the initial condition

$$\omega(0,x,\sigma,\eta) = \langle x,\eta\rangle.$$

(Here $\omega_t = (\partial/\partial t)\psi$.) It is natural to look for ω in the form $\omega(t,x,\sigma,\eta) = t\sigma + \tilde{\omega}(t,x,\eta)$, with $\tilde{\omega}$ satisfying :

$$\tilde{\omega}_t(t,x,\eta) - b(t,x,\nabla_x\tilde{\omega}(t,x,\eta)) = 0, \tag{4.2}$$

$$\tilde{\omega}(0,x,\eta) = \langle x,\eta\rangle. \tag{4.3}$$

As an additional condition for $\tilde{\omega}$ we ask for $\tilde{\omega} \in S^1_\mu$, $\mu(\eta) = (|\eta'| + |\eta''|^\delta)$, starting from second derivatives.

One can solve (4.2), (4.3) with the aid of the method of bicharacteristics. The only problem is with the size of the domain on which we can solve our equation and with the estimates. How these difficulties can be circumvented is described e.g. in the paper of [15] (in which the authors started from [8]). The bicharacteristic system associated with (4.2) admits t as a natural parameter. It is

$$t = t,$$

$$\frac{dx_j}{dt}(t,y,\eta) = -\frac{\partial b}{\partial \xi_j}(t,x(t,y,\eta),\xi(t,y,\eta)), \tag{4.4}$$

$$\frac{d\xi_j}{dt}(t,y,\eta) = \frac{\partial b}{\partial x_j}(t,x(t,y,\eta),\xi(t,y,\eta)), \tag{4.5}$$

$$\frac{d\tau}{dt}(t,y,\sigma,\eta) = \frac{\partial b}{\partial t}(t,x(t,y,\eta),\xi(t,y,\eta)), \tag{4.6}$$

to which we add the initial conditions

$$x(0,y,\eta) = y, \xi(0,y,\eta) = \eta, \tau(0,y,\eta) = \sigma. \tag{4.7}$$

It is of course due to the initial conditions that $t \to (x,\xi,\tau)$ depends also on (y,σ,η). Note that the equation (4.6) is not coupled with the equations (4.4), (4.5), so we can at first study (4.4), (4.5) with their respective initial conditions and then solve (4.6), together with the last condition in (4.7) in the end. The first step in the argument is now to show that the system (4.4),(4.5),(4.7), admits a solution for $\{t \in C, |t| < c\}$ with c independent of (y,σ,η). How this comes about is described e.g. in [15]. Let us then denote by $X(t,y,\eta)$, $\Xi(t,y,\eta)$ the solution of (4.4),(4.5),(4.7). It follows as in [15] that $|X(t,y,\eta) - y| < c|t|$, $|\Xi(t,y,\eta) - \eta| < c|t||\tilde{\eta}''|^\delta$. We want to solve here $X(t,y,\eta) = x$ for y to get $y = Y((t,x,\eta)$. That this is possible for small fixed t and for $\eta \in \{\eta \in C^n; |\eta - \tilde{\eta}| \le c|\tilde{\eta}|^\delta\}$ follows from the implicit function theorem in view of the fact that $(\partial X/\partial y)(0,y,\eta) = I$, together with the fact that $\partial^2 X/\partial y^2 = O(t)$. (Also cf. here section 3.4 in [15].) The next thing is to consider the function $L(t,x,\eta) = \Xi(t,Y(t,x,\eta),\eta)$, which associates to the x−component on some bicharacteristic curve the corresponding ξ−component. The Hamilton-Jacobi theory gives that $L(t,x,\eta) = \nabla_x \tilde{\omega}(t,x,\eta)$. We have thus a rather explicit formula for $\nabla_x \tilde{\omega}$, which shows that $\nabla_x \tilde{\omega} \in S^1_\mu$. We can now also recover $\tilde{\omega}$ from the relation

$$\tilde{\omega}(t,x,\eta) = \int_0^t b(s,x,\eta)\, ds.$$

We conclude thus that we can define $\tilde{\omega}$, and therefore also ψ, for (t,x) small and $|\eta - \tilde{\eta}| < c|\tilde{\eta}|^\delta$. We recall here that the property of $\tilde{\eta}$ was that $\tilde{\eta}' = 0$. The initial condition for $\tilde{\omega}$ at $t = 0$ however did not depend on $\tilde{\eta}$ and $\tilde{\omega}$ is locally uniquely determined by the initial condition and is analytic. It follows that actually $\tilde{\omega}$ can be defined on a set of form $\{(t,x,\eta) \in C^{n+1}_{t,x} \times C^n; |(t,x)| < c, |\eta'| < c|\eta''|^\delta, \eta'' \in G''\}$, where G'' is a complex conic neighborhood of $\lambda^{0''}$.

2. To see which is the image of the canonical transformation associated with the generating function, let us at first analyze which is the image of points of form $(0,x,0,\eta)$ under the map $(t,x,\sigma,\eta) \to T(t,x,\sigma,\eta) = (t,x,\psi_t(t,x,\sigma,\eta), \nabla_x \omega(t,x,\sigma,\eta))$. When $t = 0, \sigma = 0$, we have in fact $\omega_t(0,x,0,\eta) = b(0,x,\nabla_x \psi(0,x,0,\eta))$ and $\nabla_x \omega(0,x,\sigma,\eta) = \eta$. Thus the image point is $(0,x,b(0,x,\eta),\eta)$ which is, as expected, a point on the characteristic variety W. (Also the points $(t \neq 0, x, \sigma = 0, \eta)$ are mapped to the characteristic variety, but the image point is characterized in a less direct way.)

To obtain quantitative conclusions from this, we look at

$$\frac{\partial^2 \omega}{\partial(t,x)\partial(\sigma,\eta)}(t,x,\sigma,\eta) = I + R(t,x,\sigma,\eta),$$

(the left hand side is the mixed Hessian of ψ; the first row is $[\partial^2\omega/(\partial t \partial \sigma), \partial^2\omega/(\partial t \partial \eta_1), \ldots, \partial^2\omega/(\partial t \partial \eta_n)]$, and the first column $[\partial^2\omega/(\partial t\partial\sigma), \partial^2\omega/(\partial x_1 \partial \sigma), \ldots, \partial^2\omega/(\partial x_n \partial\sigma)]$,) where R is a matrix with entries in S^0_μ. Moreover, $R(0,x,\sigma,\eta) = 0$, so we have that

$$\frac{\partial^2 \omega}{\partial(t,x)\partial(\sigma,\eta)}(t,x,\sigma,\eta) \geq \frac{1}{2} I \tag{4.8}$$

if $|(t,x)|$ is small and I is the identity $(n+1) \times (n+1)$-matrix. As a consequence of this is also easy to see that T is injective on sets of form $A' = \{P = (t,x,\sigma,\eta);\ \mathrm{dist}_\sim(P,\tilde{P}) \leq c\}$ if $\tilde{P} = (\tilde{t},\tilde{x},\tilde{\sigma},\tilde{\eta})$ is fixed and that $T(A')$ contains $B' = \{Q = (t,x,\lambda);\ \mathrm{dist}_\sim(Q, T(\tilde{P}) \leq c'\}$ if c' is sufficiently small.

We also claim that T is globally injective if $|t|$ is small. In fact to begin with, let us note that T is clearly injective in the variables (t, x). If then, in addition $\psi_t(t, x, \sigma^1, \eta^1) = \psi_t(t, x, \sigma^2, \eta^2)$, $\nabla_x \omega(t, x, \sigma^1, \eta^1) = \nabla_x \omega(t, x, \sigma^2, \eta^2)$, then we have at first $\eta^1 + O(t)|\eta^1|^\delta = \eta^2 + O(t)|\eta^2|^\delta$, so $|\eta^1 - \eta^2| \leq c(|\eta^1| + |\eta^2|)^\delta$, with c as small as we please if t is small enough. But, as observed above, on such sets T is injective, so $\eta^1 = \eta^2$. It then also follows immediately that $\sigma^1 = \sigma^2$.

Having seen that T is injective, we want next to find a lower bound for the image of T when applied to the set

$$A = \{(t, x, \sigma, \eta); |(t, x)| \leq \varepsilon, |\sigma| + |\eta'| \leq c|\eta''|^\delta, \eta'' \in G''\}, \tag{4.9}$$

where G'' is a conic neighborhood of $(0, \ldots, 0, 1) \in R^{n''}$. It is clear in fact that $T(A)$ contains a set of form

$$B = \cup[\{(0, x, b(0, x, \eta); |\eta'| < c|\eta''|^\delta, \eta'' \in G''\}]_{c,\delta}. \tag{4.10}$$

On the other hand it is not difficult to see that B itself contains a set of form $V \cap W'_{c,\delta}$ with V a bineighborhood of $(0, (1, 0, \ldots, 0), \lambda^0)$.

We also need to perform similar considerations for the map $(t, x, \sigma, \eta) \to S(t, x, \sigma, \eta) = (t, \nabla_\eta \tilde{\omega}(t, x, \sigma, \eta), \sigma, \eta)$, which we consider again as a map on $\{|(t, x)| < \varepsilon, |\sigma| + |\eta'| < c|\eta''|^\delta\}$. Once more, S is injective on sets of type A' and $S(A')$ contains a set of type B'. To understand the global behavior of S, we note that for $t = 0$, $\nabla_\eta \omega(t, x, \sigma, \eta) = x$, so S is the identity when $t = 0$. Arguing as above, we see that S is injective on the set A defined in (4.9) and that $S(A)$ contains a set of form

$$D = \{(t, y, \sigma, \eta); |(t, y)| \leq \varepsilon', |\sigma| + |\eta'| \leq c'|\eta''|^\delta\}. \tag{4.11}$$

We can now consider, finally, the canonical transformation χ associated with ω. In the notations above it is $\chi = TS^{-1}$. It is thus defined at least on D and $\chi(D)$ contains a set of form $V \cap W'_{c,\delta}$.

5. Fourier integral operators and proof of theorem 1.11

1. We consider in this section F.I.O.'s associated with the phase function ω constructed in section 4. They will be of the following two types:

$$Au(t, x) = \int e^{i\omega(t,x,\theta)} a(t, x, \theta) \hat{u}(\theta)\, d\theta, \tag{5.1}$$

$$B^* v(t, y) = \int\!\!\int e^{it\sigma + i\langle y, \eta\rangle - i\omega(t', x, \theta)} b(t, y, t', x, \theta) v(t', x)\, dt'\, dx\, d\theta. \tag{5.2}$$

Following [15], we call A, respectively B^*, F.I.O.'s of the first, respectively the second kind associated with ω. The integration in the θ-variable is here on a set of form $\{(\sigma,\eta); |\sigma|+|\eta'|\le c|\eta''|^\delta, \eta''\in G''\}$ where G'' is a conic neighborhood in $R^{n''}$ of $(0,\dots,0,1)$.

The reason why we denote operators of the second kind with a "$*$" comes of course from the fact that the formal adjoint of an operator of the first kind is an operator of the second kind. We are interested in the study of the operator $B^*\rho(t,x,D_t,D_x)A$, where ρ is the pseudodifferential operator associated with the symbol $\tau - b(t,x,\xi)$, and in the study of the mapping properties of the operators A and B^*. Actually, we can obtain the main informations which we need from [15], the only difficulty being to describe the setting in which these results have to be applied.

2. We shall apply here the theory from [15] for the weight function $\varphi(\sigma,\eta) = |\sigma|^\delta + |\eta'|^\delta + |\eta''|$. Sometimes we just write $\varphi = \varphi(\eta)$, since we are working in the region $|\sigma|^\delta < |\eta''|$ where $|\sigma|^\delta + |\eta'|^\delta + |\eta''| \sim |\eta'|^\delta + |\eta''|$.

Starting point in our considerations is that the phase function ω is in S^1_φ. This phase function is not homogeneous and it is not possible in general to work with homogeneous versions of wave front sets. However, if we localize in the phase variables to sets of form $\{\eta; |\eta-\tilde\eta| < c|\tilde\eta|^\delta\}$, then ω has the right behavior.

We want to check next that ω satisfies the conditions needed to apply the results from [15]. We need therefore to consider the conditions called "B,C,D" and the so called "ω-compatibility" introduced in that paper.

Condition B. For the situation at hand it comes to $\varphi(\nabla_{t,x}\omega(t,x,\sigma,\eta)) \sim \varphi(\sigma,\eta)$ on the sets on which we work. Here $\nabla_{t,x}\omega(t,x,\sigma,\eta) = (\sigma + b(t,x,\nabla_x\tilde\omega(t,x,\eta)), \nabla_x\tilde\omega(t,x,\eta))$.

Condition C. Condition C asks for the fact that $|\eta-\tilde\eta| < c_1|\eta|^\delta$ implies $|\nabla_x\tilde\omega(t^0,x^0,\eta) - \nabla_x\tilde\omega(t^0,x^0,\tilde\eta)| \ge c_2(|\eta|^\delta + |\tilde\eta|^\delta)$. Also this is clear for $t=0$ and will then be valid for t small.

Condition D. For any $X' \subset\subset X$ and for any $c>0$ there is c_2 so that $|\omega_\theta(z,\theta) - \omega_\theta(z,\theta)| \ge c_2$ if $|x-x'| \ge c_1$, $z,z'\in X'$, $\theta\in\Gamma$.

Let us verify also this condition: $\omega_\theta(z,\theta) - \omega_\theta(z',\theta) = (t-t', x-x' + O(t)\tilde\omega_\eta(z,\eta) + O(t')\tilde\omega_\eta(z',\eta)$. If here $|t-t'| > c_1/C$, for some fixed large C, then we are already o.k. If $|t-t'| < c_1/C$ and t is small, also t' is small. By choosing C conveniently, we will have then that $|x-x'| > c_1/C'$.

From [15] we then obtain

Theorem 5.1. *Assume that* $\Lambda \cap WF^s_\sim u = \emptyset$. *Then* $T(\Lambda)\cap WF^s_\sim Au = \emptyset$.

Theorem 5.2. *Let Λ be such that $T(\Lambda)\cap WF^s_{\sim}v = \emptyset$. Then $\Lambda\cap WF^s_{\sim}B^*v = \emptyset$.*

Also note that the operator $q = B^*pA$ is pseudodifferential. The symbol calculus gives that it is associated with the symbol $p\circ\chi$ where p is the symbol of p and χ is the canonical transformation associated with ω. After composition with an elliptic operator, it follows that the symbol of q is $\sigma + r$, where r is a symbol in $S^{1-(J+1)/(k-1)}$.

3. Proof of theorem 1.11. First we apply assumption (1.21). Since the $p_{m-1,j}$ are real-valued for $0 \le j \le J = k-2$ it follows from theorem 3.1 that the b_j are also real-valued for the same j. If A and B^* are defined as before, the operator $q(x,D) = B^*pA$ has symbol $\sigma + r$, where $r \in S^0$. Using a standard argument (see for example lemma 4.1.5 in [15]), we may then further reduce ourselves to the case when $r = 0$, by composing everything with an appropriate elliptic operator. We are in this way reduced to propagation of singularities for the operator D_t, and that is trivial. We obtain in particular from the theorems 5.1 and 5.2 that for every solution u of $D_tu - b(t,x,D_x)u = 0$ the set $WF^s_{\sim}u$ is given by $\chi(\Lambda)$, where Λ is an union of lines which are parallel to the t-axis in the manifold $\sigma = 0$ or, equivalently, that $WF^s_{\sim}u$ is invariant under the action of the Hamiltonian flow from (4.4), ..., (4.7). At this moment, to relate this with the assumptions of theorem 1.5, we introduce the map $\tilde{\chi}$, defined by means of the reduced Hamiltonian system:

$$\frac{d\tilde{x}_j'}{dt}(t,y',y'',\eta',\eta'') = -\frac{\partial b_0}{\partial \xi_j'}(t,\tilde{x}'(t,y',y'',\eta',\eta''),y'',\tilde{\xi}'(t,y',y'',\eta',\eta''),\eta'') \tag{5.3}$$

$$\frac{d\tilde{\xi}_j'}{dt}(t,y',y'',\eta',\eta'') = \frac{\partial b_0}{\partial x_j'}(t,\tilde{x}'(t,y',y'',\eta',\eta''),y'',\tilde{\xi}'(t,y',y'',\eta',\eta''),\eta'') \tag{5.4}$$

where b_0 is the principal part of b and we add the initial conditions:

$$\tilde{x}_j'(0,y',y'',\eta',\eta'') = y_j',, \tilde{\xi}_j'(0,y',y'',\eta',\eta'') = \eta_j', \tag{5.5}$$

$j = 1,\dots,n'$. The solutions $\tilde{x}_j'(t,y',y'',\eta',\eta'')$, $\tilde{\xi}_j'(t,y',y'',\eta',\eta'')$ are quasi-homogeneous of degrees 0 and 1 respectively. We can now give the definition of $\tilde{\chi}$:

$$\tilde{\chi}(t,y',y'',\eta',\eta'') = (t,\tilde{x}'(t,y',y'',\eta',\eta''),y'',\tilde{\tau}(t,y',y'',\eta',\eta''), \tilde{\xi}'(t,y',y'',\eta',\eta''),\eta'')$$

where $\tilde{\tau}$ is quasihomogeneous of degree 1 and is computed by solving

$$\frac{d\tilde{\tau}'}{dt}(t,y',y'',\eta',\eta'') = -\frac{\partial b_0}{\partial t}(t,\tilde{x}'(t,y',y'',\eta',\eta''),y'',\tilde{\xi}'(t,y',y'',\eta',\eta''),\eta'') \tag{5.6}$$

$$\tilde{\tau}'(0,y',y'',\eta',\eta'') = \sigma. \tag{5.7}$$

From theorem 4.2.6 of [15] we have that the preceding statement of propagation remains valid if we replace χ by $\tilde{\chi}$. The proof in [15] is based on the remark that, because of the quasihomogeneity of b, the first $n+1$ components of $\chi - \tilde{\chi}$ have strictly negative degree, whereas the second $n+1$ components have degree < 1. It follows then easily that $WF^s_{\sim} v \cap \chi(\Lambda) = \emptyset$ if and only if $WF^s_{\sim} v \cap \tilde{\chi}(\Lambda) = \emptyset$; cf. definition 1.9.

The discussion of domain and image of $\tilde{\chi}$ is similar to that for χ; we have in particular that $\tilde{\chi}$ is defined at least in a set D of the form (4.11), and $\tilde{\chi}(D)$ contains a set of form $V \cap W'_{c,\delta}$. It remains to observe that the solutions of (5.3), ..., (5.7) are exactly the solutions of (1.23), once we return to the initial notations, i.e., write x instead of (t,x) and impose initial data x'^0 instead of $(0,y')$ and η' instead of (σ,η'), with parameters $y'' = x^{0''}$, $\eta'' = \xi^{0''}$, $\rho^0 = (x^0 = (x^{0'},x^{0''})$, $\xi^0 = (0,\xi^{0''})$ as in theorem 1.5. In fact, from (3.8) we have with the present notations

$$p_{m,0}(t,x'(t),x^{0''},b_0(t,x'(t),x^{0''},\xi'(t),\xi^{0''}),\tau(t),\xi'(t),\xi^{0''}) +$$
$$p_{m-1,0}(t,x'(t),x^{0''},\xi^{0''}) = 0;$$

differentiating we obtain, with $\Pi(t,x',\tau,\xi')$ defined as in (1.22):

$$\frac{\partial b_0}{\partial x'_j} = -(\partial_\tau \Pi)^{-1}(\partial_{x'_j}\Pi + \partial_{x'_j} p_{m-1,0}),$$

$$\frac{\partial b_0}{\partial t} = -(\partial_\tau \Pi)^{-1}(\partial_t \Pi + \partial_t p_{m-1,0}),$$

$$\frac{\partial b_0}{\partial \xi'_j} = -(\partial_\tau \Pi)^{-1}(\partial_{\xi'_j}\Pi)^{-1}.$$

Observing that $(\partial_\tau \Pi) \neq 0$ in the domain of $\tau - b_0$, we are reduced to (1.23).

The proof of theorem 1.11 is now concluded by fixing $\xi'^1 \in \dot{R}^{n'}$ and considering the corresponding curve $\gamma(t) = (t,x'(t),x^{0''};0,0,\xi^{0''})$. We know from the assumptions in theorem 1.5 that there is $\tilde{t}$ for which $\tilde{\rho} = \gamma(\tilde{t}) \notin WF^s u$. This means that we can take a sufficiently small conic neighborhood $\tilde{\Gamma}$ of $\tilde{\rho}$ such that $\tilde{\Gamma} \cap WF^s_{\sim} u = \emptyset$. We now choose a bineighborhood V of (x^0,ξ'^1,ξ^0) and $c > 0$ so small that if we denote (with our initial notations)

$$B = \{(x,\xi) \in V; |x| < c, \text{ dist}_{\sim}((x,\xi),W') < c\},$$

we have $\tilde{\chi}(B) \subset \tilde{\Gamma}$. Therefore $\tilde{\chi}(B) \cap WF^s_{\sim}u = \emptyset$ and, by propagation, $B \cap WF^s_{\sim}u \neq \emptyset$. This concludes the proof of theorem 1.11.

6. Characteristic manifolds of codimension 1

1. In the case when the characteristic manifold Σ has codimension $n' = 1$, we are able to give more precise results. As before we argue in a small conic neighborhood Λ of a point ρ^0 in Σ. The assumption (0.26) reads for $n' = 1$:

- In Λ we may write

$$p_m(x,\xi) = e_{m-k}(x,\xi)(a(x,\xi))^k, \tag{6.1}$$

  where $e_{m-k}(x,\xi)$ is an elliptic symbol, homogeneous of order $m-k$, and the first order term $a(x,\xi)$ is real-valued and of microlocal principal type, i.e., $d_{x,\xi}a(x,\xi)$ never vanishes and is not parallel to $\sum_{j=1}^n \xi_j\, dx_j$ on $\Sigma = \{(x,\xi) \in \Lambda; a(x,\xi) = 0\}$.

As for the subprincipal symbol, we shall begin by assuming

$$I_0(\rho) \neq 0 \text{ for all } \rho \in \Sigma. \tag{6.2}$$

Observe that Σ trivially satisfies definition 1.1 in the present case. The leaf through ρ_0 reduces to the bicharacteristic strip γ_0, i.e. the integral curve of H_a on Σ, through the same point. As for the geodesics through ρ_0 defined in section 1, they all coincide with γ_0. After multiplying $P = p(x,D)$ by an elliptic factor, which does not change the validity of (6.2) and of the properties we want to study, we may assume that

$$p(x,\xi) = a(x,\xi)^k + p_{k-1}(x,\xi) + \cdots \tag{6.3}$$

with $a(x,\xi)$ as in (6.1). In particular, we may assume without loss of generality that

$$p_m(x,\xi) \text{ is real-valued and, when } k \text{ is even, non-negative.} \tag{6.4}$$

Let us introduce the following condition, where K may run over $0, 1, \ldots, k-1$:

$$\begin{array}{l} I_0(\rho) \text{ is real-valued and, when } k \text{ is even, negative for } \rho \in \Sigma; \\ \text{moreover } I_j(\rho, X) \text{ defined in (0.35) is real-valued for} \\ (\rho, X) \in N(\Sigma) \text{ if } 1 \leq j \leq k-2-K. \end{array} \tag{6.5}$$

When $K = k-1$, it is understood that there is no assumption on the lower order terms (also (6.2) can be omitted); when $K = k-2$, the assumption only concerns $I_0(\rho)$. When $K = 0$ all the invariants in (0.35) are supposed to be real-valued, i.e. (1.10) is satisfied. An equivalent expression for (6.5) in terms of $p'_{m-1}(x,\xi)$ (see (0.28)) is the following:

$$\partial_x^\alpha \partial_\xi^\beta p'_{m-1}(x,\xi) \text{ is real-valued on } \Sigma \text{ for } |\alpha| + |\beta| \leq k - 2 - K;$$
$$\text{in addition, when } k \text{ is even } p'_{m-1}(x,\xi) \text{ is negative on } \Sigma. \qquad (6.6)$$

Theorem 6.1. *Let (6.1) ,(6.2) ,(6.4) ,(6.5) be satisfied for some K. Assume $1 \leq s < k/K$ (if $K = 0$, then $1 \leq s \leq \infty$). Then, possibly after shrinking Λ, and writing γ_0 for the bicharacteristic strip through $\rho_0 = (x^0, \xi^0)$ restricted to Λ, we have:*

i) There is $v \in \mathcal{M}^s(\Lambda)$ with $Pu = 0$ and $WF^s u = \gamma_0$.

ii) If $u \in \mathcal{M}^s(\Lambda)$ satisfies $Pu = 0$, then $\rho_0 \in WF^s u$ implies $\gamma_0 \subset WF^s u$.

iii) For every $v \in \mathcal{M}^s(\Lambda)$ there is $u \in \mathcal{M}^s(\Lambda)$ such that $Pu = v$.

Remark 6.2. *In the case $K = k$ (when we have no assumptions on the lower order terms), the theorem was already known to be valid; see [1], [25]; obviously, ii) is a particular case of theorem 1.2. In the case $K = 0$ (all the I_j are real-valued) the result was proved for $s = \infty$ by Tulovskii, [26]; the statement was however less explicit than in our theorem 6.1; as for ii), it corresponds to theorem 1.4 when $n' = 1$. Concerning the case $k = 3$, related results in Gevrey classes were obtained by Bernardi-Bove, [2].*

Theorem 6.1 does not give information on the behavior of $P = p(x, D)$ in $\mathcal{M}^s(\Lambda)$ for $k/K \leq s \leq \infty$. In fact, for these values of s the invariants $I_{k-1-K}, \ldots, I_{k-2}$ play an important role. Let us in fact fix attention on I_{k-1-K}, the first not real-valued invariant, and introduce for $K = 1, \ldots, k-1$ the new assumption:

$$Im\, I_{k-1-K}(\rho, X) \neq 0 \text{ for all } (\rho, X) \in N(\Sigma), X \neq 0. \qquad (6.7)$$

(In the case $K = k - 1$, we set $Im\, I_0(\rho) \neq 0$ for all $\rho \in \Sigma$, or else, if k is even, $I_0(\rho) \notin R_-$ for all $\rho \in \Sigma$.)

Theorem 6.3. *Let (6.1),(6.2),(6.4),(6.5),(6.7) be satisfied. Then we have for $k/K \leq s \leq \infty$:*

i) P is s-micro-hypoelliptic, i.e., $WF^s Pu = WF^s u$ for all $u \in \mathcal{M}^s(\Lambda)$.

ii) For every $v \in \mathcal{M}^s(\Lambda)$ there is $u \in \mathcal{M}^s(\Lambda)$ such that $Pu = v$.

In the case $s = \infty$, the result was proved by Tulovskii, [26], under assumptions which can be proved to be equivalent to our assumptions. For $K = k - 1$, theorem 6.3 is a particular case of theorem 1.3.

A model operator for the theorems 6.1 and 6.3 is given by the partial differential operator in $R^2_{x,y}$ with analytic coefficients:

$$\begin{aligned} P &= D_x^k + c_0(x,y)D_y^{k-1} + c_1(x,y)D_xD_y^{k-2} + \cdots + c_{k-2}(x,y)D_x^{k-2}D_y \\ &+ c_{k-1}(x,y)D_x^{k-1} + \text{ terms of order } k-2. \end{aligned} \tag{6.8}$$

We may fix $X = \partial_\xi$ and then obtain $I_j = c_j(x,y)\eta^{k-1-j}$, $0 \le j \le k-2$. The conditions (6.2), (6.5), (6.6) can then be expressed in terms of the coefficients c_j.

2. Proof of theorem 6.1. Arguing as in the sections 2, 3, 4, 5, cf. in particular the end of section 3, nr.4, we are reduced to prove i), ii), iii) in a small conic set Γ for the pseudodifferential operator with symbol

$$\tau + r(t,x,\tau,\xi),$$

where the variables are here $(t,x) \in R^{n+1}$, $(\tau,\xi) \in R^{n+1}$; the bicharacteristic γ_0 is now the parallel line to the t-axis through $\rho_0 = (t^0, x^0, 0, \xi^0)$. In view of (6.5), the symbol r belongs to $S^{1-(J+1)/(k-1)}$ with $J = k-2-K$, that is $r \in S^{K/(k-1)}$. More precisely, from theorem 3.1 we have that

$$r(t,x,\tau,\xi) = \tilde{r}(t,x,\xi) + r_0(t,x,\tau,\xi) \tag{6.9}$$

where $r_0 \in S^0$ and $\tilde{r}$ does not depend on τ. Observe now that, under our assumption $n' = 1$, all the ξ-variables have the same weight $k/(k-1)$; this means that $\tilde{r}$ can be regarded as a classical analytic symbol of order $\sigma = K/k$. As for r_0, it is also a classical symbol, with asymptotic expansion involving the τ-variable. At this moment we apply the following proposition.

Proposition 6.4. *With the preceding notations, let $P = D_t + R$, where R is a classical analytic pseudodifferential operator in Γ of order $\le \sigma$ with $0 \le \sigma < 1$. Assume moreover that $1 \le s < 1/\sigma$. Then there are two linear maps $Q, Q' : \mathcal{M}^s(\Gamma) \to \mathcal{M}^s(\Gamma)$ such that :*

- *Q, Q' are s-microlocal, i.e., $WF^sQu \subset WF^su$ and $WF^sQ'u \subset WF^su$*
- *$QQ' = Q'Q$ is the identity in $\mathcal{M}^s(\Gamma)$.*
- *$Q'PQ = D_t$ in Γ.*

For a proof of proposition 6.4 we refer to [25], where Q, Q' are constructed as pseudodifferential operators with symbols of infinite order. We are hence reduced to study the operator D_t, for which the assertions in the statement are trivial. Theorem 6.1 is therefore proved.

Remark 6.5. *A natural question is whether theorem 6.1 extends to the case of codimension $n' > 1$. An essential step in a prospective proof modeled on* *the arguments above would be a version of proposition 6.4 for quasihomogeneous symbols; this in turn would depend on a generalization to the quasihomogeneous case of the infinite order calculus in [25].*

Proof of theorem 6.3. Arguing as in the proof of theorem 6.1, we are reduced to consider in Γ a classical analytic pseudodifferential operator P with symbol of form

$$p(t,x,\tau,\xi) = \tau + r_{K/k}(t,x,\xi) + r_{(K-1)/k}(t,x,\tau,\xi), \qquad (6.10)$$

where $r_{K/k}$ is analytic, homogeneous of order K/k, with

$$Im\, r_{K/k}(t,x,\xi) \neq 0 \text{ for } (t,x,0,\xi) \in \Gamma \qquad (6.11)$$

and $r_{(K-1)/k}$ is a classical analytic symbol of order $(K-1)/k$. Writing $y = (t,x)$, $\eta = (\tau,\xi)$, we have for large $|\eta|$ and suitable positive c, C:

$$|p(y,\eta)| \geq c|\eta|^{K/k}, \qquad (6.12)$$

$$|D_y^\alpha D_\eta^\beta p(y,\eta)| \leq C^{|\alpha|+|\beta|+1}\alpha!\beta!|p(y,\eta)|(1+|\eta|)^{-K|\beta|/k}. \qquad (6.13)$$

The proof is similar to that of theorem 2.7 and we omit the details. From (6.12), (6.13) we deduce the existence of a s-microlocal inverse of P, acting on $\mathcal{M}^s(\Gamma)$ for $k/K \leq s \leq \infty$. This gives theorem 6.3.

The theorems 6.1 and 6.3 leave open the case when $Im\, I_{k-1-K}(\rho, X)$ vanishes at $\rho = \rho_0$, but $I_{k-1-K}(\rho, X)$ is not real-valued nearby. Let us only observe here that in this case much depends on the behavior of I_{k-1-K} along the bicharacteristic strip γ_0, as indicated by some results in the literature on (non) hypoellipticity and (non-) solvability: for $k = 2$, see [7], [16], [19], [20] when $s = \infty$ and [5] when $2 < s < \infty$; for $k > 2$ see [21] and previous works quoted there, [22], [23] when $s = \infty$ and [6] in the Gevrey case. A representative model for the result of Corli, [6] is the operator of form

$$P = D_x^k + c_0 D_y^{k-1} + \cdots + c_{k-2-K} D_x^{k-2-K} D_y^{K+1} + ix D_x^{k-1-K} D_y^K, \qquad (6.14)$$

which is proved to be non-solvable (locally) for $k/K < s \leq \infty$, under the assumption that $c_0, c_1, \ldots, c_{k-2-K}$ are real constants with $c_0 \neq 0$. Note that from theorem 6.1, iii), we have that P in (6.14) is microlocally solvable for $1 \leq s < k/K$, since (6.5) is satisfied.

References

1. **Aoki T.**: *Calcul exponentiel des opérateurs microdifferentiels d'ordre infini, I,II.* Ann. Inst. Fourier Grenoble, 33 (1983), 227-250 and 36 (1986).

2. **Bernardi E.-Bove A.** : *Propagation of Gevrey singularities for a class of operators with triple characteristics* I,II. Duke Math.J., 60 (1989, 1990), 187-205, 207-220.
3. **Bony J.M.-Schapira P.** : *Propagation des singularités analytiques pour les solutions des équations aux dérivées partielles.* Ann. Inst. Fourier Grenoble, 26 (1976), 81-140.
4. **Boutet de Monvel L.** : *Propagation des singularités des solutions d'équations analogues à l'équation de Schrödinger.* In "Fourier Integral Operators and Partial Differential Equations", Lecture Notes Math. Springer Verlag, vol. 459, ed. by J. Chazarain, (1975), 1-15.
5. **Corli A.**: *On local solvability in Gevrey classes of linear partial differential operators with multiple characteristics.* Comm. Partial Differential Equations, 14 (1989), 1-25.
6. **Corli A.**: *On local solvability of linear partial differential operators with multiple characteristics.* J. Differential Equations, 81 (1989), 275-293.
7. **Egorov J.V.**: *On solvability conditions for equations with double characteristics* . Dokl. A.N. SSSR, 234 (1977), 280-282; Soviet Math. Dokl., 18 (1977), 632-639.
8. **Grushin V.V. - Sananin N. A.**: *Some theorems on the singularities of solutions of differential equations with weighted principal symbol.* Math. U.S.S.R Sb., 32 (1977), 32-44.
9. **Kessab A.**: *Propagation des singularités Gevrey pour des opérateurs à caractéristiques involutives.* Tèse, Université de Paris-Sud, Centre d'Orsay, 1984.
10. **Lascar R.**: *Distributions intégrales de Fourier et classes de Denjoy-Carleman. Applications.* C.R.Acad. Sc. Paris, 284, Sér. A (1977), 485-488.
11. **Lascar R.**: *Propagation des singularités des solutions d'équations pseudodifferentielles quasi-homogènes.* Ann. Inst. Fourier, (Grenoble), 27 (1977), 79-123.
12. **Lascar R.** : *Propagation des singularités des solutions d'équations pseudo differentielles à caracteristiques de multiplicités variables* Lecture Notes Math., vol. 856, Springer Verlag, 1981.
13. **Liess O.**: *Conical refraction and higher microlocalization.* Lecture Notes Math., vol. 1555, Springer Verlag, 1993.
14. **Liess O.-Rodino L.** : *Inhomogeneous Gevrey classes and related pseudodifferential operators.* Boll. Un. Mat. Ital., 3-C (1984), 233-323.
15. **Liess, 0.-Rodino L.**: *Fourier integral operators and inhomogeneous Gevrey classes.* Annali Mat. Pura ed Appl., (IV) vol. 150 (1988), 167-262.
16. **Menikoff A.**: *On hypoelliptic operators with double characteristics.* Ann. Scuola Norm. Pisa Cl. Sci, Ser. IV, (1977), 689-724.
17. **Parenti C.- Rodino L.**: *A class of pseudodifferential operators with involutive characteristics.* Unpublished manuscript.
18. **Parenti C.- Segala F.**: *Propagation and reflection of sigularities for a class of evolution equations.* Comm. Partial Differential Equations, 6 (1981), 741-782.
19. **Popivanov P.R.**: *On the local solvability of a class of pseudodifferential equations with double characteristics.* Trudy Sem. Petrovsk., 1 (1975), 237- 278; transl. Am. Math. Soc. Transl., 118 (1982), 51-90.
20. **Popivanov P.R.** : *Microlocal properties of a class of pseudodifferential operators with double involutive characteristics.* Banach Center Publ. vol. 19 (1987), 213-224.
21. **Popivanov P.R.- Popov G.S.**: *A priori estimates and some microlocal properties of a class of pseudodifferential operators.* C.R. Acad. Bulg. Sci., 33:4 (1980), 461 -463.
22. **Roberts G.B.**: *Quasi-subelliptic estimates for operators with multiple characteristics.* Comm. Partial Differential Equations, 11 (1986), 231-230.
23. **Roberts G.B.**: *A necessary condition for the solvability of certain operators with*

multiple characteristics Comm. Partial Differential Equations, 14 (1989), 877-929.

24. **Rodino L.**: *Linear partial differential operators in Gevrey spaces,* World Scientific 1993, Singapore.
25. **Rodino L.-Zanghirati L.**: *Pseudodifferential operators with multiple characteristics and Gevrey singularities.* Comm. Partial Differential Equations, 11 (1986), 673-711. .
26. **Tulovskii V.N.** : *Propagation of singularities of operators with characteristics of constant multiplicity.* Trudy Mosc. Mat. Obsc., 39 (1979); Trans. Moscow Math. Soc., (1981), 121-144.

GEVREY AND ANALYTIC HYPOELLIPTICITY

DAVID S. TARTAKOFF
Department of Mathematics
University of Illinois at Chicago
851 S. Morgan St., m/c 249
Chicago Illinois 60607-7045, U.S.A.
e-mail: dst@uic.edu

Abstract. In these lectures we study sharp (non-isotropic) Gevrey (and analytic) hypoellipticity for partial differential operators P which are constructed as variable coefficient quadratic polynomials in real vector fields satisfying the Hörmander condition and which satisfy a maximal estimate. We also present some new sharp results obtained jointly with A. Bove.

1. Introduction

In the early 1960's, J.J. Kohn introduced the $\overline{\partial}$- Neumann problem as an important tool for solving $\overline{\partial}$ on strictly pseudo-convex domains in $\mathbf{C}^n$ [1]. The C^∞ regularity of the solution was shown in [26], cf. also [27], [24], using essentially the subellipticity of the problem: for strictly pseudo-convex domains, in the quadratic form formulation there is a loss of one half derivative in the *a priori* estimate. Reduction to the boundary (e.g., via pseudo-differential opertors,) leads to a (pseudo-)differential equation of the form of $\Box_b$, whose prototype, in turn, is the celebrated 'sum of squares' operator $\sum_{j=1}^m X_j^2$ where the X_j are real vector fields generally assumed to satisfy the 'Hörmander condition' that their iterated brackets span the whole tangent space. It had been conjectured and hoped that these problems would turn out to be analytic hypoelliptic - that with locally real analytic data, the solutions would have to be real analytic locally as well.

The now celebrated example of Baouendi and Goulaouic, from 1971 [1], simply written as $P = D_x^2 + D_t^2 + x^2 D_s^2$, which is subelliptic with loss of 1/2 derivative but whose characteristic variety is not symplectic, was shown not to be analytic hypoelliptic and this seemed to close the door

L. Rodino (ed.), Microlocal Analysis and Spectral Theory, 39–59.

on analytic hypoellipticity for non-elliptic problems. The Gevrey regularity of this problem and others was was studied in [19], [33], [12], with the general result that a loss of $1 - 1/m$ derivatives in these problems results in (isotropic) Gevrey hypoellipticity for all $s \geq m$.

In the case of a symplectic characteristic manifold where first brackets suffice to span the tangent space, the author [35] showed that one could 'break the G^2 - barrier' by utilizing a 'maximal' estimate, and by focusing less on the subellipticity. There it was proved (relatively easily) that one had hypoellipticity in *all* Gevrey classes G^s for $s > 1$, hence in their intersection (still a non-quasianalytic class) and also in a certain quasi-analytic class, $C^{\{L \log L\}}$. This last was done by considering a larger collection of non-quasianalytic classes which behave in some essential ways like the Gevrey classes (essentially that they are closed under composition), and whose intersection was quasianalytic (but not yet the analytic class) [35]. As is well known, the local real analytic hypoellipticity for the $\overline{\partial}$-Neumann problem and for $\Box_b$ even on strictly pseudo-convex domains was much harder and was finally achieved in 1978 in [40] and [36] independently. An indication of the subtlety of the local analyticity even in the symplectic case when the lower (even zero!) order terms are not appropriate, cf. [32].

More recent results have been in two directions - proving analytic hypoellipticity in more degenerate settings, globally and in some cases locally, cf. [6], [14], [7], [15], [16], [17] & [18] and counterexamples to analytic regularity of solutions when the characteristic variety does not have some particularly nice properties cf. [8], [23], [10]. Earlier work [22] and [30] pointed in these directions, though the appropriate generalizations had not been clear.

Here we recall these results and related ones and apply the same methods, though simpler than those required for local real analyticity, to show that one can often 'break the G^m barrier' in more degenerate cases as well. Recent results [9] provide sharp isotropic results on Gevrey regularity for certain sums of squares of vector fields, and here we apply our methods to these cases and prove still sharper results which include partial Gevrey hypoellipticity, where the regularity depends on the variables being examined. These results were obtained jointly with Antonio Bove. In particular, we obtain sharp non-isotropic Gevrey regularity results for the example of Baouendi and Goulaouic cited above. A very general setting is then introduced and discussed.

2. Some Definitions and Notation

Definition 1 *A function $h(w)$ belongs to the Gevrey class G^d near w_0 if* *there exists a constant C such that for all multi-indices α and w near w_0,*

$$|D^\alpha h(w)| \leq CC^{|\alpha|}\alpha!^d.$$

Definition 2 *A function $h(x,t,s)$ belongs to the Gevrey class G^{d_1,d_2,d_3} near (x_0,t_0,x_0) if there exists a constant C such that for all multi-indices α and for all (x,t,s) near (x_0,t_0,x_0),*

$$|D_x^{\alpha_1}D_t^{\alpha_2}D_s^{\alpha_3}h(x,t,s)| \leq CC^{|\alpha|}\alpha_1!^{d_1}\alpha_2!^{d_2}\alpha_3!^{d_3}.$$

We note in passing that sup norm estimates will follow from L^2 estimates of a very small number of additional derivatives of a localization of the function in question, in view of the Sobolev Lemma. In fact, it suffices, from a result of Nelson, to bound derivatives as measured by a system of vector fields that span the tangent space, and, for real analyticity, to bound even just powers of a system of vector fields that *generate* the tangent space by their brackets. However, for non-analytic results, such as Gevrey hypoellipticity in other classes, this will not suffice, as has been pointed out by [2].

3. The Elliptic Case

While the real analytic hypoellipticity for elliptic partial differential equations has been known for many decades, we sketch here a proof that will give the flavor of our later proofs. Gårding's inequality for a second order elliptic partial differential operator $P(x,D)$ in R^n reads:

$$\sum_{i,j=1}^{n} \|D_{x_i}D_{x_j}v\|_{L^2}^2 \leq C\{\|Pv\|_{L^2}^2 + \|v\|_1^2\} \tag{1}$$

for all $v \in C_0^\infty$ of small support. Assuming for simplicity that $Pu = f \in C^\omega$ with $u \in C^\infty$, we apply the 'coercive' estimate above to $v = \phi D^\alpha u$ for suitable α and $\phi \in C_0^\infty$ to be specified in a moment. Commuting P past ϕD^α to obtain Pu about which we know the effect of derivatives,

$$\sum_{i,j=1}^{n} \|D_{x_i}D_{x_j}\phi D^\alpha u\|_{L^2}^2 \leq C\|\phi D^\alpha Pu\|_{L^2}^2 + \|[P,\phi D^\alpha]u\|_{L^2}^2 + \|\phi D^\alpha u\|_1^2.$$

Writing P as a sum of terms of the form $a_\gamma D^\gamma, \gamma \leq 2$ with known real analytic coefficients we may write $[P,\phi D^\alpha]$ as a bounded sum of terms of the form $a(x)\phi' DD^\alpha, a(x)\phi'' D^\alpha$, and $\Sigma\binom{\alpha}{\alpha'}$ terms $a^{(\alpha')}\phi D^\gamma D^{\alpha-\alpha'}$, the sum being taken over $0 \neq \alpha' \leq \alpha, |\gamma| \leq 2$.

Of these terms, the last are easy to treat: since the coefficients are of known growth, analytic in this case, their derivatives combine beautifully with the binomial coefficients so that

$$\binom{\alpha}{\alpha'} |a^{(\alpha')}| \|\phi D^\gamma D^{\alpha-\alpha'} u\|_{L^2} \leq \sup_{0 \neq \beta \leq \alpha} C_a^{|\beta|+1} |\alpha|^{|\beta|} \|\phi D^{\gamma+\alpha-\beta} u\|_{L^2},$$

a power of $|\alpha|$ appearing for each 'gain' in free derivative. When all free derivatives have been used up, we find $C^{|\alpha|+1}|\alpha|^{|\alpha|}\|u\|_{L^2}$ which is bounded, by Stirling's formula, by $C_u C^{|\alpha|+1}|\alpha|!$, hence yield a 'good' term in proving analytic (or Gevrey) regularity.

The terms with derivatives on ϕ must be treated carefully by using localizing functions due to Ehrenpreis [21]: for any ω and Ω with ω compactly contained in Ω, $\text{dist}(\omega, \Omega^c) = d$, and any N and m, there exists a constant C_0 independent of N and m and a function $\phi_N \equiv 1$ on ω and in $C_0^\infty(\Omega)$ with

$$|D^\rho \phi_N| \leq C_0 (C_0 N/d)^{|\rho|}, \qquad |\rho| \leq mN. \tag{2}$$

The reason these localizing functions are so useful is that one may differentiate them a high number of times, the choice of function, but not the constants, depending on the number of derivatives one wants to estimate. And if, say, one is really interested in N derivatives, one merely invokes Stirling's formula again to show that when $|\rho| = N, |D^\rho \phi_N| \leq C_1^{|\rho|+1}|\rho|!$, which means that ϕ_N is 'as good as analytic up to order N.' The sole *caveat* is that these estimates do not combine easily with binomial coefficients in inductive arguments and great care must be taken to deal with this in some cases. Indeed, this has been the stumbling block in some proofs.

The result is that all terms that appear from the bracket in the coercive estimate above lead to good gains. The last, as we saw, gains one power of $|\alpha|$ for a gain of one in free derivatives; the same is now seen to be true (without binomial coefficients!) for the other terms. In all, when there are at most one or two free derivatives, we will have $C^{|\alpha|}$ terms, each of analytic growth $|\alpha|^{|\alpha|}$, as long as we choose N comparable to $|\alpha|$. This finishes this (sketch of a) proof of analytic hypoellipticity for elliptic equations.

To make this argument more formal, we produce a suitable inductive hypothesis. To do so, we will abusively write $D^{\alpha+2}$ to denote any $D^\alpha D^\gamma$ with $|\gamma| \leq 2$. Then we may start with ϕ differentiated:

$$\sum_{|\beta|\leq 2} \|\phi_N^{(k)} D^\beta D^\alpha u\|_{L^2}^2 \leq C\{\|\phi_N^{(k)} D^\alpha P u\|_{L^2}^2 + \|[P, \phi_N^{(k)} D^\alpha] u\|_{L^2}^2 + \|\phi_N^{(k)} D^\alpha u\|_1^2\}$$

$$\leq C\{C_0 (C_0 N/d)^{2k} C_{Pu}^{\alpha+1} |\alpha|!^2 + \sum_{\ell=1}^{2} \sum_{|\beta|\leq 2} \|\phi_N^{(k+\ell)} D^\beta D^{\alpha-\ell} u\|_{L^2}^2$$

$$+ \sum_{|\beta|\leq 2} \sum_{0\neq\alpha'}^{\alpha} C_a^{\alpha'+1} |\alpha|^{\alpha'} \|\phi_N^{(k)} D^\beta D^{\alpha-\alpha'} u\|_{L^2}^2\}$$

A suitable inductive hypothesis then would be that

$$\sum_{|\beta|\leq 2} \|\phi_N^{(\tilde{k})} D^\beta D^{\tilde{\alpha}} u\|_{L^2} \leq C_1 C_2^{|\tilde{k}|} C_3^{|\tilde{\alpha}|} N^{|\tilde{\alpha}|+2+\tilde{k}} \tag{3}$$

provided $|\tilde{\alpha}| + \tilde{k} \leq |\alpha_0| + |k_0|$ and $\tilde{\alpha} < \alpha$. Then appropriate choices of the constants, relative to one another, will complete the induction step.

4. Subelliptic Cases From Complex Analysis

Perhaps the simplest non-elliptic equations to consider are the *subelliptic* ones. For these, we find the useful form of Gårding-type inequalities are formulated with quadratic forms. That is, while the coercive estimate was stated in (1) in terms of $\|Pu\|_{L^2}$, it could equally well have been written

$$\sum_{|\beta|\leq 1} \|D^\beta v\|_{L^2}^2 \leq C\{|\,(Pv, v)_{L^2}\,| + \|v\|_{L^2}^2\} \tag{4}$$

for v of given compact support and the analysis does not change materially (see below). By a *subelliptic* operator we will mean one where the norm on the left is replaced by a Sobolev norm of fractional degree: for some positive ϵ and all v as above,

$$\|v\|_\epsilon^2 \leq C\{|Re(Pv, v)| + \|v\|_{L^2}^2\}. \tag{5}$$

As the $\bar{\partial}$-Neumann problem on a strictly pseudo-convex domain presented perhaps the first example of a subelliptic problem and motivated most of the later ones; it is a boundary value problem which we outline here for completeness, though the details will not be essential in the sequel.

Let Ω be an open, relatively compact submanifold of a complex Hermitian manifold Ω', with smooth boundary $\Gamma = \partial\Omega$. We consider

$$\bar{\partial} : C_{p,q}^\infty(\bar{\Omega}) \to C_{p,q+1}^\infty(\bar{\Omega}),$$

together with its adjoint $\bar{\partial}^*$ The $\bar{\partial}$-Neumann problem on Ω consists in finding a (p, q) form u on $\bar{\Omega}$, in the domain, $D^{p,q}$, of $\bar{\partial}^*$, with $\bar{\partial}u$ in the domain, $D^{p,q+1}$ of $\bar{\partial}^*$ on $(p, q+1)-$ forms as well, and satisfying

$$Q(u, w) \equiv (\bar{\partial}u, \bar{\partial}w)_\Omega + (\bar{\partial}^* u, \bar{\partial}^* w)_\Omega = (\alpha, w)_{L^2(\Omega)}, \forall w \in D^{p,q},$$

for a given (p, q)-form α. Thus also

$$\Box u \equiv (\overline{\partial}\overline{\partial}^* + \overline{\partial}^*\overline{\partial})u = \alpha$$

in Ω.

$\Box_b$ is analogously defined as follows: Let Γ be any real $2n-1$ dimensional compact CR manifold, i.e., a Hermitian manifold for which $CT\Gamma = T_\Gamma^{1,0} \oplus T_\Gamma^{0,1} \oplus N, \dim_R N = 1$, $T_\Gamma^{0,1} (= \overline{T_\Gamma^{1,0}})$ is integrable, and $T_\Gamma^{0,1}$ has trivial intersection with $T_\Gamma^{1,0}$. Defining $\Lambda^{p,0} = \{p\text{-forms in } CT^*\Gamma \text{ which annihilate } E = (T_\Gamma^{1,0} \oplus T_\Gamma^{0,1})^\perp \text{ and } T^{0,1}\}$, let

$$\overline{\partial}_b = \pi_{p,q+1} \circ d,$$

mapping $C^\infty(\Lambda^{p,q})$ to $C^\infty(\Lambda^{p,q+1})$, where $\pi_{p,q+1}$ = orthogonal projection onto $\Lambda^{p,q+1}$. Thus $(\overline{\partial}_b)^2 = 0$. Let $\overline{\partial}_b^*$ = the adjoint of $\overline{\partial}_b$, and set

$$\Box_b = \overline{\partial}_b \overline{\partial}_b^* + \overline{\partial}_b^* \overline{\partial}_b.$$

To define the *Levi form*, we let Γ be given by $r = 0, dr \neq 0$, and choose $\{L_j\}_{j \leq n}$ independent, real analytic vector fields, spanning $T^{1,0}$ at each point, with $L_{j<n}$ tangent to Γ, and $L_n r = 1$ on Γ, and finally set $T = (L_n - \overline{L_n})$. The Levi matrix $c_{j,k}$ on Γ is given by $[L_j, \overline{L_k}] \equiv c_{j,k} T$ mod $\{L_j, \overline{L_j}\}, j < n$, Γ is called (strictly) pseudo-convex if $c_{j,k}$ is (strictly) definite.

Note that neither the $\overline{\partial}-$ Neumann problem nor $\Box_b$ is elliptic. Although $\Box$ is an elliptic operator, the boundary conditions are not coercive, and $\Box_b$ can never be elliptic - there are only $2(n-1)$ vector fields on a $2n-1$ dimensional manifold.

Furthermore, via a (pseudodifferential) reduction to the boundary, the $\overline{\partial}-$ Neumann problem is equivalent to a (pseudo-)differential problem of the same general form as $\Box_b$ - in particular, taking real and imaginary parts of the tangential holomorphic vector fields, the principal part of both problems becomes a sum of squares of real vector fields whose iterated brackets span the whole tangent space precisely when the domain is of 'finite type'.

5. Equivalent Real Problems. Subellipticity.

To fix notations, we will consider a slight generalization of the sum of squares of real vector fields,

$$P = \sum_{k=1}^{M} a_{j,k} X_j X_k + X_0 + c(x) \tag{6}$$

with positive definite symmetric $a_{j,k}$.

And we shall make the assumption (H_2) that the $\{X_j\}, j = 1, \dots M$, together with their iterated brackets of length, say, m generate the entire tangent space. Then one has at once, for v of fixed compact support

$$\sum_{j=1}^{M} \|X_j v\|_{L^2}^2 \leq C\{|\Re(Pv, v)| + \|v\|_{L^2}^2\}$$

and, from [31], also subellipticity since the sum of just the squares on the left dominates the Sobolev norm with loss of $1 - 1/m$ derivatives on such v:

$$\sum_{j=1}^{M} \|X_j v\|_{L^2}^2 + \|v\|_{1/m}^2 \leq C\{|\Re(Pv, v)| + \|v\|_{L^2}^2\}. \tag{7}$$

Theorem 1 (Derridj-Zuily) *Let P be given as in (6) and satisfy the estimate (7). Then P is G^s hypoelliptic for any $s \geq m$.*

Proof: As in the proof of analytic hypoellipticity for elliptic operators above, we introduce $v = \phi_N D^\alpha u$ where the solution u is (assumed for convenience to be) in C^∞. Then we have, from (7),

$$\sum_{j=1}^{M} \|D_{x_j} \phi_N D^\alpha u\|_{L^2}^2 + \|\phi_N D^\alpha u\|_{1/m}^2 \leq C\{|\Re(\phi_N D^\alpha P u, \phi_N D^\alpha u)_{L^2}| +$$

$$+ |\Re([P, \phi_N D^\alpha] u, \phi_N D^\alpha u)_{L^2}| + \|\phi_N D^\alpha u\|_{L^2}^2\}.$$

Again writing P as a sum of terms of the form $a_{j,k} X_j X_k$, X_0, and $c(x)$ with known real analytic coefficients we may write $([P, \phi_N D^\alpha] u, \phi_N D^\alpha u)_{L^2}$ as a bounded sum of terms (underlining a coefficient to indicate the number of terms of a given form that may occur) of the form

$$(a(x) X_j \phi_N' D^\alpha u, \phi_N D^\alpha u)_{L^2}, \qquad (a(x) \phi_N'' D^\alpha u, \phi_N D^\alpha u)_{L^2},$$
$$\underline{\alpha}(a(x) \phi_N X_k [X_j, D] D^{\alpha-1} u, \phi_N D^\alpha u)_{L^2},$$

and

$$\underline{\alpha(\alpha-1)}(a(x) \phi_N D^\alpha u, \phi_N D^\alpha u)_{L^2}$$

plus terms such as $\Sigma \binom{\alpha}{\alpha'}$ terms $\|a^{(\alpha')} \phi(X_j) D^{\alpha-\alpha'} u\|_{L^2} \|X_j \phi_N D^\alpha u\|_{L^2}$, the sum being taken over $0 \neq \alpha' \leq \alpha$.

All of these terms have a new look to them, since now we must distinguish between X_j derivatives and unspecified derivatives, which we have denoted by D.

The Schwarz inequality will be used on all of these terms. In the first, the X_j is brought to the right side of the inner product and a small multiple of the L^2 norm of that part absorbed on the left hand side of the basic inequality. What remains is a large multiple of $\|\phi'_N D^\alpha u\|^2_{L^2}$.

In the second term, we split the weight between the two halves, assigning $|\alpha|$ to right and $|\alpha|^{-1}$ to the left.

The third term is quite benign, since we may move the X_j to the second member modulo an error of the form $\|\phi'_N D^\alpha u\|_{L^2}\|\underline{\alpha}\phi_N D^\alpha u\|_{L^2}$.

And in the fourth term, we associate one power of α with each side.

The result, modulo a small multiple of the left hand side of the basic inequality, is that one may bound $|([P, \phi_N D^\alpha]u, \phi_N D^\alpha u)_{L^2}|$ by

$$\|\phi'_N D^\alpha u\|^2_{L^2}, \quad |\alpha|^2\|\phi_N D^\alpha u\|^2_{L^2}, \quad |\alpha|^{-2}\|\phi''_N D^\alpha u\|^2_{L^2},$$

and

$$\sum_{0\neq\alpha'\leq\alpha} \underline{\binom{\alpha}{\alpha'}} \|a^{(\alpha')}\phi_N(X_j)D^{\alpha-\alpha'}u\|_{L^2}\|(X_j)\phi_N D^\alpha u\|_{L^2}$$

The terms where D^α is still present but lack the help of an X_j, such as $\phi'_N D^\alpha$ and $\underline{\alpha}\,\phi_N D^\alpha$ are new and require making use of the subellipticity. We write, for example,

$$\|\phi'_N D^\alpha u\|^2_{L^2} = \|\Lambda^{-1/m}\phi'_N D^\alpha u\|^2_{1/m}$$

where Λ is the pseudo-differential operator with $\sigma(\Lambda) = (1+|\xi|^2)^{1/2}$. While $\Lambda^{-1/m}\phi'_N D^\alpha u$ no longer has compact support, we can introduce a second cut-off function $\tilde{\phi}_N$ to the left of $\Lambda^{-1/m}$ with the same type of growth of derivatives as ϕ_N but identically equal to one on the support of ϕ_N, so that except for an error (namely $[\tilde{\phi}_N, \Lambda^{-1/m}]\phi_N$) of arbitrarily low order we may proceed as if $\Lambda^{-1/m}\phi'_N D^\alpha u$ still has compact support.

Thus modulo further brackets, which drop the order by one full derivative each time, we may iterate the whole estimate with $\Lambda^{-1/m}$ and either an extra derivative on ϕ_N or a factor of $|\alpha|$. After a total of $C^{m|\alpha|}$ iterations, all free derivatives will have been used up, and the resulting terms will either contain Pu, which will not concern us, or have the form

$$\underline{|\alpha|^{2b_1}}\|\phi_N^{(b_2)}u\|^2_{L^2}$$

with $b_1 + b_2 = m|\alpha|$. Invoking the bounds on derivatives of ϕ_N above yields a bound on these terms of $|\alpha|^{m|\alpha|}$, which is the statement of G^m hypoellipticity. For $s \geq m, G^s$ hypoellipticity follows the same lines.

6. Breaking the G^2 Barrier. Analyticity.

For what was perhaps the most interesting case, that of the problems from complex analysis in the case of strictly pseudo-convex domains, and where G^2 hypoellipticity had been proven by [19], [34], [11], etc., it had been conjectured since the end of the 1960's that real analytic hypoellipticity might still hold despite the Baouendi-Goulaouic example. Here the situation could be modelled by the real operator

$$P = \sum_{j=1}^{n-1} X_j^2 + \sum_{j=1}^{n-1} Y_j^2$$

where, in local coordinates,

$$X_j = \frac{\partial}{\partial x_j} - y_j \frac{\partial}{\partial t}$$

and

$$Y_j = \frac{\partial}{\partial y_j} + x_j \frac{\partial}{\partial t}.$$

These simple vector fields occur in the context of the Heisenberg group, though we make no use of the group structure. Clearly the 1/2 estimate holds, as well as the 'maximal' estimate with which we have been dealing all along:

$$\sum_{j=1}^{n-1} \|X_j v\|_{L^2}^2 + \sum_{j=1}^{n-1} \|Y_j v\|_{L^2}^2 + \|v\|_{1/2}^2 \le C\{|(Pv, v)| + \|v\|_{L^2}^2\}$$

but every effort to bound derivatives of the form $v = \phi_N T^k u$, with $T = \partial/\partial t$ more effectively than G^2 had failed. Of course since P is microlocally elliptic in directions other than T this would suffice. The problem was that the inevitable bracket $[X_j, \phi_N T^k]u = (X_j(\phi_N))T^k u$ could not be effectively majorized without loss of 1/2 derivative.

6.1. GLOBAL ANALYTICITY: THE STRONGLY PSEUDOCONVEX CASE

It became clear that for *global* anlyticity, such derivatives on ϕ were of no concern, and in fact one should not use the specialized ϕ_N at all - virtually any partition of unity would suffice. Whenever the localizing functions were differentiated one would bring them out of the norm and start over. What did seem to be required was a *globally defined* $T-$ vector field, but that was not a limitation in the embedded case since the (inward pointing) normal ν

could be globally defined and, via the complex structure, $T = J\nu$ could be taken as a global starting point; if this vector field did not commute perfectly with the tangential holomorphic and anti-holomorphic vector fields then it could be modified by the addition of multiples of those vector fields in a unique way to commute adequately. This was done independently in 1976 in [28] and [34], though had been found earlier by Tanaka.

6.2. HIGHER LOCAL REGULARITY

The crucial observation in breaking the G^2 barrier locally was that, utilizing the excellent commutation relations enjoyed by the X_j, Y_k, which could always be arranged by using the Darboux theorem, one could 'correct' the *localized* vector field ϕT^k in such a way that its bracket with X_j or Y_j would *not* have a derivative on ϕ without a decrease in power of T. In fact,

$$[X_j, \phi T - X_j(\phi)Y_j] = -(X_j^2(\phi))Y_j$$

is free of T. Arguing that the extra derivative on ϕ is far less troublesome than free T derivatives, the localization $\phi_N T$ of T was replaced with

$$T_{\phi_N} = \phi_N T - \sum_j (X_j(\phi_N))Y_j + \sum_j (Y_j(\phi_N))X_j,$$

and $\phi_N T^k$ by $(T_{\phi_N})^k$. While the localizing functions were still required to be able to accept large numbers of derivatives, since they would occur embedded between many T's, whenever derivatives appeared on a ϕ_N it was brought out of the norm as quickly as possible. The result [35] was hypoellipticity in all the Gevrey classes except the analytic one and in some other non-quasi analytic classes whose intersection was quasi-analytic but, it was not hard to realize, not the real analytic class.

The real analytic hypoellipticity required a more courageous construction. T_{ϕ_N} and its powers were not good enough - a construction requiring only *one* ϕ for the entire T^k seemed needed. And the idea that worked was an iterate of the construction of T_{ϕ_N}. That is, if

$$[X_k, T_{\phi_N} = \phi_N T - \sum_j (X_j(\phi_N))Y_j + \sum_j (Y_j(\phi_N))X_j] \equiv 0$$

modulo $\{\phi_N^{(\prime\prime)}X, \phi_N^{(\prime\prime)}Y\}$, (and the same bracketed with Y_k,) then could one not correct $\phi_N T^2$ in similar fashion so that the bracket of this 'correction,' $(T^2)_{\phi_N}$, when bracketed with either X_j or Y_k contained no free T derivatives at all? The answer is 'no'. The best one can do is to make a choice of the order in which the X's and Y's appear and write

$$\left(T^2\right)_{\phi_N} = T_{\phi_N}T + \sum_{|\delta_1+\delta_2|=2} (-1)^{|\delta_1|}\frac{(X^{\delta_1}Y^{\delta_2}(\phi_N))}{\delta_1!\delta_2!}X^{\delta_2}Y^{\delta_1}$$

Then it comes as no surprise that $(T^2)_{\phi_N}$ commutes beautifully with the X_j and Y_k *except for the order in which the vector fields occur.* But while $[X_k, (T^2)_{\phi_N}] \equiv 0$ modulo terms $\phi_N^{(3)} Z^2$ where Z may be an X or a Y, this is not true of the bracket with Y_k. However, a miracle does occur - modulo these terms and another expression of the form $T_{\phi_N} Y_k$ this does occur. An extremely brief calculation gives:

Proposition 1 $[Y_k, (T^2)_{\phi_N}] \equiv T_{T\phi_N} \circ Y_k$ *modulo terms* $\phi_N^{(3)} Z^2$ *where each* Z *is a* Y *or an* X.

It is this miracle, that modulo terms with only free Z's, the remaining terms combine in the perfect balance of $T_{T\phi_N} Y_k$, and which is strongly tied to the particular commutation relations enjoyed by the X's and Y's, that permits the

Definition 3

$$(T^p)_{\phi_N} = \sum_{|\delta_1+\delta_2|\le p} (-1)^{|\delta_1|} \frac{X^{\delta_1} Y^{\delta_2}(\phi_N)}{\delta_1!\delta_2!} X^{\delta_2} Y^{\delta_1} T^{p-|\delta_1+\delta_2|}$$

Proposition 2

$$[Y_k, (T^p)_{\phi_N}] \equiv \left(T^{p-1}\right)_{T\phi_N} \circ Y_k$$

and

$$[X_k, (T^p)_{\phi_N}] \equiv 0,$$

both modulo terms $\phi_N^{(p+1)} Z^p$ *where each* Z *is a* Y *or an* X.

There remain many long and difficult arguments before one reaches the analytic hypoellipticity, the most difficult being the brackets of this expression with (variable!) coefficients, but the essential result contained in this proposition, that brackets with the vector fields occurring in the differential operator under consideration lead to expressions which can be again subjected to the *a priori* estimate (the maximal estimate, in this case), bounding X's and Y's, with no sacrifice - the remaining Y_k of the proposition gives the *a priori* estimate its full power and, as is quite evident, each iteration results in trade-off of one derivative on ϕ_N for one gain in the number of T derivatives: $\left(T^{k-1}\right)_{T\phi_N}$ where one had started with $\left(T^k\right)_{\phi_N}$. This 'one-for-one' trade-off leads, as we saw in the elliptic argument included above just for this point, to analyticity.

7. Weakly Pseudo-convex Cases

Much attention has been given to cases where the Levi form is merely semi-definite, the so-called weakly pseudo-convex cases. When the domain

Ω is bounded with real analytic boundary, we know that the $\bar{\partial}-$ Neumann problem is subelliptic [20] and more generally, Catlin has shown that finite type conditions imply subellipticity more generally [5]. But beyond the conclusion that subellipticity implies Gevrey regularity as in the above Theorems, most attention has gone to study real analytic regularity when there is a maximal estimate (usually requiring a lot of symmetry - cf. [12]), e.g., [6], [7] with a series of papers [34], [13] proving local real analytic hypoellipticity for either the $\bar{\partial}-$Neumann problem or for $\square_b$ whose Levi forms degenerate in well-controlled ways). Semi-global regularity is also an active area of research currently.

In addition, some mixed results, where the characteristic manifold splits into an involutive portion and a symplectic portion were studied for pseudo-differential operators very recently by the author and Antonio Bove [4]. The class satisfies the hypotheses for C^∞ hypoellipticity with loss of one derivative, and the results state that for $1 \le s < 2$ the operator P propagates singularities, i.e., smoothness, along the leaves of the characteristic variety, if there are any, while for $s \ge 2$, the operator is G^2- hypoelliptic.

Using Métivier's technique of addition of variables and making an analytic canonical transformation we write the operator in question as

$$\langle A(x,\xi)\begin{bmatrix} X \\ Z \end{bmatrix}, \begin{bmatrix} X^* \\ Z \end{bmatrix}\rangle + \langle L(x,\xi), \begin{bmatrix} X^* \\ Y \end{bmatrix}\rangle + \tilde{p}_1(x,\xi) + \tilde{p}_0(x,\xi), \tag{8}$$

where

$$\begin{cases} X_j = \frac{1}{i}\frac{\partial}{\partial x_j} - x_{k+j}|D_{x_n}|, & 1 \le j \le k, \\ X_{k+j} = \frac{\partial}{\partial x_{k+j}} & 1 \le j \le k, \\ Z_s = \frac{\partial}{\partial x_{2k+s}} & 1 \le s \le \ell, \end{cases} \tag{9}$$

$X = (X_1, \ldots, X_k, X_{k+1}, \ldots, X_{2k})$, $Z = (Z_1, \ldots, Z_\ell)$, A is a self adjoint positive definite matrix, of size $2k + \ell$, of pseudo-differential operators of order 0, and L is a (complex) $2k+\ell$ dimensional vector of pseudo-differential operators of order 0. The characteristic variety is denoted by Σ, and these results are stated micro-locally:

Theorem 2 (Bove-Tartakoff, 1996) *Let P be as above. Let $(x_0, \xi_0) \in \Sigma$ and W be a neighborhood of (x_0, ξ_0). Suppose $1 \le s < 2$ and that $(x_0, \xi_0) \notin WF_s(Pu)$; then if $\Gamma_{(x_0,\xi_0)} \cap (W \setminus \{(x_0, \xi_0)\}) \cap WF_s(u) = \emptyset$ we have $(x_0, \xi_0) \notin WF_s(u)$.*

Theorem 3 (Bove-Tartakoff, 1996) *Under the same assumptions as in the above Theorem, let $s \ge 2$ and $(x_0, \xi_0) \notin WF_s(Pu)$. Then $(x_0, \xi_0) \notin WF_s(u)$, i.e. P is microlocally G^s-hypoelliptic.*

Remark: For $s = \infty$, this theorem is well known by [24], [3]. For $s = 1$, Theorem (2) is found in [29]. And the second of these theorems was proved using different methods in [25].

The principal part of the simplest (partial differential) prototype in $R^{2k+\ell+1}$ is

$$P = \sum_{j=1}^{2k} X_j^2 + \sum_{j=1}^{\ell} Z_j^2,$$

where the $\{X_j\}$ satisfy the Heisenberg commutation relations (microlocally), and the Z_j commute with all other vector fields, the $\{X_j\}$ and $\{Z_j\}$ being independent. Of course, together with the first brackets of the X_j they do span the tangent space. Here the results may be stated as saying that G^s– hypoellipticity for $s \geq 2$. For example, and we shall have more to say about this kind of example later, in the most concrete situation, when

$$P = \left(\frac{\partial}{\partial x_1} - x_2\frac{\partial}{\partial t}\right)^2 + \left(\frac{\partial}{\partial x_2} + x_1\frac{\partial}{\partial t}\right)^2 + \left(\frac{\partial}{\partial z}\right)^2$$

in R^4, with the obvious vector fields, the characteristic manifold consists of $\Sigma = \{\xi = (\xi_1, \xi_2) = 0, \eta = 0, x_1 = x_2 = 0\}$ and its leaves are free in y. Thus the first theorem states that if $(0,0,0,0;0,0,0,1) \notin WF_s(Pu)$ and $(0,0,y,0;0,0,0,1) \notin WF_s(u)0$ for $0 \neq |y| < \epsilon$, for some $\epsilon > 0$ then $(0,0,0,0;0,0,0,1) \notin WF_s(u)$ if $1 \leq s < 2$.

The proof of this result employs the analytic hypoellipticity proof of [38] which extended [36], [37] to prove Métivier's Theorem (analytic hypoellipticity for higher codimension symplectic characteristic variety with loss of 1/2 derivative) [29] together with propagation techniques which in this simple setting are easy to state. For, localizing $T = \partial/\partial t$, the localizing function $\phi(x_1, x_2, y, t)$ must behave as before in (x_1, x_2, t) since one expects fine hypoellipticity in those variables, but in y can be quite general, since one does not expect to have analyticity in any case in that variable. Thus whenever a derivative in y lands on the localizing function, we have no way to produce better than G^2 results *unless* we make the *a priori* assumption that where such a derivative is non zero, the solution is already smooth. Thus if there is no singularity in y derivatives in a band $\epsilon_1 \leq |y| \leq \epsilon_2$, there will be none at $y = 0$ either.

8. Breaking the G^s– Barrier; Other Rational Exponents

Recently there has been study of Gevrey hypoellipticity when s takes on values other than $1/m$. Christ [9] has very recently obtained sharp isotropic results in rational Gevrey classes that are better than those predicted by the subellipticity index. In this section and those that follow we improve on those regularity results by considering non-isotropic classes and show that all of these results are accessible with L^2 methods alone.

We consider here the particular, though apparently fairly typical, example

$$P = \frac{\partial^2}{\partial x^2} + (x\frac{\partial}{\partial t})^2 + (x^2\frac{\partial}{\partial s})^2 \qquad (10)$$
$$= X_1^2 + X_2^2 + X_3^2.$$

Theorem 4 *The operator P is G^d hypoelliptic for all $d \geq 3/2$.*

Remark: This theorem is due to M. Christ. However our proof is both elementary and allows more precise results about partial regularity, where derivatives in different directions grow at different rates.

Theorem 5 (Bove-Tartakoff, 1996) *The operator P is G^{d_1,d_2,d_3} hypoelliptic for $d_1 \geq 7/6, d_2 \geq 1$, and $d_3 \geq 3/2$, and microlocally as well.*

Proof: We shall use the *a priori* estimate, for $v \in C^\infty$, any vector field W, and the localizing function $\phi = \phi_N$, as above:

$$\sum_j \|X_j\phi W^p v\|_{L^2}^2 + \sum_j \|\phi X_j W^p v\|_{L^2}^2 + \|\phi W^p v\|_{1/3}^2 \qquad (11)$$

$$\leq C\{|(P\phi W^p v, \phi W^p v)_{L^2}| + \sum_j \|X_j(\phi)W^p v\|_{L^2}^2\} + \|\phi W^{p-1}v\|_{L^2}^2\}, v \in C^\infty.$$

(Note that we distinguish the different derivatives on ϕ since the X_j carry coefficients which are powers of x, a fact that will be crucial in the sequel.) We have used several facts in writing down this estimate: the form of P clearly allows us to bound the basic vector fields X_j on v, and thus the 1/3 norm since second brackets suffice to span the tangent space, we have used a symmetric form on the left, with the X_j either before or after the localizing functions, since both will occur as errors. It follows that if $Pw = f \in G^s$ for some $s \geq 3$ the same is true of w locally.

We shall obtain bounds of the solution locally (in L^2 norms) of the type

$$\|\phi D_x^{\alpha_1} D_t^{\alpha_2} D_s^{\alpha_3} u\|_{L^2} \leq CC^{|\alpha|}\alpha_1!^{d_1}\alpha_2!^{d_2}\alpha_3!^{d_3}, |\alpha| \leq N.$$

and to do this it suffices, by integration by parts and a simple inductive hypothesis, to treat pure powers of D_x, D_t, and D_s.

We start with $W = D_t$, as this turns out to be the simplest. Then

$$[P, \phi_N D_t^p] = D_x(\phi_N)_x D_t^p + (\phi_N)_x D_x D_t^p + xD_t x(\phi_N)_t D_t^p$$

$$+x^2(\phi_N)_{tt}D_t^p + x^2 D_s x^2(\phi_N)_s D_t^p + x^4(\phi_N)_{ss}D_t^p.$$

Now the (t,s) derivatives on ϕ_N are serious, (recall that $x-$ derivatives leave us in the elliptic region where the result is known) but even these derivatives will not be harmful to a proof even of analyticity (in the variable t) *if* there is a corresponding gain in powers of D_t without losing the two good X's.

These two good X's may come either from P (one X comes from P in the first, third and fifth terms on the right here, though in the first two terms the presence of an x derivative on ϕ_N lands us in the elliptic region) or are *created* by combining powers of the variable x with copies of D_t, namely once in the third and fifth terms and twice in the fourth and sixth, thus reducing the power p on D_t. Note that these (one or two) powers of D_t must still be commuted past ϕ_N to be in a useful position, and they may land on the localizing function. If it they do, we must try to bring another D_t out, etc. But to simplify this exposition, we include only the principal term, ignoring these second-order brackets. Once we have X's on the left, one of these X's will be moved to the right (by integration by parts). We obtain:

$$([P,\phi_N D_t^p]u,\phi_N D_t^p u)_{L^2} = (D_x(\phi_N)_x D_t^p u,\phi_N D_t^p u)_{L^2}$$

$$+((\phi_N)_x D_x D_t^p u,\phi_N D_t^p u)_{L^2} + 2\left(xD_t(\phi_N)_t D_t^{p-1}u, xD_t\phi_N D_t^p u\right)_{L^2}$$

$$+\left(xD_t(\phi_N)_{tt}D_t^{p-2}u, xD_t\phi_N D_t^p u\right)_{L^2} + 2\left(x^2D_t(\phi_N)_s D_t^{p-1}u, x^2 D_s\phi_N D_t^p u\right)_{L^2}$$

$$+\left(x^2 D_t(\phi_N)_{ss}D_t^{p-2}u, x^2 D_t\phi_N D_t^p u\right)_{L^2}$$

so that (recalling the norms are squared) for any $\epsilon > 0$,

$$|\,([P,\phi_N D_t^p]u,\phi_N D_t^p u)_{L^2}\,| \le C\|\phi_N D_t^p u\|_{L^2}^2 + CC^{2|\alpha|}(2|\alpha|)!$$

$$+\epsilon\|xD_t\phi_N D_t^p u\|_{L^2}^2 + C_\epsilon\{\|xD_t(\phi_N)_t D_t^{p-1}u\|_{L^2}^2 + \|xD_t(\phi_N)_{tt}D_t^{p-2}u\|_{L^2}^2\}$$

$$+C_\epsilon\{\|x^2 D_t(\phi_N)_s D_t^{p-1}u\|_{L^2}^2 + \|x^2 D_t(\phi_N)_{ss}D_t^{p-2}u\|_{L^2}^2\}.$$

After replacing the L^2 norm first on the right by again ϵ times the $1/3$ norm modulo a large constant times the -1 norm and using this to absorb one power of D_t, this leads, upon iteration p times, starting from (11) with $W = D_t$, to the bounds

$$\sum_j \|X_j\phi_N D_t^p u\|_{L^2}^2 + \sum_j \|\phi_N X_j D_t^p v\|_{L^2}^2 + \|\phi_N D_t^p u\|_{1/3}^2$$

$$\le CC_{u,Pu}^{2|\alpha|}(2|\alpha|)! + C^{2|\alpha|+2}(2|\alpha|)!\|u\|_{L^2}^2$$

$$= CC_{u,Pu}^{2|\alpha|}(2|\alpha|)! + C_u^{2|\alpha|+2}(2|\alpha|)!$$

which yields analytic growth in the D_t derivatives.

Next we tackle D_s derivatives, which are not nearly as simple:

$$[P, \phi_N D_s^p] = D_x(\phi_N)_x D_s^p + (\phi_N)_x D_x D_s^p + 2xD_t x(\phi_N)_t D_s^p$$

$$+x^2(\phi_N)_{tt} D_s^p + 2x^2 D_s x^2(\phi_N)_s D_s^p + x^4(\phi_N)_{ss} D_s^p$$

Again, the x derivatives on ϕ_N are not serious, and the last two terms may be treated precisely as we did with high powers of D_t: combining x^2 with D_s 'creates' a 'good' derivative (i.e., an X_j) which may be integrated to the right in the inner product. This exchange - a derivative on ϕ_N for a gain in p - i.e., a gain in power of D_s derivatives is what has led to optimal (i.e., analytic) regularity all along.

But it is the third and fourth terms that are more troublesome; and where the new features arise. For combining only one power of x with D_s will not generate a 'good' vector field sufficient to balance a derivative (D_t) falling on ϕ_N . Two powers of x are required. However a little patience will produce two, even in this 'worst case scenario'. For if we are repeatedly so unlucky (and all cases do occur) as to bracket with xD_t, (and we ignore other contributions) we find that after *two* such brackets we may decrease p by one.

But something even better happens. Consider: we start with $X\phi_N D_s^p u$ and after the first use of the basic *a priori* estimate we have come up with $x(\phi_N)_t D_s^p u$. As with the proof in the general subelliptic case above, we do have a gain of 1/3 derivative to make use of. That is, for the next iteration we write

$$\|x(\phi_N)_t D_s^p u\|_{L^2} = \|\Lambda^{-1/3} x(\phi_N)_t D_s^p u\|^2_{1/3},$$

and treat $\Lambda^{-1/3}x(\phi_N)_t D_s^p u$ as the new version of $\phi_N D_s^p u$ whose X_j derivatives as well as whose 1/3 norm is bounded in the estimate. (See the proof of the first Theorem for justification of using $\Lambda^{-1/3}x(\phi_N)_t D_s^p u$ as if it had compact support in introducing it into the *a priori* estimate again.) In the next iteration, we will be led to analogous 'errors', such as $\Lambda^{-1/3}x^2(\phi_N)_{tt} D_s^p u$ with one more power of x and another (D_t) derivative of ϕ_N.

But now something very exciting happens. We *may* treat the x^2 together with D_s as a *good* vector field, namely as one of the X_j's. After these two iterations we have gone from $\|X_j \phi_N D_s^p u\|_{L^2}$ to $\|X_j \Lambda^{-1/3}(\phi_N)_{tt} D_s^{p-1} u\|_{L^2}$. After *three* of *these* iterations we will find $\|X_j \Lambda^{-1}(\phi_N)_{tttttt} D_s^{p-3} u\|_{L^2} \sim \|X_j(\Lambda^{-1}D_s)(\phi_N)_{tttttt} D_s^{p-4} u\|_{L^2} \sim \|(X_j)(\phi_N)_{tttttt} D_s^{p-4} u\|_{L^2}$, which exhibits a trade of *four* D_s derivatives for *six* derivatives on the localizing function ϕ_N.

This is the 'trade-off' that leads to the Gevrey class $G^{3/2}$.

Finally, we turn to x derivatives, where the computations are simpler, but seem to depend on the preceeding ones: This time $W = D_x$ and the *a priori* estimate reads essentially

$$\sum_j \|X_j \phi_N D_x^p u\|_{L^2}^2 + \sum_j \|\phi_N X_j D_x^p u\|_{L^2}^2 + \|\phi_N D_x^p u\|_{1/3}^2 \leq$$

$$\leq C\{|(\phi_N D_x^p Pu, \phi_N D_x^p u)_{L^2}| + \sum_{j=1}^{3} |\left([\phi_N D_x^p, X_j^2]u, \phi_N D_x^p u\right)_{L^2}|$$

$$+ \sum_j \|\phi_N^{(l)} D_x^p u\|_{L^2}^2\}.$$

In the bracket (second term on the right) the problem is now not keeping good vector fields - the D_x themselves are good. The difficult term that arises is when $x^2 D_s[D_x^p, x^2 D_s]$ (or $x D_t[D_x^p, x D_t]$ or double brackets) enter. All terms contribute p terms where one of the D_x differentiates x. Thus the error that arises is of the form $p x D_s D_x^{p-1}$ (and $p D_t D_x^{p-1}$) after the other X_j has been moved to the second member in the inner product. From the double brackets we may also have $p(p-1) D_s D_x^{p-2}$ etc. That is, starting with $X\phi_N D_x^p$ we now have essentially $p X x \phi_N D_s D_x^{p-2}$ or $p^2 X \phi_N D_s D_x^{p-3}$, and similar terms with D_t.

Now the former, $p X x \phi_N D_s D_x^{p-2}$, turns out to be the worst in terms of managing growth of derivatives - there has been a 'gain' of two D_x's, but a factor of p and a new D_s derivative. When this continues, what started as $X\phi_N D_x^p$ becomes, after the next iteration, $p(p-1) x^2 \phi_N D_s^2 D_x^{p-3}$ or $p \phi_N D_s D_x^{p-2}$ or $p^3 X \phi_N x D_s^2 D_x^{p-5}$, etc., where, again, in the first term we may call $x^2 D_s$ an X. That is, this term is of the form $p^2 X \phi_N D_s D_x^{p-3}$. *This*, iterated $p/3$ times, will lead to terms such as $p^{2p/3} D_s^{p/3}$. Since we know that derivatives in s of the solution grow like the Gevrey class $G^{3/2}$, we will get

$$\|X\phi_N D_x^p u\|_{L^2} \leq C^p p^{2p/3} p!^{(3/2)(1/3)} \leq C^p p!^{7/6}.$$

This is the origin of the $G^{7/6}$ behavior in x. Finally, the microlocalization poses no additional problems.

9. Other Special Cases, Leading to a General Conjecture

An analysis of the above proof, which certainly includes the statement of $G^{3/2}$ hypoellipticity, shows that it applies equally well to the somewhat more general operator

$$P = \frac{\partial^2}{\partial x^2} + (x^{p-1}\frac{\partial}{\partial t})^2 + (x^{q-1}\frac{\partial}{\partial s})^2 \qquad (12)$$

for $1 \leq p \leq q$.

Theorem 6 *The operator P in (12) is G^d hypoelliptic for all $d \geq q/p$.*

Remark 1 *This theorem is also due to M. Christ. However our proof is both elementary and allows more precise results about partial regularity, where derivatives in different directions grow at different rates.*

Theorem 7 (Bove-Tartakoff, 1996) *The operator P in (12) is G^{d_1,d_2,d_3} hypoelliptic for any $d_1 \geq 1 + 1/p - 1/q, d_2 \geq 1$, and $d_3 \geq q/p$.*

Theorem 8 *When $p = q = 1$ this yields analytic hypoellipticity, but in all other cases yields new examples of Gevrey classes of solutions to subelliptic partial differential equations.*

Theorem 9 (Bove-Tartakoff, 1996) *The Baouendi-Goulaouic example,*

$$P = D_x^2 + D_t^2 + x^2 D_s^2$$

is G^{d_1,d_2,d_3} hypoelliptic for any $d_1 \geq 3/2, d_2 \geq 1$, and $d_3 \geq 2$ but not hypoelliptic in any smoother Gevrey class.

Proof: The first part is a special case of the above result. The sharpness comes by studying the function

$$u_\epsilon(x,t,s) = \int_0^\infty exp[i\rho^2 s - t\rho - \rho^2 x^2/2 - \rho^\epsilon]\, d\rho$$

for $\epsilon > 1$ which solves $Pu_\epsilon = 0$ yet brief calculations show that u_ϵ satisfies

$$|\partial_t^k u_\epsilon(0)| = |\int_0^\infty e^{-\rho^\epsilon} \rho^k d\rho| \sim C^k k!^{1/\epsilon},$$

$$|\partial_s^k u_\epsilon(0)| = |\int_0^\infty e^{-\rho^\epsilon} \rho^{2k} d\rho| \sim C^k k!^{2/\epsilon},$$

and

$$|\partial_x^{2k} u_\epsilon(0)| = |\int_0^\infty e^{-\rho^\epsilon} \rho^{2k} k! d\rho| \sim C^k k!^{1+2\epsilon} \sim C^k (2k)!^{1/2+1\epsilon},$$

showing that for any $\epsilon > 1, u_\epsilon \in G^{1/2+1/\epsilon, 1/\epsilon, 2/\epsilon}$ and no better.

More general situations clearly pose no additional difficulties, such as those studied (isotropically) in [9].

$$P = \frac{\partial^2}{\partial x^2} + a_1(x,t,s)(x^{p-1}\frac{\partial}{\partial t})^2 + a_2(x,t,s)(x^{q-1}\frac{\partial}{\partial s})^2 \tag{13}$$

where both $a_1(x,t,s)$ and $a_2(x,t,s)$ are strictly positive and belong to the Gevrey classes under consideration.

10. The General Conjecture and Result

The general conjecture suggested by these results, and voiced by Trèves, concerns the iterated brackets of a set of real vector fields in the operator

$$P = \sum_{i=1}^{r} X_i^2$$

where the X_j satisfy the Hörmander condition that their iterated brackets span the tangent space. Call $A_1 = \{X_1, \dots X_N\}$, $A_2 = A_1 \cup \{[X_i, X_j], i \neq j\}$, $\dots, A_N = \bigcup_{i=1}^{N-1} A_i \cup \{[X_{i_1}, [X_{i_2}, [\dots, [X_{i_{N-1}}, X_{i_N}] \dots]]]\}$. We agree to call A_j, $j = 1, \dots, N$ the "layers" of the Lie algebra of the tangent vector fields.

Call now $\Sigma_1 = \{(x, \xi) | X_j(x, \xi) = 0, j = 1, \dots, r\}$, the multiple characteristic set of the operator P, $\Sigma_2 = \{(x, \xi) | Y(x, \xi)| = 0 \forall Y \in A_2\}$, i.e. the characteristic set of the second layer of the Lie algebra. Proceeding in this way we get to defining $\Sigma_N = \{(x, \xi) | Y(x, \xi)| = 0 \forall Y \in A_N\}$. By the Hörmander assumption Σ_N coincides with the null section of the cotangent bundle.

Of course the Σ_j are a decreasing finite sequence of subsets of the cotangent. We need not to assume that they are manifolds; if they are not just think of them as of stratified varieties, whose strata are smooth manifolds.

Tentative statement 1: If all the Σ_j's are symplectic—i.e. every layer of each Σ_j is symplectic—then P is analytic hypoelliptic.

Tentative statement 2: Assume that Σ_ℓ is the first of the Σ_j that is not symplectic, i.e. there is at least one stratum with an involutive leaf. Then the operator P is G^s hypoelliptic if $s \geq N/\ell$.

The proofs above yield the second statement in the cases studied as well as in others under consideration by E. Bernardi, A. Bove and the author. The first conjecture remains open.

References

1. Baouendi, M.S. and Goulaouic, C. (1971) Analyticity for Degenerate Elliptic Equations and Applications, *Proc. Symp. in Pure Math.* **Vol. no. 23**, pp. 79-84.
2. Bolley, C., Camus, J., & Rodino, L., (1987) Hypoellipticit analytique-Gevrey et itrs d'oprateurs. *Rend. Sem. Mat. Univ. Politec. Torino* **Vol. no. 45(3)**, pp. 1-61.
3. Boutet de Monvel, L., Grigis, A., & Helffer, B., (1976) Paramétrices d'opérateurs pseudo-différentiels à caracteristiques multiples *Societé Math. de France, Astérisque* **Vol. no. 34-35**.
4. Bove, A., & Tartakoff, D., (1996) Propagation of Gevrey Regularity for a Class of Hypoelliptic Equations *Transactions of the A.M.S.* **Vol. no. 348(7)**, pp. 2533-2575.
5. Catlin, D. (1987) Subelliptic Estimates for the $\bar{\partial}$-Neumann Problem on Pseudoconvex Domains *Annals of Math.* **Vol. no. 126**, pp. 131-191.
6. Chen, S.C. (1988) Global Real Analyticity of Solutions to the $\bar{\partial}$-Neumann Problem on Reinhardt Domains, *Indiana U. Math. J.* **Vol. no. 37(2)**, pp. 421-430.
7. Chen, S.C. (1991) Global Analytic Hypoellipticity of $\Box_b$ on Circular Domains., *Pac. J. Math.* **Vol. no. 148(2)**, pp. 225-235.

8. Christ, M. (1991)Certain Sums of Squares of Vector Fields Fail to Be Analytic Hypoelliptic, *Comm. in P.D.E.* **Vol. no. 10**, pp. 1695-1707.
9. Christ, M. (1996) Intermediate Gevrey Exponents Occur, *to appear, Comm. P.D.E.*.
10. Christ, M. & Geller, D., (1992) Counterexamples to Analytic Hypoellipticity for Domains of Finite Type, *Annals of Math.* **Vol. no. 135**, pp. 551-566.
11. Derridj, M. (1975) Gevrey regularity up to the boundary for the $\overline{\partial}$– Neumann problem, *Proceedings of Symposia in Pure Math.* **Vol. no. XXX**, pp. 123-126.
12. Derridj, M. (1978) Régularité pour $\overline{\partial}$ dans Quelques Domaines Faiblement Pseudoconvexes., *Journal of Differential Geometry* **Vol. no. 13(4)**, pp. 559-576.
13. Derridj, M. & Tartakoff, D.S., (1976) On the Global Real Analyticity for the $\overline{\partial}$-Neumann Problem. *Comm. P. D. E.* **Vol. 5** pp. 401-435.
14. Derridj, M. & Tartakoff, D.S., (1988) Local Analyticity for $\Box_b$ and the $\overline{\partial}$-Neumann Problem at Certain Weakly Pseudo-Convex Points. *Comm. P. D. E.* **Vol. 13(12)** pp. 1521-1600.
15. Derridj, M. & Tartakoff, D.S., (1991) Local Analyticity in the $\overline{\partial}$-Neumann Problem and for $\Box_b$ in Some Model Domains Without Maximal Estimates. *Duke Mathematical Journal* **Vol. 64(2)** pp. 377-402.
16. Derridj, M. & Tartakoff, D.S., (1993) Global Analyticity for $\Box_b$ on Three Dimensional Pseudoconvex CR Manifolds *Comm. P. D. E.* **Vol. 18(11)** pp. 1847-1868.
17. Derridj, M. & Tartakoff, D.S., (1993) Local Analyticity in the $\overline{\partial}$-Neumann Problem for a Class of Totally Decoupled Weakly Pseudoconvex Domains. *Journal of Geometric Analysis* **Vol. 3(2)** pp. 141-151.
18. Derridj, M. & Tartakoff, D.S., (1994) Microlocal Analyticity for the Canonical Solution to $\overline{\partial}_b$ on Strictly Pseudoconvex CR Manifolds of Real Dimension Three. *Comm. P. D. E.* **Vol. 20(9&10)** pp. 1647-1667.
19. Derridj, M., and Zuily, C. (1973) Régularité analytique et Gevrey pour des classes d'opérateurs élliptiques paraboliques dégénérées du second ordre, *Astérisque* **Vol. no. 2,3**, pp. 371-381.
20. Diederich, K., and Fornaess, J.E. (1978) Pseudoconvex Domains with Real-Analytic Boundaries *Ann. of Math.* **Vol. no. 107**, pp. 371-384.
21. Ehrenpreis, L. (1960) Solutions of some Problems of Division IV *Amer. J. Math.* **Vol. no. 82**, pp. 522-588.
22. Grušin(1971) A certain class of elliptic pseudodifferential operators that are degenerate on a submanifold, *Mat. Sbornik (=Math. USSR Sbornik)* **Vol. no. 84(13)**, pp. 163-195(155-185).
23. Hanges, N., & Himonas, A., (1991) Singular Solutions for Sums of Squares of Vector Fields., *Comm. in P.D.E.* **Vol. no. 16(8,9)**, pp. 1503-1511.
24. Hörmander, L.(1967) Hypoelliptic Second Order Differential Equations *Acta Math.* **Vol. no. 119** pp. 147-171.
25. Kajitani, K. and Wakabayashi, S. (1988) Hypoelliptic Operators in Gevrey Classes *Recent Developments in Hyperbolic Equations, L. Cattabriga et al. Ed.s, Pitman Research Notes in Math.* **Vol. no. 183**, pp. 115-134.
26. Kohn,J.J. (1963) Harmonic Integrals on Strongly Pseudo-Convex Manifolds, I and II. *Ann. of Math* **Vol. no. 78** pp. 112-148 and **Vol. no. 79** pp. 450-472.
27. Kohn,J.J. and Nirenberg, L.(1967) Non-Coercive Boundary Value Problems *Comm. Pure Appl. Math.* **Vol. no. 18** pp. 443-492.
28. Komatsu, G. (1976) Global Analytic Hypoellipticity of the $\overline{\partial}$-Neumann Problem *Tohoku Math. J.* **Vol. no. 28**, pp. 145-156.
29. Metivier, G. (1981) Analytic hypoellipticity for operators with multiple characteristics *Comm. in P.D.E.* **Vol. no. 6**, pp. 1-90.
30. Oleinik, O., & Radkevich, R., (1974) Conditions for the analyticity of all solutions of a second order linear equation. (Russian) no. 3(177), 221–222. *Uspehi Mat. Nauk* **Vol. no. 177(3)**, pp. 221-222.
31. Rothschild, L.P. and Stein, E.M. (1977) Hypoelliptic Differential Operators and

Nilpotent Groups *Acta Math.* **Vol. no. 137**, pp. 248-315.

32. Stein, E.M., (1982) An Example on the Heisenberg Group Related to the Lewy Operator *Inventiones mathematicae* **Vol. no. 69**, pp. 209-216.
33. Tartakoff D.S. (1973) Gevrey Hypoellipticity for Subelliptic Boundary Value Problems *Communications on Pure and Applied Math.* **Vol. no. 26**, pp. 251-312.
34. Tartakoff D.S. (1976) On the Global Real Analyticity of Solutions to $\Box_b$ on Compact Manifolds *Comm. in P.D.E.* **Vol. no. 1**, pp. 283-311.
35. Tartakoff D.S. (1977) On the Local Gevrey and Quasianalytic Hypoellipticity for $\Box_b$. *Comm. in P.D.E.* **Vol. no. 2**, pp. 699-712.
36. Tartakoff D.S. (1978) Local Analytic Hypoellipticity for $\Box_b$ on Non-Degenerate Cauchy Riemann Manifolds *Proc. Nat. Acad. Sci. U.S.A.* **Vol. no. 75**, pp. 3027-3028.
37. Tartakoff D.S. (1980) On the Local Real Analyticity of Solutions to $\Box_b$ and the $\bar{\partial}$-Neumann Problem *Acta Math.* **Vol. no. 145**, pp. 117-204.
38. Tartakoff D.S. (1983) Operators with Multiple Characteristics - An L^2 Proof of Analytic Hypoellipticity *Conference on Linear Partial and Pseudodifferential Operators, Rend. Sem. Mat. Univ. Politec. Torino 1983.*, pp. 251-282.
39. Tartakoff D.S. (1996) Global (and local) analyticity for second order operators constructed from rigid vector fields on products of tori *Transactions of the A.M.S.* **Vol. no. 348(7)**, pp. 2577-2583.
40. Treves, F. (1978) Analytic Hypo-ellipticity of a Class of Pseudo-Differential Operators with Double Characteristics and Application to the $\bar{\partial}$-Neumann Problem *Comm. in P.D.E.* **Vol. no. 3 (6-7)**, pp. 475-642.

HIGHER MICROLOCALIZATION AND PROPAGATION OF SINGULARITIES

OTTO LIESS
Dipartimento di Matematica
Università di Bologna
40127 Bologna, ITALIA

1. Inverse Fourier transforms for functions of infraexponential growth.

1. A good point to start is to discuss the range of applicability of the Fourier transform, or rather of its inverse. We define the inverse Fourier transform formally by

$$h(x) = (2\pi)^{-n} \int_{R^n} e^{i\langle x,\xi\rangle} f(\xi)\, d\xi \tag{1}$$

and the problem is to see for which classes of functions (or distributions) we can give a reasonable meaning to (1). In the last century f would have been assumed to be integrable; nowadays we would write the corresponding condition in terms of Lebesgue integrals and ask for $f \in L^1(R^n)$. There is a natural extension of this to $f \in L^2(R^n)$ using Plancherel's theorem. A far reaching extension was achieved by L.Schwartz, who took $f \in S'(R^n)$.

Every now and then it has been attempted to extend the range of these definitions further: cf. e.g. [6], [13]. In particular, there should be some intersections between the results described here and the results in [13], but the present theory is in my opinion simpler. I want to show here that (1) has a simple natural meaning if we assume that

$$f \in L^1_{loc}(R^n),\ |f(\xi)| \leq c\, e^{\varphi(\xi)}, \tag{2}$$

where φ is sublinear. (We call a function $\varphi : A \to R_+$ sublinear if $\forall \varepsilon > 0\ \exists c_\varepsilon$ so that $\varphi(t) \leq \varepsilon|t| + c_\varepsilon$.) We shall say that f has infraexponential growth (at infinity) then.

To discuss (1), assume at first that supp $f \subset G$, for some convex cone $G \subset \dot{R}^n$. ($\dot{R}^n = R^n \setminus \{0\}$.) Even with this additional assumption it is not

L. Rodino (ed.), Microlocal Analysis and Spectral Theory, 61–90.

possible to give a direct meaning to (1) for real x, but we can give a meaning to it when we replace x by $z = x + iy \in C^n$, $y \in G^\perp$. ($G^\perp$ denotes the polar of G.) Indeed, for $y^0 \in G^\perp$ fixed, we will have $\langle y^0, \xi \rangle \geq c|\xi|$, for some $c > 0$ if $\xi \in G$, and in fact we can find a neighborhood V of y^0 so that

$$\langle y, \xi \rangle \geq c|\xi|, \text{ if } \xi \in G \text{ and } y \in V.$$

It follows that

$$|e^{i\langle x+iy,\xi\rangle}| = e^{-\langle y,\xi\rangle} \leq e^{-c'|\xi|} \text{ if } \xi \in G, y \in V,$$

so that $|f(\xi)\exp[i\langle x + iy, \xi\rangle]| \leq c\exp[\varphi(\xi) - c'|\xi|]$. Therefore $f(\xi)e^{i\langle x+iy,\xi\rangle} \in L^1(R^n)$ uniformly in $(x, y) \in R^n \times V$. We conclude that

$$h(z) = h(x + iy) = \int_{R^n} e^{i\langle x+iy,\xi\rangle} f(\xi)\, d\xi$$

makes sense and defines an analytic function, at first on $R^n \times iV$, but then of course on $R^n \times iG^\perp$, since $y^0 \in G^\perp$ was arbitrary. Thus, $h(z) \in \mathcal{A}(R^n \times iG^\perp)$. (For Ω open in C^n we denote by $\mathcal{A}(\Omega)$ the holomorphic functions on Ω.)

Actually, in this way we have not given a direct meaning to (1) for real x, but, assuming that the support of f lied in G, we were able to associate some $h(z) \in \mathcal{A}(R^n \times iG^\perp)$ with the integral there. Let us next drop the assumption that supp $f \subset G$ and return to (1). Let us then choose a finite collection of open convex cones $G_j \subset \dot{R}^n$, $j = 1, \ldots, k$, so that $\cup_j \Gamma_j = \dot{R}^n$. Also fix some bounded open neighborhood G_0 of the origin and write f in the form

$$f = f_0 + \sum_{j=1}^{k} f_j \tag{3}$$

with supp $f_j \subset G_j$, $j = 0, 1, \ldots, k$. We can in addition assume that the f_j have infraexponential growth and are in $L^1_{loc}(R^n)$ for $j \geq 1$, the f_0 being bounded. The integrals $h_j(z) = (2\pi)^{-n} \int_{R^n} \exp[i\langle z, \xi\rangle] f_j(\xi)\, d\xi$ have then a natural meaning for $Im\, z \in G_j^\perp$ when $j \geq 1$, respectively for all $z \in C^n$ when $j = 0$. At this moment we have thus associated with the right hand side of (1) the formal collection of analytic functions h_j, $j = 0, 1, \ldots, k$. Let us also recall that in the theory of hyperfunctions one associates with analytic functions defined on sets of type $R^n \times i\Gamma$, Γ some open cone in $\dot{R}^n$ a "formal boundary value" on the edge "R^n" of the wedge "$R^n \times i\Gamma$". (Actually the situation considered in hyperfunctions is somewhat more general and we recall it for the convenience of the reader in section 2.) We denote the formal boundary value of some analytic function defined on some wedge by $b(h)$. It is thus a hyperfunction and is to be regarded as an object living on R^n. In particular, we can calculate the sum $\sum_{j=0}^{k} b(h_j)$ as a hyperfunction on R^n and we shall set

$$(2\pi)^{-n} \int_{R^n} e^{i\langle x,\xi\rangle} f(\xi)\, d\xi = \sum_{j=0}^{k} b(h_j). \tag{4}$$

It is not difficult to see that the hyperfunction $u = \sum_{j=0}^{k} b(h_j)$ depends only on f and not on the decomposition of f in the form (3). It is this hyperfunction which we shall call the "inverse Fourier transform" of f. We shall also say that this u is the regularization of the right hand side of (1).

2. It might be instructive to see what happens when rather than having "infraexponential growth" as in (2), we assume that we have the more restrictive assumption $|f(\xi)| \le c(1+|\xi|)^b$, for some non-negative real number b. Assume as in the beginning of our argument that supp $f \subset G$ and fix $\Gamma' \subset\subset G^{\perp}$. (If Γ_1, Γ_2 are two cones in $\dot{R}^n$ or $\dot{C}^n$, we shall write $\Gamma_1 \subset\subset \Gamma_2$ if the closure of Γ_1 in $\dot{R}^n$ (or in $\dot{C}^n$) lies in the interior of Γ_2.) We now also want to be a little bit more careful with our estimate for the exponential $\exp i\langle z,\xi\rangle$ and therefore observe that there is some constant $c' > 0$, so that $\langle y,\xi\rangle \ge c'|y||\xi|$ if $\xi \in G$ and $y \in \Gamma'$. Since $|\xi|^b \exp[-c'|y|\,|\xi|] \le c''|y|^{-b} \exp[-c'|y|\,|\xi|/2]$, it follows that $h(z) = \int \exp[i\langle z,\xi\rangle] f(\xi)\, d\xi$ has tempered growth when $y \to 0$ as long as $y \in \Gamma'$. (In other words, there are constants c_1, b', so that $|h(z)| \le c_1|y|^{-b'}$ if $y \in \Gamma'$.) We conclude from this that $b(h)$ makes sense in distributions. When f has polynomial growth at infinity, the meaning we give to (1) is therefore precisely the one it has in standard distribution theory in $S'(R^n)$. Indeed, with appropriate changes in notations, we could have replaced the assumption that f be a function with polynomial growth by the condition that $f \in S'(R^n)$. We do not make this here more explicit.

2. A brief review of hyperfunctions (from a local point of view)

1. Let me make a paranthesis on what hyperfunctions "locally" are. We want here to consider hyperfunctions as equivalence classes of formal boundary values of analytic functions defined on tuboids. (But we shall call such "tuboids" "wedges" henceforth.) Starting point is that distributions themselves appear often as distributional boundary values of such analytic functions. Before we continue, we define the notion of a wedge: let us then consider at first U open in R^n. We call "wedge over U" , or simply "wedge", a set of type

$$\omega = (U \times i\Gamma) \cap \tilde{U} \subset C^n$$

where Γ is an open connected cone in R^n and $\tilde{U}$ is a complex neighborhood of U in C^n. (Sets of this form are sometimes called "tuboids" in the literature. I myself find the notion "wedge", although perhaps less accurate, more intuitive. The set U is often called the "edge" of ω.)

Let us now consider a finite number of wedges ω_j over U and let h_j be holomorphic functions on ω_j. Wedges will "always" be over the same U. With the h_j we associate the formal sum $\sum h_j$. It is then clear that in the space of such formal sums we have a natural commutative addition. Let us denote the space of all such formal sums by $\mathcal{M}$. In $\mathcal{M}$ we shall now introduce an equivalence relation by saying that

$$\sum h_j \sim \sum h'_k,$$

for $h_j \in \mathcal{A}((U + i\Gamma_j) \cap \tilde{U}_j)$, $h'_k \in \mathcal{A}((U + i\Gamma'_k) \cap \tilde{U}'_k)$, if we can find $f_{ls} \in \mathcal{A}((U + i \text{ co } (\Gamma_l \cup \Gamma'_s)) \cap \tilde{U}_{ls})$, ("co" is "convex hull") such that

$$h_j = \sum_s f_{js}, h'_k = \sum_l f_{lk}. \tag{5}$$

It is not difficult to see that $\sum h'_j \sim \sum h''_k$, $\sum f'_l \sim \sum f''_s$ implies $\sum h'_j + \sum f'_l \sim \sum h''_k + \sum f''_s$. Further, if h_1, h_2 are in $\mathcal{A}(\omega)$ for the same wedge ω, then the pointwise sum $h_3 \in \mathcal{A}(\omega)$ defined by $h_3(z) = h_1(z) + h_2(z)$ is easily seen to be equivalent with $h_1 + h_2$, the latter sum being performed in $\mathcal{M}$. Also note that a convenient way to characterize $\sum_{j=1}^m h_j \sim 0$, $h_j \in \mathcal{A}((U + i\Gamma_j) \cap \tilde{U}_j)$ is, that we can find $f_{jk} \in \mathcal{A}((U \times i \text{ co } (\Gamma_j \cup \Gamma_k)) \cap \tilde{U}_{jk})$ with

$$f_{jk} = -f_{kj} \text{ and } h_j = \sum_k f_{jk}.$$

We shall denote the factor space $\mathcal{M}/\sim$ by $B(U)$. The projection $\mathcal{M} \to \mathcal{M}/\sim$ is called "hyperfunctional boundary value" and is denoted by b. (It is the "b" we had already had above on analytic functions defined on wedges.)

Let us also consider explicitly the case $n = 1$. If u is a hyperfunction, we can write it (locally this is clear) as $u = b(h_1) + b(h_2)$ where $h_1 \in \mathcal{A}(U \times i(0, e))$ and $h_2 \in \mathcal{A}(U \times i(-e, 0))$, $e > 0$. The fact that u is zero as a hyperfunction on an open subset U' subset of U means that there is an open set $V' \subset C$, which intersects R on U' and a holomorphic function h defined on V' so that h_1 is the restriction of h to $V' \cap \{Imz > 0\}$ and h_2 the restriction of $-h$ to $V' \cap \{Imz < 0\}$. It is then easy to believe that one can identify for example the hyperfunctions on some interval $W \subset R$ with the factor space $\mathcal{A}(V \setminus W)/\mathcal{A}(V)$ where V is some open set in C which intersects R on W. Since $\mathcal{A}(V)$ is dense in $\mathcal{A}(V \setminus W)$, one sees from this that $\mathcal{B}(W)$ has no natural separable topology,

3. L^2 estimates on C^n rather than sup-norm estimates on R^n

1. We now return to (1). Unfortunately, it is not easy to develop a non-trivial theory of integrals of type (1) while working in the real domain. We shall in fact replace integration on R^n with integration on C^n. Moreover, in order to be able to apply classical results from the theory of the $\bar{\partial}$-operator in weighted L^2-spaces, we shall replace the L^∞-type estimate in (3) by some related L^2-type estimate. More precisely, we shall assume that $f \in L^2_{loc}(C^n)$ and that

$$f(\zeta)e^{-\varphi(Re\,\zeta)+\varepsilon|Im\,\zeta|} \in L^2(C^n).$$

We have then to give a meaning to

$$(1/2i)^n \int_{C^n} e^{i\langle x,\zeta\rangle} f(\zeta)\, d\zeta \wedge d\bar{\zeta}, \tag{6}$$

where $(1/2i)^n\, d\zeta \wedge d\bar{\zeta}$ is the Lebesgue measure on C^n. We can then regularize the integral in the following way: again we consider open convex cones G_j, $j = 1, \ldots, k$, and a bounded open neighborhood of the origin G_0 such that $\cup_{j\geq 1} G_j = \dot{R}^n$. We then split f into a sum $f = \sum_j f_j$ with supp $f_j \in G_j \times iR^n$, and notice that (if the f_j have the same type of behaviour at infinity which we had for f)

$$h_j(z) = (2\pi)^{-n}(1/2i)^n \int_{C^n} e^{i\langle z,\zeta\rangle} f_j(\zeta)\, d\zeta \wedge d\bar{\zeta}$$

is analytic for $|Re\, z| < \varepsilon$, $Im\, z \in G_j$, when $j \geq 1$, respectively that $h_0 \in \mathcal{A}(z; |Re\, z| < \varepsilon)$. It follows that the integral (6) has a natural meaning as a hyperfunction u on $|x| < \varepsilon$. We denote with the letter T the map which associates with f this hyperfunction. We shall then also say that the hyperfunction u has been represented with the aid of f and call f a "representation function" of u. (Since we are working on $|x| < \varepsilon$ our representations are only local.) In a related way we say that u is the inverse Fourier transform of f. As in our discussion of the case of inverse Fourier transforms of functions defined on R^n, it is useful to observe that we need not necessarily work with "functions" on C^n, but may replace "functions" by "measures" or "distributions", which satisfy appropriate growth conditions at infinity. Moreover, these growth type conditions can in principle be formulated in the form of L^2-type conditions, or also in the form of L^∞-type conditions. That this is so is related to the fact that on spaces of analytic functions locally all reasonable norms are equivalent. We do not make this here more precise, but refer to [22] for explicit statements.

2. The following questions are now natural:

- which hyperfunctions on $|x| < \varepsilon$ can be represented in the form above?
- what is the degree of non-unicity in such a representation?
- how can one recognize properties of the hyperfunction u in terms of properties of a representation function by which it can be defined?

The answer to our first question is: any hyperunction on $|x| < \varepsilon$ can be represented as the inverse Fourier of some f. That this is so has been known, in another form, since quite some time. Indeed, let us at first recall the following result,which is a consequence of results of Ehrenpreis (cf. [3]):

Theorem 3.1. *Let $\tilde{\varepsilon} > 0$, $d > 0$ and an open convex cone $\Gamma \subset R^n$ be given. Denote by $H_{\Gamma,d}$, the support-functions of the set $\{y \in \Gamma, |y| \leq d\}$. (Recall that the support-function H_K of a compact set K in R^n is defined by $H_K(\xi) = \sup_{y \in K}\langle y, \xi\rangle$.) Consider an analytic function h defined on*

$$\Omega = \{z \in C^n; |Re\, z| < \tilde{\varepsilon}, Im\, z \in \Gamma, |Im\, z| < d\}.$$

Also fix $\tilde{\varepsilon}' < \tilde{\varepsilon}$, $\Gamma' \subset\subset \Gamma$. Then we can find a sublinear function φ and a Radon measure ν on C^n with the following properties:

$$\int_{C^n} d|\nu(\zeta)| < \infty, \tag{7}$$

$$h(z) = \int_{C^n} \exp\left[i\langle z, \zeta\rangle + \varphi(-Re\, \zeta) - H_{\Gamma',d/2}(-Re\, \zeta) - \tilde{\varepsilon}'|Im\, \zeta|\right] d\nu(\zeta), \tag{8}$$

if $|Re\, z| < \tilde{\varepsilon}'$, $Im\, z \in \Gamma'$, $|Im\, z| < d/2$.

(For a direct proof of this result we refer to [22]. Also cf. [3], [31].) To relate this result to our problem of writing some given hyperfunction u on $|x| < \varepsilon$ as an inverse Fourier transform, we first extend u to some hyperfunction v defined on $|x| < \varepsilon'$ with $\varepsilon' > \varepsilon$. We then write v as a sum of formal boundary values of analytic functions h_j defined on wedges over $|x| < \varepsilon'$ and apply the preceding result for each h_j and for some $\tilde{\varepsilon}' < \tilde{\varepsilon}$ which satisfy $\varepsilon \leq \tilde{\varepsilon}' < \tilde{\varepsilon} < \varepsilon'$.

3. Before we now say a few words on non-unicity, let us introduce a notation. For A measurable in C^n and $\varepsilon > 0$, we shall in fact denote by $L^2(A, \mathcal{F}, -\varepsilon)$ the space of measurable functions f on A such that $f \exp[-\varphi(Re\, \zeta) + \varepsilon|Im\, \zeta|] \in L^2(A)$ for some sublinear function φ.

We turn to non-unicity. In fact it is quite obvious that our representation functions (or measures) will not be unique. Let us give two examples which show what can happen. Assume then at first that μ is of the form $(\partial/\partial\bar{\zeta}_j)\nu$ for some j and some ν in $C_0^\infty(C^n)$. Then (8) makes sense and defines the zero hyperfunction. We shall see that non-unicity is indeed deeply related in local representations to the fact that μ has the form $\mu = \sum_{j=1}^n (\partial/\partial\bar{\zeta}_j)\nu_j$. Indeed, the following theorem is one of the main results in the theory:

Theorem 3.2. *(Cf. [22].) Fix $0 < \varepsilon < \varepsilon'' \leq \varepsilon'$ and let $f \in L^2(C^n, \mathcal{F}, -\varepsilon')$ be such that $T(f) = 0$ on $|x| < \varepsilon''$. Then there are $g_j \in L^2(C^n, \mathcal{F}, -\varepsilon)$, $j = 1, \ldots, n$, so that*

$$f = \sum_j (\partial/\partial\bar{\zeta}_j) g_j. \tag{9}$$

Conversely, if (9) is valid, then $T(f) = 0$ on $|x| < \varepsilon$.

It is clear that non-uniqueness as in our first example is somehow related to the fact that we are working with representations on C^n rather than with representations on R^n. However, our second example is with representation functions in R^n: consider $g \in C_0^\infty(R^n)$, $g \not\equiv 0$, so that $g(x) = 0$ for $|x| < \varepsilon$. Then it follows that $\hat{g}$ is a function on R^n which defines the zero-function on $|x| < \varepsilon$. Since $\hat{g}$ is a representation function for g, we have thus represented the zero hyperfunction with the non-trivial $\hat{g}$. We might here recall that it is not easy to characterize holes in the support of some C_0^∞-function in terms of its Fourier transform. Theorem 3.2 (which has to be reformulated to include representation measures) then gives a result in this direction.

Let us finally say something on our third problem. In fact, as is the case for the standard Fourier transform, regularity properties for a given hyperfunction are often reflected in decay properties of the representation function. Let us give two examples: when u is a distribution on $|x| < \varepsilon$ and if we fix $\varepsilon' < \varepsilon$, then we can represent u on $|x| < \varepsilon'$ in the form:

$$u = \int_{C^n} e^{i\langle x,\zeta\rangle} f(\zeta)\, d\zeta \wedge d\bar{\zeta},$$

where $f \exp[\varepsilon'|Im\,\zeta| - b\ln(1 + |\zeta|)] \in L^2(C^n)$ for some suitable $b \in R$. (Thus, f has polynomial, rather than infraexponential, growth in $Re\,\zeta$.) Another interesting case is when $f \exp[\varepsilon|\zeta|] \in L^2(C^n)$. (I.e., f is exponentially decaying in the L^2-sense in $|\zeta|$.) In fact, it is immediate then that the hyperfunction u associated with this f is real-analytic near zero and the converse is also true, as is easy to see. Indeed, this is the simplest instance of a theorem of the type of theorem 3.1, and we give a few details to explain how the argument goes. Let us then assume that u admits a holomorphic extension to the set $\{z \in C^n; |z| < 2\varepsilon\}$ and define a continuous linear functional L on the space $\mathcal{A}(C^n, \varepsilon|\zeta|)$ of entire analytic functions h on C^n for which the norm $|h|_\varepsilon = \sup_\zeta |h(\zeta)| \exp -[\varepsilon|\zeta|]$ is finite, by $L(h) = \mathcal{F}^{-1}(h)(u)$. $\mathcal{F}^{-1}$ is of course the inverse Fourier-Borel transform. It is easy to see that we also have $|L(h)| \leq c||h||_{\varepsilon/2}$ where $||h||_{\varepsilon/2} = \sqrt{|\int_{C^n} |h(\zeta)|^2 \exp[-\varepsilon|\zeta|] \, d\zeta \wedge d\bar{\zeta}|}$. Applying the Hahn-Banach and the Riesz-Fischer theorem it follows that there is a function f on C^n so that $L(h) = \int f(\zeta)h(\zeta) \, d\zeta \wedge d\bar{\zeta}$ and so that $f \exp[(\varepsilon/2)|\zeta|] \in L^2(C^n)$. (Of course, we have to assume here that $h \exp[-(\varepsilon/2)|\zeta|] \in L^2(C^n)$.) We conclude that $u(x) = \int \exp[i\langle x, \zeta\rangle f(\zeta) \, d\zeta \wedge d\bar{\zeta}$ if $|x| < \varepsilon/2$.

Since we consider later on real-analytic functions which depend on an additional parameter (this will happen when we speak about pseudodifferential operators), it is useful to have a more explicit way to obtain representation functions in this case. We are only interested in the case of germs of analytic functions near 0; let us then assume that u is of form

$$u(z) = \sum_{\alpha} a_\alpha z^\alpha,$$

where the $a_\alpha \in C$ satisfy Cauchy's inequalities: there are constants c, c' so that

$$|a_\alpha| \leq cc'^{|\alpha|}\alpha!. \tag{10}$$

Actually, we want our Fourier representation to be valid in a complex neighborhood of 0. The construction of the representation function for u will then be performed in the following way: we shall choose representation functions σ_α for z^α, i.e., σ_α will be so that

$$z^\alpha = (/2i)^n \int_{C^n} \exp[i\langle z, \zeta\rangle]\sigma_\alpha(\zeta) \; d\zeta \wedge d\bar{\zeta}. \tag{11}$$

We can then take $\sum_\alpha a_\alpha \sigma_\alpha$ as a representation function for u. Here the σ_α are of course not unique, and indeed, we will have to choose them according to the value of c' in (10). However, once c' in (10) is fixed, we can work with the same σ_α, independently of u, so that our construction is stable if we have additional parameters. The main step in the argument is the following easy

Lemma 3.3. *There are constants c, c_1, so that for any $A > 0$ we can find measurable functions σ_α on C^n with the following properties:*

a) supp $\sigma_\alpha \subset \{\zeta \in C^n; A|\alpha| - 1/2 \leq |\zeta| \leq A|\alpha| + 1/2\}$,

b) $||\sigma_\alpha||_{L^2(C^n)} \leq cc_1^{|\alpha|}A^{-|\alpha|}$,

c) for any $h \in A(C^n)$ we have $D_\zeta^\alpha h(0) = \int_{C^n} h(\zeta)\sigma_\alpha(\zeta) \, d\lambda(\zeta)$. In particular, (11) is valid. (D_ζ^α stands for $(1/i)^{|\alpha|}(\partial/\partial\zeta)^\alpha$.) (We put $h(\zeta) = e^{i\langle z, \zeta\rangle}$.)

As a consequence, we obtain:

Theorem 3.4. *Let u be of form $u(z) = \sum_\alpha a_\alpha z^\alpha$, and assume that the coefficients a_α satisfy (10). Then we can find a representation function f for u with the following properties:*

a) It is of form $f = \sum_\alpha a_\alpha \sigma_\alpha$, with σ_α as in lemma 3.3, for some suitable A, which depends only on c'.

b) There is $c'' > 0$ so that $\exp[c''|\zeta|]f \in L^2(C^n)$.

It is interesting to note that, although we are working with exponentially decreasing Fourier representations, there is no need to work with "quasi-analytic" cut-off functions. This is of course due to the fact that we work with representation functions which live on C^n.

4. We have stated theorem 3.4 in terms of L^2-type estimates: it is this type of estimates which we use in almost all of the paper. In the applications of the theorem it is however more convenient to work with L^∞-type estimates. For this purpose we consider the following strengthened form of theorem 3.4:

Theorem 3.5. *Let u be as in the assumptions of theorem 3.4. Then the conclusion there remains valid, even if we replace the condition b) by the following stronger condition*

b)' There is $c'' > 0$ so that $\exp[c''|\zeta|]f \in L^\infty(C^n)$.

4. Microfunctions and Bony's theorem on the equivalence of definitions of the analytic wave front set.

1. Until now, we have worked locally. One can microlocalize theorem 3.2. This is in so far interesting as it gives a representation of microfunctions in terms of $\bar{\partial}$-cohomology classes and it permits to analyze problems on microfunctions with the aid of methods of complex analysis. For the convenience of the reader, we recall the following definition:

Definition 4.1. *(Sato) Let u be a hyperfunction defined near x^0. We shall say that (x^0, ξ^0) is not in the analytic singular spectrum of u, and write $(x^0, \xi^0) \notin ss_A u$ if we can find open convex cones $\Gamma_j \subset \dot{R}^n$, $j = 1, \ldots, k$, and holomorphic function h_j defined on $\{x + iy \in C^n; x \in R^n, y \in R^n, |x - x^0| < \varepsilon, y \in \Gamma_j, |y| < d\}$ so that $u = \sum_j b(h_j)$ near x^0 and so that ξ^0 is not in the polar G_j of Γ_j.*

One can then characterize the analytic singular spectrum as follows:

Proposition 4.2. *$(x^0, \xi^0) \notin ss_A u$ if and only if we can find a sublinear function φ, an open cone $G \subset \dot{R}^n$ which contains ξ^0, $d > 0$, and a repre-*

sentation function ν for u so that

$$|\int_{C^n} |\nu(\zeta)|^2 \exp[-2\varphi(Re\,\zeta) + 2d\,|Re\,\zeta|_G + 2\varepsilon|Im\,\zeta|]\, d\zeta \wedge d\bar{\zeta}| < \infty. \quad (12)$$

Here we have denoted for a given set $A \subset R^n$ by $|\xi|_A$ the function defined in the following way: $|\xi|_A = |\xi|$, if $\xi \in A$, $|\xi|_A = 0$ if $\xi \notin A$. Note that definition 9.1 is "geometrical", whereas (12) is analytical. The inverse Fourier transform thus gives a relation between an analytical and a geometrical information for u.

A result similar to proposition 4.2 is valid also in distributions. Let us in fact denote by $WF_A v$ the analytic wave front set (as introduced e.g. in [7]: we shall recall the corresponding definition in a more general frame in section 6 below) of the distribution v. Then we have:

Proposition 4.3. *Let $v \in D'(U)$. Then $(0, \xi^0) \notin WF_A v$ if and only if we can find an open cone $G \subset \dot{R}^n$ which contains ξ^0, $d > 0$, $b \in R$, and a representation function ν for u so that*

$$|\int_{C^n} |\nu(\zeta)|^2 \exp[2d|Re\,\zeta|_G + 2\varepsilon|Im\,\zeta| + 2b\ln(1+|\zeta|)]\, d\zeta \wedge d\bar{\zeta}| < \infty. \quad (13)$$

To simplify the statements of the following results, we shall now introduce some (rather heavy) notation: the first is $L^2(\xi^0, \mathcal{F}, -\varepsilon)$ and it stands (if ξ_0 has been fixed in $\dot{R}^n$) for the space $\cup_W L^2(W, \mathcal{F}, -\varepsilon)$, where the union is for all open convex complex neighborhoods W of ξ^0.

Furthermore, if X is some space of distributions in C^n, we denote by $X_{(0,q)}$ the space of complex $(0,q)$-forms with coefficients in X. Moreover, if (x^0, ξ^0) is fixed we shall write for two hyperfunctions u and v, that $u \sim v$ if $(x^0, \xi^0) \notin ss_A(u - v)$. We denote by $\mathcal{B}_{(x^0,\xi^0)}$ the factor space $\mathcal{B}(U)/\sim$, where U is some open set containing x^0. We call the elements in $\mathcal{B}_{(x^0,\xi^0)}$ "microfunctions" and denote by $J_{(x^0,\xi^0)}$ the map which maps $u \in \mathcal{B}(U)$ in its residual class in $\mathcal{B}_{(x^0,\xi^0)}$.

If now $\mu \in L^2(\xi^0, \mathcal{F}, -\varepsilon)$, we can associate with it a microfunction $\dot{T}(\mu)$ in the following way: if W is the domain of definition of μ, we define $\tilde{\mu} : C^n \to C$ by setting $\tilde{\mu}(\zeta) = \mu(\zeta)$ for $\zeta \in W$ and $\tilde{\mu}(\zeta) = 0$ for $\zeta \notin W$ and then we put $\dot{T}(\mu) = J_{(x^0,\xi^0)}(T(\tilde{\mu}))$. (More precisely, $T(\tilde{\mu})$ will be a hyperfunction on $|x| < \varepsilon$ and $J_{(x^0,\xi^0)}$ maps $B(\{x \in R^n; |x| < \varepsilon\})$ onto $\mathcal{B}_{(x^0,\xi^0)}$.) It is immediate to see that $\dot{T}$ is well-defined (i.e., $\dot{T}(\mu)$ does not depend on the choice of μ, W) and it is clear that $\dot{T}$ is surjective and we want to analyze the kernel of $\dot{T}$. In analogy with theorem 3.2 we can prove:

Theorem 4.4. *Let* $\mu \in L^2(\xi^0, \mathcal{F}, -\varepsilon')$ *be given so that* $\dot{T}(\mu) = 0$. *Then there is* ε *and* $\nu_j \in L^2(\xi^0, \mathcal{F}, -\varepsilon)$ *with* $\mu = \sum_j (\partial/\partial\bar{\zeta}_j)\nu_j$ *in a complex neighborhood of* x^0. *(The converse is also true.)*

The result thus says, intuitively, that when we work with microfunctions, then we can practically argue for representation functions which are defined on a complex conic neighborhoods of ξ^0 rather than on all of C^n, and that the space of microfunctions at $(0, \xi^0)$ can (roughly speaking) be identified with

$$\bigcup_{\varepsilon'>0} L^2_{(0,n)}(\xi^0, \mathcal{F}, -\varepsilon') / \bigcup_{\varepsilon>0} \bar{\partial} L^2_{(0,n-1)}(\xi^0, \mathcal{F}, -\varepsilon).$$

That this result can be of some interest can be seen from the following theorem which gives a formulation in terms of $\bar{\partial}$-cohomology for Bony's theorem on the equivalence of Sato's and Hörmander's definitions of the analytic singular spectrum for distributions. We denote in it by $L^2(A, -\varepsilon, -b)$ the space of measurable functions f on A so that $f \exp[-\varepsilon|Im\,\zeta| - b\ln(1+|\zeta|)] \in L^2(A)$.

Theorem 4.5. *Let* $b \geq 3$ *and* $\mu \in L^2_{(0,n)}(C^n, -\varepsilon, -b)$ *be given and assume that* $J(T(\mu)) = 0$ *in* $\mathcal{B}_{(x^0,\xi^0)}$. *Then there is a complex conic neighborhood* W *of* ξ^0, $\varepsilon > 0$ *and* $\nu \in L^2_{(0,n-1)}(W, -\varepsilon, -b+3)$ *so that* $\mu = \bar{\partial}\nu$ *on* W.

Thus, with somewhat sloppy, by now self-explanatory, notations, we can write that

$$\frac{\bigcup_{\varepsilon'>0} L^2_{(0,n)}(\xi^0, -\varepsilon', -b)}{\bigcup_{\varepsilon>0} \bar{\partial} L^2_{(0,n-1)}(\xi^0, \mathcal{F}, -\varepsilon)} = \frac{\bigcup_{\varepsilon'>0} L^2_{(0,n)}(\xi^0, -\varepsilon', -b)}{\bigcup_{\varepsilon>0} \bar{\partial} L^2_{(0,n-1)}(\xi^0, -\varepsilon, -b+3)}.$$

(For proofs and comments on how this result is related to Bony's theorem, cf. [22].)

5. Representation functions and pseudodifferential operators.

1.In the sections 9, 10 below, we shall show how one can extend the preceding results to the case of higher analytic microlocalization in hyperfunctions and discuss some related arguments. One of the advantages of our approach is that one can develop then a calculus of pseudodifferential operators in higher microlocalization which is completely analoguous to the theory of pseudodifferential operators in distributions. To make the underlying ideas more transparent, we shall describe the main steps in the calculus of pseudodifferential operators only for standard first microlocalization. The theory which we describe is thus parallel to any theory of infinite order analytic pseudodifferential operators. (Cf. in particular [2] or [26].) The case of higher microlocalization can be treated in a similar way but involves much heavier notations.

2. Let U be a neighborhood of the origin and consider $\xi^0 \in \dot{R}^n$. Also let $p(x,\xi)$ be some analytic symbol of infraexponential type defined in a conic neighborhood of $(0,\xi^0)$ for $|\xi|$ large. By this we mean that there is $C > 0$, a complex neighborhood Ω of the origin, a complex open cone W which contains ξ^0 and some analytic function $\tilde{p}$ on $S = \{(z,\zeta) \in C^{2n}; z \in \Omega, \zeta \in W, |\zeta| > C\}$ which extends p and so that

$$|\tilde{p}(z,\zeta)| \leq ce^{\varphi(|\zeta|)} \text{ for } (z,\zeta) \in S,$$

for some sublinear function $\varphi : R_+ \to R_+$. We shall later on write "p" also for "$\tilde{p}$". Consider next $u \in \mathcal{B}_{(0,\xi^0)}$. We can represent u in the form $\dot{T}(\mu)$ where $\mu \in L^2(\xi^0, \mathcal{F}, -\varepsilon)$ for some $\varepsilon > 0$. It is no loss of generality to assume that in fact $\mu \in L^2(C^n, \mathcal{F}, -\varepsilon)$. We can now define $p(x,D)u$ to be the microfunction associated with

$$h(z) = (1/2i)^n \int_W \exp[i\langle z,\zeta\rangle] p(z,\zeta)\mu(\zeta)\, d\zeta \wedge d\bar{\zeta}. \tag{14}$$

In fact, when W is a convex cone, the preceding expression already defines an analytic function when $|Re\, z| < \varepsilon$, $Im\, z \in (W \cap R^n)^\perp$, and when W is not convex, we can regularize the expression in exactly the same way we regularized (6). (The growth type of $p(z,\zeta)\mu(\zeta)$ is exactly the one we had for f in (6), only that here we have the additional analytic parameter z in the symbol $p(z,\zeta)$. It is clear that this does not make the situation more complicated.) It is easy to see that the value of $p(x,D)u$ in $\mathcal{B}_{(0,\xi^0)}$ does not depend on the choice of μ to represent u (recall here theorem 4.4; actually it is in this type of situations that we need theorem 4.4) or on the way in which we regularized (14). In particular, pseudodifferential operators are pseudolocal. (We start here from an analytic information for μ and get a geometric information for h.) To study composition of operators, let us consider one more analytic symbol $q(x,\xi)$. We can apply theorem 3.5 to write $q(z,\zeta)$ in the form

$$q(z,\zeta) = \int_{C^n} \exp[i\langle z,\theta\rangle]\sigma(\theta,\zeta)\, d\theta \wedge d\bar{\theta},$$

for some σ which satisfies for suitable c_1, c_2 and some sublinear F the inequality

$$|\sigma(\theta,\zeta)| \leq c_1 \exp[-c_2|\theta| + F(|\zeta|)].$$

Let also μ be some representation function for u which satisfies

$$(1/2i)^n \int_W |\mu(\zeta)|^2 \exp[-2\varphi(Re\,\zeta) + 2\varepsilon|Im\,\zeta|]\, d\zeta \wedge d\bar{\zeta} < \infty.$$

The first result which we need to study composition of operators is then

Proposition 5.1. *$q(z, D)u$ admits*

$$\nu(\theta) = (1/2i)^n \int_W \sigma(\theta - \zeta, \zeta)\mu(\zeta)\, d\zeta \wedge d\bar{\zeta} \tag{15}$$

as a representation function.

3. To consider composition, let us next assume that two analytic symbols, p and q are given on $\{|z| < \varepsilon\} \times \{\zeta \in W; |\zeta| > C\}$. Let also $u \in \mathcal{B}_{(0,\xi^0)}$. Since we can calculate a representation function for $q(x, D)u$, it is easy to calculate $p(x, D)(q(x, D)u))$. We would like to show that this is the same with $r(x, D)u$ for some analytic pseudodifferential operator which has symbol $r(x, \zeta)$ asymptotically equal with the formal symbol: $\sum_\alpha p^\alpha(x, \xi)q_\alpha(x, \xi)/\alpha!$, where $p^{(\alpha)} = (\partial/i\partial\xi)^\alpha$ and $q_{(\alpha)} = (\partial/\partial x)^\alpha q$. The main step in the argument is the following result:

Theorem 5.2. *Consider $W' \subset\subset W$. Define $I(z, \zeta)$ by the expression*

$$I(z, \zeta) = \int_{W, |\theta| > c} \exp[i\langle z, \theta - \zeta\rangle] p(z, \theta)\sigma(\theta - \zeta, \zeta)\, d\theta \wedge d\bar{\theta}.$$

Also denote for $a \in R_+$ by $Int[a]$ the integer part of a. Then there are constants $c_1, c_2, c_3, \varepsilon' > 0$, so that $I(z, \zeta)$ is an analytic symbol on $\{(z, \zeta); |z| < \varepsilon', \zeta \in W', |\zeta| > c_3\}$, and so that

$$I(z, \zeta) = \sum_{|\alpha| \le Int[c_1|\zeta|]} p^{(\alpha)}(z, \zeta)q_{(\alpha)}(z, \zeta)/\alpha! + O(\exp[-c_2|\zeta|]), \zeta \in W', |z| < \varepsilon'. \tag{16}$$

6. Higher microlocalization

1. Second analytic microlocalization has first been considered by Kashiwara-Kawai in [11]. The theory of Kashiwara has then been developed by Laurent and is described in [17]. A theory of analytic microlocalization of arbitrary order was developed by Sjöstrand and later on extended by Lebeau: cf. [30], [18]. A third approach to high order analytic microlocalization has finally been proposed in [20]. The three theories are not equivalent among themselves, in that they refer in part to different frames. Indeed, also the approaches used in the three competing theories are very different. For applications of arguments in higher analytic microlocalization, cf. e.g. [15], [18], [30], [32], [34], [35], [20]. For the present author, higher analytic microlocalization was foremost a tool to understand phenomena of propagation of analytic singularities for solutions of partial differential equations with characteristics of highly changing multiplicity. Since propagation phenomena in the analytic category are often related to some kind of partial

analyticity, we should perhaps point out that higher order microlocalization is much closer related to partial analyticity than is first microlocalization. As an illustration for this we shall state in section 7 a result which characterizes partial analyticity with the aid of the first two analytic wave front sets, whereas it is not possible to characterize it with the aid of the first analytic wave front set alone.

2. To develop a full theory of higher microlocalization, we need higher order wave front sets, higher order pseudodifferential operators and higher order Fourier integral operators. We shall here describe the higher order wave front sets. Higher order pseudodifferential operators can easily be studied in a way parallel to what we did in section 5 in the frame of first microlocalization. We shall not give details here and refer to [20] for results on higher order analytic pseudodifferential operators acting on distributions. (With the aid of the present results one can extend the results on pseudodifferential operators proved in [20] for distributions to the case of hyperfunctions.) We have not explicitly studied Fourier integral operators in higher analytic microlocalization when we microlocalize to order higher than 3. Unfortunately, this implies that we can not show that the results which we obtain in the case of third microlocalization and higher, have an invariant meaning. (In fact, what has to be done is not difficult to see, but seems to be very technical.)

3. Let us at first recall the definition of the higher analytic wave front set as given in [20]. We start from a finite sequence of subspaces M_j, $j = 0, 1, \dots k$, in R^n such that $M_0 = R^n$, $M_j \subset M_{j-1}$, $M_j \neq M_{j-1}$, $M_k = \{0\}$. Also denote by $\Pi_j : R^n \to M_j$ the orthogonal projection on M_j and by $\dot{M}_j = M_j \ominus M_{j+1}$ the orthogonal complement of M_{j+1} in M_j. It follows in particular that

$$R^n = \oplus_{j \geq 0} \dot{M}_j.$$

Also consider $\xi^j \in M_j \ominus M_{j+1}$.

Definition 6.1. *Let U be open in R^n, consider $x^0 \in U$, $u \in D'(U)$, and let M_j and ξ^j be as above. We shall say that $(x^0, \xi^0, \xi^1, \dots, \xi^{k-1})$ is not in the analytic k-wave front set of u and write*

$$(x^0, \xi^0, \xi^1, \dots, \xi^{k-1}) \notin WF_A^k u,$$

if we can find open conic neighborhoods $G^j \subset M_j$ of $\xi^j, \varepsilon, c, c_j, \beta > 0$, (increasing) sublinear functions $F_j : R_+ \to R_+$ and a bounded sequence of distributions $\{u_i\}_{i=1}^{\infty} \subset E'(U)$ such that

$$u = u_i, \text{for } |x - x^0| < \varepsilon, \ \textit{and all } i$$

$$|\hat{u}_i(\xi)| \le c(ci/|\Pi_{k-1}\,\xi|)^i \text{ if } i = 1, 2, \ldots, \Pi_j\,\xi \in G^j, j = 0, ..., k-1,$$
$$|\Pi_j\,\xi| \ge F_j(|\Pi_{j-1}\,\xi|), j = 1, ..., k-1,$$
$$\text{and } |\Pi_{j+1}\,\xi| \ge c_j|\Pi_j\,\xi|^{\beta+1}/|\Pi_{j-1}\,\xi|^{\beta}, j = 1, ..., k-2.$$

(In the oral lectures, we only stated the definitions for the cases $k = 2$ and $k = 3$. The reader will have understood why.)

Definition 6.2. *A set G is called a multineighborhood of $(\xi^0, \xi^1, ..., \xi^{k-1})$ if we can find open conic neighborhoods $G_j \subset M_j$ of ξ_j so that $G = \{\xi; \Pi_j\xi \in G_j, j = 0, \ldots, k-1\}$.*

Remark 6.3. *a) One can of course write the conditions*

$$|\Pi_{j+1}\,\xi| \ge c_j\,|\Pi_j\,\xi|^{\beta+1}/|\Pi_{j-1}\,\xi|^{\beta}\ ,\ j = 1, ..., k-2 \tag{17}$$

in homogeneous form $c_j\,|\Pi_j\,\xi|^{\beta}/|\Pi_{j-1}\,\xi|^{\beta} \le |\Pi_{j+1}\xi|/|\Pi_j\,\xi|\ ,\ j = 1, ..., k-2$. Since the quantities $|\Pi_{j+1}\xi|/|\Pi_j\,\xi|$ correspond to Sjöstrand's "small parameters", this makes comparision with the theory of Sjöstrand-Lebeau easier. We stick here to the form (17), since it is more natural in propagation phenomena.

b) When $k \le 2$ the conditions (17) are considered void. For $k = 3$, they reduce to $|\Pi_2\,\xi| \ge c_1\,|\Pi_1\,\xi|^{\beta+1}/|\Pi_0\,\xi|^{\beta}$.

Remark 6.4. *As stated above, we have not proved that the notion WF_A^k has an invariant meaning when $k \ge 3$. The situation is more favorable for $k = 2$. One can then give an invariant meaning to WF_A^2, and in fact, the above definition just reduces to second analytic microlocalization (as considered e.g. by [17] or [30]) with respect to the involutive submanifold $\xi' = 0$, if we denote the variables from M_1 by ξ'. (On the equivalence of the various definitions cf. [5], [25], [20].)*

Let us also recall for the convenience of the reader how one can characterize the second analytic wave front set in special coordinates with the aid of the FBI-transform: we shall assume here that $M_1 = \{\xi \in R^n; \xi_{n'+1} = \cdots = \xi_n = 0\}$ and denote $(\xi_1, \ldots, \xi_{n'})$ by ξ' and $(\xi_{n'+1}, \ldots, \xi_n)$ by ξ''. It can then be shown that

$$(x^0, \xi^0, \xi^1) \notin WF_A^2 u,$$

iff we can find c, c', c'', $\Gamma \subset R^n$, $\Gamma' \subset R^{n'}$, $\alpha > 0$, $\beta > 0$, open cones and a sublinear function $F : R_+ \to R_+$ such that $\xi^0 \in \Gamma$, $\xi^1 \in \Gamma'$,

$$\left|u_x(e^{[-i\langle x, \xi\rangle - \alpha|\xi'|\,|t' - x'|^2/2 - \beta|\xi''|\,|t'' - x''|^2/2]})\right| \le c\,e^{-c'|\xi'|}.$$

if $|x^0 - t| < c'$ (localization in x) $\xi \in \Gamma$, $\xi' \in \Gamma'$, $|\xi'| \ge F(|\xi|)$. The main reason why analytic microlocalization of some order k can be interesting if one studies propagation phenomena is that the following result of "micro-Holmgren" type (for which there is no analogue in the C^∞-category, of course,) is valid :

Theorem 6.5. *Let $U \subset R^n$ be open and consider $x^1, x^2 \in U$ with $[x^1, x^2] \subset U$, $x^1 - x^2 \in M_{k-1}$. Here and later on we denote by $[x^1, x^2]$ the segment with endpoints x^1 and x^2. Consider $\xi^0, \xi^1, ..., \xi^{k-2}$ and assume that for any $\eta \in M_{k-1}$ and any $x \in [x^1, x^2]$ it follows that*

$$(x, \xi^0, \xi^1, ..., \xi^{k-2}, \eta) \notin WF_A^k u. \tag{18}$$

Moreover, assume that $(x^1, \xi^0, \xi^1, ..., \xi^{k-2}) \notin WF_A^{k-1} u$. Then it follows that

$$(x, \xi^0, \xi^1, ..., \xi^{k-2}) \notin WF_A^{k-1} u, \text{ whatever } x \in [x^1, x^2] \text{ is.}$$

It also follows from this that $(x, \xi^0, \xi^1, ..., \xi^{k-2}) \notin WF_A^{k-1} u$, for all x in the connected component of $(x^1 + M_{k-1}) \cap U$ which contains x^1.

7. Second microlocalization and partial analyticity

We want to explain here the relation between partial analyticity and higher analytic microlocalization. Let us at first explain what we mean by "partial analyticity". We will say that $u \in D'(U)$ is partially analytic in the distribution sense in the variable x_1 if we can find an open set W in R^{n+1} with $W \cap R^n = U$ and $v \in D'(W)$ (if we only ask for $v \in \mathcal{B}(W)$, we shall say that u is partially analytic in the hyperfunction sense; the two assumptions are not completely equivalent) so that

$$[(\partial/\partial x_1) + i(\partial/\partial x_{n+1})]v = 0, v_{|x_{n+1}=0} = u.$$

We can here perform the restriction in view of the fact that v satisfies an homogeneous equation which is non-characteristic with respect to $x_{n+1} = 0$.

From general results on the wave front set of a restriction we can conclude that

$$(x, \xi^0) \in WF_A u \text{ implies } \xi_1^0 = 0, \forall x \in U. \tag{19}$$

Thus we obtain complete information on the first analytic wave front set of u when $\xi_1^0 \neq 0$, but in fact nothing can be said on the first wave front set of u when $\xi_1^0 = 0$. One can however prove the following result (cf. [20]):

Proposition 7.1. *Consider $u \in D'(U)$. Then the following two conditions are equivalent:*

i) u is partially analytic in the hyperfunction sense with respect to x_1.

ii) $(x^0,\xi^0) \notin WF_A u$, *whatever* $x^0 \in U$, $\xi^0 \in \dot{R}^n$ *with* $\xi_1^0 \neq 0$ *and in addition* $(x^0,\xi^0,\xi^1) \notin WF_A^2 u$, *whatever* $x^0 \in U$, $\xi^0 \in \dot{R}^n$ *with* $\xi_1^0 = 0$ *and whatever* $\xi^1 \in \dot{R}^n$ *with* $\xi_i^1 = 0$ *for* $i \geq 2$.

Remark 7.2. *A similar result is of course also valid for partial analyticity in the distribution sense, if we replace* WF_A^2 *as we defined it above with some tempered version. (Cf. again [20].)*

8. Successive localizations; an example

When $f(\xi)$ is a (say) C^∞-function which vanishes of some order s at some point ξ^0, we call "localization" of f at ξ^0 the function: $\sum_{|\alpha|=s}(\partial/\partial\xi)^\alpha f(\xi^0)(\xi -\xi^0)^\alpha/\alpha!$. When p is an analytic symbol whose principal part vanishes of variable multiplicity on its characteristic variety, localizations of this principal part to subvarieties play an important role. It may happen that the localized symbol itself vanishes of variable multiplicty, so one will be tempted to consider localizations of the first localized symbol, etc. In a fully involutive setting, the first localized symbol is an object of the second microlocal category, the second localization an object of the theory of third order microlocalization and so on. We do not make this here more precise, but refer to [20] for examples where this idea has been implemented. We want to show here with a simple example how this same idea has led to the geometry of higher order wave front sets as we have considered it above. (A more complicated situation of this type will appear in section 11.) Let us start from the symbol $p = \xi_1\xi_2\xi_3 + a(x,\xi)\xi_1^2 + b(x,\xi)\xi_2^3$ (plus perhaps lower order terms,) where a, b are symbols of order 1, respectively 0. Thus, from the point of view of the classical theory, the principal part is $\xi_1\xi_2\xi_3 + a(x,\xi)\xi_1^2 + b(x,\xi)\xi_2^3$. I want to study this microlocally near $\xi^0 = (0,0,1)$, two-microlocally near $\xi^0 = (0,0,1)$, $\xi^1 = (0,1,0)$, and 3-microlocally near $\xi^0 = (0,0,1)$, $\xi^1 = (0,1,0)$, $\xi^2 = (1,0,0)$.

The classical, i.e. 1-microlocal principal part is of course

$$\xi_1\xi_2\xi_3 + a(x,\xi)\xi_1^2 + b(x,\xi)\xi_2^3.$$

For 2-microlocalization, we need the localization of the principal symbol to the subspace $\xi_1 = \xi_2 = 0$. The lowest order of vanishing is two, therefore the localization p_1 of p to this subspace is $p_1 = \xi_1\xi_2\xi_3 + a(x,\xi)\xi_1^2$. 3-microlocally, we calculate the localization to $\xi_1 = 0$. The lowest order of vanishing is one, the symbol of the localized becomes $p_2 = \xi_1\xi_2\xi_3$.

This should be elliptic 3-microlocally and we want $\xi_1\xi_2\xi_3$ to domiate $|\xi_1|$ is dominated by ξ_2. all the other terms. We must then in particular have that $b(x,\xi)\xi_2^2$ can be dominated by $\xi_1\xi_3$, and this is true only if we restrict our attention to regions of form $\xi_1 >> \xi_2^2/\xi_3$. This is of the form $|\xi_1| \geq c|\xi_2|^{1+\beta}/|\xi|^\beta$ with $\beta = 1$.

9. Higher order microfunctions

1. Let M_j and $(\xi^0, \xi^1, \ldots, \xi^{k-1})$ be given as in section 6. Also fix $\beta > 0$, c_j, some multineighborhood G of $(\xi^0, \xi^1, \ldots, \xi^{k-1})$ and some sublinear functions $F_j : R_+ \to R_+$. We denote by $Y = Y(\xi^0, \xi^1, \ldots, \xi^{k-1})$ the set of $\xi \in R^n$ which satisfy the following conditions:

- $\xi \in G$,
- $|\Pi_{j+1}\xi| \geq c_j |\Pi_j \xi|^{\beta+1}/|\Pi_{j-1}\xi|^\beta$, $j = 1, \ldots, k-2$,
- $|\Pi_j \xi| \geq F_j(|\Pi_{j-1}\xi|)$, $j = 1, \ldots, k-1$.

We shall often say that Y is of type "$Y(\xi^0, \xi^1, \ldots, \xi^{k-1})$" then. Moreover, if Y is given of type $Y(\xi^0, \xi^1, \ldots, \xi^{k-1})$, then we denote by Y_C the set $\{\zeta \in C^n; Re\,\zeta \in Y, |Im\,\zeta| < c|\Pi_{k-1}(Re\,\zeta)|\}$ for some suitable $c > 0$.

Consider now a hyperfunction u defined near x^0. For notational simplicity we shall assume that $x^0 = 0$. We want to extend the definitions of higher order analytic wave front sets to the case of hyperfunctions. This is in so far new that in [20] higher order wave front sets were considered only for the case of distributions.

Definition 9.1. *We shall say that $(0, \xi^0, \xi^1, \ldots, \xi^{k-1}) \notin WF_A^k u$ if there are Y, d, ε, a sublinear function φ and a representation function μ for u so that*

$$|\mu(\zeta)| e^{-[\varphi(Re\,\zeta) + d|\Pi_{k-1}(Re\,\zeta)|_Y - \varepsilon|Im\,\zeta|]} \in L^2(Y_C). \tag{20}$$

Next we note that if μ is fixed and if we define $\tilde{\mu}$ by $\tilde{\mu}(\zeta) = \mu(\zeta)$ for $\zeta \in Y_C$ and by $\tilde{\mu}(\zeta) = 0$ if $\zeta \notin Y_C$, then $(0, \xi^0, \xi^1, \ldots, \xi^{k-1}) \notin WF_A^k \mathcal{F}^{-1}(\mu - \tilde{\mu})$. Thus the microfunction associated with u at $(0, \xi^0, \xi^1, \ldots, \xi^{k-1})$ depends only on the restriction of μ to Y_C.

2. In this subsection we consider a hyperfunction u defined in a neighborhood of 0 and assume that $(0, \xi^0, \xi^1, \ldots, \xi^{k-1}) \notin WF_A^k u$. Let μ be a representation function of u which satisfies (20) and consider an additional representation function μ' for u. Since μ and μ' both represent u, we can write

$$\mu - \mu' = \sum_j \bar{\partial}_j \chi_j,$$

with $|\chi(\zeta)| \exp -[\varphi(Re\,\zeta) - \varepsilon|Im\,\zeta|] \in L^2(C^n)$. Moreover, since on Y_C, $-\varphi(Re\,\zeta) + \varepsilon|Im\,\zeta|$ is dominated by $|\Pi_{k-1} Re\,\zeta|$ and since $-|\Pi_{k-1}(Re\,\zeta)|$ is plurisubharmonic on Y_C, we can write

$$\mu = \sum_j \bar{\partial}_j \chi'_j,$$

on Y_C, with χ'_j satisfying $|\chi'_j(\zeta)|\exp[d'|\Pi_{k-1}(Re\,\zeta)|] \in L^2(Y_C)$. (We use here Hörmander's theory on the $\bar\partial$-operator in weighted L^2-spaces.) We conclude that if $(0,\xi^0,\xi^1,\ldots,\xi^{k-1}) \notin WF_A^k u$, and if μ' is a representation measure of u, then there are $\varphi',\varepsilon',b',Y'_C$ so that

$$\mu' = \sum_j \bar\partial_j \chi''_j \text{ on } Y_C,$$

with $|\chi''_j(\zeta)|\exp-[\varphi'(Re\,\zeta) - \varepsilon'|Im\,\zeta|] \in L^2(Y'_C)$.

3. Conversely, let μ be a representation function of u and assume that $\mu = \sum_j \bar\partial_j \chi_j$, on Y_C. We can cut off χ_j near $Y'_C \subset\subset Y_C$ and therefore conclude that u admits a representation measure which vanishes on Y'_C. In particular this implies that $(0,\xi^0,\xi^1,\ldots,\xi^{k-1}) \notin WF_A^k u$. Microhyperfunctions at $(0,\xi^0,\xi^1,\ldots,\xi^{k-1})$ can in particular therefore be identified with

$$L^2_{(0,n)}(\xi^0,\xi^1,\ldots,\xi^{k-1},\varphi,-\varepsilon,b)/\bar\partial L^2_{(0,n-1)}(\xi^0,\xi^1,\ldots,\xi^{k-1},\varphi,-\varepsilon,b).$$

4. Microregularity for hyperfunctions can be tested by duality. What we can prove is the following result:

Theorem 9.2. *Let u be given. Then there are equivalent:*

i) $(0,\xi^0,\xi^1,\ldots,\xi^{k-1}) \notin WF_A^k u$.

ii) Let μ be a representation function for u on Y'_C. Then there are $Y \subset\subset Y'$, c,d,ε,b and a sublinear function φ so that $|\int_{Y'_C} h(\zeta)\mu(\zeta)\,d\zeta\wedge d\bar\zeta| \le c$ for any $h \in A(Y'_C)$ which satisfies

$$|h(\zeta)| \le \exp[-\varphi(Re\,\zeta) + d|\Pi_{k-1}(Re\,\zeta)|_Y + \varepsilon|Im\,\zeta| + b\ln(1+|\zeta|)],$$

$$|h(\zeta)| \le C\exp[-\varphi(Re\,\zeta) + \varepsilon|Im\,\zeta| + b\ln(1+|\zeta|)],$$

for some constant C which may depend on h. (The second condition is needed to give a meaning to $\int h(\zeta)\mu(\zeta)\,d\zeta\wedge d\bar\zeta$.)

This result is in so far interesting that the space of hyperfunctions does not admit a natural topology. Nevertheless, we see here that for microfunctions it is possible to work with duality arguments. (This will be further developed in a future paper, which is in preparation.) With the aid of the preceding result, we can now repeat the arguments from [20] and prove the full analogue in hyperfunctions of theorem 6.5.

10. Restriction of microfunctions

One can use the theory of Fourier-inverse transforms and of higher order wave front sets to define and study restrictions of higher order microfunctions to subvarieties. Since our theory is not invariant for high order microlocalization and in order to avoid notational complications, we shall only consider the case when we want to restrict 1-microlocal microfunctions. Rather than working in R^n we shall now work in R^{n+1}. We denote the variables there by (t, x), $t \in R$, $x \in R^n$, and want to consider restrictions to $t = 0$. The Fourier-dual variables will be denoted by $\lambda = (\tau, \xi)$. Let us also assume that we are given a microfunction $\dot{u}$ defined near $(0, \lambda^0)$, where the τ component of λ^0 is zero. We shall therefore have $\lambda^0 = (0, \xi^0)$ for some $\xi^0 \in \dot{R}^n$. For notational simplicity we assume that $\xi^0 = (0, \ldots, 0, 1)$ and denote by Π_1 the map $\Pi_1(\lambda) = (\xi_1, \ldots, \xi_{n-1})$. To define the restriction of $\dot{u}$ to $t = 0$, we can argue locally and then work with the Fourier transform, i.e. representation functions. In fact, on the Fourier side, we have just to integrate the τ-variable away in $\mu(\lambda)$, if μ is a suitable representation function on C^{n+1} for $\dot{u}$. A problem could be that only the values of μ in a complex conic neighborhood of λ^0 can have a real meaning for $\dot{u}$. Let us then fix a small complex conic neighborhood $V \subset C^n$ of ξ^0 and consider for some small constant $c > 0$ the set $W = \{\lambda \in C^{n+1}; \zeta \in V, |\tau| < c|Re\,\zeta|\}$. We tentatively define a function $\nu : V \to C$ by

$$\nu(\zeta) = \frac{1}{2i} \int_{|\tau|<c|Re\,\zeta|} \mu(\tau, \zeta)\, d\tau \wedge d\bar{\tau}. \tag{21}$$

It is easy to see that the function ν satisfies the estimate

$$|\int_V |\nu(\zeta)|^2 e^{-2[\varphi'(Re\,\zeta) - \varepsilon|Im\,\zeta| + b\ln(1+|\zeta|)]}\, d\zeta \wedge d\bar{\zeta}| \leq c \tag{22}$$

where

$$\varphi'(\xi) = \sup_{|\tau| \leq c|\xi|} \varphi(\tau, \xi). \tag{23}$$

(φ' is obviously sublinear.) The estimate for ν shows in particular that ν can play the role of a representation function of some microfunction $\dot{v}$ defined near $(0, \xi^0)$. Note that we have not used any additional assumption on μ. We however recall from the classical theory, that in order to define restrictions of hyperfunctions we need some additional assumptions on the hyperfunctions and that a good choice for such an assumption is that the co-normal directions to $t = 0$ do not belong to the analytic wave front set of the distribution under consideration. In fact, although we have not used any assumption in order to define $\dot{v}$, we will not obtain a reasonable theory unless we impose some additional assumptions on $\dot{u}$. Indeed, if we do not make additional assumptions, $\dot{v}$ is not correctly defined as a microfunction, in that it depends on the choice of the representation function μ choosen to define $\dot{u}$ and on the choice of c which we had to make to define W in terms of V. We shall therefore introduce the following additional assumption:

$$(0, \lambda^0, \pm N) \notin WF_A^2 \dot{u}, \text{ where } N = (1, 0, \ldots, 0). \tag{24}$$

To define $\dot{v}$, we shall use the assumption (24) in a rather strong way. In fact, one can show that the assumption implies that there is a constant $c' > 0$ and a representation function μ for $\dot{u}$ so that

$$\mu(\lambda)e^{d|Re\,\tau|+\varepsilon|Im\,\lambda|+b\ln(1+|\lambda|)} \in L^2(A), \tag{25}$$

where

$$A = \{Re\,\lambda \in G; |Re\,\tau| \geq c'|\Pi_1 Re\,\zeta|, |Im\,\lambda| < c''|Re\,\lambda|\}, \tag{26}$$

where G is some open cone in R^{n+1} which contains λ^0.

It can now be shown that $\dot{v}$ defined with the aid of ν depends only on $\dot{u}$, provided we argue on representation functions for $\dot{u}$ which satisfy the condition (25). Let us also mention the following result:

Theorem 10.1. *Let λ^0 and $\dot{u}$ be as above, in particular we assume that (25) is valid. Also assume that for some λ^1 with $\tau^1 = 0$, $\xi_n^1 = 0$, we have $(0, \lambda^0, \theta N + \lambda^1) \notin WF_A^2\dot{u}$ whatever $\theta \in R$ is. Then it follows that $(0, \xi^0, \xi^1) \notin WF_A^2\dot{u}_{|t=0}$. (Here ξ^1 is the ξ-component of λ^1.)*

The present considerations will be used in a forthcoming paper.

11. The Kawai-Kashiwara theorem.

Let us at first recall the Kawai-Kashiwara theorem in its classical form:

Theorem 11.1 *(Kawai-Kashiwara, cf. [12].) Let $p(x, D)$ be an analytic pseudodifferential operator defined in a conic neighborhood W of (x^0, ξ^0), let $\psi : W \to R$ be a real analytic function such that $\psi(x^0, \xi^0) = 0$, and assume that p is microhyperbolic at (x^0, ξ^0) in the direction $(-(\partial\psi/\partial\xi)(x^0, \xi^0), (\partial\psi/\partial x)(x^0, \xi^0))$. Let u be a hyperfunction such that $p(x, D)u = 0$ on W (in the sense that $WF_A p(x, D)u \cap W = \emptyset$) and assume that*

$$WF_A u \cap \{(x, \xi) \in W; \psi(x, \xi) < 0\} = \emptyset.$$

Then it follows that $(x^0, \xi^0) \notin WF_A u$.

(A real analytic function f defined in a neighborhood U of y^0 in R^m is called "microhyperbolic with respect to the direction θ" if there is $\varepsilon > 0$ and a neighborhood U' of y^0 such that

$$f(y + it\theta) \neq 0, \text{ if } 0 < t < \varepsilon, y \in U'.)$$

The Kawai-Kashiwara theorem is one of the most important theorems on propagation of analytic singularities. Many proofs, variants, extensions and applications (to results on propagation of singularities or conical refraction) of this result have been considered in the literature: [9], [12], [29], [36] [37],

[23]. Our aim in this section is to state a higher order variant of this theorem. We expect this higher order version to have applications which are similar in spirit to the ones of the original theorem and to help to understand propagation phenomena in a systematic way. Actually, the original Kawai-Kashiwara theorem consists of a "regularity part", which is theorem 11.1 above, and an existence part. Much of what we have said in the previous sections is related to an attempt to prove higher order existence results of Kawai-Kashiwara type. In the sequel we shall stick to the case of third wave front sets. The reason is that the value of high order variants becomes less and less clear if the calculus is not invariant and that for third order wave front sets we can understand at least the meaning of the main conditions which one imposes from an invariant point of view. To state our result, it will be convenient to work in R^{n+1} rather than in R^n. This is due to the fact that theorems of Kawai-Kashiwara type are deeply related to the Cauchy problem in which one has a distinguished variable in R, which we shall call "t", reserving the notation "x" for the remaining variables, which will then be variables in R^n, as above. These notations are thus in fact the same with the ones in section 10. As for the Fourier-dual variables, they are denoted, again as in section 10, by $\lambda = (\tau, \xi)$ with $\xi \in R^n$. It is convenient to write λ as $(\lambda_0, \lambda_1, \dots \lambda_n)$, so that $\tau = \lambda_0$ and $\xi = (\lambda_1, \dots, \lambda_n)$. We now start from a classical analytic symbol p of order μ defined on $U \times G$ where U is a neighborhood of $0 \in R^{n+1}$ and G is an open cone in R^{n+1}. Let p_μ be its principal part. We assume that p_μ vanishes of some order s on an analytic homogenous regular involutive variety Σ in T^*U which contains the point $(0, \lambda^0)$, $\lambda^0 = (0, \dots, 0, 1)$. It will be no loss of generality in applications to assume that $\Sigma = \{(z, \lambda); \lambda' = 0\}$ for some group of variables of type $\lambda' = (\lambda_0, \lambda_1, \dots, \lambda_d)$, $\lambda = (\lambda', \lambda_{d+1}, \dots, \lambda_n)$.

Let us also compute the localization $p_{\mu,1}$ of p_μ along Σ. It is in general a function on the co-normal bundle to Σ, but in our special coordinates above, we may just write (for λ in a conic neighborhood of λ^0) that

$$p_{\mu,1}(z, \lambda) = \sum_{|\alpha| = s} (\partial / \partial \lambda')^\alpha p_\mu(z, 0, \lambda_{d+1}, \dots, \lambda_n) \lambda'^\alpha / \alpha!.$$

In particular it is clear from this that $p_{\mu,1}$ is positively homogenous of order μ in the variables λ, and, in addition, homogenous of order s in the variables λ'. It also follows that

$$p_\mu(z, \lambda) = p_{\mu,1}(z, \lambda) + O(|\lambda'|^{s+1} |\lambda|^{\mu - s - 1}).$$

If $p_{\mu,1}$ were elliptic, its order of magnitude were $|\lambda'|^s |\lambda|^{\mu - s}$. Thus in a small conic neighborhood, $p_{\mu,1}$ would dominate p_μ. Consider $d' < d$, denote $\lambda'' = (\lambda_0, \lambda_1, \dots, \lambda_{d'})$ and fix $\lambda^1 \neq 0$ with $\lambda^{1''} = 0$ and $\lambda^1_i = 0$ for $i > d$. Also assume that $p_{\mu,1}$ vanishes of some order m on $\{(z,\lambda); \lambda''=0\}$. We denote by $p_{\mu,2}$ the localization of $p_{\mu,1}$ along $\lambda''=0$. It is thus given by the relation

$$p_{\mu,2}(z,\lambda)=\sum_{|\beta|=m}(\partial/\partial\lambda'')^{\beta}p_{\mu,1}(z,0,\lambda_{d'+1},\ldots,\lambda_n)\lambda''^{\beta}/\beta!.$$

It follows that, in addition to the homogeneities inherited from $p_{\mu,1}$, $p_{\mu,2}$ is homogeneous of order m in λ''. It is possible to give an invariant meaning also to these conditions in terms of the bi-homogeneous and bi-symplectic structures of the normal bundle to Σ; we refer to [17] or [20] for details. We have not studied the invariant meaning for the statements which follow hereafter. In any case, $p_{\mu,2}$ is of form $\sum_{|\gamma|=m}a_\gamma(z,\lambda_{d'+1},\ldots,\lambda_n)\lambda''^{\gamma}$ with $a_\gamma(z,\lambda_{d'+1},\ldots,\lambda_n)$ positively homogeneous of order $\mu-m$ in λ and homogeneous of order $s-m$ in λ' and the relation between $p_{\mu,1}$ and $p_{\mu,2}$ is

$$p_{\mu,1}(z,\lambda)=p_{\mu,2}(z,\lambda)+O(|\lambda''|^{m+1}|\lambda'|^{s-m-1}|\lambda|^{\mu-s}).$$

We shall now write the variables λ'' as $\lambda''=(\tau,\zeta'')$, where again $\tau=\lambda_0$. Similarily, $\lambda'=(\tau,\zeta')$. We also fix $\lambda^2\neq 0$ in R^{n+1} with $\lambda^2_0=0$, $\lambda^2_i=0$ for $i>d'$. We moreover assume that $p_{\mu,2}$ satisfies the following conditions:

a) the coefficient of τ^m in $p_{\mu,2}$ does not vanish at $(z=0,\zeta^{2\prime\prime},\lambda^1_{d'+1},\ldots,\lambda^1_{d},\lambda^0_{d+1},\ldots,\lambda^0_n)$. Recall that this coefficient has homogeneity $\mu-m$ in λ and $s-m$ in λ'. It has therefore order of magnitude $O(|\lambda'|^{s-m}|\lambda|^{\mu-s})$.

b) $p_{\mu,2}$ vanishes of order m at $(z=0,\lambda^{2\prime\prime},\lambda^1_{d'+1},\ldots,\lambda^1_{d},\lambda^0_{d+1},\ldots,\lambda^0_n)$.

c) $p_{\mu,2}$ is micro-hyperbolic with respect to $t=0$ at $(z=0,\lambda^0,\lambda^1,\lambda^2)$. By this we mean that there is a real neighborhood U' of $z=0$, a real tri-neighborhood G' of (ξ^0,ξ^1,ξ^2) ("tri" ="multi", for $k=3$.) , and $c>0$ so that $p_{\mu,2}(z,\tau,\xi)=0$, $z\in U'$, $\xi\in G'$ together with $|\tau|\leq c|\xi''|$ implies $\operatorname{Im}\tau\leq 0$.

Note that by assumption a), the coefficient of τ in $p_{\mu,2}$ is elliptic in the third order microlocal calculus near $(0,\lambda^0,\lambda^1,\lambda^2)$. 3-microlocally near $(0,\lambda^0,\lambda^1,\lambda^2)$, it is therefore no loss of generality to assume (if we compose everything with the inverse of the coefficient of τ) that we have

$$p_\mu(z,\lambda)=\tau^m+\sum_{|\alpha|+j=m,j<m}a_{\alpha,j}(z,\lambda_{d'+1},\ldots,\lambda_n)\zeta''^{\alpha}\tau^j+O(|\lambda''|^{m+1}/|\lambda'|)$$

$$+O(|\lambda'|^{m+1}/|\lambda|), \tag{27}$$

with coefficients $a_{\alpha,j}$ which are positively homogeneous of order zero in λ and λ'. This is, actually, 3-microlocally, the model on which we work.

Even in regions where $p_{\mu,2}$ is 3-microlocally elliptic, it will have at most the order of magnitude $|\lambda''|^m$. It can therefore dominate the remainder term $O(|\lambda'|^{m+1}/|\lambda|)$ only in regions of form $|\lambda''| \geq c|\lambda'|^{1+\beta}/|\lambda|^\beta$ with $\beta < 1/m$. (We shall work with $\beta = 1/(2m)$, to make a choice.) This is the justification why we restrict our attention to such regions in the definition of WF_A^3.

Remark 11.2. *It seems that we have lost orders of vanishing in λ', since the model p_μ vanishes only of order m when $\lambda' = 0$, if it has form (27). Fact is that we are working 3-microlocally, in a region away from $\lambda' = 0$, and that the representation (27) can be used only there.*

We can now state the following result:

Theorem 11.3 *Assume that under the above assumptions u is a distribution defined in a neighborhood of U and that it satisfies the following conditions for some tri-neighborhood W of $(0, \lambda^0, \lambda^1, \lambda^2)$:*

$$WF_A^3 p(z, D)u \cap W = \emptyset, \tag{28}$$

$$WF_A^3 u \cap W \cap \{t < 0\} = \emptyset. \tag{29}$$

Then it follows that $(0, \lambda^0, \lambda^1, \lambda^2) \notin WF_A^3 u$.

Remark 11.4 *Although we have stated theorem 11.3 for the case of 3-microlocalization, the argument works as well for the case of standard wave front sets, respectively for the case of second microlocalization. In particular, one thus obtains a new proof for theorem 11.1. As far as the case of two-microlocalization is concerned, I was told by prof. N.Tose that he is also aware of the fact that a result of the type of theorem 11.3 is true. I also think that the result remains valid in arbitrary microlocalization, but I have not checked all details. Note that the result presented here, as well as its analogue for the case of two-microlocalization, refer to a highly involutive setting. The proof of theorem 11.3 will be given elsewhere.*

We give an example in which one sees that the theorems above give the possibility to study questions of propagation of singularities in a rather systematic way. Let $q(x, \xi_1, \xi_2, \xi_3)$ be an elliptic polynomial of order two in (ξ_1, ξ_2, ξ_3), with real analytic coefficients which depend on $x \in R^4$ (much less is needed) and denote by $p(x, \xi)$ a symbol of form:

$$(\xi_1 + ix_1^2\xi_2)q(x, \xi_1, \xi_2, \xi_3) + O(|(\xi_1, \xi_2, \xi_3)|^4/\xi_4) + O(|\xi|^2)$$

in four variables at $(0, \xi^0)$ where $\xi^0 = (0, 0, 0, 1)$. Singularities of solutions of $p(x, D)u = 0$ then propagate in planes parallel to the (x_1, x_2)-plane. This can be obtained in the following way:

Lemma 11.5. *If $n' = 3$, $\xi' = (\xi_1, \xi_2, \xi_3) \in R^3$, $\xi^1 \in \dot{R}^3$, then $WF_A^2 u$ at (ξ^0, ξ^1) is either void or propagates in planes parallel to the (x_1, x_2)-plane.*

Proof. Here $\mu = 3$ and p_μ vanishes of order $s = 3$ on $\xi_1 = \xi_2 = \xi_3 = 0$. The localization polynomial on $\xi_1 = \xi_2 = \xi_3 = 0$ is $p_{\mu,1}(x,\xi) = (\xi_1 + ix_1^2\xi_2)q(x,\xi_1,\xi_2,\xi_3)$. Let $\xi^1 = (\xi_1^1, \xi_2^1, \xi_3^1)$. If $\xi_1^1 \neq 0$, then $p_{\mu,1}$ is elliptic at (x, ξ^0, ξ^1) in the two-microlocal calculus, so $(x, \xi^0, \xi^1) \notin WF_A^2 u$. The same is true if $\xi_2^1 \neq 0$ and $x_1 \neq 0$. When $\xi_1^1 = 0, \xi_2^1 \neq 0$, then $p_{\mu,1}$ is microhyperbolic at $x_1 = 0$. Thus we obtain once more that $(x, \xi^0, \xi^1) \notin WF_A^2 u$. (We apply the second order version of theorem 11.3.) Thus $|\xi_1^1| + |\xi_2^1| \neq 0$ implies $(x, \xi^0, \xi^1) \notin WF_A^2 u, \forall x$. It remains to study the case $\xi^1 = (0, 0, \pm 1)$. We propagate $WF_A^2 u$ along the (x_1, x_2)-planes in this case by showing that $WF_{A,s}^3 u$ at (x, ξ^0, ξ^1, ξ^2) is void for any choice of $\xi^2 = (\xi_1^2, \xi_2^2)$. Here $p_{\mu,2}(x,\xi) = p_{\mu,1}(x,\xi)$. If $\xi_1^2 \neq 0$ or if $\xi_2^2 \neq 0$ and simultaneously $x_1 \neq 0$, this is again ellipticity. When $\xi_1^2 = 0$, $\xi_2^2 \neq 0, x_1 = 0$, we can use again theorem 11.3 to see that we have no $WF_{A,s}^3 u$ at (x, ξ^0, ξ^1, ξ^2).

12. The theorem of Bony-Schapira

1. We want to explain some of the ideas involved in the proof of theorem 11.3 by sketching a proof of a theorem of Bony-Schapira, which was the first in a number of results which were generalized in [12]. We recall at first the Cauchy-Kowalewska theorem. Let us in fact consider some linear partial differential operator with analytic coefficients of order m defined in a neighborhood U of $0 \in R^{n+1}$ of form

$$p(t, x, D_t, D_x) = D_t^m + \sum_{|\alpha|+j \leq m, j<m} a_{\alpha j}(t,x) D_t^j D_x^\alpha.$$

We assume that the coefficients $a_{\alpha j}$ admit analytic extensions to a complex neighborhood Ω of $0 \in C^{n+1}$ (Here $t \in R$, $x \in R^n$, $(t,x) \in U$.) We also consider the Cauchy problem

$$p(t, x, , D)u = g \text{ on } |z| < \varepsilon_2, (i\partial/\partial t)^j_{|t=0} = g_j, j = 0, ..., m-1. \tag{30}$$

Theorem 12.1. *For any $\varepsilon_1 > 0$ which is sufficiently small, we can find $\varepsilon_2 > 0$ so that if $f \in \mathcal{A}(z \in C^{n+1}; |z| < \varepsilon_1)$, $g_j \in \mathcal{A}(x \in C^n; |x| < \varepsilon_1)$, then we can find $u \in f \in \mathcal{A}(z \in C^{n+1}; |z| < \varepsilon_2)$ so that (30) is valid.*

By duality we obtain from this

Theorem 12.2. *For any $\varepsilon_1 > 0$ which is sufficiently small, we can find $\varepsilon_2 > 0$ so that if $v \in \mathcal{A}'(z \in C^{n+1}; |z| < \varepsilon_2)$ is given, then there are $w \in \mathcal{A}'(z \in C^{n+1}; |z| < \varepsilon_1)$ and $w_j \in \mathcal{A}'(z \in C^{n+}; |z| < \varepsilon_1)$ such that $v(u) = w(f) + \sum_{j=0}^{m-1} w_j(g_j)$, if u, f, g_j are as in theorem 12.1. This can* *also be written as*

$$v = {}^tp(z, D_z)w + \sum_{j=0}^{m-1} D_t^j \delta \otimes w_j. \tag{31}$$

Here we have denoted by ${}^tp$ the formal adjoint of p and by δ the Dirac distribution at 0 in the variable t.

The interesting thing is now that from (31) we can obtain by direct means rather explicit information on the map ${}^tT$, in that one can obtain an almost explicit formula which gives the Fourier-Borel transform of w. We explain this at first in the case of constant coefficients. In fact, if we take the Fourier-Borel transform of (31), we get

$$\hat{v}(\lambda) = p(-\lambda)\hat{w}(\lambda) + \sum_{j=0}^{m-1} \hat{w}_j(\zeta)\tau^j, \tag{32}$$

which shows that $\hat{w}$ and the $\hat{w}_j$ are just the quotient and remainder terms in a Weierstrass-type decomposition of $\hat{v}$. (This also shows that we may view (31) as a non-commutative version of the Weierstrass preparation theorem.) One can compute $\hat{w}$ explicitly from (32) using contour integration formulas. Note in fact that it follows from (32) that

$$\frac{1}{2\pi i}\frac{\hat{w}(\tau+\sigma,\zeta)}{\sigma} = \frac{1}{2\pi i}\frac{\hat{v}(\tau+\sigma,\zeta)}{\sigma p(-\tau-\sigma,\zeta)} - \frac{1}{2\pi i}\sum_{j=0}^{m-1} \hat{w}_j(\zeta)\frac{(\tau+\sigma)^j}{\sigma p(-\tau-\sigma,\zeta)},$$

whenever σ and $p(-\tau-\sigma,\zeta)$ are different from zero. Integrating this in the complex σ- plane over a contour of form $|\sigma| = c(1+|\zeta|)$ for some sufficiently large c, (contours in contour integrals are always with counterclockwise orientation,) we obtain

$$\hat{w}(\tau,\zeta) = \frac{1}{2\pi i}\int_{|\sigma|=c(1+|\zeta|)} \frac{\hat{v}(\tau+\sigma,\zeta)}{\sigma p(-\tau-\sigma,-\zeta)}d\sigma, \tag{33}$$

since

$$\int_{|\sigma|=c(1+|\zeta|)} \frac{(\zeta_n+\sigma)^k}{\sigma p(-\tau-\sigma,\zeta)}d\sigma = 0, \text{ for } k \leq m-1, \tag{34}$$

if c is large enough. (We can apply the residuum theorem at ∞.) All this is of course standard. For the variable coefficient case, the situation is only slightly more involved. Indeed, let us denote by $\sum_j q_j$ a formal analytic symbol inverse to p. It is defined e.g. on the set $|\tau| > c(1+|\zeta|)$ if c is large. Also consider the map $\tilde{T}$ which associates with v the function

$$g(\lambda) = \frac{1}{2\pi i}\int_{\Lambda(\lambda)} v[e^{-i\langle z,\, \lambda+\sigma N\rangle}\frac{1}{\sigma}\sum_{j\leq c'|\lambda|} q_j(z, -\lambda-\sigma N)]\, d\sigma\,.$$

One can then prove with arguments which are only slightly more complicated than in the constant coefficient case that

$$\tilde{T}({}^t p(z,D)w)(\lambda) \sim \hat{w}(\lambda), \tilde{T}(D_t^j \delta_t \otimes w_j)(\lambda) \sim 0.$$

Let me in fact sketch a proof of the first assertion. We start with a preparation

Proposition 12.3. *There is χ^0 such that for any fixed χ, $\chi \leq \chi^0$, we can find $C > 0$, $d > 0$, such that*

$$|1 - e^{i\langle z,\lambda\rangle} p(z,D)[e^{-i\langle z,\lambda\rangle} \sum_{j<\chi|\lambda|} q_j(z,\lambda)]| \leq Ce^{-d|\lambda|}, \text{ if } |z| < \varepsilon, -\lambda \in G.$$

(This is based on

$$e^{i\langle z,\lambda\rangle} p(z,D)[e^{-i\langle z,\lambda\rangle} \sum_{j<\chi|\lambda|} q_j(z,\lambda)] = p(z, -\lambda + D) \sum_{j<\chi|\lambda|} q_j(z,-\lambda) =$$

$$\sum_{|\alpha|\leq m} (1/\alpha!) p^{(\alpha)}(z,-\lambda) D_z^\alpha \sum_{j<\chi|\lambda|} q_j(z,-\lambda) = S(z,\lambda) \sim 1,$$

etc.)

We return to the proof of the first assertion: we have

$$\tilde{T}({}^t p(z,D)w)(\lambda) = (1/2\pi i) w\{\int_{\Lambda(\lambda)} (1/\sigma) p(z,D)[e^{-i\langle z,\lambda+\sigma N\rangle}$$

$$\sum_{j<\chi|\lambda|} q_j(z, -\lambda - \sigma N)]\, d\sigma\}.$$

Applying the proposition, we remain with

$$w[\int_{\Lambda(\lambda)} e^{-i\langle z,\lambda+\sigma N\rangle} (1/\sigma)\, d\sigma] = \hat{w}(\lambda).$$

To state the theorem of Bony-Schapira let us denote by p_m the principal symbol of p. Thus

$$p_m(t,x,\tau,\xi) = \tau^m + \sum_{|\alpha|+j\leq m, j<m} a_{\alpha j}(t,x)\tau^j \xi^\alpha.$$

We assume that p_m is hyperbolic with respect to $t = c$. By this we mean that $(t,x) \in U$, $\xi \in R^n$ and $p_m(t,x,\tau,\xi) = 0$ together, imply $\tau \in R$. Let us also recall the following result of Bony-Schapira (cf. [1]):

Theorem 12.4. *Let $\tilde{\varepsilon}$ be given small enough, fix $\varepsilon < \tilde{\varepsilon}$ and $C' > 0$. Then there is $C > 0$ so that if*

$$g_j \in \mathcal{A}(x \in R^n; |x| < \tilde{\varepsilon}),\ f \in \mathcal{A}((t,x) \in R^{n+1}; C'|t| + |x| < \tilde{\varepsilon}),$$

are given, then we can find $u \in \mathcal{A}((t,x) \in R^{n+1}; C|t| + |x| < \varepsilon)$, so that (30)is valid.

We want to sketch a new proof of this results which if suitably adapted will be used to prove theorem 11.3. The proof is by duality. One of the main preliminary ingredients which we want to mention is an inequality for the roots of the characteristic equation if the principal part is hyperbolic. More precisely, one has

$$p_m(z,\tau,\xi) = 0 \text{ implies } |Im\,\tau| \le c(|Im\,z||Re\,\zeta| + |Im\,\zeta|), \tag{35}$$

if z is in a complex neighborhood of $0 \in C^{n+1}$ and $\zeta \in C^n$.

Remark 12.5. *(35) is often obtained from hyperbolicity using a so called "local form of Bochner's tube theorem". (Cf. again [1], or [14]). It is therefore perhaps interesting to mention that it is also a consequence of a suitable form of the Phragmén-Lindelöf principle applied to the plurisubharmonic functions $\rho(z,\zeta) = \sup Im\,\tau(z,\zeta)$, $\tilde{\rho}(z,\zeta) = \sup\ [-Im\,\tau(z,\zeta)]$, where the supremum is over all roots $\tau(z,\zeta)$ of $p_m(z,\tau,\zeta) = 0$.*

(The form of the Phragmén-Lindelöf principle which we have in mind is the following:

Theorem 12.6.*(Cf. e.g.[24]) Consider $A = \{y \in C^s, |y| < 1\}$ and let $\chi : A \to R$ be a plurisubharmonic function so that $\chi(y) \le 0$ if $y \in A \cap R^s$ and $\chi(y) \le c_1$. Then there are constants c_2, c_3 which depend only on c_1 so that*

$$\chi(y) \le c_2|Im\,y| \text{ if } y \in C^s, |y| \le c_3.$$

By applying this for the function ρ, we obtain at first $\rho(z,\zeta) \le c_2(|Im\,z| + |Im\,\zeta|)$ for small (z,ζ), and from this we obtain (35) after homogeneization.)

We next introduce some notations. When $C, C', \varepsilon, \varepsilon'$ are given, we denote by

$$K = \{(t,x) \in R^{n+1}; C|t| + |x| \le \varepsilon\}, \tag{36}$$

$$K' = \{(t,x) \in R^{n+1}; C'|t| + |x| \le \varepsilon'\}, \tag{37}$$

and by H_K, $H_{K'}$, the supporting functions of K and K' in C^{n+1}. Thus

$$H_K(\lambda) = \sup_{y \in K} Im\,\langle y, \lambda \rangle,$$

and similarly for K'.

It is now clear that theorem 12.4 is a consequence of the following result:

Proposition 12.7. *Consider $0 < \varepsilon < \varepsilon' < \tilde{\varepsilon}$ with $\tilde{\varepsilon}$ fixed sufficiently small, but no other condition on $\varepsilon, \varepsilon'$. Also fix $C' > 0$, $\nu' > 0$. Then there are $\nu > 0, C > 0$ (with C independent of ν'!) so that if v is a Radon measure on $\{z \in C^{n+1}; C|Re\, t| + |Re\, x| < \varepsilon, |Im\, z| < \nu\}$, which satisfies $\int d\,|v(z)| < 1$, then we can find analytic functionals $(w, w_0, \ldots, w_{m-1})$ related to v by (31) so that (with the notation for K' explained in (37)),*

$$|\hat{w}(\lambda)| \leq c\exp[\nu'|Re\,\lambda| + H_{K'}(Im\,\lambda)], \tag{38}$$

$$|\hat{w}_j(\zeta)| \leq c\exp[\nu'|Re\,\zeta| + \varepsilon''|Im\,\zeta|]. \tag{39}$$

How will we prove this? Well, we have a precise geometric information on supp v and want a precise analytic information on $\hat{w}$ and $\hat{w}_j$. Here we can calculate $\hat{w}$ approximatively with our contour integral formulas, whereas the $\hat{w}_j$ can be estimated with the aid of (31). The geometric information on supp v tells us which z are to be taken into account. Hyperbolicity gives as an estimate on where the residua of the contour integral are located. We can now deform the integration contour in such a way as to make the exponential harmless. (To some extent, the situation is inverse to the one we had in proposition 4.2: we have a geometric information on v and we obtain an analytic information on $\hat{w}$.)

References

1. J.M. Bony- P. Schapira: Solutions hyperfunctions du problème de Cauchy. In "Hyperfunctions and pseudodifferential equations", Springer LNM **vol. 287**, Springer Verlag Berlin-Heidelberg, 1973.
2. L.Boutet de Monvel: *Opérateurs pseudo-differentiels analytiques et opérateurs d'ordre infini.* Ann. Inst. Fourier,Grenoble, **22:3**, (1972), 229-268.
3. L.Ehrenpreis: *Fourier analysis in several complex variables.* Interscience Publ.Comp., 1970.
4. P. Esser- P. Laubin: Second microlocalization on involutive submanifolds. Séminaire d'analyse supérieure, Univ. de Liège, Institut de Mathematique, Liege 1987.
5. Second analytic wave front set and boundary values of holomorphic functions. *Applicable Analysis,* **vol. 25** (1987), 1-27.
6. I.M. Gelfand-G.E.Shilov : *Generalized functions.* Academic Press.
7. L. Hörmander: Uniqueness theorems and wave front sets for solutions of linear differential equations with analytic coefficients. *C.P.A.M.*, **24**, (1971), 671-704.
8. *The analysis of linear partial differential operators* I. Springer Verlag, Grundlehren der mathematischen Wissenschaften, **vol.251**, Berlin-New York, 1983.
9. K. Kajitani- S. Wakabayashi: Microhyperbolic operators in Gevrey classes. *Publ. R.I.M.S. Kyoto Univ.*,**25**(1989), 169-221.
10. The hyperbolic mixed problem in Gevrey classes. *Japan J. Math.* **15** (1989), 315-383.
11. M. Kashiwara M.- T. Kawai: *Deuxième microlocalisation.* Proc. Conf. Les Houches 1976, Lecture Notes in Physics **vol. 126**, Springer Verlag, Berlin Heidelberg New York.
12. Microhyperbolic pseudodifferential operators I. *J. Math. Soc. Jap.* **27** (1975), 359-404.

13. T. Kawai: *On the theory of Fourier hyperfunctions and its applications to partial differential operators with constant coefficients.* J. Fac. Sci Tokyo IA, 17:3 (1970), 467-519.
14. H. Komatsu: A local version of Bochner's tube theorem. *J.Fac.Sci. Tokyo,* **IA,19** (1972), 201-214.
15. P. Laubin: Propagation of the second analytic wave front set in conical refraction. *Proc. Conf. on hyperbolic equations and related topics,* Padova, 1985.
16. Etude 2-microlocale de la diffraction. *Bull. Soc. Royale de Science de Liège,* **56:4** (1987), 296-416.
17. Y. Laurent: *Theorie de la deuxième microlocalisation dans le domaine complexe.* Birkhäuser Verlag, Basel, Progress in Math., **vol.53**, 1985.
18. G. Lebeau: Deuxième microlocalisation sur les sous-varietés isotropes. *Ann. Inst. Fourier Grenoble,* **XXXV:2**, (1985), 145-217.
19. O. Liess: The Cauchy problem in inhomogeneous Gevrey classes. *C.P.D.E.,* **11**, (1986), 1379-1439.
20. *Conical refraction and higher microlocalization* Springer LNM **1555**, 1993, Springer Verlag, Berlin Heidelberg.
21. Higher microlocalization and propagation of analytic singularities. K̂okyoroku Series of the R.I.M.S. in Kyoto, **1996:2**, 60-72.
22. $\bar{\partial}$ – *cohomology* with bounds and hyperfunctions. Preprint nr.1, University of Bologna, 1996.
23. A. Martinez: Lectures in these proceedings.
24. R. Meise- B.A.Taylor-D. Voigt: Phragmén-Lindelöf principle on algebraic varieties. To appear.
25. Y. Okada-N. Tose : FBI-transformation and microlocalization- equivalence of the second analytic wave front sets and the second singular spectrum. *Journal de Math. Pures et Appl.,* **t. 70:4**, (1991), 427-455.
26. M. Sato- T. Kawai- M. Kashiwara: Hyperfunctions and pseudodifferential operators. Lecture Notes in Math., **vol. 287**, Springer Verlag, Berlin Heidelberg New York, 1973, 265-529.
27. J. Sjöstrand: Propagation of analytic singularities for second order Dirichlet problems I. *C.P.D.E.,* **5:1**, (1980), 41-94.
28. *Singularities in Boundary Value Problems.* Proc. of the Nato ASI, Maratea 1980, Ed. by H.G.Garnir, Reidel Publ. Comp., Dordrecht Boston-London 1981, 235-271.
29. Analytic singularities and microhyperbolic boundary values problems. *Math. Ann.* **254** (1980), 211-256.
30. *Singularités analytiques microlocales.* Astérisque **vol.95**, 1982, Soc.Math. France.
31. B.A. Taylor: *Analytically uniform spaces of infinitely differentiable functions.* C.P.A.M., vol. XXIV (1971), 39-51.
32. N. Tose: On a class of microdifferential operators with involutory double characteristics- as an application of second microlocalization. *J. Fac. Sci. Univ. Tokyo,* Sect. IA, Math. **33**, (1986), 619-634.
33. The 2-microlocal canonical form for a class of microdifferential equations and propagation of singularities. *Publ. R.I.M.S. Kyoto,* **23-1**, (1987), 101-116.
34. Second microlocalisation and conical refraction. *Ann. Inst. Fourier Grenoble,* **37:2**, (1987), 239-260.
35. Second microlocalisation and conical refraction, II. "Algebraic analysis", Vol.II. Volumes in honour of Prof. M.Sato, edited by T.Kawai and M.Kashiwara, Academic Press, 1989, 867-881.
36. K. Uchikoshi: Construction of the solutions of microhyperbolic pseudodifferential equations. *J.Math. Soc. Japan.* **vol. 40**, (1988), 289-318.
37. S. Wakabayashi: A classical approach to studies on propagation of analytic singularities. *Kokyoroku Series of the R.I.M.S. in Kyoto,* **1996:2**, 60-72.

CONORMALITY AND LAGRANGIAN PROPERTIES IN DIFFRACTIVE BOUNDARY VALUE PROBLEMS

P. LAUBIN
Institut de Mathématique
Université de Liège
Avenue des Tilleuls, 15
4000 Liège, Belgique

1. Introduction

Our main purpose is to study the lagrangian structure of the solution of a strictly diffractive boundary value problem at the transition from the shadow to the illuminated region. If the incoming data or the boundary data are conormal then two lagrangian submanifolds are involved there. Because of the geometry of the diffractive rays, their intersection is not clean. We try to describe the solution with phase functions and oscillatory integrals.

Let M be a real manifold with boundary and P a second order differential operator with smooth coefficients and real principal symbol p. We assume that p is of real principal type and not characteristic on the boundary. Let us consider the classical Dirichlet problem

$$Pu = 0 \quad \text{in} \quad M, \quad u_{|\partial M} = 0.$$

If the equation of the boundary is $f = 0$ with $f > 0$ in M, the diffractive region is defined by

$$\mathcal{G}_{+} = \{\rho \in \dot{T}^{*}\partial M : p(\rho) = 0, \quad \frac{\{p, \{p, f\}\}_{\rho}}{\{\{p, f\}, f\}_{\rho}} > 0\}$$

and corresponds to rays tangent to the boundary. The propagation of singularities of C^{∞}, Gevrey and analytic singularities is known in this setting, see [11], [15], [8], [9]. However, very few lagrangian properties are preserved along diffractive rays. In [10], Lebeau proves that, far away from the data, the operator mapping the Dirichlet data to the normal derivative of the solution belongs to a class of lagrangian Gevrey 3 distributions with weight.

L. Rodino (ed.), Microlocal Analysis and Spectral Theory, 91–113.

We first study the properties of the solution at the transition from the shadow to the illuminated region in the C^∞ framework. Using the canonical invariance, we prove that the solution belongs to a class of lagrangian distributions associated to a pair of lagrangian submanifolds. As a consequence, we see that, for a conormal data, the second wave front lies in a lagrangian submanifold.

We next investigate the same problem in the analytic category. Here we use the geometry of complex canonical transforms and the H_φ spaces of Sjöstrand. Our main tool is the parametrix of Lebeau. We generalize the definition of bilagrangian distributions in this framework and describe the FBI transform of the solution of the boundary value problem.

2. Pairs of lagrangian submanifolds

2.1. MICROLOCAL PHASE

Let X be a C^∞ manifold of real dimension n and with local coordinates $x_1, \ldots, x_n$. On the cotangent bundle T^*X, we consider the canonical 2-form

$$\sigma = \sum_{j=1}^{n} d\xi_j \wedge dx_j$$

where the dual coordinates are defined by $d\xi_j(D_{x_k}) = \delta_{jk}$. This manifold is conic for the multiplication $M_t : (x, \xi) \mapsto (x, t\xi)$. We denote by $\dot{T}^*X = T^*X \setminus \{0\}$ the cotangent bundle with the zero section removed.

A submanifold Λ of $\dot{T}^*X$ of dimension n is lagrangian if $\sigma_{|\Lambda} = 0$. It is said conic if it is invariant through T_t for every $t > 0$.

The classical definition of a phase function for a conic lagrangian submanifold is the following, [2]. For simplicity, we restrict ourself to the case of a real non-degenerate phase function.

Definition 1 *Let X be a C^∞ manifold and φ be a C^∞ real valued function in an open conic subset Γ of $X \times \mathbb{R}^N \setminus \{0\}$ which is homogeneous of degree 1. The function φ is called a local phase function of X if $d\varphi \neq 0$ in Γ and $\mathrm{rk}(\varphi''_{\theta x}, \varphi''_{\theta\theta}) = N$ in the set*

$$C_\varphi = \{(x, \theta) \in \Gamma : \varphi'_\theta(x, \theta) = 0\}.$$

If φ is a local phase function then the differential of the map

$$j_\varphi : C_\varphi \to \dot{T}^*X : (x, \theta) \mapsto (x, \varphi'_x(x, \theta))$$

is of rank n. If it is an embedding then φ is called a *phase function*. Since

$$j_\varphi^*\sigma = j_\varphi^* d(\xi dx) = d(\varphi'_x dx) = d(d\varphi_{|C_\varphi}) = 0,$$

its image $\Lambda_\varphi = j_\varphi(C_\varphi)$ is a lagrangian submanifold of $\dot{T}^*X$.

2.2. 2-MICROLOCAL PHASE

The second wave front set along a lagrangian submanifold Λ is defined as a subset of the cotangent bundle of Λ. To define lagrangian distributions associated to this geometric setting, we introduce new phase functions.

If Λ is a conic lagrangian submanifold of $\dot{T}^*X$, then we have the identification

$$\dot{T}^*\Lambda \sim T_\Lambda \dot{T}^*X$$

where the right hand side is the normal bundle of Λ. Indeed, if k is a normal to Λ at a point ρ then $T_\rho\Lambda \ni h \mapsto \sigma(h,k)$ is a well-defined 1-form.

Moreover this manifold has two homogeneities: one inherited from Λ and another one as a cotangent bundle. A lagrangian submanifold of $\dot{T}^*\Lambda$ is said *conic bilagrangian* if it is conic for both homogeneities. We introduce phase functions that parameterize such a manifold.

Let Γ_0 be an open subset of $X \times \mathbb{R}^N \setminus \{0\} \times \mathbb{R}^M \setminus \{0\}$ such that $(x,\theta,\eta) \in \Gamma_0$ and $s,t > 0$ imply $(x, t\theta, st\eta) \in \Gamma_0$. Such an open set is called a *profile*. An open subset Γ of $X \times \mathbb{R}^N \setminus \{0\} \times \mathbb{R}^M \setminus \{0\}$ is said *biconic with profile* Γ_0 if

- $(x,\theta,\eta) \in \Gamma$ and $t > 0$ imply $(x, t\theta, t\eta) \in \Gamma$,
- for each compact subset K of Γ_0, there is $\epsilon > 0$ such that $(x,\theta,s\eta) \in \Gamma$ if $(x,\theta,\eta) \in K$ and $0 < s < \epsilon$.

If Γ is biconic with respect to a family of profiles, it is also biconic with respect to their union. The *profile* of Γ is the largest profile Γ_0 such that the last condition is satisfied.

We also introduce

$$\Gamma_1 = \{(x,\theta) : \exists\, \eta \text{ such that } (x,\theta,\eta) \in \Gamma\}.$$

This is an open conic subset of $X \times \mathbb{R}^N \setminus \{0\}$.

Let $p, q \in \mathbb{R}$ and $r \in \mathbb{N}_0$. A C^∞ function $f : \Gamma \to \mathbb{R}^m$ is said *bihomogeneous of degree* $(p,q;r)$ if

- $f(x,t\theta,t\eta) = t^p f(x,\theta,\eta)$ if $(x,\theta,\eta) \in \Gamma$, $t > 0$,
- for every $(x_0,\theta_0,\eta_0) \in \Gamma_0$, there is a neighborhood V of (x_0,θ_0,η_0) and a C^∞ function F in $V \times]-\epsilon,\epsilon[$ satisfying

$$f(x,\theta,s\eta) = s^q F(x,\theta,\eta,s^{1/r})$$

if $(x,\theta,\eta,s) \in V \times]0,\epsilon[$.

The integer r is inserted here essentially for technical reasons. In the application, it does not affect the 2-microlocal geometry but has some effects on the microlocal lagrangian submanifolds involved. We say that f has the regularity r.

Definition 2 *Let*

- Λ *be a conic lagrangian submanifold of* $\dot{T}^*X$,
- φ *be a* C^∞ *real valued function which is homogeneous of degree* 1 *in* Γ_1,
- ψ *be a* C^∞ *real valued function which is bihomogeneous of degree* $(1,1;r)$ *in* Γ

and

$$C_{\varphi,\psi} = \{(x,\theta,\eta) \in \Gamma_0 : \varphi'_\theta(x,\theta) = 0, \psi'_{1,\eta}(x,\theta,\eta) = 0\}.$$

The pair (φ,ψ) *is a local 2-phase function of* Λ *(with regularity* r*) if*

- φ *is a local phase function that parameterizes* Λ,
- *at each point of* $C_{\varphi,\psi}$, *the vector* $(\psi'_{1,x},\psi'_{1,\theta})$ *is different from* 0 *and*

$$\operatorname{rk}\begin{pmatrix} \psi''_{1,\eta x} & \psi''_{1,\eta\theta} & \psi''_{1,\eta\eta} \\ \varphi''_{\theta x} & \varphi''_{\theta\theta} & 0 \end{pmatrix} = N + M.$$

If φ is a phase function, the last condition means that the map $(\rho,\eta) \mapsto \psi_1(j_\varphi^{-1}(\rho),\eta)$ is a local phase function of Λ. This definition has the following consequences.

a) *The map*

$$j_{\varphi,\psi} : C_{\varphi,\psi} \to \dot{T}^*\Lambda : (x,\theta,\eta) \mapsto ((x,\varphi'_x), j_{\varphi *}((\psi'_{1,x},\psi'_{1,\theta})_{|TC_\varphi})).$$

is a lagrangian immersion.

Assume that (h,k,u) is in the kernel of the differential of this map. It follows that $h=0$, $\varphi''_{x\theta}.k=0$, $\varphi''_{\theta\theta}.k=0$, $\psi''_{1,\eta\theta}.k+\psi''_{1,\eta\eta}.u=0$ and there is v such that

$$\begin{pmatrix} \tilde{k} & \tilde{u} \end{pmatrix}\begin{pmatrix} \psi''_{1,\theta x} & \psi''_{1,\theta\theta} \\ \psi''_{1,\eta x} & \psi''_{1,\eta\theta} \end{pmatrix} = v\begin{pmatrix} \varphi''_{\theta x} & \varphi''_{\theta\theta} \end{pmatrix}.$$

We have $\tilde{k}(\varphi''_{\theta x},\varphi''_{\theta\theta}) = 0$ hence $k=0$. Moreover

$$(\tilde{u},-v)\begin{pmatrix} \psi''_{1,\eta x} & \psi''_{1,\eta\eta} & \psi''_{1,\eta\theta} \\ \varphi''_{\theta x} & 0 & \varphi''_{\theta\theta} \end{pmatrix} = 0$$

hence $u=0$. This proves that $j_{\varphi,\psi}$ is an immersion.

Let α be the canonical 1-form of $T^*\Lambda$, $\pi : T^*\Lambda \to \Lambda$ the projection on the base, (x, θ, η) a point of $C_{\varphi,\psi}$ and $j = j_{\varphi,\psi}$. If (h, k, u) is tangent to $C_{\varphi,\psi}$ at (x, θ, η) and $\sigma = j(x, \theta, \eta)$ then

$$\begin{aligned} j^*(\alpha).(h,k,u) &= \alpha(j_*(h,k,u)) = \sigma(\pi_* j_*(h,k,u)) = \sigma \begin{pmatrix} h \\ \varphi''_{xx}.h + \varphi''_{x\theta}.k \end{pmatrix} \\ &= \psi'_{1,x}.h + \psi'_{1,\theta}.k = d\psi_1.(h,k) = 0 \end{aligned}$$

since ψ_1 is equal to 0 on $C_{\varphi,\psi}$.

Following the identification $\dot{T}^*\Lambda \sim T_\Lambda \dot{T}^* X$, *the map* $j_{\varphi,\psi}$ *can be identified with*

$$C_{\varphi,\psi} \to \dot{T}_\Lambda T^* X : (x, \theta, \eta) \mapsto ((x, \varphi'_x), (h, \tilde{\psi}'_{1,x} + \varphi''_{xx}.h + \varphi''_{x\theta}.k))$$

where h, k *satisfy*

$$\varphi''_{\theta x}.h + \varphi''_{\theta\theta}.k + \tilde{\psi}'_{1,\theta} = 0.$$

Indeed, if $\varphi''_{\theta x}.h + \varphi''_{\theta\theta}.k + \tilde{\psi}'_{1,\theta} = 0$ and (u, v) is tangent to C_φ then

$$\begin{aligned} &\omega\left(\begin{pmatrix} h \\ \tilde{\psi}'_{1,x} + \varphi''_{xx}.h + \varphi''_{x\theta}.k \end{pmatrix}, \begin{pmatrix} u \\ \varphi''_{xx}.u + \varphi''_{x\theta}.v \end{pmatrix}\right) \\ &= \psi'_{1,x}.u + \tilde{k}.\varphi''_{\theta x}.u - \tilde{h}.\varphi''_{x\theta}.v \\ &= \psi'_{1,x}.u + \tilde{k}.\varphi''_{\theta x}.u + \psi'_{1,\theta}v + \tilde{k}.\varphi''_{\theta\theta}v = (\psi'_{1,x}, \psi'_{1,\theta}) \begin{pmatrix} u \\ v \end{pmatrix}. \end{aligned}$$

b) Let (φ, ψ) be a local 2-phase function (with regularity r) in a biconic set Γ and $(x_0, \theta_0, \eta_0) \in C_{\varphi,\psi}$. By the definition, φ is a local phase function in Γ_1 and there is a biconic open subset $\tilde{\Gamma}$ of Γ whose profile contains (x_0, θ_0, η_0) such that $(x, (\theta, \eta)) \mapsto \varphi(x, \theta) + \psi(x, \theta, \eta)$ is a local phase function in $\tilde{\Gamma}$. A local 2-phase function (φ, ψ) is called a *2-phase function* if j_φ, $j_{\varphi+\psi}$ and $j_{\varphi,\psi}$ are embeddings.

One can verify that *if* (φ, ψ) *is a local 2-phase function in* Γ *and* $(x_0, \theta_0, \eta_0) \in C_{\varphi,\psi}$ *then there is a biconic open set* $\tilde{\Gamma}$ *whose profile contains* (x_0, θ_0, η_0) *such that* (φ, ψ) *is a 2-phase function in* $\tilde{\Gamma}$.

Hence, *if* (φ, ψ) *is a 2-phase function then*

$$\{((x, \varphi'_x), (h, \psi'_{1,x} + \varphi''_{xx}.h + \varphi''_{x\theta}.k)) : (x, \theta) \in C_{\varphi,\psi}, \psi'_{1,\theta} + \varphi''_{\theta x}.h + \varphi''_{\theta\theta}.k = 0\}$$

is a conic bilagrangian submanifold of $\dot{T}^*\Lambda_\varphi$. It is denoted $\Lambda_{\varphi,\psi}$.

c) *If* (φ, ψ) *is a 2-phase function, then*

$$n - \mathrm{rk}(\pi_{\Lambda_\varphi, X}) = N - \mathrm{rk}(\varphi''_{\theta\theta}) \quad , \quad n - \mathrm{rk}(\pi_{\Lambda_{\varphi,\psi}, \Lambda_\varphi}) = M - \mathrm{rk}(\psi''_{1,\eta\eta}),$$

and

$$n - \mathrm{rk}(\pi_{\Lambda_{\varphi,\psi},X}) = N + M - \mathrm{rk}\begin{pmatrix} \psi''_{1,\eta\eta} & \psi''_{1,\eta\theta} \\ 0 & \varphi''_{\theta\theta} \end{pmatrix}.$$

The first two equalities are known from the study of microlocal phase functions. Consider the map $C_{\varphi,\psi} \ni (x,\theta,\eta) \mapsto x \in X$. A vector (h,k,u) is in the kernel of the differential of this map if and only if $h=0$, $\varphi''_{\theta\theta}.k = 0$ and $\psi''_{1,\eta\theta}.k + \psi''_{1,\eta\eta}.u = 0$. This proves the equality.

2.3. PAIRS OF LAGRANGIAN SUBMANIFOLDS

We now describe the geometric setting associated to a 2-phase. If Y is a submanifold of a C^∞ manifold X, the blowup of X along Y is

$$\hat{X}_Y = (X \setminus Y) \cup \dot{T}_Y X.$$

The sets

$$\bigcap_{1\le j\le p} \Big(\{x \in \omega : f_j(x) > 0\} \cup \{(x,h) \in \dot{T}_Y X : x \in \omega, df_j(x).h > 0\}\Big)$$

where ω is an open subset of X and $f_j \in C^\infty(\omega)$, $f_{j|Y\cap\omega} = 0$ for all j, form a basis of topology of $\hat{X}_Y$. For this topology, the projection $\pi : \hat{X}_Y \to X$ is continuous.

Definition 3 A pair (Λ_0, Λ_1) is a *2-microlocal pair* of lagrangian submanifolds of $\dot{T}^*X$ if

- Λ_0 is a conic lagrangian submanifolds of $\dot{T}^*X$, $\Lambda_1 \subset (\dot{T}^*X)^{\wedge}_{\Lambda_0}$,
- $\Lambda_1 \cap (\dot{T}^*X \setminus \Lambda_0)$ is a conic lagrangian submanifold of $\dot{T}^*X$,
- for each $(\rho, h) \in \Lambda_1 \cap \dot{T}_{\Lambda_0} T^*X$, there is an open neighborhood V of (ρ,h) in $(\dot{T}^*X)^{\wedge}_{\Lambda_0}$ and a 2-phase function (φ,ψ) such that

$$\Lambda_0 \cap \pi(V) = \Lambda_\varphi \quad \text{and} \quad \Lambda_1 \cap V = \Lambda_{\varphi+\psi} \cup \Lambda_{\varphi,\psi}.$$

In this situation, we say that the 2-phase function (φ,ψ) defines (Λ_0,Λ_1). Let $T_{\Lambda_0}\Lambda_1 = \Lambda_1 \cap \dot{T}_{\Lambda_0}(T^*X)$. This is a conic bilagrangian submanifold of $\dot{T}^*\Lambda_0$.

Example 4 *In $\dot{T}^*\mathbb{R}^n$, consider*

$$\varphi(x,\xi) = x.\xi \quad , \quad \psi(x,\xi,\eta') = \frac{\eta'.\xi'}{\xi_n} - H(\eta',\xi_n).$$

where $\xi = (\xi', \xi_n)$ and H is bihomogeneous of degree $(1,1;r)$. We have

$$\Lambda_\varphi = \{(0,\xi) : \xi_n \neq 0\}$$

and

$$\Lambda_{\varphi+\psi} = \{((-\frac{\eta'}{\xi_n}, \frac{\eta'.H'_{\eta'}}{\xi_n} + H'_{\xi_n}), (\xi_n H'_{\eta'}, \xi_n)) : \xi_n \neq 0\}.$$

If $H(\eta', \xi_n) = \eta_1^3/\eta_2^2$ in $\mathbb{R}^3$, the projection of $T_{\Lambda_\varphi}\Lambda_{\varphi+\psi}$ on Λ_φ is the cusp

$$\{(0,\xi) : (\frac{\xi_1}{3})^3 = (\frac{\xi_2}{2})^2\xi_3 : \xi_3 \neq 0\}.$$

It can be shown, see [6], that the property of being a microlocal pair of lagrangian submanifolds is preserved by an homogeneous canonical transformation.

Let us describe the equivalence of 2-phase functions.

Two 2-phase functions (φ, ψ) and $(\tilde{\varphi}, \tilde{\psi})$ defined in biconic open subsets Γ and $\tilde{\Gamma}$ of $X \times \mathbb{R}^N \setminus \{0\} \times \mathbb{R}^M \setminus \{0\}$ are said *equivalent* if there is a C^∞ diffeomorphism $\Gamma \to \tilde{\Gamma} : (x,\theta,\eta) \mapsto (x, f(x,\theta,\eta), g(x,\theta,\eta))$ such that

- $\varphi(x, f(x,\theta,\eta)) + \psi(x, f(x,\theta,\eta), g(x,\theta,\eta)) = \tilde{\varphi}(x,\theta) + \tilde{\psi}(x,\theta,\eta)$,
- f is strictly bihomogeneous of degree $(1,0;r)$ and g is bihomogeneous of degree $(1,1;r)$,
- $D_\theta f_0$ and $D_\eta g_1$ are invertible in Γ_0.

These two pairs define the same 2-microlocal pair.

If Δ is a diagonal real invertible matrix, the pair of phases

$$\varphi(x,\theta) = \tilde{\varphi}(x,\theta'') + \frac{\langle \Delta\theta', \theta' \rangle}{2|\theta''|} \; , \quad \psi(x,\theta'',\eta) = \tilde{\psi}(x,\theta'',\eta)$$

defines the same lagrangian submanifolds as $\tilde{\varphi}$ and $\tilde{\psi}$. In the same way,

$$\varphi(x,\theta) = \tilde{\varphi}(x,\theta) \; , \quad \psi(x,\theta,\eta) = \tilde{\psi}(x,\theta,\eta'') + \frac{\langle \Delta\eta', \eta' \rangle}{2|\eta''|}$$

defines the same lagrangian submanifolds as $\tilde{\varphi}$ and $\tilde{\psi}$.

It can be shown that the transition between two 2-phase functions defining the same 2-microlocal pair of lagrangian submanifolds can be obtained by a composition of the previous reductions.

3. Bilagrangian distributions

3.1. SYMBOLS

We use only classical symbols. This is enough for the applications that we consider here.

Definition 5 *If $m, p \in \mathbb{R}$ and X is an open subset of $\mathbb{R}^n$, we denote by $S^{m,p}(X, \mathbb{R}^N, \mathbb{R}^M)$ the set of all $a \in C^\infty(X \times \mathbb{R}^N \times \mathbb{R}^M)$ such that for every compact subset K of X and all multiorders α, β, γ there is a $C > 0$ satisfying*

$$|D_x^\alpha D_\theta^\beta D_\eta^\gamma a(x, \theta, \eta)| \leq C(1 + |\theta| + |\eta|)^{m-|\beta|}(1 + |\eta|)^{p-|\gamma|}$$

for all $(x, \theta, \eta) \in K \times \mathbb{R}^N \times \mathbb{R}^M$.

Write

$$S_2^\infty = \bigcup_{m,p\in\mathbb{R}} S^{m,p} \quad , \quad S^{m,-\infty} = \bigcap_{p\in\mathbb{R}} S^{m,p}.$$

It is clear that $S^{m,p}$ is a Fréchet space with semi-norms given by the smallest constants which can be used in the definition.

Oscillatory integrals can be defined using symbols in $S^{m,p}$ and 2-phase functions.

Theorem 6 *Let (φ, ψ) be a 2-phase function in an open biconic set Γ and let F be a closed conic subset of Γ such that $F \ll \Gamma$. For every $u \in C_0^\infty(X)$, the linear form*

$$a \mapsto \iiint e^{i(\varphi(x,\theta)+\psi(x,\theta,\eta))} a(x, \theta, \eta) u(x)\, dx d\theta d\eta$$

defined in the set of all $a \in S^{-\infty}(X; \mathbb{R}^N \times \mathbb{R}^M)$ satisfying $\operatorname{supp}(a) \subset F$, *can be extended on S_2^∞ in a unique way such that it is continuous on the set of $a \in S^{m,p}(X, \mathbb{R}^N, \mathbb{R}^M)$ satisfying* $\operatorname{supp}(a) \subset F$ *for every m, p.*

3.2. DISTRIBUTION CLASS

Let X be a C^∞ manifold of dimension n and let (Λ_0, Λ_1) be a 2-microlocal pair of lagrangian submanifolds of $\dot{T}^*X$.

Definition 7 *The space $I^{m,p}(X, \Lambda_0, \Lambda_1)$ is the set of all locally finite sums of an element of $I^m(X, \Lambda_0)$, an element of $I^{m+p}(X, \Lambda_1 \cap T^*X)$ and distributions of the form*

$$I_{\varphi,\psi,a}(u) = (2\pi)^{-(n+2(N+M))/4} \iiint e^{i(\varphi(x,\theta)+\psi(x,\theta,\eta))} a(x, \theta, \eta) u(x)\, dx d\theta d\eta$$

where (U, χ) is a chart of X, $u \in C_0^\infty(X)$, (φ, ψ) is a 2-phase function of (Λ_0, Λ_1) defined in an open biconic subset Γ of $\chi(U) \times \mathbb{R}^N \setminus \{0\} \times \mathbb{R}^M \setminus \{0\}$ and

$$a \in S^{m+(n-2N)/4, p-M/2}(\chi(U), \mathbb{R}^N, \mathbb{R}^M)$$

satisfies $\operatorname{supp}(a) \ll \Gamma$.

It can be shown that this space is invariant by composition with a Fourier integral operators. Moreover, any 2-phase function defining the pair (Λ_0, Λ_1) near a point $\rho_0 \in \Lambda_0$ can be used to define any element of $I^{m,p}(X, \Lambda_0, \Lambda_1)$ near ρ_0.

The singularities of an element of $I^{m,p}(X, \Lambda_0, \Lambda_1)$ are included in the lagrangian submanifolds involved, [6].

Theorem 8 *If* $u \in I^{m,p}(X, \Lambda_0, \Lambda_1)$ *then*

$$WF(u) \subset \Lambda_0 \cup \Lambda_1 \ , \quad WF^{(2)}_{\Lambda_0}(u) \subset T_{\Lambda_0}\Lambda_1.$$

4. Application to diffraction

Let us consider the boundary value problem

$$\begin{cases} (-\Delta + (1+x_n)\partial_t^2)u = 0 \\ u_{|x_n=0} = \delta_0 \quad , \quad u_{|t<0} = 0 \end{cases}$$

where we use the decomposition $(t, x', x_n) \in \mathbb{R} \times \mathbb{R}^{n-1} \times \mathbb{R}_+$. This is a model for the strictly diffractive problems in the C^∞ category, see [14].

Let

$$p(x_n, \tau, \xi) = |\xi|^2 - (1+x_n)\tau^2$$

be the principal symbol of the operator and $r(\tau, \xi') = |\xi'|^2 - \tau^2$ be the boundary hamiltonian. Two lagrangian submanifolds are involved here. On one hand, we consider the flowout $\Lambda_0 = \Lambda_{0,+} \cup \Lambda_{0,-}$ of

$$\{((0,0),(\tau,\xi)) : \tau = \pm|\xi'| \neq 0, \, \xi_n = 0\}$$

through H_r on the boundary and followed by H_p intersected with $t > 0$ and $x_n > 0$. On the other hand, the flowout $\Lambda_1 = \Lambda_{1,+} \cup \Lambda_{1,-}$ of

$$\{((0,0),(\tau,\xi)) : \tau = \pm|\xi|, \xi_n \neq 0\}$$

through H_p intersected with $t > 0$ and $x_n > 0$. These two manifolds are smooth but are tangent at their intersection.

It can be checked that $(\Lambda_{0,\pm}, \Lambda_{1,\pm})$ is a 2-microlocal pair of lagrangian submanifolds with

$$\begin{aligned} T_{\Lambda_{0,\pm}}\Lambda_{1,\pm} \ = \ & \{(((\frac{2}{3}x_n^{3/2} + 2\sqrt{x_n}, x', x_n), (\pm|\xi'|, \xi', \mp|\xi'|\sqrt{x_n})), \\ & ((0,0,0), (\pm\frac{1}{2}\sigma, 0, \mp\frac{1}{2}\sigma(\sqrt{x_n} + \frac{1}{\sqrt{x_n}}))) : \sigma, x_n > 0, \xi' \neq 0\}. \end{aligned}$$

A 2-phase function $(\varphi_\pm, \psi_\pm)$ of $(\Lambda_{0,\pm}, \Lambda_{1,\pm} \cup T_{\Lambda_{0,\pm}}\Lambda_{1,\pm})$ is given by

$$\varphi_\pm(t,x,\xi') = x'.\xi' \pm |\xi'|(t - \frac{2}{3}x_n^{3/2})$$

and

$$(\varphi_\pm+\psi_\pm)(t,x,\sigma,\xi') = x'.\xi' \pm |\xi'|(1-\frac{\sigma}{|\xi'|})^{-1/2}(t-\frac{2}{3}((x_n+\frac{\sigma}{|\xi'|})^{3/2}-(\frac{\sigma}{|\xi'|})^{3/2})).$$

This 2-phase function has the regularity 2.

We denote by $I_\rho^m(X,\Lambda_0)$ the set of all lagrangian distributions on Λ_0 with symbol in S_ρ^m. This means that the symbol satisfies the following inequalities

$$|D_x^\alpha D_\theta^\beta a(x,\theta)| \le C_{\alpha,\beta}(1+|\theta|)^{m-|\beta|+(1-\rho)(|\alpha|+|\beta|)}.$$

An analysis of the solution of the initial boundary value problem given in [3] leads to the following result.

Theorem 9 *The solution u of the previous boundary value problem belongs to*

$$I^{\frac{n}{4}-1,\frac{3}{4}}(\mathbb{R}\times\mathbb{R}^{n-1}\times\mathbb{R}_+,\Lambda_0,\Lambda_1\cup T_{\Lambda_0}\Lambda_1))$$
$$+\ I_{2/3}^{\frac{n}{4}-\frac{1}{2}}(\mathbb{R}\times\mathbb{R}^{n-1}\times\mathbb{R}_+,\Lambda_0).$$

5. The geometry in the complex domain

Our purpose is to define the phase functions used to characterize the bi-lagrangian distributions in the formalism of the Fourier-Bros-Iagolnitzer transform. In the microlocal case, we closely follow [7] and collect some material from [10], see also [16].

As usual, we identify

- $\mathbb{C}^n$ with $\mathbb{R}^n \times \mathbb{R}^n$ and write $z = x + iy$,
- $\zeta \in T_z^*\mathbb{C}^n$ with $(\zeta_1,\ldots,\zeta_n) \in \mathbb{C}^n$ using $\zeta(h) = \sum_j \zeta_j h_j$,
- $T_z^*\mathbb{C}^n$ with $T_{(x,y)}^*\mathbb{R}^{2n}$ by mapping the $\mathbb{C}$-linear form $\zeta \in T_z^*\mathbb{C}^n$ to the $\mathbb{R}$-linear form $h \mapsto -\mathcal{I}\zeta(h)$.

This map is symplectic if $T^*\mathbb{R}^{2n}$ is endowed with the usual canonical 2-form and $T^*\mathbb{C}^n$ with the 2-form $-\mathcal{I}\sigma$ defined below.

It follows that if f is a holomorphic function, $\partial f \in T_z^*\mathbb{C}^n$ is identified with $d(-\mathcal{I}f) \in T_{(x,y)}^*\mathbb{R}^{2n}$ since $d(-\mathcal{I}f) = -\mathcal{I}(df) = -\mathcal{I}(\partial f)$.

In the same way, if φ is a real function then $d\varphi \in T_{(x,y)}^*\mathbb{R}^{2n}$ is identified with $\frac{2}{i}D_z\varphi \in \mathbb{C}^n$.

All the constructions described in this section are local even this is not stated explicitly.

5.1. FBI TRANSFORM

Writing $z = x + iy$ and $\zeta = \xi + i\eta$, the canonical 2-form on $T^*\mathbb{C}^n$ is

$$\sigma = \sum_j d\zeta_j \wedge dz_j.$$

Its real and imaginary parts

$$\mathcal{R}\sigma = \sum_j (d\xi_j \wedge dx_j - d\eta_j \wedge dy_j), \quad \mathcal{I}\sigma = \sum_j (d\eta_j \wedge dx_j + d\xi_j \wedge dy_j)$$

are symplectic forms on $\mathbb{R}^{2n}$.

Let φ be a real C_1 function defined in a neighborhood of $z_0 \in \mathbb{C}^n$ and

$$\Lambda_\varphi = \{(z, \frac{2}{i} D_z\varphi(z)) : z \in \mathbb{C}^n\}.$$

This manifold is $\mathcal{I}$-lagrangian since it is identified with

$$\{(z, d\varphi(z)) : z \in \mathbb{C}^n\} \subset T^*\mathbb{R}^{2n}.$$

If j_φ denotes the immersion $z \mapsto (z, \frac{2}{i} D_z\varphi(z))$ then

$$j_\varphi^*(\mathcal{R}\sigma) = j_\varphi^*(\sigma) = j_\varphi^*(d(\zeta dz)) = d(\frac{2}{i}\partial\varphi) = \frac{2}{i}\overline{\partial}\partial\varphi.$$

It follows that, if $\overline{\partial}\partial\varphi$ is non degenerate, j_φ is a symplectic map from $(\mathbb{C}^n, \frac{2}{i}\overline{\partial}\partial\varphi)$ onto $(\Lambda_\varphi, \mathcal{R}\sigma)$. Its inverse is the projection.

Let us remark that

$$\frac{2}{i}\overline{\partial}\partial\varphi(z)(u,v) = \frac{2}{i}\sum_{j,k} D_{z_j} D_{\overline{z}_k}\varphi(z)(\overline{u}_k v_j - u_j \overline{v}_k)$$

is real for every $u, v \in \mathbb{C}^n$.

The tangent space to Λ_φ at the point $j_\varphi(z)$ is given by

$$\{(h, \frac{2}{i}(\partial_z^2\varphi(z).h + \partial_{\overline{z}}\partial_z\varphi(z).\overline{h})) : h \in \mathbb{C}^n\}.$$

This shows that if $\overline{\partial}\partial\varphi$ is non degenerate then Λ_φ is a totally real submanifold and that its complexification is $T^*\mathbb{C}^n$.

Let us denote

$$\begin{aligned}\mathcal{L}_\varphi(u,v) &= \frac{2}{i}\overline{\partial}\partial\varphi(u,iv)\\ &= 2\sum_{j,k} D_{z_j}D_{\overline{z}_k}\varphi(z)(\overline{u}_k v_j + u_j\overline{v}_k)\\ &= 4\mathcal{R}\,\langle H_\varphi(z)u,v\rangle\end{aligned}$$

the Levi 2-form of φ. This form defines a hermitian quadratic form $\mathcal{L}_\varphi(u,u)$. It is positive definite if φ is strictly plurisubharmonic.

We now assume that φ is strictly plurisubharmonic and endow Λ_φ with the symplectic form $\mathcal{R}\sigma$.

If $z \in \mathbb{C}^n$, we denote by $u \mapsto \overline{u}$ the unique antilinear bijection from $T_{j_\varphi(z)}T^*\mathbb{C}^n$ onto itself which is the identity on $T_{j_\varphi(z)}\Lambda_\varphi$. Let us consider the hermitian form

$$q : T_{j_\varphi(z)}\mathbb{C}^n \times T_{j_\varphi(z)}\mathbb{C}^n \to \mathbb{C} : (u,v) \mapsto \frac{1}{i}\,\sigma(u,\overline{v}).$$

The signature of q is (n,n) and q is negative definite on the tangent space to the fiber.

Indeed, since σ is not degenerate, the rank of q is $2n$. If L is a complex subspace of $T_{j_\varphi(z)}\mathbb{C}^n$ such that $q(u,u) > 0$ for every $u \in L\setminus\{0\}$ then

$$q(\overline{u},\overline{u}) = \frac{1}{i}\,\sigma(\overline{u},u) = -q(u,u) < 0$$

for every $u \in L\setminus\{0\}$. This shows that the signature of q is (n,n). If $(0,u) \neq 0$ is tangent to the fiber at $j_\varphi(z)$ then

$$\overline{(0,u)} = (h, \frac{2}{i}\,D_z^2\varphi(z).h) \quad\text{with}\quad u = \frac{2}{i}D_{\overline{z}}D_z\varphi(z).\overline{h}.$$

Hence

$$q((0,u),(0,u)) = -2\,\langle H_\varphi(z)h,h\rangle < 0.$$

The following result is proven in [8], see also [5].

Theorem 10 *Let φ be a strictly plurisubharmonic function near $z_0 \in \mathbb{C}^n$ and $\chi : \dot{T}^*\mathbb{R}^n \to \Lambda_\varphi$ a canonical transform defined near (y_0,η_0) such that $\chi(y_0,\eta_0) = (z_0, \frac{2}{i}D_z\varphi(z_0))$. Here Λ_φ is endowed with the 2-form $\mathcal{R}\sigma$. There is a unique holomorphic function $g(z,y)$ near (z_0,y_0), such that*

- *the complexification of χ is*

$$\chi^{\mathbb{C}} : T^*\mathbb{C}^n \to T^*\mathbb{C}^n : (y, -D_y g(z,y)) \mapsto (z, D_z g(z,y)),$$

- *$ig(z_0, y_0) = \varphi(z_0)$, $-D_y g(z_0, y_0) = \eta_0$,*
- *the function $y \mapsto -\mathcal{I}g(z,y)$ has a non degenerate critical point $y(z)$ with signature $(0,n)$ and critical value $\varphi(z)$. Moreover, we have*

$$(y(z), -D_y g(z, y(z)) = \chi^{-1}(z, \frac{2}{i} D_z\varphi(z)).$$

For example, if

$$\chi : (x, \xi) \mapsto (x - i\xi, \xi), \quad \varphi(z) = \frac{1}{2} |\mathcal{I}z|^2,$$

then

$$g(z,y) = \frac{i}{2} (z-y)^2.$$

The FBI transform associated to φ, χ near the points (y_0, η_0), z_0 is

$$T_\chi u(z, \lambda) = \int e^{i\lambda g(z,y)} a(z, y, \lambda) u(y)\, dy$$

where a is a classical symbol.

5.2. ISOTROPIC AND INVOLUTIVE SUBMANIFOLDS

The following result links the isotropic submanifolds of $(\mathbb{C}^n, \frac{2}{i}\overline{\partial}\partial\varphi)$ to the complex structure.

Proposition 11 *Let φ be an analytic strictly plurisubharmonic function near $z_0 \in \mathbb{C}^n$. If Γ is an isotropic submanifold of $(\mathbb{C}^n, \frac{2}{i}\overline{\partial}\partial\varphi)$ then Γ is totally real and there is a unique pluriharmonic function h on $\Gamma^{\mathbb{C}}$ such that*

$$\varphi_{|\Gamma^{\mathbb{C}}} - h \geq 0 \quad \textit{and} \quad (\varphi - h)_{|\Gamma} = 0.$$

Moreover, there is $C > 0$ such that

$$\varphi_{|\Gamma^{\mathbb{C}}} - h \geq C\, d(z, \Gamma)^2.$$

Proof. Let us prove that Γ is totally real. If $u, iu \in T_z\Gamma$, it follows that

$$0 = \frac{2}{i} \overline{\partial}\partial\varphi(u, iu) = 2 \sum_{j,k} D_{z_j} D_{\overline{z}_k}\varphi(z)\, u_j \overline{u}_k.$$

Since φ is strictly plurisubharmonic, u is 0.

Denote by ρ the injection from Γ to $\Gamma^{\mathbb{C}}$. Since Γ is isotropic, we have

$$0 = \rho^*(\overline{\partial}\partial(\varphi_{|\Gamma^{\mathbb{C}}})) = d\rho^*(\partial(\varphi_{|\Gamma^{\mathbb{C}}})).$$

Hence there is an analytic function $H : \Gamma \to \mathbb{C}$ such that

$$\rho^*(\partial(\varphi_{|\Gamma^{\mathbb{C}}})) = dH.$$

Let us denote θ the holomorphic extension of H in $\Gamma^{\mathbb{C}}$. If $z \in \Gamma$, we have

$$\partial(\varphi_{|\Gamma^{\mathbb{C}}})(z) = d\theta(z)$$

on $T_z\Gamma^{\mathbb{C}}$ since the equality holds on $T_z\Gamma$ and both sides are $\mathbb{C}$-linear. We get

$$d(\varphi_{|\Gamma^{\mathbb{C}}}) = \partial(\varphi_{|\Gamma^{\mathbb{C}}}) + \overline{\partial}(\varphi_{|\Gamma^{\mathbb{C}}}) = 2d(\mathcal{R}\theta)$$

on $T_z\Gamma^{\mathbb{C}}$ at the points of Γ. Modifying θ by a constant, we can take $h = 2\mathcal{R}\theta$.

The inequality follows from the fact that $\varphi_{|\Gamma^{\mathbb{C}}} - h$ is strictly plurisubharmonic and vanishes to the second order on the maximal totally real submanifold Γ of $\Gamma^{\mathbb{C}}$.

The function h is unique. Indeed, a plurisubharmonic function vanishing on Γ and having a null differential on $T_z\Gamma^{\mathbb{C}}$ at the points of Γ is equal to 0. □

Let us now consider an involutive submanifold V of $(\mathbb{C}^n, \omega = \frac{2}{i}\overline{\partial}\partial\varphi)$. If $z \in V$ then

$$T_zV \oplus i(T_zV)^\omega = \mathbb{C}^n.$$

Indeed, if $u \in T_zV$ and $iu \in (T_zV)^\omega$ we have

$$0 = \frac{2}{i}\overline{\partial}\partial\varphi(u, iu) = 2\sum_{j,k} D_{z_j}D_{\overline{z}_k}\varphi(z)\, u_j\overline{u}_k$$

hence $u = 0$.

This shows that the union of the complexifications of the bicharacteristic leaves of V can be locally identified with $\mathbb{C}^n$. It follows from the proposition 11 that there is a unique real analytic function φ_V equal to φ on V, pluriharmonic on the complexification of the bicharacteristic leaves of V and such that $\varphi_V \leq \varphi$.

Proposition 12 *Let φ be an analytic strictly plurisubharmonic function near $z_0 \in \mathbb{C}^n$. If V is an involutive submanifold of $(\mathbb{C}^n, \frac{2}{i}\overline{\partial}\partial\varphi)$ then*

$$\Lambda_{\varphi_V} = \{(z, \frac{2}{i}D_z\varphi_V(z)) : z \in \mathbb{C}^n\}$$

is the union of the complexification of the bicharacteristic leaves of $j_\varphi(V)$.

Proof. The submanifold $j_\varphi(V)$ is involutive in $(\Lambda_\varphi, \mathcal{R}\sigma)$ and totally real since it is included in Λ_φ. Let us denote by W the union of the complexification of the bicharacteristic leaves of $j_\varphi(V)$.

This submanifold is $\mathcal{I}$-lagrangian. Indeed, let $\rho = j_\varphi(z)$ and let Γ be the leaf of V containing z. If $h \in T_\rho j_\varphi(\Gamma)$ and $k \in T_\rho j_\varphi(V)$, we have $\mathcal{I}\sigma(h,k) = 0$ since Λ_φ is $\mathcal{I}$-lagrangian and also $\mathcal{I}\sigma(ih,k) = \mathcal{R}\sigma(h,k) = 0$ since $j_\varphi(V)$ is $\mathcal{R}$-involutive.

Let us show that W is transversal to the fibers. If

$$u = v + iw \in T_\rho j_\varphi(V) \oplus iT_\rho j_\varphi(\Gamma),$$

it follows that

$$q(u,u) = \frac{1}{i}\,\sigma(u,\overline{u}) = \frac{1}{i}\,\sigma(v+iw, v-iw) = 2\,\sigma(w,v) = 0$$

since $j_\varphi(V)$ is isotropic with respect to $\mathcal{R}\sigma$ and $\mathcal{I}\sigma$. We know that q is negative definite on the tangent to the fiber, hence W is transversal to the fibers.

There is a real analytic function ψ such that

$$W = \Lambda_\psi = \{(z, \frac{2}{i}\,D_z\psi(z) : z \in \mathbb{C}^n\}.$$

If $z \in V$, we have $D_z\varphi_V(z) = D_z\varphi(z) = D_z\psi(z)$. Hence we can assume that $\psi = \varphi$ on V. To see that $\psi = \varphi_V$, we have to show that ψ is pluriharmonic on the complexification of the leaves of V.

Let Γ be a leaf of V and

$$j_\Gamma : \Gamma^{\mathbb{C}} \to T^*\mathbb{C}^n : z \mapsto (z, \frac{2}{i}\,D_z\psi(z)).$$

We have $j_\Gamma(\Gamma^{\mathbb{C}}) = j_\varphi(\Gamma)^{\mathbb{C}}$. Indeed, if π is the projection $T^*\mathbb{C}^n \to \mathbb{C}^n$, we have $\pi \circ j_\Gamma = \mathrm{id}_{\Gamma^{\mathbb{C}}}$. Moreover, π is holomorphic hence $\pi(j_\varphi(F)^{\mathbb{C}}) = F^{\mathbb{C}}$ for all the leaves F of V. This implies $j_\Gamma(\Gamma^{\mathbb{C}}) \subset j_\varphi(\Gamma)^{\mathbb{C}}$.

On the other hand $j_\varphi(F)$ is $\mathcal{R}$ and $\mathcal{I}$-isotropic. Its complexification $j_\varphi(\Gamma)^{\mathbb{C}}$ is isotropic for σ. Hence

$$0 = j^*\sigma = d(j^*(\zeta dz)) = d(\frac{2}{i}\partial[\psi_{\Gamma^{\mathbb{C}}}]) = \frac{2}{i}\,\overline{\partial}\partial[\psi_{\Gamma^{\mathbb{C}}}].$$

This proves that ψ is pluriharmonic on $\Gamma^{\mathbb{C}}$. □

Remark that j_Γ is holomorphic since it has a holomorphic inverse.

5.3. LAGRANGIAN SUBMANIFOLDS

For a lagrangian submanifold, we can use a holomorphic function.

Proposition 13 *Let Λ be a lagrangian submanifold of $\dot{T}^*\mathbb{R}^n$, h be a phase function of Λ near ρ_0 and χ be a local canonical map from $\dot{T}^*\mathbb{R}^n$ to Λ_φ mapping ρ_0 to z_0. If g the FBI phase defined in theorem 10 and*

$$\phi_\Lambda(z) = \mathrm{cv}_{(x,\theta)}(g(z,x) + h(x,\theta))$$

then $\varphi_\Lambda = -\mathcal{I}\phi_\Lambda$. The critical points are given by

$$(x,\theta) = j_{\mathbb{C}}^{-1} \circ \chi_{\mathbb{C}}^{-1}(z, D_z\phi_\Lambda(z)).$$

Here j is the immersion $(x,\theta) \mapsto (x, h'_x)$ and $j_{\mathbb{C}}$ is its complexification.

Proof. The critical points are given by $g'_x + h'_x = 0$ and $h'_\theta = 0$. The hessian matrix

$$\begin{pmatrix} g''_{xx} + h''_{xx} & h''_{x\theta} \\ h''_{\theta x} & h''_{\theta\theta} \end{pmatrix}$$

is invertible since $\mathcal{I}g''_{xx} > 0$ and $\mathrm{rk}(h''_{\theta x}, h''_{\theta\theta})$ is maximal. This implies that the function ϕ_Λ is holomorphic. Moreover, if $z \in \pi \circ \chi(\Lambda)$, we have

$$(z, \frac{2}{i}D_z\varphi(z)) = (z, D_zg(z,y(z)) = \chi_{\mathbb{C}}(y(z), -D_yg(z,y(z))) \in \chi(\Lambda).$$

It follows that there is a θ such that

$$D_yg(z,y(z)) + D_yh(y,\theta) = 0, \quad D_\theta h(y,\theta) = 0.$$

This shows that the critical point is $(y(z),\theta)$ and we get $-\mathcal{I}\phi_\Lambda(z) = \varphi(z)$. Since the first derivative are also the same at these points, the proposition follows from the uniqueness in proposition 11. □

We have

$$\chi^{\mathbb{C}}(\Lambda^{\mathbb{C}}) = \{(z, D_z\phi_\Lambda(z)) : z \in \mathbb{C}^n\}$$

and

$$\varphi_\Lambda(z) \leq \varphi(z).$$

The equality holds if and only if $(z, \frac{2}{i}D_z\varphi(z)) \in \chi(\Lambda)$.

In this formalism, the lagrangian distributions are defined in the following way.

Definition 14 *Let u be a distribution in an open subset Ω of $\mathbb{R}^n$, Λ a lagrangian submanifold of $\dot{T}^*\Omega$. With the notations of proposition 13, u is said lagrangian at ρ_0 if, in a neighborhood of z_0, we have*

$$(T_\chi u)(z,\lambda) = e^{i\lambda\phi_\Lambda(z)}b(z,\lambda)$$

where b is a classical analytic symbol.

This is equivalent to the fact that u can be written $u = u_1 + u_2$ with $\rho_0 = j_h(x_0, \theta_0)$ not in the singular spectrum of u_2 and

$$u_1(x) = \int_\Gamma e^{ih(x,\theta)} a(x,\theta)\, d\theta$$

where Γ is a conic neighborhood of θ_0 and a is a classical analytic symbol near (x_0, θ_0).

5.4. PAIRS OF LAGRANGIAN SUBMANIFOLDS

Let us consider the FBI transform of a 2-phase function. For simplicity, we restrict ourself to the case of one 2-microlocal parameter.

Proposition 15 *Let (Λ_0, Λ_1) be a 2-microlocal pair of lagrangian submanifolds and (h, ψ) be a 2-phase function for the pair (Λ_0, Λ_1) near a point $\rho_0 \in \Lambda_0$. We assume that h is analytic and that ψ is an analytic function of $(x, \theta, \sigma^{1/2})$,*

$$\psi(x,\theta,\sigma) = \psi_1(x,\theta)\sigma + \psi_{3/2}(x,\theta)\sigma^{3/2} + \psi_2(x,\theta)\sigma^2 + \mathcal{O}(\sigma^{5/2}).$$

If g is an FBI phase function associated to a local canonical map χ such that $\chi(\rho_0) = z_0 \in \mathbb{C}^n$, we have

$$\begin{aligned} \phi(z,\sigma) &= \operatorname{cv}_{(x,\theta)}\Big(g(z,x) + h(x,\theta) + \psi(x,\theta,\sigma)\Big) \\ &= \Phi_{\Lambda_0}(z) + \Phi_1(z)\sigma + \Phi_{3/2}(z)\sigma^{3/2} + \Phi_2(z)\sigma^2 + \mathcal{O}(\sigma^{5/2}). \end{aligned}$$

Here Φ_1 and $\Phi_{3/2}$ are real on $\pi \circ \chi(\Lambda_0)$, $\Phi_1(z_0) = 0$, $D_z\Phi_1(z_0) \neq 0$ and $\mathcal{I}\Phi_2(z_0) > 0$.

Proof. The critical points are the solutions of

$$\begin{cases} h'_x + \psi'_x + g'_x = 0 \\ h'_\theta + \psi'_\theta = 0 \end{cases}$$

and are given by

$$(x,\theta) = j_{h,\mathbb{C}}^{-1} \circ \chi_{\mathbb{C}}^{-1}(z, D_z\phi_{\Lambda_0}(z))$$

if $\sigma = 0$. So the critical points are not degenerate for small σ. Let $\sigma = s^2$. An easy computation gives

$$\begin{aligned} \phi'_s(z,\sigma) &= \psi'_s(x,\theta,\sigma) \\ \phi''_{ss}(z,\sigma) &= \psi''_{ss} + \psi''_{sx}x'_s + \psi''_{s\theta}\theta'_s. \end{aligned}$$

Hence

$$\Phi'_s(z,0) = 0, \quad \Phi''_{ss}(z,0) = 2\psi_1(x,\theta).$$

Moreover, if $s = 0$,

$$\begin{cases} h''_{xx}x'_s + h''_{x\theta}\theta'_s + g''_{xx}x'_s = 0 \\ h''_{\theta x}x'_s + h''_{\theta\theta}\theta'_s = 0 \end{cases}$$

This implies $(x'_s, \theta'_s) = 0$ on $s = 0$. Two more derivations give

$$\begin{aligned} \Phi'''_{sss}(z,0) &= 6\psi_{3/2}(x,\theta) \\ \Phi^{(iv)}_{s(4)}(z,0) &= 24\psi_2(x,\theta) + 6(\psi'_{1,x}x''_{ss} + \psi'_{1,\theta}\theta''_{ss}). \end{aligned}$$

The critical points are real if and only if $z \in \pi \circ \chi(\Lambda_0)$. This shows that Φ_1 and $\Phi_{3/2}$ are real on this totally real submanifold. Since the differential of ψ_1 does not vanish on C_h, we have $D_z\Phi_1(z_0) \neq 0$.

We also get

$$\begin{cases} h''_{xx}x''_{ss} + h''_{x\theta}\theta''_{ss} + g''_{xx}x''_{ss} + 2\psi'_{1,x} = 0 \\ h''_{\theta x}x''_{ss} + h''_{\theta\theta}\theta''_{ss} + 2\psi'_{1\theta} = 0. \end{cases}$$

It follows that

$$\begin{aligned} \phi^{(iv)}_{s(4)}(z,0) &= 24\,\psi_2(x,\theta) - 3\left\langle \begin{pmatrix} h''_{xx} + iI & h''_{x\theta} \\ h''_{\theta x} & h''_{\theta\theta} \end{pmatrix} \begin{pmatrix} x''_{ss} \\ \theta''_{ss} \end{pmatrix}, \begin{pmatrix} x''_{ss} \\ \theta''_{ss} \end{pmatrix} \right\rangle \\ &= 12\left(2\psi_2(x,\theta) - \left\langle \begin{pmatrix} h''_{xx} + iI & h''_{x\theta} \\ h''_{\theta x} & h''_{\theta\theta} \end{pmatrix}^{-1} \begin{pmatrix} \psi'_{1,x} \\ \psi'_{1,\theta} \end{pmatrix}, \begin{pmatrix} \psi'_{1,x} \\ \psi'_{1,\theta} \end{pmatrix} \right\rangle\right) \end{aligned}$$

Since (h, ψ) is a 2-phase function, the matrix

$$\begin{pmatrix} \psi'_{1,x} & h''_{x\theta} \\ \psi'_{1,\theta} & h''_{\theta\theta} \end{pmatrix}$$

has full rank. Using the lemma 16, we obtain $\mathcal{I}\phi''_{\sigma\sigma}(z_0) > 0$. □

Lemma 16 *Let S, H be hermitian matrices with H semi-positive definite. If $A = S + iH$ is invertible and $x \in \mathbb{C}^n$ does not belong to the image by A of the kernel of H then $\mathcal{I}\langle A^{-1}x, x\rangle < 0$.*

Proof. We have

$$\left\langle A^{-1}x, x\right\rangle = \left\langle A^*A^{-1}x, A^{-1}x\right\rangle = \left\langle SA^{-1}x, A^{-1}x\right\rangle - i\left\langle HA^{-1}x, A^{-1}x\right\rangle.$$

In the right hand side, the first term is real and $\langle HA^{-1}x, A^{-1}x\rangle > 0$. □

With the notations of the proposition 15, a distribution u is said *analytic bilagrangian* at ρ_0 with respect to (Λ_0, Λ_1) if, in a neighborhood of z_0, we have

$$(T_\chi u)(z,\lambda) = \int_0^\delta e^{i\phi(z,\sigma)}a(z,\sigma,\lambda)\,d\sigma$$

where a is holomorphic in an open set of the form

$$\{(z,\sigma)\in \mathbb{C}^n\times\mathbb{C} : |z-z_0|<\epsilon, |\mathcal{I}\sigma|<c\mathcal{R}\sigma\}$$

and is bounded by $C\lambda^m$ for $\lambda>1$.

Since $\mathcal{I}\Phi_2(z_0)>0$ and $\Phi_1(z_0)$, $\Phi_{3/2}(z_0)$ are real, we can choose $\delta>0$ small such that

$$-\mathcal{I}\phi(z_0,\delta)<-\mathcal{I}\varphi_{\Lambda_0}(z_0).$$

For example, if

$$\Lambda_0=\{((0,x_n),(\xi',0))\},\quad \Lambda_1=\{((0,0),(\xi',\xi_n))\}$$

and $g(z,y)=i(z-y)^2/2$, we have

$$\Phi_{\Lambda_0}(z)=\frac{i}{2}z'^2,\quad \Phi_{\Lambda_1}(z)=\frac{i}{2}z^2$$

and

$$\phi(z,\sigma)=\frac{iz'^2}{2}+\sigma z_n+\frac{i\sigma^2}{2}.$$

6. Bilagrangian structure of the parametrix

We review the construction of the parametrix in the analytic case, see [8], and show that, at the transition of the shadow and the illuminated region, it defines a bilagrangian distribution if the boundary data is conormal.

We consider a differential operator with real analytic coefficients

$$P(x,D)=D^2_{x_n}+R(x,D_{x'})$$

Its principal symbol is

$$p(x,\xi)=\xi_n^2+r(x,\xi').$$

Let $r_0(x',\xi')=r(x',0,\xi')$. We assume that the point (x_0',ξ_0') is diffractive. This means that $r_0(x_0',\xi_0')=0$ and $dr_0\neq 0$, $\partial_{x_n}r<0$.

Following [8], we first perform a complex canonical transform. We choose the weight function $\varphi_0(z')=|\mathcal{I}z'|^2/2$ and a canonical map

$$\chi_0:\dot{T}^*\mathbb{R}^{n-1}\to(\Lambda_{\varphi_0},\mathcal{R}\sigma)$$

mapping (x_0',ξ_0') to $(0,0)$ and the glancing region $\{r_0=0\}$ to $\{\mathcal{I}z_1=0\}$. To this canonical map is associated a FBI transform.

After this transform, we obtain a pseudodifferential operator

$$P(x,\tilde{D},\lambda)=\tilde{D}_{x_n}^2+R(x,\tilde{D}_{x'},\lambda)$$

near $(0,0)$ on Λ_{φ_0}. Its principal symbol $p(x,\xi)=\xi_n^2+r(x,\xi')$ is real on Λ_{φ_0} and $p(x,\xi)=0$ is equivalent to $x_n+q(x',\xi)=0$ with

$$q(x',\xi)=\xi_1-e(x',\xi')\xi_n^2+\mathcal{O}(\xi_n^4),\quad e(0,0)>0.$$

In the H_φ space, the problem is reduced to find an outgoing solution to

$$P(x,\tilde{D},\lambda)u(x,\lambda)=0,\quad u_{|x_n=0}=g. \tag{1}$$

Denote by G the solution of

$$\begin{cases} \xi_n^2+r(x',-\partial_{\xi_n}G,\partial_{x'}G)=0\\ G_{|\xi_n=0}=x'.\xi' \end{cases}$$

We have

$$G(x',\xi)=x'.\xi'+\xi_n\xi_1-\frac{\xi_n^3}{3}e(x',\xi')-\frac{\xi_n^4}{12}\partial_{x_1}e(x',\xi')+\mathcal{O}(\xi_n^5).$$

Let C be the canonical relation of $G(x',\xi_n,\xi')-y'.\xi'$. Lebeau shows that there is a $C^{1,1}$ function $\psi_0(y')$ real analytic for $\mathcal{I}y_1\neq 0$ such that C maps Λ_{ψ_0} to Λ_{φ_0}.

For a suitable choice of the symbol a, the operator

$$Jf(x,\lambda)=\int_{\xi_n^-}^{\xi_n^+}d\xi_n\int_{\Sigma_0}e^{i\lambda(G(x',\xi)-y'.\xi'+x_n\xi_n)}a(x',\xi,\lambda)\,f(y',\lambda)\,dy'd\xi'$$

defines asymptotic solutions to P and maps H_{ψ_0} to H_{φ_0} for $x_n\geq 0$, see [8] for the choice of contours. Note that ξ_n^+ and ξ_n^- are respectively close to $\delta e^{-5i\pi/6}$ and $\delta e^{i\pi/2}$.

We have the critical values

$$\text{c.v.}_{(\xi_n)}G(x',\xi_n,\xi')=x'.\xi'-\xi_1^2\gamma(x',\xi')\pm\frac{2}{3}(\xi_1\rho(x',\xi'))^{3/2}$$

with $\rho=e^{-1/3}$ on $\xi_1=0$. Introduce a new parameter t such that $t^2=\xi_1$ and $\arg t\in[-\pi/2,0]$ if ξ_1 is real. In the definition of Jf, the critical value corresponding to the critical point $\xi_n=-te^{-1/2}+\mathcal{O}(t^2)$ is

$$(x'-y').\xi'-\frac{2}{3}t^3\rho(x',t^2,\xi'')^{3/2}-t^4\gamma(x',t^2,\xi'').$$

It is natural to invert the trace of Jf by an operator of the form

$$Ig(y',\lambda) = \int_{z_1^-}^{z_1^+} dz_1 \int_{t-}^{t+} 2t\,dt \int_{\Sigma''} e^{i\lambda((y_1-z_1)t^2+(y''-z'').\eta''+F(z',t,\eta''))} \sigma_\tau(z',t,\eta'',\lambda) f(z',\lambda)\, dz'' d\eta''.$$

Here

$$F(z',t,\eta'') = \frac{2}{3}t^3\rho(z',t^2,\eta'')^{3/2} + t^4\gamma(z',t^2,\eta'').$$

The symbol σ_τ is chosen to avoid the ramification of the asymptotic behavior of the Airy function. If $\tau > 0$, the function

$$f(z) = (z+\tau)^{1/2}(z-\frac{\tau}{2}) - z^{3/2} = \frac{(z(z+\tau))^{3/2}(z-\tau/2) - z^2}{\sqrt{z}}$$

is holomorphic and bounded on the 2-sheets covering of $[-\tau, 0]$. For a large but fixed τ, take

$$\sigma_\tau(z',t,\eta'',\lambda) = (w+\tau)^{1/4} e^{\frac{2}{3}[(w+\tau)^{1/2}(w-\frac{\tau}{2})-w^{3/2}]}$$

with

$$w = e^{i\pi/3}\lambda^{2/3}t^2\rho(z',t^2,\eta'').$$

It follows that the symbol is not holomorphic in a fixed neighborhood of 0 but in a set of the form

$$\{(z',\eta') : |z'|, |\eta'| < r, \ \ |\arg(\eta_1 - \tau\lambda^{-2/3}e^{2i\pi/3}) - \frac{2\pi}{3}| > \epsilon\}.$$

For a good choice of contours, the operator I maps H_{φ_0} on H_{ψ_0}. Lebeau shows how to invert the trace of $J \circ I$ by an operator that propagates the singular support only on one side.

In $x_n > 0$, we get the phase

$$G(x',\xi) - y'.\xi' + x_n\xi_n + (y'-z').\eta' + F(z',\sqrt{\eta_1},\eta'').$$

Using the theorem of the stationary phase function, we can reduce evaluate the integrals involving y', ξ' and we get the phase

$$G(x',\eta',\xi_n) + x_n\xi_n - z'.\eta' + F(z',\sqrt{\eta_1},\eta'').$$

Since we study the solution near a point where $x_n > 0$, the function $\xi_n \mapsto G(x',\eta',\xi_n) + x_n\xi_n$ has exactly one critical point whose argument is close to π and satisfying

$$x_n + \eta_1 - \xi_n^2 e(x',\eta') + O(\xi_n^3) = 0.$$

Let

$$H(x', \eta', \sqrt{x_n + \eta_1}) = x'.\eta' + \frac{2}{3}(x_n + \eta_1)^{3/2} e^{-1/2}(x', \eta') + \mathcal{O}((x_n + \eta_1)^2)$$

be the critical value.

Define, as above, Λ_0 as the flowout of the set of diffractive points through the boundary hamiltonian H_r followed by H_p and Λ_1 as the flowout of all the characteristic points at $x = 0$ through H_p.

In the boundary value problem (1), we consider the boundary data $g(x', \lambda) = \exp(i\lambda z'^2)$ corresponding to a Dirac mass.

Theorem 17 *The function*

$$\begin{aligned} \varphi(z, \sigma) &= \mathrm{cv}_{(x,\eta'')} \Big(\frac{i}{2}(z_n - x_n)^2 + H(z', \sigma, \eta'', \sqrt{x_n + \sigma}) \\ &\qquad - x_1\sigma - x''.\eta'' + F(x', \sqrt{\sigma}, \eta'') + \frac{ix'^2}{2} \Big) \end{aligned}$$

satisfies the conditions of proposition 15. Moreover, the solution u of the boundary value problem (1) can be written $u_1 + u_2$ where u_1 is analytic bilagrangian and

$$|u_2(z, \lambda)| \leq C_\epsilon e^{\lambda(\varphi_{\Lambda_0}(z) + Cd(z, \pi\circ\chi(\Lambda_0))^3) + \epsilon\lambda}$$

near 0 *for every* $\epsilon > 0$.

References

1. Delort, J-M., *Deuxième microlocalisation simultanée et front d'onde de produits*, Ann. scient. Ec. Norm. Sup., 23, 1990, 257-310.
2. Hörmander, L., *The analysis of linear partial differential operators I-IV*, Springer-Verlag, 1983-85.
3. Friedlander, F.G. and Melrose, R.B., *The wave front set of the solution of a simple initial-boundary value problem with glancing rays II*, Math. Proc. Camb. Phil. Soc., 87, 1977, 97-120.
4. Lafitte O., *The kernel of the Neumann operator for a strictly diffractive analytic problem*, Comm. in Part. Diff. Eq., 20, 1995, 419-483.
5. Laubin, P., *Etude 2-microlocale de la diffraction*, Bull. Soc. Roy. Sc. Liège, 4, 1987, 295-416.
6. Laubin P., Willems B., *Distributions associated to a 2-microlocal pair of lagrangian manifolds*, Comm. in Part. Diff. Eq., 19, 1994, 1581-1610.
7. Lebeau, G., *Deuxième microlocalisation sur les sous-variétés isotropes*, Ann. Inst. Fourier, Grenoble, 35, 1985, 145-216.
8. Lebeau, G., *Régularité Gevrey 3 pour la diffraction*, Comm. in Part. Diff. Eq., 9(15), 1984, 1437-1494.
9. Lebeau, G., *Propagation des singularités Gevrey pour le problème de Dirichlet*, Advances in microlocal analysis, Nato ASI, C168, 1986, 203-223.
10. Lebeau, G., *Scattering frequencies and Gevrey 3 singularities*, Invent. math., 90, 1987, 77-114.

11. Melrose R.B., Sjöstrand J., *Singularities in boundary value problem I,* Comm. Pure Appl. Math., 1978, 31, 593-617.
12. Melrose R.B., Sa Barreto A., Zworski M., *Semi-linear diffraction of conormal waves,* preprint.
13. Melrose, R. B., *Local Fourier-Airy integral operators,* Duke Math. J., 42, 1975, 583-604.
14. Melrose, R. B., *Transformation of boundary value problems,* Acta Math. J., 147, 1981, 149-236.
15. Sjöstrand, J., *Propagation of analytic singularities for second order Dirichlet problems, I-III,* Comm. Part. Diff. Eq., 5(1), 1980, 41-94; 5(2), 1980, 187-207; 6(5), 1981, 499-567.
16. Sjöstrand, J., *Singularités analytiques microlocales,* Astérisque, 95, 1982, 1-166.

PARAMETRIZED PSEUDODIFFERENTIAL OPERATORS AND GEOMETRIC INVARIANTS

GERD GRUBB
Copenhagen University Mathematics Department
Universitetsparken 5, DK-2100 Copenhagen, Denmark

Abstract. This is based on joint work with R. T. Seeley. The introduction presents the problem of parameter-dependent calculi for ψdo's and the question of trace asymptotics for Atiyah-Patodi-Singer operators. Chapter 2 establishes relations between the three operator functions: resolvent, heat operator and power operator (zeta function). Chapter 3 explains our parameter-dependent ψdo calculus with weak polyhomogeneity, showing how logarithmic terms appear in trace formulas. In Chapter 4, the APS problem is treated in the case with a product structure near the boundary, where functional calculus on the cylinder leads to precise formulas for heat trace expansions and zeta function pole structure. Finally, Chapter 5 treats the APS problem in the non-product case where the weakly polyhomogeneous ψdo calculus is used to get asymptotic trace expansions generalizing those in the product case.

1. Introduction

1.1. PARAMETER-DEPENDENT CALCULI

A typical case of an interesting parameter-dependent *pseudodifferential operator* (henceforth abbreviated to ψdo) is the resolvent $R_\lambda = (P-\lambda)^{-1}$ of a, say, strongly elliptic operator P on a compact manifold. Let the symbol of P (in a local coordinate system) be

$$p(x,\xi) = p_m(x,\xi) + p_{m-1}(x,\xi) + \dots,$$

where each term $p_{m-j}(x,\xi)$ is homogeneous of degree $m-j$ (for a positive integer m), then we write $-\lambda$ as

$$-\lambda = e^{i\theta}\mu^m, \quad \mu = |\lambda|^{1/m}, \theta \in [0, 2\pi]$$

L. Rodino (ed.), Microlocal Analysis and Spectral Theory, 115–164.

(where i is the imaginary unit $\sqrt{-1}$), and assign to $P - \lambda$ the *principal* symbol

$$\overline{p}_m(x, \xi, \lambda) = p_m(x, \xi) + e^{i\theta}\mu^m$$

(also denoted $\overline{p}_m(x, \xi, \theta, \mu)$), and the full symbol $\overline{p} + e^{i\theta}\mu^m$ where the lower order terms are the same as those for P. The inverse of this principal symbol,

$$q_m(x, \xi, \lambda) = \overline{p}_m(x, \xi, \lambda)^{-1}$$

will then be the principal symbol of the resolvent.

Here μ can be considered as one more "cotangent variable" in addition to $\xi_1, \xi_2, \dots, \xi_n$, and $\overline{p}_m$ is homogeneous of degree m in (ξ, μ).

There is a marked difference between the case of a differential operator and that of a ψdo. In the first case, $\overline{p}_m$ is polynomial in (ξ, μ), hence homogeneous and C^∞ in $(\xi, \mu) \in \overline{\mathbf{R}}_+^{n+1}$. In the second case, the homogeneous symbol $p_m(x, \xi)$ usually has a lack of smoothness at $\xi = 0$ (it has only m bounded derivatives), so $\overline{p}_m$ will have this lack of smoothness on the whole halfline $\{(0, \mu) \mid \mu \geq 0\}$. (Alternatively, if p_m is modified in a bounded neighborhood of 0 to be C^∞, the ensuing modification of $\overline{p}_m$ takes place in an unbounded set.)

This also has an effect on the *estimates* of q_m. Here one has (with $\langle x \rangle = (|x|^2 + 1)^{1/2}$):

$$\begin{aligned} D_\xi^\alpha q_m &= O(\langle(\xi, \mu)\rangle^{-m-|\alpha|}), \text{ for } |\alpha| \leq m, \\ D_\xi^\alpha q_m &= O(\langle(\xi, \mu)\rangle^{-2m}\langle\xi\rangle^{m-|\alpha|}), \text{ for } |\alpha| \geq m, \end{aligned} \tag{1.1}$$

where the first line extends to all α *if and only if* p_m is polynomial in ξ. In the polynomial case one can apply the usual symbolic calculus, just in one more variable, getting simple and straightforward results, whereas in the general case the fact that only the first m estimates are standard (the so-called *regularity number* is m), gives severe trouble.

For boundary value problems there are similar phenomena. In the differential operator case, the resolvent parameter enters as another cotangent variable, on a par with the others, whereas for a *pseudodifferential boundary operator*, a resolvent parameter, when considered as a cotangent variable, gives symbolic estimates where only finitely many of them are "good". Again one assigns a regularity number to the operator, this will now be different for the different types (trace operators, Poisson operators, singular Green operators).

This phenomenon is one of the main subjects of the book Grubb [12]. It is shown there that in the application to obtain trace formulas for resolvents (and heat kernels), one get finitely many well-defined terms in an asymptotic expansion, namely as many the regularity number indicates. For resolvents in the case without boundary, there is a trick to extend the analysis to get full trace expansions with infinitely many terms, some of them logarithmic; also this is explained in [12].

More recently, we have developed a somewhat more special calculus in collaboration with Robert Seeley [14], which allows a systematic construction of full asymptotic expansions for a class of ψdo's containing the resolvents: the calculus of *weakly polyhomogeneous* operators. It is completely described for the boundaryless case (whereas the additional details needed for general pseudodifferential boundary problems only exist in a sketched form).

For differential operators with pseudodifferential boundary conditions, one can however use the weakly polyhomogeneous ψdo calculus in cases where the trace formula in question can be reduced to one for an operator *in the boundary* of the weakly polyhomogeneous kind.

The calculus was developed for, and applies in particular to, the general Atiyah-Patodi-Singer problem. We describe this in detail below.

1.2. THE ATIYAH-PATODI-SINGER PROBLEM

On a compact n-dimensional C^∞ manifold X with boundary $\partial X = X'$, consider a first-order elliptic differential operator

$$P : C^\infty(E_1) \to C^\infty(E_2)$$

between sections of vector bundles over X. E_1 and E_2 have Hermitian metrics, and X has a smooth volume element, defining Hilbert spaces structures on the sections (primarily the spaces of L_2-sections, denoted $L_2(E_i)$, and more generally the Sobolev spaces $H^s(E_i)$, $s \in \mathbf{R}$).

The restrictions of the E_i to the boundary X' are denoted E_i'. A neighborhood of X' in X has the form $X_c = X' \times [0, c]$, and there the E_i are isomorphic to the pull-backs of the E_i'. Let x_n denote the coordinate in $[0, c]$, x' the coordinate in X'. Then we assume that P is represented in X_c as

$$P = \sigma(\tfrac{\partial}{\partial x_n} + A + x_n P_1 + P_0), \tag{1.2}$$

where σ is a unitary morphism from E_1' to E_2', independent of x_n, and A is a fixed elliptic first-order differential operator on $C^\infty(E_1')$, selfadjoint with respect to the Hermitian metric in E_1' and the volume element $v(x', 0)dx'$ on X' induced by the element $v(x', x_n)dx'dx_n$ on X. The P_j are smooth differential operators (in all variables) of order $\leq j$; they can be taken arbitrary near X', but for larger x_n, P_1 is subject to the requirement that P be elliptic. All morphisms are assumed C^∞.

In comparison with completely general elliptic first-order operators, the assumption means (modulo homotopies) that we have restricted the attention to operators such that when the principal symbol is written near X' as $\sigma_1(x',x_n)(i\xi_n I + a_1(x',x_n,\xi'))$ (with a bundle isomorphism σ_1 from E_1 to E_2 in front), then $a_1(x',0,\xi')$ is *symmetric;* cf. Grubb [13], p. 2036. The case considered by Atiyah, Patodi and Singer in [2] is the case where, furthermore, $P_1 = P_0 = 0$ in (1.2); this is often called the *product case.* Important examples are the Dirac operator and its generalizations.

We denote $u|_{X'} = \gamma_0 u$ and observe the Green's formula:

$$(Pu,w)_X - (u,P^*w)_X = -(\gamma_0 u, \sigma^*\gamma_0 w)_{X'}. \tag{1.3}$$

Since P is a first-order system, it may not be possible to formulate a well-posed boundary value problem in terms of a *differential* boundary condition (a Dirichlet condition is too much, no boundary condition is too little, and the boundary bundle structure will not in general allow putting a Dirichlet condition on some "half" of the boundary data). But using ψdo's, one can get well-posedness:

Definition 1.1 The APS boundary problem consists of finding $u \in H^1(E_1)$ for a given $f \in L_2(E_2)$, so that

$$Pu = f \text{ on } X, \quad B\gamma_0 u = 0 \text{ on } X'. \tag{1.4}$$

Here B is an orthogonal projection in $L_2(E_1')$ of the form $B = \Pi_{\geq} + B_0$, where $\Pi_{\geq}$ ($\Pi_<$, Π_R, ...) denotes the orthogonal projection onto $V_{\geq}$ ($V_<$, V_R, ...), the sum of eigenspaces for A with eigenvalues $\lambda \geq 0$ ($\lambda < 0$, $|\lambda| \leq R$, ...), and B_0 commutes with A and ranges in V_R for some $R \geq 0$.

The associated realization P_B is defined as the operator from $L_2(E_1)$ to $L_2(E_2)$ acting like P and with domain

$$D(P_B) = \{ u \in H^1(E_1) \mid B\gamma_0 u = 0 \}; \tag{1.5}$$

it is a Fredholm operator called **the APS operator**, and **the APS index problem** consists of determining its index.

This type of boundary condition is often called a *spectral boundary condition.* The Fredholm property of P_B was shown by Seeley in [23], where it was moreover shown that the adjoint of P_B is of a related type (in view of (1.4)):

$$(P_B)^* = (P^*)_{B'}, \text{ with } B' = B^\perp\sigma^*, \quad B^\perp = I - B. \tag{1.6}$$

One of the ways to study the index of P_B is to consider the "Laplacians"

$$\Delta_1 = P_B{}^*P_B, \quad \Delta_2 = P_BP_B{}^*, \tag{1.7}$$

and search for asymptotic expansions for $t \to 0$ (with $\varepsilon > 0$):

$$\operatorname{Tr} e^{-t\Delta_i} = c_{-n,i} t^{-n/2} + \cdots + c_{-1,i} t^{-1/2} + c_{0,i} + O(t^{\varepsilon}),\ i = 1, 2. \tag{1.8}$$

When (1.8) holds, the index is determined by

$$\operatorname{index} P_B = \operatorname{Tr} e^{-t\Delta_1} - \operatorname{Tr} e^{-t\Delta_2} = c_{0,1} - c_{0,2}. \tag{1.9}$$

Remark 1.2 The systems $\begin{pmatrix} P \\ B\gamma_0 \end{pmatrix}$ and $\begin{pmatrix} P^* \\ B'\gamma_0 \end{pmatrix}$ are injectively elliptic (also called overdetermined elliptic or left-elliptic). The operators Δ_1 and Δ_2 are realizations of truly elliptic systems (two-sided elliptic) such as

$$\begin{pmatrix} P^*P \\ A'B\gamma_0 + B'\gamma_0 P \end{pmatrix}, \text{ resp. } \begin{pmatrix} PP^* \\ A'B'\gamma_0 + B\gamma_0 P^* \end{pmatrix}. \tag{1.10}$$

(We here use that B and B' map into complementing subspaces of $L_2(E_1')$, and we have inserted the invertible ψdo $A' = A + \Pi_0(A)$ in order to make the boundary conditions first-order. The operators are *principally* the same as in the case where $B = \Pi_{\geq}$, discussed in detail in [13].) Another truly elliptic system incorporating P_B and $P_B{}^*$ is discussed below in Section 5.1 (and in [14]).

Remark 1.3 If $\sigma^* = -\sigma$ and $A\sigma = -\sigma A$, then in the product case, P is formally selfadjoint. Then if furthermore $B\sigma = \sigma(I - B)$, P_B is selfadjoint. This holds in many geometrically interesting cases, see e.g. Gilkey [10].

In [2] it was shown in the product case, with $B = \Pi_{\geq}$, that

$$\operatorname{index} P_B = \int_X \alpha(x) - \tfrac{1}{2}\eta_A; \quad \eta_A = \eta(A, 0) + \dim \ker A; \tag{1.11}$$

where $\alpha(x)$ is a certain form defined from the symbol of P, and $\eta(A, 0)$ is the value at $s = 0$ of the eta function

$$\eta(A, s) = \operatorname{Tr}(A|A|^{-s-1}). \tag{1.12}$$

(Here A^{-s-1} is defined as 0 on the nullspace of A, and meromorphic extension is used for $\operatorname{Re} s < n$.) Formula (1.11) was extended to the non-product case in [13] as

$$\operatorname{index} P_B = \int_X \alpha(x) + \int_{X'} \beta(x') - \tfrac{1}{2}\eta_A, \tag{1.13}$$

with a boundary form $\beta(x')$ defined from the symbols of P and B at X'. The forms defined from the symbols are regarded as *local contributions*, whereas the term η_A depends on the full set-up in a global way.

Actually, [2] did not calculate the two expressions $\operatorname{Tr} e^{-t\Delta_1}$ and $\operatorname{Tr} e^{-t\Delta_2}$ separately, but only their difference. They showed for the product case that this has the same asymptotic expansion as

$$\operatorname{Tr}(e^{-t\widetilde{\Delta}_1}|_X) - \operatorname{Tr}(e^{-t\widetilde{\Delta}_2}|_X) + \operatorname{Tr}(e^{-t\Delta_1^0} - \sigma^* e^{-t\Delta_2^0}\sigma); \tag{1.14}$$

here $\widetilde{\Delta}_1 = \widetilde{P}^*\widetilde{P}$ and $\widetilde{\Delta}_2 = \widetilde{P}\widetilde{P}^*$, where $\widetilde{P}$ is a certain extension of P to bundles $\widetilde{E}_1$ and $\widetilde{E}_2$ over the double manifold $\widetilde{X}$ (cf. [2], p. 55, where the roles of E_1 and E_2 are switched on $\widetilde{X}\setminus X$); the Δ_i^0 are x_n-independent extensions of the Δ_i on X_c to the cylinder $X^0 = X' \times \mathbf{R}_+$. The first difference is well-known, and the second can be analyzed by use of functional calculus for the selfadjoint operator A; this sufficed to get the index formula in the product case.

In [13] the separate expansions (1.8) were proved with $\varepsilon = \frac{3}{8}$ in the non-product case with $B = \Pi_{\geq}$, by a combination of the general treatment of parameter-dependent ψdo boundary problems [12] with the special results from [2]. It was shown that the global term $-\frac{1}{2}\eta_A$ enters in $c_{0,i}$ for both expansions, as $-\frac{1}{4}\eta_A$ for $i = 1$, resp. $\frac{1}{4}\eta_A$ for $i = 2$.

Now the index is just one special geometric invariant connected with the APS problem. More generally, one can ask about the value of the general coefficient $c_{j-n,i}$ in (1.7), and one can ask whether there is a more detailed structure of the $O(t^\varepsilon)$ term, giving a full asymptotic expansion $\sum_{j=0}^{\infty} c_{j-n,i} t^{(j-n)/2}$ for the trace $\operatorname{Tr}\exp(-t\Delta_i)$.

These questions have been answered in two papers written in cooperation with Seeley, [14] and [15]. It is shown there that there does exist a full asymptotic expansion, which however includes also logarithmic terms $c\,t^{(j-n)/2}\log t$ for $j - n > 0$. For the product case, a precise description of the coefficients in terms of the zeta and eta functions of A is given, when B_0 ranges in the nullspace of A.

In the following we shall give an account of these results, explaining the highlights of the methods.

2. The three operator-functions

2.1. DEFINITION OF THE OPERATOR FUNCTIONS

One can associate several interesting operator-functions with an elliptic operator Q. The following have been studied extensively:

• The resolvent $(Q-\lambda)^{-1}$ and its asymptotic behaviour for $\lambda \to \infty$ on rays in $\mathbf{C}$.

• The heat operator e^{-tQ} $(t \in \mathbf{R}_+)$ and its asymptotic behavior for $t \to 0+$.

• The power operator Q^{-s} and the pole structure of associated functions of $s \in \mathbf{C}$.

For the questions we address here, there are essentially equivalent formulations in terms of each of the three operator functions, and one can pass from one formulation to another by suitable transformations. Very briefly stated, the heat operator and the resolvent are related to one another by the Laplace transformation, and the heat operator and power operator are related to one another by the Mellin transformation. One can also define the heat operator and the power operator from the resolvent by suitable Cauchy integral formulas (Dunford integrals), and there is another complex integration formula involving a reciprocal sinus function going from the power function to the resolvent. (In the proofs of Theorems 2.1 and 2.3 below, we also relate the formulas to the Fourier transformation.) In the following we collect the facts on these operator functions that we need.

Much of this has been known in the literature for a long time (but not always explained as generally as here). Applications to trace asymptotics have been made earlier e.g. in Seeley [22], Duistermaat and Guillemin [8], Grubb [12], Agranovič [1], Branson and Gilkey [5]. The explanation in the following is essentially copied from [15], and is given here with full details since it may be of interest also for other purposes.

Suppose that Q is a closed operator in a Hilbert space having a resolvent $(Q-\lambda)^{-1}$ which is holomorphic in some sector $|\arg(-\lambda)| < \alpha$, with $\|(Q-\lambda)^{-1}\| = O(|\lambda|^{-1})$, and is meromorphic at 0 (in the sense that $(Q-\lambda)^{-1} - (-\lambda)^{-1}\Pi_0(Q)$ is holomorphic at 0, where $\Pi_0(Q)$ is the orthogonal projection onto the nullspace of Q). Then the power function $Z(Q,s)$ is defined for $\operatorname{Re} s > 0$ by

$$Z(Q,s) = \tfrac{i}{2\pi} \textstyle\int_C \lambda^{-s}(Q-\lambda)^{-1} d\lambda, \tag{2.1}$$

where C is a curve

$$\begin{aligned} C_{\theta,r_0} &= \{\, \lambda = re^{i\theta} \mid \infty > r \geq r_0 \,\} + \{\, \lambda = r_0 e^{i\theta'} \mid \theta \geq \theta' \geq -\theta \,\} \\ &\quad + \{\, \lambda = re^{i(2\pi-\theta)} \mid r_0 \leq r < \infty \,\}, \end{aligned} \tag{2.2}$$

with $\pi - \alpha < \theta \leq \pi$ and $r_0 > 0$ chosen so that $(Q-\lambda)^{-1}$ is holomorphic for $0 < |\lambda| \leq r_0$. If Q is invertible then $Z(Q,s) = Q^{-s}$ (further details are found e.g. in Seeley [22] or Shubin [24]); in any case, $Z(Q,s)$ is zero on the nullspace of Q, since $\int_C \lambda^{-s-1} d\lambda = 0$. We can also write

$$Z(Q,s) = \tfrac{i}{2\pi} \textstyle\int_{C_{\theta,0}} \lambda^{-s}(Q-\lambda)^{-1}\Pi_0^{\perp}(Q)\, d\lambda, \tag{2.3}$$

where $\Pi_0^\perp(Q) = I - \Pi_0(Q)$.

If $Z(Q,s)$ is trace class for some s, then Q has a zeta function

$$\zeta(Q,s) = \operatorname{Tr} Z(Q,s), \tag{2.4}$$

and, for appropriate operators D and values s, a "modified zeta function"

$$\zeta(D,Q,s) = \operatorname{Tr} DZ(Q,s). \tag{2.5}$$

Similarly, under appropriate conditions, we define

$$\begin{aligned} Y(Q,s) = QZ(Q^*Q, \tfrac{s+1}{2}) &= \tfrac{i}{2\pi} \textstyle\int_{\mathcal{C}} \lambda^{-(s+1)/2} Q(Q^*Q - \lambda)^{-1} d\lambda \\ &= \tfrac{i}{2\pi} \textstyle\int_{\mathcal{C}_{\theta,0}} \lambda^{-(s+1)/2} Q(Q^*Q - \lambda)^{-1} d\lambda \end{aligned} \tag{2.6}$$

(since $\Pi_0(Q^*Q) = \Pi_0(Q)$ and $Q\Pi_0(Q) = 0$, we can leave out the nullspace projection), and the eta functions

$$\eta(Q,s) = \operatorname{Tr} Y(Q,s), \quad \eta(D,Q,s) = \operatorname{Tr} DY(Q,s). \tag{2.7}$$

When Q is selfadjoint,

$$\sum_{\lambda \in \operatorname{sp}(Q)\setminus\{0\}} |\lambda|^{-s} = \zeta(Q^2, \tfrac{s}{2}), \qquad \sum_{\lambda \in \operatorname{sp}(Q)\setminus\{0\}} \operatorname{sign} \lambda\, |\lambda|^{-s} = \eta(Q,s), \tag{2.8}$$

with summation over the eigenvalues, repeated according to multiplicities.

In order to move the trace inside the integral, we may represent the power function by use of a derivative of the resolvent. Note that

$$\partial_\lambda^m (Q - \lambda)^{-1} = m!(Q-\lambda)^{-m-1}. \tag{2.9}$$

If Q is a ψdo of order $r > 0$ on a compact manifold M, say, then the mth derivative of $(Q-\lambda)^{-1}$ is a ψdo of order $-(1+m)r$ and hence is trace class when $(m+1)r > \dim M$. By an integration by parts, one can replace (2.1) by

$$Z(Q,s) = \tfrac{1}{(s-1)\cdots(s-m)} \tfrac{i}{2\pi} \textstyle\int_{\mathcal{C}} \lambda^{m-s} \partial_\lambda^m (Q-\lambda)^{-1} d\lambda, \tag{2.10}$$

whereby (2.4) can be written

$$\zeta(Q,s) = \operatorname{Tr} Z(Q,s) = \tfrac{1}{(s-1)\cdots(s-m)} \tfrac{i}{2\pi} \textstyle\int_{\mathcal{C}} \lambda^{m-s} \operatorname{Tr} \partial_\lambda^m (Q-\lambda)^{-1} d\lambda, \tag{2.11}$$

for sufficiently large m. Similar modifications can be made when there is a factor D as in (2.5) and when eta functions as in (2.7) are studied; and the integral can be replaced by an integral over $\mathcal{C}_{\theta,0}$ when $\Pi_0^\perp(Q)$ is inserted in front of $d\lambda$. There are similar formulas for the symbols and kernels of the operators.

When Q is lower bounded selfadjoint, the heat operator e^{-tQ} (also called the exponential function or the semigroup generated by $-Q$) can be defined by

$$e^{-tQ} = \tfrac{i}{2\pi} \int_{C'} e^{-t\lambda}(Q-\lambda)^{-1}\, d\lambda, \quad t>0; \tag{2.12}$$

where C' is a curve encircling the full spectrum in the positive direction and such that $e^{-t\lambda}$ falls off for $|\lambda| \to \infty$ on the curve (e.g. one can let C' begin with a ray with argument $\in]0, \frac{\pi}{2}[$ and end with a ray with argument $\in]-\frac{\pi}{2}, 0[$). This is well-known from the literature, see e.g. Hille-Phillips [16], Friedman [9] or Kato [20].

The exponential function and the power function of an operator $Q \geq 0$ with resolvent as above are related to one another by the formulas:

$$\begin{aligned} Z(Q,s) &= \tfrac{1}{\Gamma(s)} \int_0^\infty t^{s-1} e^{-tQ} \Pi_0^\perp(Q)\, dt, \quad \operatorname{Re} s > 0, \\ e^{-tQ}\Pi_0^\perp(Q) &= \tfrac{1}{2\pi i} \int_{\operatorname{Re} s = c} t^{-s} Z(Q,s)\Gamma(s)\, ds, \quad c>0, \end{aligned} \tag{2.13}$$

that follow e.g. from Theorem 2.3 below, with $e(t) = e^{-tQ}\Pi_0^\perp(Q)$, $\varphi(s) = \Gamma(s)Z(Q,s)$.

Taking $Q = S^*S$ for suitable operators S, we have accordingly (cf. (2.6)):

$$\begin{aligned} Z(S^*S,s) &= \tfrac{1}{\Gamma(s)} \int_0^\infty t^{s-1} e^{-tS^*S} \Pi_0^\perp(S)\, dt, \\ e^{-tS^*S}\Pi_0^\perp(S) &= \tfrac{1}{2\pi i} \int_{\operatorname{Re} s = c} t^{-s} Z(S^*S,s)\Gamma(s)\, ds, \\ Y(S,2s) = SZ(S^*S, s+\tfrac12) &= \tfrac{1}{\Gamma(s+\frac12)} \int_0^\infty t^{s-\frac12} S e^{-tS^*S}\, dt, \\ Se^{-tS^*S} &= \tfrac{1}{2\pi i} \int_{\operatorname{Re} s = c} t^{-s} Y(S, 2s-1)\Gamma(s)\, ds. \end{aligned} \tag{2.14}$$

(Also here we can omit mention of the nullspace projection in the last two formulas.)

Again, these formulas can be composed with a suitable operator D. When the expressions are trace class (usually for $\operatorname{Re} s$ resp. c sufficiently large) one can take the trace on both sides in (2.14) (composed with D), obtaining the formulas relating zeta and eta functions to exponential function traces:

$$\begin{aligned} \zeta(D, S^*S, s) &= \tfrac{1}{\Gamma(s)} \int_0^\infty t^{s-1} \operatorname{Tr} De^{-tS^*S} \Pi_0^\perp(S)\, dt, \\ \operatorname{Tr} De^{-tS^*S}\Pi_0^\perp(S) &= \tfrac{1}{2\pi i} \int_{\operatorname{Re} s = c} t^{-s} \zeta(D, S^*S, s)\Gamma(s)\, ds, \\ \eta(D,S,2s) = \zeta(DS, S^*S, s+\tfrac12) &= \tfrac{1}{\Gamma(s+\frac12)} \int_0^\infty t^{s-\frac12} \operatorname{Tr} DSe^{-tS^*S} dt, \\ \operatorname{Tr} DSe^{-tS^*S} &= \tfrac{1}{2\pi i} \int_{\operatorname{Re} s = c} t^{-s} \eta(D, S, 2s-1)\Gamma(s)\, ds. \end{aligned} \tag{2.15}$$

There are similar transition formulas for the symbols and kernels of the operators.

2.2. RELATIONS BETWEEN THE RESOLVENT AND THE POWER FUNCTION

Let us first consider the passage between properties of the resolvent and properties of the power and zeta functions. In order to handle operator functions defined not only as in (2.1), but also as in (2.10), we include functions with higher order poles at 0. We denote $\{0, 1, 2, \dots\} = \mathbf{N}$.

Theorem 2.1 1° *Suppose that f is meromorphic at 0 with Laurent expansion*

$$f(\lambda) = \sum_{j=-k}^{\infty} h_j(-\lambda)^j, \ |\lambda| \le \rho, \tag{2.16}$$

that f is holomorphic in the open sector $S_{\delta_0} = \{\lambda \in \mathbf{C} \mid |\arg\lambda - \pi| < \delta_0\}$ (for some $\delta_0 \le \pi$), and that $f(\lambda) = O(|\lambda|^{-\alpha})$ for some $\alpha \in]0,1]$ as $\lambda \to \infty$, uniformly in each sector S_δ for $\delta < \delta_0$. Let C be a curve C_{π,r_0} as in (2.2) (a Laurent loop, since $\theta = \pi$), with $0 < r_0 < \varrho$. Set $f_0(\lambda) = f(\lambda) - \sum_{-k}^{-1} h_j(-\lambda)^j$, and

$$\zeta(s) = \tfrac{i}{2\pi} \int_C \lambda^{-s} f(\lambda)\, d\lambda, \quad \operatorname{Re} s > 1 - \alpha, \tag{2.17}$$

with $\lambda^{-s} = r^{-s}e^{-is\theta}$, $r > 0$ and $|\theta| \le \pi$. Then ζ and f_0 are interrelated by:

$$\zeta(s) = \tfrac{\sin \pi s}{\pi} \int_0^\infty r^{-s} f_0(-r)\, dr, \quad 1 - \alpha < \operatorname{Re} s < 1, \tag{2.18}$$

$$f_0(-\lambda) = \tfrac{1}{2i} \int_{\operatorname{Re} s = \sigma} \lambda^{s-1} \tfrac{\zeta(s)}{\sin \pi s}\, ds, \quad 1 - \alpha < \sigma < 1. \tag{2.19}$$

The function $\frac{\pi\zeta(s)}{\sin \pi s}$ is meromorphic for $\operatorname{Re} s > 1 - \alpha$, having simple poles at $s = j + 1$ with residues $(-1)^{j+1}\zeta(j+1) = -h_j$, $j \in \mathbf{N}$.

2° *Moreover, the following properties* a) *and* b) *are equivalent:*

a) *f has an asymptotic expansion as λ goes to infinity*

$$f(-\lambda) \sim \sum_{j=0}^{\infty} \sum_{l=0}^{m_j} a_{j,l} \lambda^{-\alpha_j} (\log \lambda)^l, \quad 0 < \alpha_j \nearrow +\infty \tag{2.20}$$

(with $m_j \in \mathbf{N}$), uniformly for $-\lambda$ in S_δ, for each $\delta < \delta_0$.

b) *$\psi(s) = \frac{\pi\zeta(s)}{\sin \pi s}$ is meromorphic on $\mathbf{C}$ with the singularity structure*

$$\frac{\pi\zeta(s)}{\sin \pi s} \sim - \sum_{j=-k}^{\infty} \frac{h_j}{s - j - 1} + \sum_{j=0}^{\infty} \sum_{l=0}^{m_j} \frac{a_{j,l}\, l!}{(s + \alpha_j - 1)^{l+1}} \tag{2.21}$$

(in the sense that for large N, the left hand side minus the sums for $j \le N$ in the right hand side is holomorphic for $1-\alpha_N < \operatorname{Re} s < N+1$); and for each real C_1, C_2 and each $\delta < \delta_0$,

$$|\psi(s)| \le C(C_1, C_2, \delta) e^{-\delta|\operatorname{Im} s|}, \text{ for } |\operatorname{Im} s| \ge 1,\ C_1 \le \operatorname{Re} s \le C_2. \tag{2.22}$$

In particular, the singularities of $\psi(s)$ in $\operatorname{Re} s < 1$ are determined by the expansion (2.20) and the singular Laurent terms of $f(\lambda)$ at $\lambda = 0$, and vice versa.

3° Let f take values in a Banach space, and be holomorphic in S_{δ_0}, and meromorphic at 0 in the sense that there is a function $\sum_{j=-k}^{-1}(-\lambda)^j H_j$ with bounded operators H_j such that $f_0(\lambda) = f(\lambda) - \sum_{-k}^{-1}(-\lambda)^j H_j$ is holomorphic for $|\lambda| < \varrho$, some $\varrho > 0$. Let $\|f(\lambda)\|$ be $O(|\lambda|^{-\alpha})$ for $\lambda \to \infty$ in subsectors S_δ with $\delta < \delta_0$. Then with $\zeta(s)$ defined by (2.17), the formulas (2.18)–(2.19) are valid.

Proof: 1°. For $j \le -1$ and $\operatorname{Re} s > 0$, $\int_C \lambda^{j-s} d\lambda = 0$, since the contour can be closed at ∞ in $\{|\arg\lambda| < \pi\}$. So the singular part of f, $\sum_{-k}^{-1} h_j(-\lambda)^j$, is "killed" by the integral over C in (2.17). For the remaining part f_0, the circular part of C can be reduced to the origin if $\operatorname{Re} s < 1$, reducing (2.17) to (2.18) (note that f_0 is $O(|\lambda|^{-\alpha})$ too).

The inversion (2.19) requires growth estimates for $\zeta(s)$. Replacing the integration curve by $C(\delta) := C_{\pi-\delta,0}$, $0 < \delta < \delta_0$, we have that

$$|\zeta(s)| = \left|\tfrac{i}{2\pi} \textstyle\int_{C(\delta)} \lambda^{-s} f_0(\lambda)\, d\lambda\right| = O(e^{(\pi-\delta)|\operatorname{Im} s|}),$$
$$1-\alpha < C_1 \le \operatorname{Re} s \le C_2 < 1. \tag{2.23}$$

For, when $\lambda = re^{i(\pi-\delta)}$, we can use the estimate

$$\left|\int_0^\infty r^{-s} e^{i(\pi-\delta)(1-s)} O((1+r)^{-\alpha})\, dr\right| \le Ce^{(\pi-\delta)|\operatorname{Im} s|}, \tag{2.24}$$

and there is a similar estimate on the other half of $C(\delta)$.

Now let

$$\psi(s) = \int_0^\infty r^{-s} f_0(-r)\, dr = \frac{\pi\zeta(s)}{\sin \pi s}. \tag{2.25}$$

Since $(\sin\pi s)^{-1}$ is $O(e^{-\pi|\operatorname{Im} s|})$ for $|\operatorname{Im} s| \ge 1$, we have by (2.23) that $\psi(\sigma + i\tau) = O(e^{-\delta|\tau|})$ for $1-\alpha < C_1 \le \sigma \le C_2 < 1$. Also, $\psi(\sigma+i\tau)$ is the Fourier transform $\hat{F}(\tau)$ of the function $F(x) = e^{(1-\sigma)x} f_0(-e^x)$.

Since $f_0(\lambda) = O(\langle\lambda\rangle^{-\alpha})$, $F(x)$ decays exponentially as $x \to \pm\infty$, for $1-\alpha < \sigma < 1$. By Fourier inversion, $F(x) = \frac{1}{2\pi}\int_{-\infty}^{\infty} e^{ix\tau}\psi(\sigma + i\tau)d\tau$, giving (2.19), for $\lambda > 0$. It extends to $|\arg\lambda| < \delta_0$ by analytic continuation.

It is seen from (2.17) that $\zeta(s)$ is holomorphic for $\operatorname{Re} s > 1-\alpha$; and since $\zeta(j+1) = \frac{-i}{2\pi}\int_{|\lambda|=r_0} \lambda^{-j-1} f(\lambda)\,d\lambda = (-1)^j h_j$ for $j \in \mathbf{N}$, $\psi(s)$ is meromorphic for $\operatorname{Re} s > 1-\alpha$, having simple poles with residues $-h_j$.

2°. Now suppose that a) holds; then

$$f_0(-\lambda) = \sum_{j=0}^{N-1}\sum_{l=0}^{m_j} a_{j,l}\lambda^{-\alpha_j}(\log\lambda)^l - \sum_{j=-k}^{-1} h_j\lambda^j + O(|\lambda|^{-\alpha_N+\varepsilon}) \text{ for } \lambda \to \infty, \tag{2.26}$$

for $\alpha_N \geq k$, any $\varepsilon > 0$. Note that

$$\int_0^1 r^{j-s}\,dr = \frac{-1}{s-j-1} \quad \text{for } \operatorname{Re} s < j+1,$$

$$\int_1^\infty r^{\beta-s}(\log r)^l\,dr = \frac{l!}{(s-\beta-1)^{l+1}} \quad \text{for } \operatorname{Re} s > \beta+1$$

(the cases $l > 0$ follow from the case $l = 0$ by application of ∂_s^l); the right hand sides extend meromorphically to $\mathbf{C}$. Then we get from (2.25), for arbitrarily large N:

$$\begin{aligned}\psi(s) &= \int_0^1 \Big[\sum_{j=0}^{N-1} h_j r^{j-s} + r^{-s}O(r^N)\Big]dr \\ &\quad + \int_1^\infty \Big[\sum_{j=0}^{N-1}\sum_{l=0}^{m_j} a_{j,l} r^{-\alpha_j-s}(\log r)^l - \sum_{j=-k}^{-1} h_j r^{j-s} + r^{-s}O(r^{-\alpha_N+\varepsilon})\Big]dr \\ &= -\sum_{j=-k}^{N-1} \frac{h_j}{s-j-1} + \sum_{j=0}^{N-1}\sum_{l=0}^{m_j} \frac{a_{j,l}\,l!}{(s+\alpha_j-1)^{l+1}} + h_N(s)\end{aligned}$$

where h_N is holomorphic for $1-\alpha_N+\varepsilon < \operatorname{Re} s < N+1$, and the other terms are meromorphic on $\mathbf{C}$. This gives the singularities (2.21).

To show the decay, we use the integral in (2.23) and expand on each piece of $\mathcal{C}(\delta)$:

$$\begin{aligned}\zeta(s) &= -\tfrac{i}{2\pi}\Big(\textstyle\int_0^1 + \int_1^\infty (re^{i(\pi-\delta)})^{-s} f_0(re^{i(\pi-\delta)})e^{i(\pi-\delta)}\,dr\Big) \\ &\quad + \tfrac{i}{2\pi}\Big(\textstyle\int_0^1 + \int_1^\infty (re^{i(-\pi+\delta)})^{-s} f_0(re^{i(-\pi+\delta)})e^{i(-\pi+\delta)}\,dr\Big).\end{aligned} \tag{2.27}$$

The first integral from 0 to 1 is written as

$$\begin{aligned}&\tfrac{-i}{2\pi}\textstyle\int_0^1 (re^{i(\pi-\delta)})^{-s} f_0(re^{i(\pi-\delta)})e^{i(\pi-\delta)}\,dr \\ &\quad = \tfrac{-i}{2\pi}\textstyle\int_0^1 \sum_{j=0}^{N-1} e^{i(j+1-s)(\pi-\delta)} h_j r^{j-s}\,dr + \int_0^1 r^{-s}e^{i(\pi-\delta)(1-s)}O(r^N)\,dr \\ &\quad = \textstyle\sum_{j=0}^{N-1} \frac{-e^{i(j+1-s)(\pi-\delta)}h_j}{j+1-s} + e^{i(\pi-\delta)(1-s)}\int_0^1 r^{-s}O(r^N)\,dr.\end{aligned} \tag{2.28}$$

Let $|\operatorname{Im} s| \geq 1$. The sum over j extends meromorphically to $\mathbf{C}$, and its terms are $O(e^{(\pi-\delta)|\operatorname{Im} s|})$ for $-\infty < C_1 \leq \operatorname{Re} s \leq C_2 < \infty$. The last term exists and is $O(e^{(\pi-\delta)|\operatorname{Im} s|})$ when $\operatorname{Re} s < N+1$. Similar considerations hold for the other integral from 0 to 1. In the integrals from 1 to ∞ we expand as in (2.26), obtaining functions that are $O(e^{(\pi-\delta)|\operatorname{Im} s|})$ for $\operatorname{Re} s > 1-\alpha_N+\varepsilon$. We conclude that the estimate in (2.23) extends to $1-\alpha_N < \operatorname{Re} s < N+1$, $|\operatorname{Im} s| \geq 1$, for arbitrarily large N. Dividing by $\sin \pi s$ we find that $\psi(s)$ satisfies (2.22). This shows a) $\Longrightarrow$ b).

Conversely, assume b). Then $f_0(-\lambda)$ is given by (2.19), and we obtain the expansion (2.20) by shifting the contour of integration past the poles of $\psi(s)$. The remainder after all terms up to the singularity $s = 1-\alpha_N$ is given by the integral (2.19) but with $\sigma < 1-\alpha_N$; it is $O(|\lambda|^{-\alpha_N+\varepsilon})$ on S_δ.

3°. The proof under 1° is generalized straightforwardly to Banach spaces, with the relevant estimates valid for the norms. □

In this analysis, the poles in (2.21) may very well be considered in a general sense where we allow some of the coefficients $a_{j,l}$ to be 0; this is practical for the applications where vanishing coefficients often occur, and we shall use this point of view in the following. (So we can e.g. speak of a simple pole with residue 0 — this is usually not called a pole.)

Corollary 2.2 *When $f(\lambda)$ and $\zeta(s)$ are as in Theorem* 2.1 1°-2°, *then $\Gamma(s)\zeta(s)$ is meromorphic on* $\mathbf{C}$ *with the singularity structure*

$$\Gamma(s)\zeta(s) \sim \sum_{j=-k}^{j=-1} \frac{-\tilde{h}_j}{s-j-1} + \sum_{j=0}^{\infty}\sum_{l=0}^{m_j} \frac{\tilde{a}_{j,l}\, l!}{(s+\alpha_j-1)^{l+1}},$$
$$\tilde{h}_j = \frac{h_j}{\Gamma(-j)},\quad \tilde{a}_{j,l} = \frac{a_{j,l}}{\Gamma(\alpha_j)}. \tag{2.29}$$

Thus the singularity structure (2.29) *of $\Gamma(s)\zeta(s)$ is determined from the asymptotic expansion* (2.20) *of f together with the singular part of the Laurent expansion* (2.16) *(the coefficients h_j with $-k \leq j \leq -1$), and vice versa.*

When $\delta_0 > \frac{\pi}{2}$, one has moreover, for any $\delta' < \delta_0 - \frac{\pi}{2}$, any real C_1 and C_2:

$$|\Gamma(s)\zeta(s)| \leq C'(C_1, C_2, \delta)e^{-\delta'|\operatorname{Im} s|}, \quad \textit{for } |\operatorname{Im} s| \geq 1,\ C_1 \leq \operatorname{Re} s \leq C_2. \tag{2.30}$$

Proof: Since $\pi(\sin \pi s)^{-1} = \Gamma(s)\Gamma(1-s)$, (2.29) results from (2.21) by multiplication by $\Gamma(1-s)^{-1}$, whose zeros cancel the poles $h_j/(s-j-1)$, $j \geq 0$. If $\delta - \pi/2 = \delta' > 0$, the estimate $|\zeta(s)| \leq Ce^{(\pi-\delta)|\operatorname{Im} s|}$ shown in the proof of Theorem 2.1 (and assured by (2.22)) implies (2.30), since $\Gamma(s)$ is $O(e^{(-\frac{\pi}{2}+\varepsilon)|\operatorname{Im} s|})$ for $|\operatorname{Im} s| \geq 1$, $-\infty < C_1 \leq \operatorname{Re} s \leq C_2 < \infty$, any $\varepsilon > 0$. (Cf. e.g. the assertion in Bourbaki [3], p. 182:

$$|\Gamma(s)| \sim \sqrt{2\pi}\,|\operatorname{Im} s|^{\operatorname{Re} s-\frac{1}{2}} e^{-\frac{\pi}{2}|\operatorname{Im} s|} \text{ for } |\operatorname{Im} s| \to \infty, \tag{2.31}$$

valid for fixed $\operatorname{Re} s$ or $\operatorname{Re} s$ in compact intervals of $\mathbf{R}$.) □

Note in particular that a case $m_j = 1$ in (2.20) corresponds to a double pole of $\Gamma(s)\zeta(s)$ at $s = 1 - \alpha_j$ (in the strict sense if $a_{j,m_j} \neq 0$).

2.3. RELATIONS BETWEEN THE POWER FUNCTION AND THE EXPONENTIAL FUNCTION

Now we shall investigate the relation between properties of exponential functions and of power and zeta functions. The general transition goes as follows:

Theorem 2.3 1° *Let $e(t)$ be a function holomorphic in a sector V_{θ_0} (for some $\theta_0 \in]0, \frac{\pi}{2}[$),*

$$V_{\theta_0} = \{t = re^{i\theta} \mid r > 0, |\theta| < \theta_0\}, \tag{2.32}$$

such that $e(t)$ decreases exponentially for $|t| \to \infty$ and is $O(|t|^a)$ for $t \to 0$ in V_δ, any $\delta < \theta_0$, for some $a \in \mathbf{R}$. Let φ be the Mellin transform of e,

$$\varphi(s) = (\mathcal{M}e)(s) := \int_0^\infty t^{s-1} e(t)\, dt, \tag{2.33}$$

for $\operatorname{Re} s > -a$. Then $\varphi(s)$ is holomorphic for $\operatorname{Re} s > -a$ and $\varphi(c + i\xi)$ is $O(e^{-\delta|\xi|})$ for $|\xi| \to \infty$, when $c > -a$ (uniformly for c in compact intervals of $]-a, \infty[$); and $e(t)$ is recovered from $\varphi(s)$ by the formula

$$e(t) = \tfrac{1}{2\pi i} \textstyle\int_{\operatorname{Re} s = c} t^{-s} \varphi(s)\, ds. \tag{2.34}$$

2° *Moreover, the following properties* a) *and* b) *are equivalent:*
a) *$e(t)$ has an asymptotic expansion for $t \to 0$,*

$$e(t) \sim \sum_{j=0}^{\infty} \sum_{l=0}^{m_j} b_{j,l} t^{\beta_j} (\log t)^l, \quad \beta_j \nearrow +\infty,\ m_j \in \mathbf{N}, \tag{2.35}$$

uniformly for $t \in V_\delta$, for each $\delta < \theta_0$.
b) *$\varphi(s)$ is meromorphic on $\mathbf{C}$ with the singularity structure*

$$\varphi(s) \sim \sum_{j=0}^{\infty} \sum_{l=0}^{m_j} \frac{(-1)^l l! b_{j,l}}{(s + \beta_j)^{l+1}}, \tag{2.36}$$

and for each real C_1, C_2 *and each* $\delta < \theta_0$,

$$|\varphi(s)| \leq C(C_1, C_2, \delta)e^{-\delta|\operatorname{Im} s|}, \ |\operatorname{Im} s| \geq 1, \ C_1 \leq \operatorname{Re} s \leq C_2. \tag{2.37}$$

3° *Let* $f(\lambda)$ *take values in a Banach space, and be holomorphic in* $S_{\delta_0} = \{|\pi - \arg\lambda| < \delta_0\}$ *for some* $\delta_0 \in]\frac{\pi}{2}, \pi]$ *and meromorphic at* $\lambda = 0$ *(holomorphic for* $0 < |\lambda| < \varrho$*). Assume that as* $\lambda \to \infty$ *in* S_δ *(for* $\delta < \delta_0$*), some derivative* $\partial_\lambda^m f(\lambda)$ *is* $O(|\lambda|^{-1-\varepsilon})$ *for some* $\varepsilon > 0$ *(so that* $f(\lambda)$ *is* $O(|\lambda|^{m-1})$*). Let* θ_0 *and* θ *be such that* $]\theta - \theta_0, \theta + \theta_0[\subset]\pi - \delta_0, \frac{\pi}{2}[$, *let* $C = C_{\theta, r_0}$ *as in* (2.2) *with* $r_0 \in]0, \varrho[$, *and let*

$$e(t) = \tfrac{i}{2\pi} \int_C e^{-t\lambda} f(\lambda)\, d\lambda, \quad \varphi(s) = \Gamma(s) \tfrac{i}{2\pi} \int_C \lambda^{-s} f(\lambda)\, d\lambda, \tag{2.38}$$

for $t \in V_{\theta_0}$ *resp.* $\operatorname{Re} s > m - \varepsilon$. *Then* $e(t)$ *is exponentially decreasing for* $t \to \infty$ *in sectors* V_δ *with* $\delta < \theta_0$, *and is* $O(|t|^{-m})$ *for* $t \to 0$, *and* $\varphi(s)$ *and* $e(t)$ *correspond to one another by* (2.33), (2.34). *Here, when* $f(\lambda) = (Q - \lambda)^{-1}$, *then* $e(t) = e^{-tQ}\Pi_0^\perp(Q)$ *and* $\varphi(s) = \Gamma(s)Z(Q, s)$.

Proof: 1°. Note first that replacing $e(t)$ by $t^b e(t)$ replaces $\varphi(s)$ by $\varphi(s+b)$, so we can assume that $a > 0$ and then consider $c \geq 0$. The function $\varphi(s)$ is holomorphic for $\operatorname{Re} s \geq 0$ since the integrand $t^{s-1}e(t)$ is so and has an integrable majorant there.

By a change of variables $t = e^x$, we see that $\varphi_1(\xi) = \varphi(i\xi)$ is the conjugate Fourier transform of $e_1(x) = e(e^x) \in L_2(\mathbf{R})$:

$$\varphi_1(\xi) = \varphi(i\xi) = \int_0^\infty t^{i\xi} e(t) \frac{dt}{t} = \int_{-\infty}^\infty e^{ix\xi} e(e^x)\, dx = \int_{-\infty}^\infty e^{ix\xi} e_1(x)\, dx,$$

so by Fourier's inversion formula,

$$e(t) = e_1(x) = \tfrac{1}{2\pi} \int_{-\infty}^\infty e^{-ix\xi} \varphi_1(\xi)\, d\xi = \tfrac{1}{2\pi i} \int_{\operatorname{Re} s = 0} t^{-s} \varphi(s)\, ds. \tag{2.39}$$

Similarly, $\varphi(c + i\xi)$ is the conjugate Fourier transform of $e(e^x)e^{xc}$ for $c > 0$.

The hypothesis on exponential decrease of $e(t)$ in the sectors V_δ allows us to shift the path of integration in (2.33) from $t \in \mathbf{R}_+$ to $t \in e^{i\delta}\mathbf{R}_+$ for $|\delta| < \theta_0$ (corresponding to a shift to $x \in \mathbf{R} + i\delta$); this gives:

$$\begin{aligned} \varphi(c + i\xi) &= \int_0^\infty (re^{i\delta})^{c+i\xi} e(re^{i\delta}) \frac{dr}{r} \\ &= e^{-\delta\xi} \int_0^\infty r^{i\xi} e(re^{i\delta})(re^{i\delta})^c \frac{dr}{r} = e^{-\delta\xi} g(\delta, \xi, c), \end{aligned}$$

where g is bounded as a function of $\xi \in \mathbf{R}$, locally uniformly in $c \geq 0$. Taking $\delta > 0$ for $\xi > 0$ and $\delta < 0$ for $\xi < 0$, we see that $\varphi(c+\mathrm{i}\xi)$ decreases exponentially (like $e^{-\delta|\xi|}$) for $|\xi| \to \infty$, in any vertical strip $\{s = c + \mathrm{i}\xi \mid C_1 \leq c \leq C_2, \xi \in \mathbf{R}\}$ with $0 \leq C_1 \leq C_2$. Then we can also shift the integration path in (2.39) from $\operatorname{Re} s = 0$ to $\operatorname{Re} s = c$, $c \geq 0$. This shows 1°.

2°. Assume now in addition (2.35). Let us first write $\varphi(s)$ as

$$\varphi(s) = \int_0^1 t^{s-1} e(t)dt + \int_1^\infty t^{s-1} e(t)dt. \tag{2.40}$$

The second integral defines an entire function of s. The expansion (2.35) means that

$$e(t) = \sum_{j=0}^{N-1} \sum_{l=0}^{m_j} b_{j,l} t^{\beta_j} (\log t)^l + \varrho_N(t),$$
$$\varrho_N(t) = O(|t|^{\beta_N - \varepsilon}) \text{ for } t \to 0 \text{ in } V_\delta, \tag{2.41}$$

for $\varepsilon > 0$ and any positive integer N; we insert this in the first integral. Observe the formulas, valid for $\operatorname{Re} s > -\beta$,

$$\int_0^1 t^{s-1+\beta} (\log t)^l \, dt = \frac{(-1)^l l!}{(s+\beta)^l},$$
$$\int_0^\infty t^{s-1+\beta} (\log t)^l \, e^{-t} \, dt = \partial_s^l \Gamma(s+\beta), \tag{2.42}$$

where the cases $l > 0$ follow from the cases $l = 0$ by application of ∂_s^l. The remainder $\varrho_N(t)$ in (2.41) gives a function holomorphic for $\operatorname{Re} s > -\beta_N + \varepsilon$, and for the powers of t we use (2.42); this shows (2.36).

To show the exponential decrease of $\varphi(s)$ on general vertical strips, one can shift the contour in (2.33) and proceed much as in the proof of Theorem 2.1. Another instructive method is to insert the expansion $e^t = \sum_{\nu \geq 0} \frac{1}{\nu!} t^\nu$, that gives

$$e^t t^{\beta_j} (\log t)^l = \sum_{\nu=0}^{M-1} \tfrac{1}{\nu!} t^{\beta_j + \nu} (\log t)^l + O(t^{\beta_j + M - \varepsilon}),$$

for any $\varepsilon > 0$ and positive integer M. Then we can write

$$e(t) = e(t) e^t e^{-t} = \Big(\sum_{\beta_j + \nu < M} \sum_{l \leq m_j} b_{j,l} \tfrac{1}{\nu!} t^{\beta_j + \nu} (\log t)^l \Big) e^{-t} + \tilde{\varrho}_M(t),$$
$$\text{with } \tilde{\varrho}_M(t) = O(|t|^{M-\varepsilon}) \text{ for } t \to 0 \text{ in } V_\delta,$$

where $\tilde{\varrho}_M(t)$ is exponentially decreasing for $|t| \to \infty$ in V_δ since the other terms are so, and hence

$$\varphi(s) = \int_0^\infty t^{s-1}\Big(\sum_{\beta_j+\nu<M} \sum_{l\le m_j} b_{j,l}\tfrac{1}{\nu!}t^{\beta_j+\nu}(\log t)^l\Big)e^{-t}dt$$

$$+ \int_0^\infty t^{s-1}\tilde{\varrho}_M(t)\,dt. \quad (2.43)$$

The last integral defines a function that is holomorphic for $\operatorname{Re} s > -M+\varepsilon$ and exponentially decreasing (like $e^{-\delta|\operatorname{Im} s|}$) on strips $-M+\varepsilon < C_1 \le \operatorname{Re} s \le C_2$, by 1°. For the contributions from the first integral we use the second formula in (2.42) together with the fact that the gamma function $\Gamma(s)$ and its derivatives are $O(e^{(-\frac{\pi}{2}+\varepsilon')|\operatorname{Im} s|})$, any $\varepsilon' > 0$, for $|\operatorname{Im} s| \ge 1$, $-\infty < C_1 \le \operatorname{Re} s \le C_2 < \infty$, cf. e.g. [3], pp. 181–182. This gives (2.37), completing the proof of a) $\Longrightarrow$ b).

Conversely, assume b). Then $e(t)$ is given by (2.34), and we obtain the expansion (2.35) by shifting the contour of integration past the poles of $\varphi(s)$. The remainder after all terms up to and including the singularity $s = -\beta_N$ is given by an integral like (2.34) but with $c < -\beta_N$; it is $O(|t|^{\beta_N-\varepsilon})$.

3°. That $e(t)$ defined here is exponentially decreasing for $|t| \to \infty$ in V_δ, $\delta < \theta_0$, follows since $|e^{-\lambda t}| \le e^{-\gamma|t|}$ with $\gamma > 0$ on the integration curve. The estimate for $t \to 0$ follows since

$$\int_{\mathcal{C}} e^{-\lambda t}f(\lambda)\,d\lambda = (-t)^{-m}\int_{\mathcal{C}}(\partial_\lambda^m e^{-\lambda t})f(\lambda)\,d\lambda = t^{-m}\int_{\mathcal{C}} e^{-\lambda t}\partial_\lambda^m f(\lambda)\,d\lambda$$

for $t \in V_\delta$, where $e^{-\lambda t}\partial_\lambda^m f(\lambda)$ has a fixed integrable majorant for $t \to 0$. The formula (2.33) for φ is shown by a complex change of variables, where we replace t by u/λ for each λ; when $\arg\lambda \in]0, \frac{\pi}{2}[$, the ray $\mathbf{R}_+$ is transformed to a ray Λ_λ with argument $-\arg\lambda \in]-\frac{\pi}{2}, 0[$, and vice versa. The integral of $u^{s-1}e^{-u}$ on such a ray is again equal to $\Gamma(s)$, as noted above. Thus (recall that $f(\lambda)$ is $O(|\lambda|^{m-1})$)

$$\int_0^\infty t^{s-1}\tfrac{i}{2\pi}\int_{\mathcal{C}} e^{-t\lambda}f(\lambda)\,d\lambda dt = \tfrac{i}{2\pi}\int_{\mathcal{C}}\int_{\Lambda_\lambda} u^{s-1}\lambda^{-s}e^{-u}f(\lambda)\,dud\lambda$$

$$= \Gamma(s)\tfrac{i}{2\pi}\int_{\mathcal{C}}\lambda^{-s}f(\lambda)\,d\lambda. \qquad \square$$

3. Weakly polyhomogeneous symbols

3.1. POLYHOMOGENEOUS SYMBOL CLASSES

We here sketch the properties of the symbol class used to get trace expansions for the general APS problem; details are given in [14].

Consider symbols $p(x,\xi,\mu)$, where x and $\xi \in \mathbf{R}^n$, $\mu \in \Gamma$ (a sector of $\mathbf{C} \setminus \{0\}$). We shall say that:

p is **strongly homogeneous** of degree m, when

$$p(x,t\xi,t\mu) = t^m p(x,\xi,\mu) \text{ for } |\xi|^2 + |\mu|^2 \geq 1,\ t \geq 1,\quad (\xi,\mu) \in \mathbf{R}^n \times (\Gamma \cup \{0\}). \tag{3.1}$$

p is **weakly homogeneous** of degree m, when

$$p(x,t\xi,t\mu) = t^m p(x,\xi,\mu) \text{ for } |\xi|, t \geq 1,\ (\xi,\mu) \in \mathbf{R}^n \times \Gamma. \tag{3.2}$$

Example 3.1 Let $a(x,\xi)$ be positive and C^∞ on $\mathbf{R}^n \times \mathbf{R}^n$, and homogeneous in ξ of degree $r \in \mathbf{N}$ for $|\xi| \geq 1$. Then $a(x,\xi)+\mu^r$ and $(a(x,\xi)+\mu^r)^{-1}$ extend to:

strongly homogeneous symbols of degree r, resp. $-r$, if a is **polynomial** in ξ (it is the symbol of a differential operator);

weakly homogeneous symbols of degree r, resp. $-r$, if a is **not** polynomial in ξ (it is the symbol of a genuine ψdo).

If for example $r = n = 2$, $a(x,\xi) = \xi_1^2 + \xi_2^2$ enters in the first case, and $a(x,\xi) = (\xi_1^4 + \xi_2^4)/(\xi_1^2 + \xi_2^2)$ (for $|\xi| \geq 1$) enters in the second case.

Both cases can be shown to belong to the following symbol classes (where $(a(x,\xi) + \mu^r)^{-1} \in S^{-r,0} \cap S^{0,-r}$):

Definition 3.2 $S^{m,0}(\mathbf{R}^n, \mathbf{R}^n, \Gamma)$ consists of the functions $p(x,\xi,\mu)$ that are holomorphic in μ for $|(\xi,\mu)| \geq \varepsilon$, $\mu \in \Gamma$, and satisfy, denoting $\frac{1}{\mu} = z$,

$$\partial_z^j p(\cdot,\cdot,\tfrac{1}{z}) \text{ is in } S^{m+j}(\mathbf{R}^n,\mathbf{R}^n) \text{ for } \tfrac{1}{z} \in \Gamma, \text{ with uniform estimates for } |z| \leq 1,\ \tfrac{1}{z} \in \text{closed subsectors of } \Gamma. \tag{3.3}$$

Moreover, we set $S^{m,d}(\mathbf{R}^n, \mathbf{R}^n, \Gamma) = \mu^d S^{m,0}(\mathbf{R}^n, \mathbf{R}^n, \Gamma)$.

Here $S^m(\mathbf{R}^n, \mathbf{R}^n)$ denotes the standard ψdo symbol space consisting of the functions $p(x,\xi) \in C^\infty(\mathbf{R}^n \times \mathbf{R}^n)$ such that $\partial_x^\beta \partial_\xi^\alpha p$ is $O(\langle\xi\rangle^{m-|\alpha|})$ for all $\alpha, \beta \in \mathbf{N}^n$. The rules of calculus for such symbols are well-known, see e.g. Hörmander [18], Seeley [23], Shubin [24], Hörmander [19] for various set-ups with local or global estimates in x. We call the symbols in $S^m(\mathbf{R}^n \times \mathbf{R}^n)$ *classical*, when they moreover have expansions in series of homogeneous terms (in ξ, $|\xi| \geq 1$) of degrees $m - j$, $j \in \mathbf{N}$.

When symbols $p(x,\xi)$ of order m are considered as depending on one more variable μ, they lie in $S^{m,0}$:

$$S^m(\mathbf{R}^n, \mathbf{R}^n) \subset S^{m,0}(\mathbf{R}^n, \mathbf{R}^n, \Gamma), \text{ any } \Gamma. \tag{3.4}$$

The symbols in $S^{m,d}(\mathbf{R}^n, \mathbf{R}^n, \Gamma)$ define ψdo's $P = \mathrm{OP}(p)$ (which depend on the parameter μ) by the usual formula:

$$\mathrm{OP}(p)f(x) = \int e^{ix\cdot\xi} p(x,\xi,\mu)\hat{f}(\xi)\,đ\xi, \quad f \in \mathcal{S}(\mathbf{R}^n), \tag{3.5}$$

with $đ\xi = (2\pi)^{-n}d\xi$. The definition extends to more general functions and distributions f as in the nonparametrized case. When $m < -n$, $\mathrm{OP}(p)$ is an integral operator with continuous kernel $K_p(x,y,\mu)$;

$$\begin{aligned} K_p(x,y,\mu) &= \int e^{i(x-y)\cdot\xi} p(x,\xi,\mu)\,đ\xi, \\ \text{in particular, } K_p(x,x,\mu) &= \int p(x,\xi,\mu)\,đ\xi. \end{aligned} \tag{3.6}$$

The operators have good composition rules, since $S^{m,d} \cdot S^{m',d'} \subset S^{m+m',d+d'}$, and since one can refer to the standard rules for S^m symbol classes, which must here hold uniformly in z as in (3.3). One finds for example that

$$P \in \mathrm{OP}(S^{m,d}),\ P' \in \mathrm{OP}(S^{m',d'}) \implies PP' \in \mathrm{OP}(S^{m+m',d+d'}) \tag{3.7}$$

(under the usual precautions on supports or global estimates), and the resulting symbol is described by the usual formula

$$(p \circ p')(x,\xi,\mu) \sim \sum_{\alpha\in\mathbf{N}^n} \tfrac{1}{\alpha!}\partial_\xi^\alpha p(x,\xi,\mu)(-i\partial_x)^\alpha p'(x,\xi,\mu) \text{ in } S^{m+m',d+d'}. \tag{3.8}$$

The expansion in (3.8) is an expansion in terms with decreasing m-exponents $m + m' - j$, $j \to \infty$ ($j = |\alpha|$). Such expansions enter in the theory as follows:

When $p_j \in S^{m_j,d}$ for a sequence $m_j \searrow -\infty$ (for $j \to \infty$, $j \in \mathbf{N}$), and $p \in S^{m_0,d}$, we say that $p \sim \sum_{j\in\mathbf{N}} p_j$ in $S^{m_0,d}$ if

$$p - \sum_{j<J} p_j \in S^{m_J,d} \text{ for any } J \in \mathbf{N}. \tag{3.9}$$

For any given sequence $p_j \in S^{m_j,d}$ with $m_j \searrow -\infty$, there exists a p such that (3.9) holds.

For the present special symbols there is *another* type of expansion that is of great interest:

Theorem 3.3 *When $p \in S^{m,d}(\mathbf{R}^n, \mathbf{R}^n, \Gamma)$, then p has an expansion in terms $\mu^{d-k}p_{(d,k)}(x,\xi)$ with $p_{(d,k)} \in S^{m+k}(\mathbf{R}^n, \mathbf{R}^n)$, such that for any N,*

$$p(x,\xi,\mu) - \sum_{0\le k<N} \mu^{d-k}p_{(d,k)}(x,\xi) \in S^{m+N,d-N}(\mathbf{R}^n, \mathbf{R}^n, \Gamma). \tag{3.10}$$

In the proof one reduces to the case $d = 0$ by multiplication by μ^{-d}; then the expansion is essentially a Taylor expansion in $z = \frac{1}{\mu}$ at $z = 0$.

Note that in (3.10), the order of $p_{(d,k)}$ *increases* with increasing k, whereas the power of μ *decreases*. A very simple example is

$$(1 + |\xi|^2 + \mu^2)^{-1} = \mu^{-2}(1 - \mu^{-2}\langle\xi\rangle^2 + \mu^{-4}\langle\xi\rangle^4 - \dots).$$

Corollary 3.4 *When $p \in S^{-\infty,d}$, the kernel $K_p(x, y, \mu)$ of* $\mathrm{OP}(p)$ *has an expansion*

$$K_p(x, y, \mu) \sim \sum_{k \in \mathbf{N}} \mu^{d-k} K_{p,k}(x, y), \quad K_{p,k} \in C^\infty. \tag{3.11}$$

Definition 3.2 contains no homogeneity requirements, but we now define a polyhomogeneous subspace:

Definition 3.5 A symbol $p \in S^{m_0-d,d}$ is called **weakly polyhomogeneous**, when $p \sim \sum_{j \in \mathbf{N}} p_j$, with $p_j \in S^{m_j-d,d}$, $m_j \searrow -\infty$ for $j \to \infty$, $j \in \mathbf{N}$, such that the p_j are weakly homogeneous of degrees m_j (cf. (3.2)).

This will be compared with:

Definition 3.6 A function $p(x, \xi, \mu) \in C^\infty(\mathbf{R}^n \times \mathbf{R}^n \times (\Gamma \cup \{0\})$ is called **strongly polyhomogeneous** of degree m if there is a sequence of functions $p_j \in C^\infty(\mathbf{R}^n \times \mathbf{R}^n \times (\Gamma \cup \{0\})$ that are strongly homogeneous of degree $m - j$ (cf. (3.1)) such that

$$\partial_x^\beta \partial_\xi^\alpha \partial_\mu^k (p - \sum_{j<J} p_j) = O(\langle(\xi, \mu)\rangle^{m-J-|\alpha|-k}), \tag{3.12}$$

for all indices, uniformly for μ in closed subsectors of $\Gamma \cup \{0\}$.

Then one has in fact:

Theorem 3.7 *When p is strongly polyhomogeneous of degree $m \in \mathbf{Z}$, then it is also weakly polyhomogeneous, with degrees $m - j$, $j \in \mathbf{N}$, and with*

$$\begin{gathered} p \in S^{m,0} + S^{0,m} \ \textit{if}\ m \geq 0, \quad p \in S^{m,0} \cap S^{0,m} \ \textit{if}\ m \leq 0, \\ \partial_x^\beta \partial_\xi^\alpha \partial_\mu^k p \in S^{m-|\alpha|-k,0} \cap S^{0,m-|\alpha|-k} \ \textit{for}\ |\alpha| + k \geq m,\ \textit{all}\ \beta. \end{gathered} \tag{3.13}$$

As a consequence, classical symbols of order $m \in \mathbf{Z}$ in $n + 1$ cotangent variables give strongly polyhomogeneous symbols in n cotangent variables, when one cotangent variable is replaced by μ (here $\Gamma = \mathbf{R}_+ \cup \mathbf{R}_-$).

The type of parameter-dependence entering in Theorem 3.7 was used by Agmon and by Agranovič and Vishik in resolvent studies for differential operators; for ψdo's this is the kind of parameter-dependence studied e.g. in Shubin [24] and many other works. It is a mild generalization that does not cover resolvents $(P-\lambda)^{-1}$ and parabolic operators such as $\partial/\partial t + P$ when P is truly pseudodifferential (as treated in [12]).

3.2. APPLICATIONS TO KERNEL AND TRACE EXPANSIONS

Both the expansion in Theorem 3.3 and the expansion in Definition 3.5 enter in the proof of:

Theorem 3.8 *Let p be weakly polyhomogeneous as in Definition 3.5, with $m_0 - d < -n$. Then $\mathrm{OP}(p)$ has a continuous kernel $K_p(x,y,\mu)$ with an expansion on the diagonal*

$$K_p(x,x,\mu) \sim \sum_{j=0}^{\infty} a_j(x)\mu^{m_j+n} + \sum_{k=0}^{\infty}[c_k(x)\log\mu + c'_k(x)]\mu^{d-k}\,, \tag{3.14}$$

for $|\mu| \to \infty$, uniformly for μ in closed subsectors of Γ. The coefficients $a_j(x)$ and $c_{d-m_j-n}(x)$ are determined by $p_j(x,\xi,\mu)$ for $|\xi| \geq 1$ (are "local"), while the $c'_k(x)$ are "global."

Details of proof are given in [14]. A brief explanation: One uses the general principle that "remainders contribute to c'_k terms," by Corollary 3.3. The p_j contribute with (cf. (3.6))

$$\begin{aligned} K_{p_j}(x,x,\mu) &= \int_{\mathbf{R}^n} p_j(x,\xi,\mu)\,đ\xi \\ &= \int_{|\xi|\geq|\mu|} p_j\,đ\xi + \int_{|\xi|\leq 1} p_j\,đ\xi + \int_{1\leq|\xi|\leq|\mu|} p_j\,đ\xi = I_1 + I_2 + I_3, \end{aligned} \tag{3.15}$$

where I_1 gives part of the a_j term, I_2 gives c'_k terms, and I_3 gives the rest of a_j and c_{d-m_j-n} (if $d-m_j-n \in \mathrm{N}$) and some c'_k terms. One has of course to show that the contributions to the c'_k pile up in a controlled way.

When the operator acts on a compact boundaryless manifold, integration of $K_p(x,x,\mu)$ in x gives a similar expansion of the *trace*:

Corollary 3.9 *Let P be a μ-dependent ψdo on a compact manifold M of dimension n, with symbol satisfying the hypotheses of Theorem 3.8 in local coordinates. Then it is trace class, the trace satisfying*

$$\operatorname{Tr} P \sim \sum_{j=0}^{\infty} a_j\mu^{m_j+n} + \sum_{k=0}^{\infty}[c_k\log\mu + c'_k]\mu^{d-k}\,, \tag{3.16}$$

for $|\mu| \to \infty$, uniformly for μ in closed subsectors of Γ. The coefficients are derived from those in (3.14) *for coordinate patches by integration over M.*

The result applies in particular to expressions containing a differentiated resolvent:

$$P = S\partial_\lambda^m (Q-\lambda)^{-1}, \tag{3.17}$$

where Q is a classical elliptic ψdo of positive integer order r on a compact boundaryless manifold M of dimension n_1, with principal symbol $q_r(x,\xi)$ having no eigenvalues on $\mathbf{R}_-$, S is a classical ψdo of order d, and m is chosen so that $d - r(1+m) < -n_1$. With $\mu = (-\lambda)^{1/r}$ for λ in a narrow sector Γ around $\mathbf{R}_-$, the symbol is in $S^{d-r(1+m),0} \cap S^{0,d-r(1+m)}$ and weakly polyhomogeneous. Then Theorem 3.8 and its corollary lead to an expansion of the diagonal kernel and the trace, generalizing the result of Agranovič [1] for $S = I$ (cf. [14] for details). The kernel $K(x, y, P)$ satisfies on the diagonal:

$$\begin{aligned} &K(x,x,S\partial_\lambda^m(Q-\lambda)^{-1}) \\ &\sim \sum\nolimits_{j=0}^{\infty} a_j(x)\lambda^{\frac{n_1+d-j}{r}-m-1} + \sum_{k=0}^{\infty}(c_k(x)\log\lambda + c_k'(x))\lambda^{-k-m-1}, \end{aligned} \tag{3.18}$$

for $|\lambda| \to \infty$, uniformly in closed subsectors of Γ. Consequently, one has for the trace:

$$\operatorname{Tr} S\partial_\lambda^m(Q-\lambda)^{-1} \sim \sum\nolimits_{j=0}^{\infty} a_j\lambda^{\frac{n_1+d-j}{r}-m-1} + \sum_{k=0}^{\infty}(c_k\log\lambda + c_k')\lambda^{-k-m-1}\,; \tag{3.19}$$

where the coefficients are the integrals over M of the fiber traces of the coefficients defined in (3.18).

If S is a differential operator (in particular if $S = I$), then $c_0(x) = 0$ and the complete coefficient of λ^{-m-1} is locally determined.

If S and Q are both differential operators, we are in the well-known case where no logarithms occur, and all coefficients are locally determined (cf. [22]). This is shown by a simpler version of the above proof where the decomposition (3.15) is not needed since the symbols are smooth and homogeneous at $\xi = 0$, and it gives an expansion we may write as follows (denoting S by D):

$$\begin{aligned} K(x,x,D\partial_\lambda^m(Q-\lambda)^{-1}) &\sim \sum\nolimits_{j=0}^{\infty} b_j(x,D,Q)(-\lambda)^{\frac{n_1+d-j}{r}-m-1}, \\ \operatorname{Tr} D\partial_\lambda^m(Q-\lambda)^{-1} &\sim \sum\nolimits_{j=0}^{\infty} b_j(D,Q)(-\lambda)^{\frac{n_1+d-j}{r}-m-1}, \end{aligned} \tag{3.20}$$

for $\lambda \to \infty$ in suitable sectors. Let r be even. Then since the symbol terms homogeneous of odd degree satisfy $p(x,-\xi,\mu) = -p(x,\xi,\mu)$, their contributions to the diagonal kernel and the trace vanish (cf. (3.6)); hence

$$b_j(x,D,Q) \text{ and } b_j(D,Q) \text{ are zero for } d-j \text{ odd.} \tag{3.21}$$

For later reference we list the formula for the zeta function that follows from (3.20) by Corollary 2.2, in the case where the differential operator Q is selfadjoint ≥ 0 and of order 2. We have to take the singularity resulting from the nullspace $V_0(Q)$ (of finite dimension $\nu_0(Q)$) into account; in fact,

$$D\partial_\lambda^m(Q-\lambda)^{-1} - D\Pi_0(Q)\partial_\lambda^m(-\lambda)^{-1} \tag{3.22}$$

is holomorphic at 0. Here $\Pi_0(Q)$ is an integral operator with C^∞ kernel $K(x,y,\Pi_0(Q)) = \sum_{1\leq l\leq \nu_0} u_l(x)\otimes \overline{u}_l(y)$, where the u_l are a smooth orthonormal basis of V_0. The kernel of $D\Pi_0(Q)$ is $\sum_{1\leq l\leq\nu_0}(Du_l(x))\otimes\overline{u}_l(y)$. Then the singularity at 0 of $K(x,x,D\partial_\lambda^m(Q-\lambda)^{-1})$, resp. $\operatorname{Tr} D\partial_\lambda^m(Q-\lambda)^{-1}$, is

$$K(x,x,D\Pi_0(Q))\partial_\lambda^m(-\lambda)^{-1}, \text{ resp. } \operatorname{Tr} D\Pi_0(Q)\partial_\lambda^m(-\lambda)^{-1}. \tag{3.23}$$

In this case (3.20) is seen to correspond, by (2.10) and Corollary 2.2, to the following pole descriptions of the diagonal kernel and trace of $\Gamma(s)DZ(Q,s)$:

$$\begin{aligned}\Gamma(s)K(x,x,DZ(Q,s)) &\sim \sum_{j=0}^{\infty}\frac{c_j(x,D,Q)}{s+\frac{j-n_1-d}{2}} - \frac{K(x,x,D\Pi_0(Q))}{s},\\ \Gamma(s)\operatorname{Tr}(DZ(Q,s)) &\sim \sum_{j=0}^{\infty}\frac{c_j(D,Q)}{s+\frac{j-n_1-d}{2}} - \frac{\operatorname{Tr}(D\Pi_0(Q))}{s},\end{aligned} \tag{3.24}$$

with

$$\begin{gathered} c_j(x,D,Q) = \frac{b_j(x,D,Q)}{\Gamma(m+1+\frac{j-n_1-d}{2})},\quad c_j(D,Q) = \int_M \operatorname{tr} c_j(x,D,Q)\,dx,\\ c_j = 0 \text{ for } j-d \text{ odd.}\end{gathered} \tag{3.25}$$

In particular, if Q is the square of a selfadjoint first-order differential operator A, and D is multiplication by $\psi(x)$, we get for the pole structure of the zeta and eta functions, taking the vanishing of coefficients into account:

$$\begin{aligned}\Gamma(s)\zeta(\psi,A^2,s) = \Gamma(s)\operatorname{Tr}(\psi Z(A^2,s)) &\sim \sum_{k=0}^{\infty}\frac{c_{2k}(\psi,A^2)}{s+k-\frac{n_1}{2}} - \frac{\operatorname{Tr}(\psi\Pi_0(A))}{s},\\ \Gamma(s)\eta(\psi,A,2s-1) = \Gamma(s)\operatorname{Tr}(\psi AZ(A^2,s)) &\sim \sum_{k=0}^{\infty}\frac{c_{2k+1}(\psi A,A^2)}{s+k-\frac{n_1}{2}}\end{aligned} \tag{3.26}$$

Let us also list the consequences for the heat kernel and trace, by Theorem 2.3:

$$\begin{aligned} K(x,x,De^{-tQ}\Pi_0^\perp(Q)) &\sim \sum_{j=0}^{\infty} c_j(x,D,Q)t^{\frac{j-n_1-d}{2}} - K(x,x,D\Pi_0(Q)), \\ K(x,x,De^{-tQ}) &\sim \sum_{j=0}^{\infty} c_j(x,D,Q)t^{\frac{j-n_1-d}{2}}, \\ \operatorname{Tr}(De^{-tQ}) &\sim \sum_{j=0}^{\infty} c_j(D,Q)t^{\frac{j-n_1-d}{2}}; \end{aligned} \tag{3.27}$$

note that the effects of the nullspace projection have been cancelled out in the second and third lines. (In Theorem 2.3, a simple pole at $s=0$ for $\varphi(s)$ corresponds to a constant term for $e(t)$.)

4. The APS resolvent in the product case

4.1. GENERALITIES ON RESOLVENTS

We now return to the APS operator on a manifold with boundary, as described in Section 1.2.

One auxiliary tool is to consider an extension of P to a larger manifold without boundary. As mentioned after (1.14), one can choose a specific extension $\widetilde{P}$ to the double $\widetilde{X}$ in the product case. However, in the final formulas, the choice of extension really plays no role, since all operators are restricted back to X (more comments in [15], Remark 3.6), so we can let $\widetilde{P}$ stand for any extension satisfying the ellipticity requirements. In the general case, we just extend to a neighboring open manifold $\widetilde{X} = X \cup (X' \times]-1,0[)$ preserving the ellipticity hypotheses there.

We denote the extended "Laplacians" $\widetilde{\Delta}_1 = \widetilde{P}^*\widetilde{P}$ and $\widetilde{\Delta}_2 = \widetilde{P}\widetilde{P}^*$, and set

$$Q_{i,\lambda} = (\widetilde{\Delta}_i - \lambda)^{-1}. \tag{4.1}$$

In the product case where $\widetilde{\Delta}_i$ is selfadjoint ≥ 0 on the compact manifold $\widetilde{X}$, this is well-defined for $\lambda \in \mathbf{C}$ outside a discrete subset of $\overline{\mathbf{R}}_+$, and the zeta and eta functions as well as the heat trace for the $\widetilde{\Delta}_i$ behave as described at the end of the preceding section.

In the general case, $Q_{i,\lambda}$ is, to begin with, just defined in a parametrix sense, but it can be modified such that for sufficiently large λ

$$(\widetilde{\Delta}_{i,+} - \lambda)Q_{i,\lambda,+} = I \text{ on } X \tag{4.2}$$

(as explained in detail in [14], p. 508–9). Here we use the convention of defining, for an operator S on $\widetilde{X}$, the truncation S_+ to X by

$$S_+u = r^+Se^+u, \tag{4.3}$$

where e^+u denotes the extension of u with $e^+u(x', x_n) = 0$ for $x_n < 0$, and r^+ denotes restriction to $\{x_n > 0\}$. We shall also write

$$\operatorname{Tr}_+ S = \operatorname{Tr}[S_+]. \tag{4.4}$$

(The plus-subscript is often omitted when one deals with *differential* operators, since they act locally.)

The $Q_{i,\lambda}$ enter as pseudodifferential parts of the resolvents we are looking for:

$$R_{i,\lambda} = Q_{i,\lambda,+} + G_{i,\lambda}, \tag{4.5}$$

where the $G_{i,\lambda}$ are *singular Green operators* (in the notation of Boutet de Monvel [4]); s.g.o.s..

Remark 4.1 One of the well-known ways to describe the resolvent of a given boundary value problem is the following: Consider a problem

$$(P - \lambda)u = f \text{ on } X, \quad Tu = \varphi \text{ on } X', \tag{4.6}$$

where P is elliptic of order d in a bundle E over X, and T is a trace operator (from $H^d(X, E)$ to a suitable Sobolev space $H_T(X', F)$ over the boundary X'). The resolvent R_λ is the solution operator $R_\lambda : f \mapsto u$ for the problem (4.6) with $\varphi = 0$. *Assume* that $P - \lambda$, extended to a larger manifold $\widetilde{X}$, has an inverse Q_λ such that $(P - \lambda)Q_{\lambda,+} = I$ on X, where $Q_{\lambda,+}$ maps $L_2(X, E)$ into $H^d(X, E)$. *Assume* moreover that the problem (4.6) with $f = 0$ has a solution operator $K_\lambda : \varphi \mapsto u$ (such that $(P - \lambda)K_\lambda = 0$, $TK_\lambda = I$), mapping $H_T(X', F)$ into $H^d(X, E)$. Such an operator going from the boundary to the interior is called a *Poisson operator* in [4].

Then the full problem (4.6) has at most one solution for any data $\{f, \varphi\}$ in $L_2(X, E) \times H_T(X', F)$, since null-data give the null-solution. Moreover, the resolvent equals

$$R_\lambda = Q_{\lambda,+} - K_\lambda T Q_{\lambda,+}, \tag{4.7}$$

for this operator verifies $(P - \lambda)R_\lambda = I$ and $TR_\lambda = 0$ and is defined on all of $L_2(X, E)$ so it must be the unique solution operator.

In (4.7) we see the structure of the resolvent as the sum of a ψdo term and a term composed of a Poisson operator K_λ and a general type of trace operator $TQ_{\lambda,+}$; here $K_\lambda TQ_{\lambda,+}$ is an example of a singular Green operator.

Another auxiliary tool in the analysis of the inverse (4.5) is to compare it with the inverse on the cylinder $X^0 = X' \times \mathbf{R}_+$. Define

$$P^0 = \sigma(\partial_n + A), \quad P^{0\prime} = (-\partial_n + A)\sigma^*, \text{ so}$$
$$P^{0\prime}P^0 = D_n^2 + A^2, \quad P^0P^{0\prime} = \sigma(D_n^2 + A^2)\sigma^*. \tag{4.8}$$

They have a meaning on X^0, where P^0 goes from E_1^0 to E_2^0, the respective liftings of E_1' and E_2', and $P^{0\prime}$ is the formal adjoint of P^0 with respect to the product measure. They can be extended to bundles $\widetilde{E}_i^0$ over $\widetilde{X}^0 = X' \times \mathbf{R}$; the simplest choice is to take the $\widetilde{E}_i^0$ as the liftings of E_i' and extend the formulas in (4.8). We denote the extensions $\widetilde{P}^0$, $\widetilde{\Delta}_1^0 = (\widetilde{P}^0)'\widetilde{P}^0$, $\widetilde{\Delta}_2^0 = \widetilde{P}^0(\widetilde{P}^0)'$.

On the cylinder X^0 we consider the realization P_B^0 of P^0 defined by the boundary condition $B\gamma_0 u = 0$, with the Laplacians $\Delta_1^0 = P_B^{0*}P_B^0$ and $\Delta_2^0 = P_B^0 P_B^{0*}$. The resolvents are:

$$R_{i,\lambda}^0 = Q_{i,\lambda,+}^0 + G_{i,\lambda}^0, \text{ with}$$
$$Q_{1,\lambda}^0 = (D_n^2 + A^2 - \lambda)^{-1}, \quad Q_{2,\lambda}^0 = \sigma(D_n^2 + A^2 - \lambda)^{-1}\sigma^*, \tag{4.9}$$

the $G_{i,\lambda}^0$ being singular Green operators (as in Remark 4.1).

In the product case one can show that the true resolvent $R_{i,\lambda}$ is, near X', very closely related to $R_{i,\lambda}^0$, in such a way that the singular Green contributions to the asymptotic expansions we are looking for are essentially the same. In the general case, $R_{i,\lambda}^0$ is a first order approximation in some sense, so we can take it as a point of departure for the construction of the true resolvent $R_{i,\lambda}$.

4.2. DECOMPOSITION FORMULAS IN THE PRODUCT CASE

In the product case, very precise information will be obtained for the asymptotic expansions, on the basis of exact formulas for the operators involved. Let

$$A_\lambda = (A^2 - \lambda)^{1/2}, \text{ for } \lambda \in \mathbf{C} \setminus \overline{\mathbf{R}}_+; \quad A' = A + \Pi_0(A). \tag{4.10}$$

We shall here describe the results for the case $B = \Pi_\geq$ (i.e., $B_0 = 0$) in detail. [15] moreover treats $B = \Pi_\geq + B_0$ with B_0 ranging in $\overline{V}_0(A)$. In a recent manucript [6], Brüning and Lesch treat certain other boundary conditions for problems as in Remark 1.3, see Remark 4.14 below.

Using the cylindrical structure, we shall write the s.g.o. terms in (4.9) explicitly in terms of the special operator

$$(G_\lambda u)(x', x_n) = \int_0^\infty e^{-(x_n + y_n)A_\lambda} u(x', y_n)\, dy_n. \tag{4.11}$$

When G is an operator defined by $Gu = \int_0^\infty \mathcal{G}(x_n, y_n)u(x', y_n)\,dy_n$, where $\mathcal{G}$ is a function of x_n, y_n valued in operators on x'-space, we call $\mathcal{G}(x_n, y_n)$ the *normal kernel* of G, and define its *normal trace* as

$$\operatorname{tr}_n G = \int_0^\infty \mathcal{G}(x_n, x_n)\,dx_n, \tag{4.12}$$

when it exists. The normal kernel of G_λ is $e^{-(x_n+y_n)A_\lambda}$, and the normal trace is

$$\operatorname{tr}_n G_\lambda = \int_0^\infty e^{-2x_nA_\lambda}\,dx_n = (2A_\lambda)^{-1}. \tag{4.13}$$

Example 4.2 To explain how G_λ enters, consider the Dirichlet problem for $D_n^2 + A^2 - \lambda$ on X^0,

$$(D_n^2 + A^2 - \lambda)u = f, \quad \gamma_0 u = \varphi, \tag{4.14}$$

from the point of view of Remark 4.1. The Poisson operator solving (4.14) with $f = 0$ is

$$K^0_{\mathrm{Dir},\lambda}\varphi = e^{-x_nA_\lambda}\varphi, \tag{4.15}$$

and the composition $\gamma_0 Q^0_{1,\lambda,+}$ acts like

$$\begin{aligned}\gamma_0(D_n^2 + A^2 - \lambda)^{-1}e^+ f &= \gamma_0\tfrac{1}{2A_\lambda}\int_0^\infty e^{-|x_n-y_n|A_\lambda}e^+ f(x', y_n)dy_n \\ &= \int_0^\infty \tfrac{1}{2A_\lambda}e^{-y_nA_\lambda}f(x', y_n)dy_n, \end{aligned}\tag{4.16}$$

so the singular Green operator term as in (4.7) equals the composed operator

$$\begin{aligned}G^0_{\mathrm{Dir},\lambda}f &= -K^0_{\mathrm{Dir},\lambda}\gamma_0 Q^0_{1,\lambda,+}f \\ &= \tfrac{-1}{2A_\lambda}\int_0^\infty e^{-(x_n+y_n)A_\lambda}f(x', y_n)\,dy_n = \tfrac{-1}{2A_\lambda}G_\lambda f. \end{aligned}\tag{4.17}$$

Thus the resolvent equals

$$(\Delta^0_{\mathrm{Dir}} - \lambda)^{-1} = R^0_{\mathrm{Dir},\lambda} = Q^0_{1,\lambda,+} - \tfrac{1}{2A_\lambda}G_\lambda. \tag{4.18}$$

For a Robin-type boundary condition $\gamma_0(\partial_n + S)u = 0$, where S commutes with A^2 and $A_\lambda - S$ is invertible, one finds in a similar way that the singular Green operator term in the resolvent is

$$G^0_{\mathrm{Rob},\lambda} = \tfrac{1}{2A_\lambda}\tfrac{A_\lambda+S}{A_\lambda-S}G_\lambda. \tag{4.19}$$

In particular for the Neumann condition, $G^0_{\mathrm{Neu},\lambda} = \tfrac{1}{2A_\lambda}G_\lambda$.

The actual boundary conditions mix boundary values and normal derivatives in a more complicated way; for example, Δ_1^0 has the boundary condition (cf. (1.6))

$$\Pi_{\geq}\gamma_0 u = 0, \quad \Pi_{<}\gamma_0(\partial_n + A)u = 0, \tag{4.20}$$

where we have used that $\Pi_{<} = I - \Pi_{\geq}$ and that $\sigma^*\sigma = I$. This is a Dirichlet condition on the functions of x_n valued in $V_{\geq}$ and a Robin-type condition on the functions of x_n valued in $V_{<}$, so by applying Example 4.2 to each component, we find that the singular Green term in $(\Delta_1^0 - \lambda)^{-1}$ has the form

$$G_{1,\lambda}^0 = \left(\tfrac{-1}{2A_\lambda}\Pi_{\geq} + \tfrac{1}{2A_\lambda}\tfrac{A_\lambda + A}{A_\lambda - A}\Pi_{<}\right)G_\lambda \tag{4.21}$$

(which has a good sense since $-A$ is positive on $V_{<}$). Along with the corresponding formula for $G_{2,\lambda}^0$, this may be written as in [13], [15]:

$$\begin{aligned}
G_{1,\lambda}^0 &= G_{e,\lambda} + G_{o,\lambda} - \tfrac{\Pi_0(A)}{2\sqrt{-\lambda}}G_\lambda, \\
G_{2,\lambda}^0 &= \sigma(G_{e,\lambda} - G_{o,\lambda} + \tfrac{\Pi_0(A)}{2\sqrt{-\lambda}}G_\lambda)\sigma^*, \text{ where} \\
G_{e,\lambda} &= \tfrac{-|A|}{2A_\lambda(|A|+A_\lambda)}G_\lambda = \left(\tfrac{-A^2}{2\lambda A_\lambda} + \tfrac{|A|}{2\lambda}\right)G_\lambda, \\
G_{o,\lambda} &= \tfrac{-1}{2(|A|+A_\lambda)}\tfrac{A}{|A'|}G_\lambda = \left(-\tfrac{A}{2\lambda} + \tfrac{A_\lambda}{2\lambda}\tfrac{A}{|A'|}\right)G_\lambda.
\end{aligned} \tag{4.22}$$

Recall (4.10); for the last expressions it is used that $1/(|A| + A_\lambda) = (|A| - A_\lambda)/(A^2 - (A^2 - \lambda)) = |A|/\lambda - A_\lambda/\lambda$. The indices e and o refer to the evenness and oddness of the principal symbols with respect to ξ'. (The parity alternates between even and odd in the sequences of lower order symbols.) All the operators are defined and holomorphic for $\lambda \in \mathbf{C} \setminus \overline{\mathbf{R}}_+$. Moreover, $G_{e,\lambda}$ and $G_{o,\lambda}$ are holomorphic at 0 because of the factors $|A|$ and A that vanish on the nullspace.

From (4.13), (4.22) follow:

$$\operatorname{tr}_n G_{e,\lambda} = \tfrac{-A^2}{4\lambda A_\lambda^2} + \tfrac{|A|}{4A_\lambda\lambda}, \quad \operatorname{tr}_n G_{o,\lambda} = \tfrac{-A}{4\lambda A_\lambda} + \tfrac{1}{4\lambda}\tfrac{A}{|A'|}. \tag{4.23}$$

Example 4.3 The corresponding expressions for the Dirichlet and Robin-type problems considered in Example 4.2 are

$$\operatorname{tr}_n G_{\mathrm{Dir},\lambda}^0 = \tfrac{-1}{4A_\lambda^2} = -\tfrac{1}{4}\tfrac{1}{A^2-\lambda}, \quad \operatorname{tr}_n G_{\mathrm{Rob},\lambda}^0 = \tfrac{1}{4A_\lambda^2}\tfrac{A_\lambda+S}{A_\lambda-S}. \tag{4.24}$$

Now one can show that near X', the true s.g.o $G_{i,\lambda}$ is very similar to the cylindrical version $G_{i,\lambda}^0$.

Lemma 4.4 *(Product case.) Let $\chi \in C_0^\infty(\mathbf{R})$ with $\chi(x_n) = 1$ for $|x_n| < \frac{c}{3}$, $\chi(x_n) = 0$ for $|x_n| > \frac{2c}{3}$. Then $G_{1,\lambda} - \chi G_{1,\lambda}^0 \chi$ is trace class in $L_2(E_1)$ with norm $O(|\lambda|^{-N})$ for $|\lambda| \to \infty$ with $\arg\lambda \in [\delta, 2\pi - \delta]$, any $\delta > 0$. The same is true of $\partial_\lambda^k[G_{1,\lambda} - \chi G_{1,\lambda}^0 \chi]$ for $k = 1, 2, \dots$, and of expressions $DG_{1,\lambda} - \chi D' G_{1,\lambda}^0 \chi$, where D is a differential operator, constant in x_n near X' and equal to D' there.*

Similar estimates hold for $G_{2,\lambda} - \chi G_{2,\lambda}^0 \chi$ in $L_2(E_2)$, and for the operators $(1-\chi)G_{i,\lambda}^0$ and $G_{i,\lambda}^0 - \chi G_{i,\lambda}^0 \chi$ in $L_2(E_i^0)$. Here the $G_{i,\lambda}^0$ can be replaced by $G_{\mathrm{e},\lambda}$ or $G_{\mathrm{o},\lambda}$.

All these functions are holomorphic in $\lambda \in \mathbf{C} \setminus \overline{\mathbf{R}}_+$.

The proof is given in detail in [15], using elements of [13]. It extends to show that the operators also map into H^s, any s, with $O(|\lambda|^{-N})$ estimates for any N.

We now construct the zeta functions. For this, we integrate $\lambda^{-s} R_{i,\lambda}$ along an appropriate curve $\mathcal{C}$ as in Theorem 2.1, running along the negative axis and around a small circle of radius

$$r_0 < \min\{\lambda_1(\Delta_i), \lambda_1(\tilde{\Delta}_i), \lambda_1(A^2)\}, \tag{4.25}$$

where λ_1 denotes the smallest positive eigenvalue. ($\mathcal{C}$ could also be taken to be a curve in $\operatorname{Re}\lambda > 0$ closer to the spectra.) Then (cf. (4.3))

$$\begin{aligned} Z(\Delta_i, s) &= \tfrac{i}{2\pi} \textstyle\int_{\mathcal{C}} \lambda^{-s} R_{i,\lambda}\, d\lambda = \tfrac{i}{2\pi} \int_{\mathcal{C}} \lambda^{-s} Q_{i,\lambda,+}\, d\lambda + \tfrac{i}{2\pi} \int_{\mathcal{C}} \lambda^{-s} G_{i,\lambda}\, d\lambda \\ &= Z(\tilde{\Delta}_i, s)_+ + G_{Z,i,s}, \text{ where we have set} \\ G_{Z,i,s} &= \tfrac{i}{2\pi} \textstyle\int_{\mathcal{C}} \lambda^{-s} G_{i,\lambda}\, d\lambda. \end{aligned} \tag{4.26}$$

In the trace calculations in Theorem 4.6 below, we shall replace $G_{i,\lambda}$ by $G_{i,\lambda}^0$ by use of Lemma 4.4. Define the transforms

$$G_{Z,\mathrm{e},s} = \tfrac{i}{2\pi} \textstyle\int_{\mathcal{C}} \lambda^{-s} G_{\mathrm{e},\lambda}\, d\lambda, \quad G_{Z,\mathrm{o},s} = \tfrac{i}{2\pi} \int_{\mathcal{C}} \lambda^{-s} G_{\mathrm{o},\lambda}\, d\lambda. \tag{4.27}$$

To describe the various G_Z, we use the function defined for $\operatorname{Re}(-t) < \operatorname{Re} s < 0$ by

$$\begin{aligned} F_t(s) &= \tfrac{i}{2\pi} \textstyle\int_{\mathcal{C}_{\pi,r_0}} \tau^{-s-1}(1-\tau)^{-t}\, d\tau \\ &= \tfrac{i}{2\pi}\left(e^{(-s-1)i\pi} - e^{(s+1)i\pi}\right) \textstyle\int_0^\infty u^{-s-1}(1+u)^{-t}\, du \\ &= \tfrac{1}{\pi} \sin\pi(s+1)\, \frac{\Gamma(-s)\Gamma(s+t)}{\Gamma(t)} = \frac{\Gamma(s+t)}{\Gamma(t)\Gamma(s+1)}; \end{aligned} \tag{4.28}$$

$\mathcal{C}_{\pi,r_0}$ is taken with $r_0 \in]0,1[$, cf. (2.2). $F_t(s)$ coincides with the binomial coefficient $\binom{s+t-1}{t-1}$, also equal to $(sB(t,s))^{-1}$, where B is the beta function.

$F_t(s)$ extends meromorphically to general s and $t \in \mathbf{C}$. In particular,

$$\begin{aligned} &F_{\frac12}(s) = \tfrac{\Gamma(s+\frac12)}{\sqrt{\pi}\,\Gamma(s+1)} = \tbinom{s-\frac12}{-\frac12}, \quad F_1(s) = 1, \\ &F_0(s) = 0 \text{ if } s \neq 0, \quad F_t(0) = 1 \text{ if } \Gamma(t) \neq \infty. \end{aligned} \tag{4.29}$$

That $F_1(s) = 1$ follows directly from the first integral in (4.14), and the formula for $F_0(s)$ follows from the fact that $\frac{i}{2\pi}\int_C \tau^{-s-1}\,d\tau = 0$ for $\operatorname{Re} s > 0$.

The formulas for the singular Green operator terms are greatly simplified when we take normal traces.

Proposition 4.5 *Define* $G_{Z,\mathrm{e},s}$ *and* $G_{Z,\mathrm{o},s}$ *by* (4.27), *cf. also* (4.13), (4.10). *Then*

$$\begin{aligned} \operatorname{tr}_n G_{Z,\mathrm{e},s} &= \tfrac14 (F_{\frac12}(s) - 1) Z(A^2, s), \\ \operatorname{tr}_n G_{Z,\mathrm{o},s} &= -\tfrac14 F_{\frac12}(s) Y(A, 2s). \end{aligned} \tag{4.30}$$

Proof: Expand the operators on X' with respect to the orthogonal eigenprojections $\{\Pi_\mu\}_{\mu\in\mathrm{sp}(A)}$ for A. Our $G_{Z,\mathrm{e},s}$ and $G_{Z,\mathrm{o},s}$ are both 0 in the zero eigenspace. Using (4.23) we find, by replacing λ by $\mu^2\tau$ for each μ,

$$\begin{aligned} \operatorname{tr}_n G_{Z,\mathrm{e},s} &= \operatorname{tr}_n \tfrac{i}{2\pi}\int_C \lambda^{-s} G_{\mathrm{e},\lambda}\,d\lambda = \tfrac{i}{2\pi}\int_C \lambda^{-s}\big(\tfrac{-A^2}{4\lambda A_\lambda^2} + \tfrac{|A|}{4A_\lambda \lambda}\big)\,d\lambda \\ &= \sum_\mu \tfrac14 \tfrac{i}{2\pi}\int_C \lambda^{-s-1}\big(-\tfrac{\mu^2}{\mu^2-\lambda} + \tfrac{|\mu|}{(\mu^2-\lambda)^{\frac12}}\big)\,d\lambda\cdot\Pi_\mu \\ &= \sum_\mu \tfrac14 |\mu|^{-2s}\tfrac{i}{2\pi}\int_C \tau^{-s-1}\big(\tfrac{-1}{1-\tau} + \tfrac{1}{(1-\tau)^{\frac12}}\big)\,d\tau\cdot\Pi_\mu \\ &= \tfrac14(-F_1(s) + F_{\frac12}(s))Z(A^2,s) = \tfrac14(-1 + F_{\frac12}(s))Z(A^2,s); \end{aligned} \tag{4.31}$$

$$\begin{aligned} \operatorname{tr}_n G_{Z,\mathrm{o},s} &= \operatorname{tr}_n \tfrac{i}{2\pi}\int_C \lambda^{-s} G_{\mathrm{o},\lambda}\,d\lambda = \tfrac{i}{2\pi}\int_C \lambda^{-s}\big(\tfrac{-A}{4\lambda A_\lambda} + \tfrac{1}{4\lambda}\tfrac{A}{|A'|}\big)\,d\lambda \\ &= \sum_\mu \tfrac14 \tfrac{i}{2\pi}\int_C \lambda^{-s-1}\big(-\tfrac{\mu}{(\mu^2-\lambda)^{\frac12}} + \tfrac{\mu}{|\mu|}\big)\,d\lambda\cdot\Pi_\mu \\ &= \sum_\mu \tfrac14 \mu|\mu|^{-2s-1}\tfrac{i}{2\pi}\int_C \tau^{-s-1}\big(\tfrac{-1}{(1-\tau)^{\frac12}} + 1\big)\,d\tau\cdot\Pi_\mu \\ &= \tfrac14(-F_{\frac12}(s) + F_0(s))Y(A,2s) = -\tfrac14 F_{\frac12}(s) Y(A,2s). \quad \square \end{aligned} \tag{4.32}$$

Note that the *even part* produces a function derived from the *zeta* function of A, and the *odd part* produces a function derived from the *eta* function of A. This is the fundamental observation for the following, relating the power functions of the boundary value problem to those of A.

Now we combine this with the interior contribution, taken from the doubled manifold $\widetilde{X}$. This leads to the key result:

Theorem 4.6 *(Product case with $B_0 = 0$.) The zeta functions have the following decompositions:*

$$\Gamma(s)\zeta(\Delta_i, s) = \Gamma(s)[\zeta_+(\widetilde{\Delta}_i, s) + \tfrac{1}{4}(F_{\frac{1}{2}}(s) - 1)\zeta(A^2, s) + (-1)^i \tfrac{1}{4} F_{\frac{1}{2}}(s)\eta(A, 2s)]$$
$$+ \tfrac{1}{s}[\mathrm{Tr}_+(\Pi_0(\widetilde{\Delta}_i)) - \nu_0(\Delta_i) + (-1)^i \tfrac{1}{4}\nu_0(A)] + h_i(s), \quad (4.33)$$

where the h_i are entire. Moreover, $\Gamma(s)\zeta(\Delta_i, s)$ is $O(e^{(-\frac{\pi}{2}+\varepsilon)|\operatorname{Im} s|})$ for $|\operatorname{Im} s| \geq 1$, $-\infty < C_1 \leq \operatorname{Re} s \leq C_2 < \infty$, any $\varepsilon > 0$.

Here $\zeta_+(\widetilde{\Delta}_i, s) = \mathrm{Tr}_+ Z(\widetilde{\Delta}_i, s)$ (cf. (4.4)).

The basic idea in the proof goes as follows: By Lemma 4.4, the resolvent $(\Delta_i - \lambda)^{-1} = (\widetilde{\Delta}_i - \lambda)_+^{-1} + G_{i,\lambda}$ has the same asymptotic behavior for λ going to infinity as $(\widetilde{\Delta}_i - \lambda)_+^{-1} + \chi G_{i,\lambda}^0 \chi$, and the last term behaves like $G_{i,\lambda}^0$. Here the contribution from $\widetilde{\Delta}_i$ is well-known; and the contributions from $G_{e,\lambda}$ and $G_{o,\lambda}$ in $G_{i,\lambda}^0$ have been dealt with in Proposition 4.5; they give the terms involving $F_{\frac{1}{2}}(s)$. What remains is some adjustments due to the Laurent expansions of the resolvents at $\lambda = 0$ and the trace of $G_{i,\lambda}^0$ restricted to the nullspace of A, plus the contribution from an $O(|\lambda|^{-N})$ term; these adjustments yield the coefficient of $\frac{1}{s}$ in (4.33) and the entire function. The explanation is slightly technical because of the need to consider differentiated resolvents as in (2.11). We leave out further details; they are given in [15].

Example 4.7 For the Dirichlet realization Δ_{Dir}^0 of $D_n^2 + A^2$, a calculation as in (4.31) gives, by (4.24),

$$\mathrm{tr}_n\, G_{Z,\mathrm{Dir},s} = \mathrm{tr}_n \tfrac{i}{2\pi} \int_{\mathcal{C}} \lambda^{-s} G_{\mathrm{Dir},\lambda}^0 \, d\lambda = \tfrac{i}{2\pi} \int_{\mathcal{C}} \lambda^{-s} \tfrac{-1}{4(A^2-\lambda)} \, d\lambda = -\tfrac{1}{4} Z(A^2, s).$$

Then the zeta function for the Dirichlet realization Δ_{Dir} of P^*P has the decomposition (with $h(s)$ entire):

$$\Gamma(s)\zeta(\Delta_{\mathrm{Dir}}, s) = \Gamma(s)\zeta_+(\widetilde{\Delta}_1, s) - \tfrac{1}{4}\Gamma(s)\zeta(A^2, s) + h(s). \quad (4.34)$$

For the Neumann case one gets this formula with $-\frac{1}{4}$ replaced by $+\frac{1}{4}$.

A similar analysis applies to the eta functions associated with P_B, and to functions with differential operators inserted in front. Consider e.g. the eta function $\Gamma(s)\,\mathrm{Tr}(\varphi P \Delta_1^{-s})$, where φ is a bundle morphism from E_2 to E_1, equal to $\varphi^0 = \varphi|_{X'}$ on $X' \times [0, c]$. (Some morphism is needed in order to allow taking the trace in $L_2(E_1)$; e.g., σ^* can be used for φ.)

Theorem 4.8 *(Product case with $B_0 = 0$.) The eta function $\Gamma(s)\eta(\varphi, P_B, 2s-1)$ has the following decomposition:*

$$
\begin{aligned}
\Gamma(s)\eta(\varphi, P_B, 2s-1) &\equiv \Gamma(s)\operatorname{Tr}(\varphi P \Delta_1^{-s}) \\
&= \Gamma(s)\big[\operatorname{Tr}_+(\varphi P \widetilde{\Delta}_1^{-s}) + \tfrac{1}{4}(F_{\frac{1}{2}}(s-1)-1)\eta(\varphi^0\sigma, A, 2s-1)\big] \\
&\quad + \tfrac{1}{4\sqrt{\pi}}\operatorname{Tr}(\varphi^0\sigma\Pi_0(A))(s-\tfrac{1}{2})^{-1} + h_1(s), \quad (4.35)
\end{aligned}
$$

where $h_1(s)$ is entire. Moreover, $\Gamma(s)\eta(\varphi, P_B, 2s-1)$ is $O(e^{(-\frac{\pi}{2}+\varepsilon)|\operatorname{Im} s|})$ for $|\operatorname{Im} s| \geq 1$, $-\infty < C_1 \leq \operatorname{Re} s \leq C_2 < \infty$, any $\varepsilon > 0$.

There is a similar result for $\Gamma(s)\operatorname{Tr}(\varphi P^* \Delta_2^{-s})$, where φ is a morphism from E_1 to E_2.

4.3. PRECISE TRACE FORMULAS IN THE PRODUCT CASE

It is shown in Theorems 4.6 and 4.8 how the zeta and eta functions of the APS operator arise by simple addition of known zeta and eta functions with factors defined from $F_{\frac{1}{2}}(s)$ in front.

This makes it easy to determine the pole structure! We know the pole structure of the zeta and eta functions of the operators $\widetilde{P}$, $\widetilde{\Delta}_i$ and A, and we also know the pole structure of $F_{\frac{1}{2}}(s)$ from its gamma function components. The result is that we get from each decomposition a meromorphic function with poles where those functions have them; and the poles will be double when there are coincidences. Accordingly, there will be heat trace expansions with powers t^β corresponding to the simple poles $-\beta$, and powers t^β plus $t^\beta \log t$ terms corresponding to double poles $-\beta$.

We list the precise result below. An interesting aspect is that it shows a difference between the cases n even and n odd.

In the case n even, coincidences between poles give rise to double poles (hence log-terms in the heat operator formulation). At a double pole $-\beta$, the singular part consists both of a coefficient c times $(s+\beta)^{-2}$ and another coefficient c' times $(s+\beta)^{-1}$. The first coefficient c is determined from the symbols of the operators in a well-known local way, whereas the second coefficient c' is usually just globally determined.

In the case n odd, there are no coincidences, hence no double poles. But here the poles of $F_{\frac{1}{2}}$ force us to evaluate the zeta and eta functions at the points midways between their well-known poles; also this gives new global coefficients. (These are the points where the poles according to (3.24) have vanishing residue (3.25), so the value can also be regarded as the second coefficient where the first one is 0.)

Now comes the detailed description:

We denote the second coefficient in the Laurent series for $\Gamma(s)\zeta(D,Q,s)$ at a pole $s_j = \frac{-j+n_1+d}{2}$ by $c'_j(D,Q)$:

$$c'_j(D,Q) = \lim_{s\to s_j}[\Gamma(s)\zeta(D,Q,s) - \tfrac{c_j(D,Q)}{s-s_j}] = \operatorname{Res}_{s=s_j}\tfrac{\Gamma(s)\zeta(D,Q,s)}{s-s_j}; \tag{4.36}$$

here $\operatorname{Res}_{s=s'}$ means the residue at s'. (In case $c_j(D,Q)=0$, $c'_j(D,Q)$ is the *value* of $\Gamma(s)\zeta(D,Q,s)$ at the point.)

We also need to define some universal constants:

$$\begin{aligned}
\beta_m &= \operatorname{Res}_{s=-\frac12-m}\tfrac14 F_{\frac12}(s) = \tfrac{(-1)^m}{4m!\sqrt{\pi}\,\Gamma(\frac12-m)},\\
\beta'_m &= \operatorname{Res}_{s=-\frac12-m}\tfrac14 F_{\frac12}(s)(s+\tfrac12-m)^{-1},\\
\gamma_k &= \tfrac14(F_{\frac12}(\tfrac{k}{2})-1) = \tfrac14\Big[\tfrac{\Gamma(\frac{k+1}{2})}{\sqrt{\pi}\,\Gamma(1+\frac{k}{2})}-1\Big],\\
\varepsilon_m &= \operatorname{Res}_{s=-\frac12-m}\tfrac14 F_{\frac12}(s)\Gamma(s) = \tfrac{(-1)^{m+1}}{4m!\sqrt{\pi}\,(m+\frac12)},\\
\delta_m &= \operatorname{Res}_{s=\frac12-m}\tfrac14 F_{\frac12}(s-1)\Gamma(s) = \operatorname{Res}_{s=\frac12-m}\tfrac{1}{4\sqrt{\pi}}\Gamma(s-\tfrac12) = \tfrac{(-1)^m}{4\sqrt{\pi}\,m!};
\end{aligned} \tag{4.37}$$

here $m\in\mathbb{N}$, and the k are integers avoiding negative odd numbers. (The explicit expressions are found by use of the formula $\Gamma(s)=\pi\Gamma(1-s)/\sin\pi s$. Also β'_m can be written more explicitly, departing from the fact that $-\Gamma'(1)$ equals Euler's constant.)

From (3.24) we find, omitting vanishing coefficients,

$$\Gamma(s)\zeta_+(\widetilde{\Delta}_1,s) \sim \sum_{k=0}^{\infty}\frac{c_{2k,+}(\widetilde{\Delta}_1)}{s+k-\frac{n}{2}} - \frac{\operatorname{Tr}_+(\Pi_0(\widetilde{\Delta}_1))}{s}, \tag{4.38}$$

where $c_{j,+}(\widetilde{\Delta}_1) = \int_X \operatorname{tr} c_j(x,\widetilde{\Delta}_1)\,dx$; cf. also (4.4). Since A acts on X' of dimension $n-1$, we get from (3.26):

$$\Gamma(s)\zeta(A^2,s) \sim \sum_{k=0}^{\infty}\frac{c_{2k}(A^2)}{s+k-\frac{n-1}{2}} - \frac{\nu_0(A)}{s}, \tag{4.39}$$

and, for example, when ψ is a morphism in E_1,

$$\begin{aligned}
\Gamma(s)F_{\frac12}(s)\eta(\psi,A,2s) &= \frac{1}{\sqrt{\pi}\,s}\Gamma(s+\tfrac12)\zeta(\psi A,A^2,s+\tfrac12)\\
&\sim \sum_{0\le k}\frac{c_{2k+1}(\psi A,A^2)}{\sqrt{\pi}\,(\frac{n}{2}-k-1)(s+k+1-\frac{n}{2})} + \frac{\eta(\psi,A,0)}{s} \quad \text{if } n \text{ is odd},\\
&\sim \sum_{0\le k\ne\frac{n}{2}-1}\frac{c_{2k+1}(\psi A,A^2)}{\sqrt{\pi}\,(\frac{n}{2}-k-1)(s+k+1-\frac{n}{2})}\\
&\qquad + \frac{c_{n-1}(\psi A,A^2)}{\sqrt{\pi}\,s^2} + \frac{c'_{n-1}(\psi A,A^2)}{\sqrt{\pi}\,s} \quad \text{if } n \text{ is even},
\end{aligned} \tag{4.40}$$

where $c'_{n-1}(\psi A, A^2)$ is defined as in (4.36). When $\psi = I$ then $c_{n-1}(A, A^2) = 0$ and $c'_{n-1}(A, A^2) = \sqrt{\pi}\,\eta(A, 0)$.

Insertion of these expansions in our decompositions gives:

Corollary 4.9 *The zeta function $\Gamma(s)\zeta(\Delta_i, s)$ is meromorphic on* **C**, *with the following singularity structure:*

For n even:

$$\Gamma(s)\zeta(\Delta_i, s) \sim \sum_{k\geq 0} \frac{c_{2k,+}(\tilde{\Delta}_i)}{s+k-\frac{n}{2}} - \frac{\mathrm{Tr}_+ \Pi_0(\tilde{\Delta}_i)}{s} + \sum_{0\leq k<\frac{n}{2}} \frac{\gamma_{n-1-2k} c_{2k}(A^2)}{s+k-\frac{n-1}{2}}$$

$$+ \sum_{k\geq\frac{n}{2}} \Big[\frac{\beta_{k-\frac{n}{2}} c_{2k}(A^2)}{(s+k-\frac{n-1}{2})^2} + \frac{\beta_{k-\frac{n}{2}} c'_{2k}(A^2) + (\beta'_{k-\frac{n}{2}} - \frac{1}{4})c_{2k}(A^2)}{s+k-\frac{n-1}{2}}\Big]$$

$$+ (-1)^i \tfrac{1}{4}\Big[\sum_{0\leq k\neq\frac{n}{2}-1} \frac{c_{2k+1}(A, A^2)}{\sqrt{\pi}\,(\frac{n}{2}-k-1)(s+k+1-\frac{n}{2})} + \frac{\eta(A,0)+\nu_0(A)}{s}\Big]. \tag{4.41}$$

For n odd:

$$\Gamma(s)\zeta(\Delta_i, s) \sim \sum_{k\geq 0} \frac{c_{2k,+}(\tilde{\Delta}_i)}{s+k-\frac{n}{2}} - \frac{\mathrm{Tr}_+ \Pi_0(\tilde{\Delta}_i)}{s} + \sum_{k\geq 0} \frac{\gamma_{n-1-2k} c_{2k}(A^2)}{s+k-\frac{n-1}{2}}$$

$$+ \sum_{m\geq 0} \frac{\varepsilon_m \zeta(A^2, -m-\frac{1}{2})}{s+m+\frac{1}{2}}$$

$$+ (-1)^i \tfrac{1}{4}\Big[\sum_{k\geq 0} \frac{c_{2k+1}(A, A^2)}{\sqrt{\pi}\,(\frac{n}{2}-k-1)(s+k+1-\frac{n}{2})} + \frac{\eta(A,0)+\nu_0(A)}{s}\Big]. \tag{4.42}$$

[15] moreover shows the formulas where a morphism is included. The terms β'_m were missing in the Preprint version of [15].

Corollary 4.10 *The eta function $\Gamma(s)\eta(\varphi, P_B, 2s-1) = \Gamma(s)\,\mathrm{Tr}(\varphi P \Delta_1^{-s})$ is meromorphic on* **C**, *with the following singularity structure:*

For n even:

$$\Gamma(s)\,\mathrm{Tr}(\varphi P \Delta_1^{-s}) \sim \sum_{k\geq 0} \frac{c_{2k+1,+}(\varphi P, \tilde{\Delta}_1)}{s+k-\frac{n}{2}}$$

$$+ \sum_{0\leq k<\frac{n}{2}-1} \frac{\gamma_{n-3-2k} c_{2k+1}(\varphi^0 \sigma A, A^2)}{s+k-\frac{n-1}{2}} + \sum_{k\geq\frac{n}{2}-1} \Big[\frac{\beta_{k+1-\frac{n}{2}} c_{2k+1}(\varphi^0 \sigma A, A^2)}{(s+k-\frac{n-1}{2})^2}$$

$$+ \frac{\beta_{k+1-\frac{n}{2}} c'_{2k+1}(\varphi^0 \sigma A, A^2) + (\beta'_{k+1-\frac{n}{2}} - \frac{1}{4}) c_{2k+1}(\varphi^0 \sigma A, A^2)}{s+k-\frac{n-1}{2}}\Big]$$

$$+ \frac{\mathrm{Tr}(\varphi^0 \sigma \Pi_0(A))}{4\sqrt{\pi}(s-\frac{1}{2})}. \tag{4.43}$$

For n odd:

$$\Gamma(s)\,\mathrm{Tr}(\varphi P\Delta_1^{-s}) \sim \sum_{k\geq 0} \frac{c_{2k+1,+}(\varphi P, \widetilde{\Delta}_1)}{s+k-\frac{n}{2}} + \sum_{k\geq 0} \frac{\gamma_{n-3-2k} c_{2k+1}(\varphi^0\sigma A, A^2)}{s+k-\frac{n-1}{2}}$$
$$+ \sum_{m\geq 0} \frac{\delta_m \eta(\varphi^0\sigma A, -2m)}{s+m-\frac{1}{2}} + \frac{\mathrm{Tr}(\varphi^0\sigma\Pi_0(A))}{4\sqrt{\pi}(s-\frac{1}{2})}. \quad (4.44)$$

There are similar formulas for $\mathrm{Tr}(\varphi P^*\Delta_2^{-s})$, *with* $\varphi^0\sigma$ *replaced by* $\sigma^*\varphi^0$.

Let us finally list the consequences for heat traces, derived from Corollary 4.9–4.10 by use of Theorem 2.3:

Corollary 4.11 *The exponential trace* $\mathrm{Tr}(e^{-t\Delta_i})$ *has the following behavior for* $t \to 0$.

For n even:

$$\mathrm{Tr}(e^{-t\Delta_i}) \sim \sum_{k\geq 0} c_{2k,+}(\widetilde{\Delta}_i)\, t^{k-\frac{n}{2}} + \sum_{0\leq k<\frac{n}{2}} \gamma_{n-1-2k} c_{2k}(A^2)\, t^{k-\frac{n-1}{2}}$$
$$+ \sum_{k\geq\frac{n}{2}} [-\beta_{k-\frac{n}{2}} c_{2k}(A^2)\log t + \beta_{k-\frac{n}{2}} c'_{2k}(A^2) + (\beta'_{k-\frac{n}{2}} - \tfrac{1}{4})c_{2k}(A^2)]\, t^{k-\frac{n-1}{2}}$$
$$+ (-1)^i \tfrac{1}{4}[\sum_{0\leq k\neq\frac{n}{2}-1} \frac{c_{2k+1}(A, A^2)}{\sqrt{\pi}\,(\frac{n}{2}-k-1)}\, t^{k+1-\frac{n}{2}} + \eta(A,0) + \nu_0(A)]. \quad (4.45)$$

For n odd:

$$\mathrm{Tr}(e^{-t\Delta_i}) \sim \sum_{k\geq 0} c_{2k,+}(\widetilde{\Delta}_i)\, t^{k-\frac{n}{2}}$$
$$+ \sum_{k\geq 0} \gamma_{n-1-2k} c_{2k}(A^2)\, t^{k-\frac{n-1}{2}} + \sum_{m\geq 0} \varepsilon_m \zeta(A^2, -m-\tfrac{1}{2})\, t^{m+\frac{1}{2}}$$
$$+ (-1)^i \tfrac{1}{4}[\sum_{k\geq 0} \frac{c_{2k+1}(A, A^2)}{\sqrt{\pi}\,(\frac{n}{2}-k-1)}\, t^{k+1-\frac{n}{2}} + \eta(A,0) + \nu_0(A)]. \quad (4.46)$$

Corollary 4.12 *The associated exponential trace* $\mathrm{Tr}(\varphi P e^{-t\Delta_1})$ *has the following behavior for* $t \to 0$.

For n even:

$$\mathrm{Tr}(\varphi P e^{-t\Delta_1}) \sim \sum_{k\geq 0} c_{2k+1,+}(\varphi P, \widetilde{\Delta}_1)\, t^{k-\frac{n}{2}}$$
$$+ \sum_{0\leq k<\frac{n}{2}-1} \gamma_{n-3-2k} c_{2k+1}(\varphi^0\sigma A, A^2)\, t^{k-\frac{n-1}{2}} +$$

$$+\sum_{k\geq\frac{n}{2}-1}\Big[-\beta_{k+1-\frac{n}{2}}c_{2k+1}(\varphi^0\sigma A,A^2)\,t^{k-\frac{n-1}{2}}\log t$$

$$+\big(\beta_{k+1-\frac{n}{2}}c'_{2k+1}(\varphi^0\sigma A,A^2)+(\beta'_{k+1-\frac{n}{2}}-\tfrac14)c_{2k+1}(\varphi^0\sigma A,A^2)\big)\,t^{k-\frac{n-1}{2}}\Big]$$

$$+\frac{1}{4\sqrt{\pi}}\operatorname{Tr}(\varphi^0\sigma\Pi_0(A))\,t^{-\frac12}. \quad (4.47)$$

For n odd:

$$\operatorname{Tr}(\varphi Pe^{-t\Delta_1})\sim\sum_{k\geq0}c_{2k+1,+}(\varphi P,\widetilde{\Delta}_1)\,t^{k-\frac{n}{2}}$$

$$+\sum_{k\geq0}\gamma_{n-3-2k}c_{2k+1}(\varphi^0\sigma A,A^2)\,t^{k-\frac{n-1}{2}}$$

$$+\sum_{m\geq0}\delta_m\eta(\varphi^0\sigma A,-2m)\,t^{m-\frac12}+\frac{1}{4\sqrt{\pi}}\operatorname{Tr}(\varphi^0\sigma\Pi_0(A))\,t^{-\frac12}. \quad (4.48)$$

There are similar formulas for $\operatorname{Tr}(\varphi P^*e^{-t\Delta_2})$*, with* $\varphi^0\sigma$ *replaced by* $\sigma^*\varphi^0$*.*

The proof shows the advantage of working with the power functions, where the contributions from the boundary condition appear as simple multiplicative formulas involving the zeta and eta functions of A; this allows an exact analysis of the pole coefficients which can then be carried over to the heat expansions by Theorem 2.3. If working directly in the heat operator framework (a point of view taken up in [6]), one has to deal with convolution-type integrals.

Gilkey and Grubb [11] show that all terms, in particular the logarithmic ones, are nontrivial in general. Dowker, Apps, Kirsten and Bordag [7] find no logarithms for the Dirac operator on the ball; this is due to special symmetries and does not contradict the above since it is not a product case.

Example 4.13 For the Dirichlet problem considered in Examples 4.2, 4.3 and 4.7, formula (4.34) implies in a similar way:

$$\operatorname{Tr}(e^{-t\Delta_{\mathrm{Dir}}})\sim\sum_{k\geq0}c_{2k,+}(\widetilde{\Delta}_1)\,t^{k-\frac{n}{2}}-\frac14\sum_{k\geq0}c_{2k}(A^2)\,t^{k-\frac{n-1}{2}}; \quad (4.49)$$

note that all the integer and half-integer powers enter here too. There is a similar formula for the Neumann problem, with $-\frac14$ replaced by $+\frac14$.

Remark 4.14 In a recent study of the gluing problem for the eta-invariant, [6], Brüning and Lesch treat boundary conditions of a somewhat different nature than those considered here and in [14], [15]; moreover they depend on a parameter and the variation in this parameter is studied. We show below how those new boundary conditions can be handled in the present framework: Restrict the attention to selfadjoint operators P satisfying $\sigma^* = -\sigma$, $\sigma A = -A\sigma$ as in Remark 1.3. Let B be an orthogonal projection in $L_2(E_1')$ commuting with A^2 and satisfying

$$\begin{aligned} &\text{(i) } \sigma B = (I-B)\sigma, \\ &\text{(ii) } BAB = \alpha|A|B \text{ for some } \alpha > -1. \end{aligned} \tag{4.50}$$

([6] gives special examples of the form $B = \sigma_1\Pi_> + \sigma_2\Pi_< + B_0$ with morphisms or zero order ψdo's σ_1 and σ_2.) Because of (i), P_B is selfadjoint, and $\Delta_B = {P_B}^2$ is the realization of P^2 under the boundary condition (where $B\gamma_0$ is written $\gamma_0 B$)

$$\gamma_0 Bu = 0, \quad \gamma_0 B\sigma(\partial_n + A)u = 0. \tag{4.51}$$

For the second equation we note that when $\gamma_0 Bu = 0$, then in view of (i), $\gamma_0 B\sigma(\partial_n + A)u = \sigma\gamma_0(I-B)(\partial_n + A)(I-B)u = \sigma\gamma_0(\partial_n + (I-B)A)(I-B)u$. Here, by (i) and (ii), $(I-B)A(I-B) = -\alpha|A|(I-B)$. Thus the boundary condition may be written:

$$\gamma_0 Bu = 0, \quad \gamma_0(\partial_n - \alpha|A|)(I-B)u = 0. \tag{4.52}$$

This is a Dirichlet condition for the functions of x_n valued in $R(B)$, and a Robin-type condition as in Example 4.2 with $S = -\alpha|A|$ for the functions valued in $R(I-B)$. Then by the calculations in Example 4.2, the resolvent on X^0 is $(\Delta_B^0 - \lambda)^{-1} = Q_{1,\lambda,+}^0 + G_{B,\lambda}^0$ with

$$\begin{aligned} G_{B,\lambda}^0 &= \big(\tfrac{-1}{2A_\lambda}B + \tfrac{1}{2A_\lambda}\tfrac{A_\lambda - \alpha|A|}{A_\lambda + \alpha|A|}(I-B)\big)G_\lambda \\ &= \big(\tfrac{-1}{2A_\lambda} + \tfrac{A_\lambda - \alpha|A|}{2((1-\alpha^2)A^2-\lambda)} + \tfrac{1}{2(A_\lambda + \alpha|A|)}(I-2B)\big)G_\lambda. \end{aligned} \tag{4.53}$$

Now Lemma 4.4 can be extended to this case. Therefore we have as in the proof of Theorem 4.6,

$$\Gamma(s)\zeta(\Delta_B, s) = \Gamma(s)\zeta_+(\tilde{\Delta}_1, s) + \Gamma(s)\operatorname{Tr}_X \tfrac{i}{2\pi}\int_C \lambda^{-s} G_{B,\lambda}^0\, d\lambda + h(s), \tag{4.54}$$

with $h(s)$ entire; and here

$$\begin{aligned} &\operatorname{Tr}_X \tfrac{i}{2\pi}\int_C \lambda^{-s} G_{B,\lambda}^0\, d\lambda = \operatorname{Tr}_{X'} \operatorname{tr}_n \tfrac{i}{2\pi}\int_C \lambda^{-s} G_{B,\lambda}^0\, d\lambda = \\ &\operatorname{Tr}_{X'} \tfrac{i}{2\pi}\int_C \lambda^{-s}\big(\tfrac{-1}{2A_\lambda} + \tfrac{A_\lambda - \alpha|A|}{2((1-\alpha^2)A^2-\lambda)} + \tfrac{1}{2(A_\lambda + \alpha|A|)}(I-2B)\big)\tfrac{1}{2A_\lambda}\, d\lambda. \end{aligned} \tag{4.55}$$

The term $\frac{1}{2(A_\lambda+\alpha|A|)}(I-2B)\frac{1}{2A_\lambda}$ contributes with zero, for by (i) and the fact that σ and B commute with A^2,

$$\begin{aligned}\tfrac{1}{2(A_\lambda+\alpha|A|)}(I-2B)\tfrac{1}{2A_\lambda} &= \tfrac{1}{4A_\lambda(A_\lambda+\alpha|A|)}(I-B) - \tfrac{1}{4A_\lambda(A_\lambda+\alpha|A|)}\sigma^*\sigma B\\ &= \tfrac{1}{4A_\lambda(A_\lambda+\alpha|A|)}(I-B) - \sigma^*\tfrac{1}{4A_\lambda(A_\lambda+\alpha|A|)}(I-B)\sigma; \quad (4.56)\end{aligned}$$

here since the trace is invariant under circular perturbations (that we can use in a reformulation with sufficiently high λ-derivatives as in (2.10)), the contributions from these two terms will cancel each other. The remaining terms are treated as in Proposition 4.5 (we give the details for $\alpha < 1$; the case $\alpha > 1$ is similar and the case $\alpha = 1$ is simpler):

$$\begin{aligned}&\tfrac{i}{2\pi}\int_C \lambda^{-s}\big(\tfrac{-1}{4A_\lambda^2} + \tfrac{A_\lambda-\alpha|A|}{4A_\lambda((1-\alpha^2)A^2-\lambda)}\big)\,d\lambda\\ &= \sum_{\mu\in\mathrm{sp}(A)} \tfrac{1}{4}\tfrac{i}{2\pi}\textstyle\int_C \lambda^{-s}\big(\tfrac{-1}{\mu^2-\lambda} + \tfrac{1}{(1-\alpha^2)\mu^2-\lambda} - \tfrac{\alpha|\mu|}{(\mu^2-\lambda)^{\frac12}((1-\alpha^2)\mu^2-\lambda)}\big)\,d\lambda\cdot\Pi_\mu\\ &= \sum_\mu \tfrac14|\mu|^{-2s}\tfrac{i}{2\pi}\textstyle\int_C \tau^{-s}\big(\tfrac{-1+(1-\alpha^2)^{-s}}{1-\tau} - \tfrac{\alpha}{(1-\tau)^{\frac12}(1-\alpha^2-\tau)}\big)\,d\tau\cdot\Pi_\mu \qquad (4.57)\\ &= \tfrac14(-1+e^{-s\log(1-\alpha^2)} + \widetilde{F}_\alpha(s))Z(A^2,s);\end{aligned}$$

with

$$\widetilde{F}_\alpha(s) = \tfrac{i}{2\pi}\int_C \tau^{-s}\frac{-\alpha}{(1-\tau)^{\frac12}(1-\alpha^2-\tau)}\,d\tau. \qquad (4.58)$$

This is a hypergeometric function whose pole structure is easily determined by use of Theorem 2.1. In fact, $\widetilde{F}_\alpha(s)$ is of the form (2.17) with $f(\tau) = -\alpha(1-\tau)^{-\frac12}(1-\alpha^2-\tau)^{-1}$. It is holomorphic on $\mathbf{C}\setminus[1,\infty[$ and has the asymptotic expansion for $-\tau\to\infty$ in closed subsectors:

$$\begin{aligned}f(-\tau) &= -\alpha\tau^{-\frac32}(1+\tfrac{1}{\tau})^{-\frac12}(1+\tfrac{1-\alpha^2}{\tau})^{-1}\\ &\sim -\alpha\tau^{-\frac32}\sum_{k\in\mathbf{N}}\tbinom{-\frac12}{k}\tau^{-k}\sum_{l\in\mathbf{N}}(\alpha^2-1)^l\tau^{-l} = \sum_{j\in\mathbf{N}}\omega_j\tau^{-\frac32-j}. \quad (4.59)\end{aligned}$$

An application of Theorem 2.1 carries the terms $\omega_j t^{-\frac32-j}$ over into simple poles at $s=-j-\frac12$ for $\frac{\pi}{\sin\pi s}\widetilde{F}_\alpha(s)$ with residues ω_j. The poles at integers $j+1$ stemming from the Taylor expansion at 0 are removed when we multiply by $\pi^{-1}\sin\pi s$. Consequently, $\widetilde{F}_\alpha(s)$ is meromorphic on $\mathbf{C}$ with simple poles at the points $-j-\frac12$, $j\in\mathbf{N}$, with residues $\pi^{-1}(-1)^{j+1}\omega_j$.

Finally,

$$\begin{aligned}\Gamma(s)\zeta(\Delta_B,s) &= \Gamma(s)\zeta_+(\widetilde{\Delta}_1,s)\\ &\quad + \tfrac14(-1+e^{-s\log(1-\alpha^2)} + \widetilde{F}_\alpha(s))\Gamma(s)\zeta(A^2,s) + h(s), \quad (4.60)\end{aligned}$$

which is meromorphic on $\mathbf{C}$ with poles at the points $(n-k)/2$, $k \in \mathbf{N}$; here the poles at the negative half-integers $-j-\frac{1}{2}$ are in general double when n is even; otherwise the poles are simple. A heat trace expansion in terms of $t^{(k-n)/2}$ and $t^{l+\frac{1}{2}}\log t$ $(k,l \in \mathbf{N})$ follows as usual by Theorem 2.3.

Note that (4.58) also implies: 1) $\widetilde{F}_\alpha(s)$ equals $\pi^{-1}\sin\pi s$ times the Mellin transform of $-\alpha(1+\tau)^{-\frac{1}{2}}(1-\alpha^2+\tau)^{-1}$ at $s-1$; cf. (2.18), (2.33).

2) $(1-\alpha^2)\widetilde{F}_\alpha(s) - \widetilde{F}_\alpha(s-1) = -\alpha F_{\frac{1}{2}}(s-1)$; cf. (4.28).

5. The general case

5.1. A GENERAL RESOLVENT CONSTRUCTION

In the non-product case the results will be more qualitative. A useful trick here is to replace the separate consideration of P_B and $P_B{}^*$ by the study of the skew-selfadjoint operator

$$\mathcal{P}_{\mathcal{B}} = \begin{pmatrix} 0 & -P_B{}^* \\ P_B & 0 \end{pmatrix}; \tag{5.1}$$

this is the realization of

$$\mathcal{P} = \begin{pmatrix} 0 & -P^* \\ P & 0 \end{pmatrix} \tag{5.2}$$

under the following boundary condition on $u = \{u_1, u_2\}$ (cf. (1.5)):

$$\mathcal{B}\gamma_0 u = 0, \text{ where } \mathcal{B} = \begin{pmatrix} B & B' \end{pmatrix} : \begin{matrix} L_2(E_1') \\ \times \\ L_2(E_2') \end{matrix} \to L_2(E_1'). \tag{5.3}$$

The advantage of taking P_B and $P_B{}^*$ together in this way is that $\mathcal{P}_{\mathcal{B}}$ is two-sided elliptic, and $\mathcal{R}_\mu = (\mathcal{P}_{\mathcal{B}} + \mu)^{-1}$, defined for $\mu \in \pm\Gamma_0$, $\Gamma_0 = \{\mu \in \mathbf{C}\setminus\{0\} \mid |\arg\mu| < \pi/2\}$, satisfies

$$\mathcal{R}_\mu = (\mathcal{P}_{\mathcal{B}} + \mu)^{-1} = \begin{pmatrix} \mu(\Delta_1+\mu^2)^{-1} & P_B{}^*(\Delta_2+\mu^2)^{-1} \\ -P_B(\Delta_1+\mu^2)^{-1} & \mu(\Delta_2+\mu^2)^{-1} \end{pmatrix}, \tag{5.4}$$

where $(\Delta_i+\mu^2)^{-1} = R_{i,-\mu^2}$ are the resolvents we are looking for (cf. (1.7)). The diagonal terms give back the individual resolvents, and the off-diagonal terms can be used to describe eta functions instead of zeta functions.

This allows us to stay working with first-order systems (instead of passing to second order), at the cost of doubling up the size of the matrix.

We shall denote $E_1 \oplus E_2 = E$ and $E_1' \oplus E_2' = E'$.

We let $\widetilde{\mathcal{P}} = \begin{pmatrix} 0 & -\widetilde{P}^* \\ \widetilde{P} & 0 \end{pmatrix}$, where $\widetilde{P}$ is an elliptic extension of P to a bundle $\widetilde{E} = \widetilde{E}_1 \oplus \widetilde{E}_2$ over $\widetilde{X} = X \cup (X' \times]-1, 0[)$. Then $\widetilde{\mathcal{P}} + \mu$ has a parametrix Q_μ (of strongly polyhomogeneous type) for $\mu \in \pm\Gamma_0$, and as shown in detail in [14], p. 508–9, it can be modified such that for large μ in closed subsectors of $\pm\Gamma_0$,

$$(\mathcal{P} + \mu) Q_{\mu,+} = I \text{ on } X. \tag{5.5}$$

Also here, a comparison with the cylinder case (cf. (4.8)) plays a role. We denote $\begin{pmatrix} 0 & -P^{0\prime} \\ P^0 & 0 \end{pmatrix} = \mathcal{P}^0$, acting in $E^0 = E_1^0 \oplus E_2^0$. We extend $\mathcal{P}^0$ to $\widetilde{X}^0$ simply by extending the formulas (4.8) to $x_n \in \mathbf{R}$, letting $\widetilde{E}^0 = \widetilde{E}_1^0 \oplus \widetilde{E}_2^0$ be the lifting of $E' = E_1' \oplus E_2'$. Then the extended operator $\widetilde{\mathcal{P}}^0$ is skew-selfadjoint, and the resolvent is

$$Q_\mu^0 = (\widetilde{\mathcal{P}}^0 + \mu)^{-1} =$$
$$\begin{pmatrix} \mu(D_n^2 + A^2 + \mu^2)^{-1} & (-\partial_n + A)(D_n^2 + A^2 + \mu^2)^{-1}\sigma^* \\ -\sigma(\partial_n + A)(D_n^2 + A^2 + \mu^2)^{-1} & \mu\sigma(D_n^2 + A^2 + \mu^2)^{-1}\sigma^* \end{pmatrix}. \tag{5.6}$$

In particular,

$$(\mathcal{P}^0 + \mu) Q_{\mu,+}^0 = I \text{ on } X^0. \tag{5.7}$$

Along with $\mathcal{P}_\mathcal{B}$, we study the realization $\mathcal{P}_\mathcal{B}^0$, acting like $\mathcal{P}^0$ on X^0 and with the same boundary condition (5.3) as $\mathcal{P}$. With a slight abuse of notation, we now denote

$$A_\mu = (A^2 + \mu^2)^{\frac{1}{2}}, \text{ for } \mu \in \pm\Gamma_0. \tag{5.8}$$

Lemma 5.1 *Define the ψdo from sections of E_1' to sections of $E_1' \oplus E_1'$:*

$$\mathcal{S}_{\mathcal{B},\mu} = \begin{pmatrix} B + \mu^{-1}(A_\mu + A)B^\perp \\ \mu^{-1}(A_\mu - A)B + B^\perp \end{pmatrix}; \tag{5.9}$$

and the Poisson operator from sections of E_1' to sections of E^0:

$$K_{\mathcal{B},\mu}^0 = \begin{pmatrix} 1 & 0 \\ 0 & \sigma \end{pmatrix} e^{-x_n A_\mu} \mathcal{S}_{\mathcal{B},\mu}. \tag{5.10}$$

Then $K_{\mathcal{B},\mu}^0$ satisfies

$$\mathcal{B}\gamma_0 K_{\mathcal{B},\mu}^0 = I \text{ on } X', \quad (\mathcal{P}^0 + \mu) K_{\mathcal{B},\mu}^0 = 0 \text{ on } X^0. \tag{5.11}$$

The proof is a direct verification, using that B commutes with A.
In other words, $K^0_{\mathcal{B},\mu} : \psi \mapsto u$ solves the problem

$$\begin{aligned}(\mathcal{P}^0 + \mu)u &= g \text{ on } X^0,\\ \mathcal{B}\gamma_0 u &= \psi \text{ on } X',\end{aligned} \tag{5.12}$$

when $g = 0$. We note that by (5.7), the full solution operator for (5.12) is

$$\begin{pmatrix}\mathcal{R}^0_\mu & K^0_{\mathcal{B},\mu}\end{pmatrix}, \text{ where } \mathcal{R}^0_\mu = (\mathcal{P}^0_{\mathcal{B}} + \mu)^{-1} = Q^0_{\mu,+} - K^0_{\mathcal{B},\mu}\mathcal{B}\gamma_0 Q^0_{\mu,+}; \tag{5.13}$$

cf. also Remark 4.1.

Now $\mathcal{R}^0_\mu$ is principally like the true resolvent $\mathcal{R}_\mu$ at X'. However, we prefer to use a better adapted approximate resolvent, namely

$$\mathcal{R}'_\mu = Q_{\mu,+} - G_1 \text{ with } G_1 = \chi K^0_{\mathcal{B},\mu}\mathcal{B}\gamma_0 Q_{\mu,+}, \tag{5.14}$$

where $Q_{\mu,+}$ satisfies (5.5) and χ is a cut-off function as in Lemma 4.4. By (5.11), $\mathcal{R}'_\mu$ maps into the domain of $\mathcal{P}_{\mathcal{B}}$, and by (5.5), we have for large enough μ,

$$\begin{aligned}(\mathcal{P} + \mu)\mathcal{R}'_\mu &= (\mathcal{P} + \mu)Q_{\mu,+} - (\mathcal{P} + \mu)\chi K^0_{\mathcal{B},\mu}\mathcal{B}\gamma_0 Q_{\mu,+}\\ &= I - ([\mathcal{P},\chi] + \chi(\mathcal{P} - \mathcal{P}^0))K^0_{\mathcal{B},\mu}\mathcal{B}\gamma_0 Q_{\mu,+}\\ &= I - G_2, \quad \text{with} \quad G_2 = (x_n\mathcal{P}_1 + \mathcal{P}_0)K^0_{\mathcal{B},\mu}\mathcal{B}\gamma_0 Q_{\mu,+},\end{aligned} \tag{5.15}$$

the $\mathcal{P}_j$ denoting differential operators of order j with smooth coefficients vanishing for $x_n > \frac{2}{3}c$. (G_1 and G_2 are μ-dependent, and so are many other auxiliary operators in the following, where we do not indicate the μ-dependence explicitly.)

The exact inverse $\mathcal{R}_\mu$ of $\mathcal{P}_{\mathcal{B}} + \mu$ can then be described by

$$\mathcal{R}_\mu = \mathcal{R}'_\mu(I - G_2)^{-1} = (Q_{\mu,+} - G_1)(I - G_2)^{-1}, \tag{5.16}$$

whenever $I - G_2$ is invertible. The main point is now to show that this holds for large μ and leads to a constructive expression for $\mathcal{R}_\mu$.

For this purpose, we analyze the various factors in (5.14) and (5.15). Let us denote

$$\begin{aligned}K_0 &= e^{-x_n A_\mu}, \quad T_0 = \gamma_0 Q_{\mu,+}, \quad S_0 = S_{\mathcal{B},\mu}\mathcal{B},\\ K_1 &= \chi\left(\begin{smallmatrix}1 & 0\\ 0 & \sigma\end{smallmatrix}\right)K_0, \quad K_2 = (x_n\mathcal{P}_1 + \mathcal{P}_0)\left(\begin{smallmatrix}1 & 0\\ 0 & \sigma\end{smallmatrix}\right)K_0,\end{aligned} \tag{5.17}$$

here K_0 goes from $C^\infty(E'_1)$ to $C^\infty(E^0_1)$, K_1 and K_2 go from $C^\infty(E'_1 \oplus E'_1)$ to $C^\infty(E)$, T_0 goes from $C^\infty(E)$ to $C^\infty(E')$, and S_0 goes from from $C^\infty(E')$ to

$C^\infty(E'_1 \oplus E'_1)$. (They also define mappings beween suitable Sobolev spaces.) Then

$$G_1 = K_1 S_0 T_0, \quad G_2 = K_2 S_0 T_0. \tag{5.18}$$

In the terminology of Boutet de Monvel [4] and Grubb [12], the K_j are parameter-dependent Poisson operators and T_0 is a parameter-dependent trace operator of class 0 (trace operators of class 0 are well-defined on L_2), but their usage entered elliptic theory much earlier, cf. Seeley [21], Hörmander [17]. For the considerations of these operators, we do *not* need to introduce new and complicated symbol classes and composition rules for boundary operators, for in fact they are of the *strongly polyhomogeneous type*: When the parameter μ runs on a ray $\{\mu = \varrho e^{i\theta_0} \mid \varrho \geq 0\}$, ϱ enters like another cotangent variable on a par with $\xi_1, \ldots, \xi_{n-1}$, in the sense that the standard estimates described in [4] are satisfied with $\{\xi_1, \ldots, \xi_{n-1}, \varrho\}$ as the boundary cotangent variable. This is similar to the situation described in Theorem 3.7, now for boundary operators.

Let us refrain from further details (that presuppose a lengthy introdution to the calculi described in [4], [12], summarized in the appendix of [14]), but just mention a consequence we need:

Lemma 5.2 *With K_1, K_2 and T_0 defined above, and φ a morphism in E, the compositions $T_0\varphi K_j$ are strongly polyhomogeneous ψdo's on X' of order -1. Moreover, the compositions $T_0\varphi Q_{\mu,+}K_j$ are strongly polyhomogeneous ψdo's on X' of order -2.*

An important trick in the following is to reduce considerations of the singular Green operators G_j to considerations of ψdo's in the boundary. This is done on several levels; one is in the study of inverses that uses Lemma 5.3 below, another is in the study of traces in Section 5.2, where a cyclic permutation brings operators of the form TK into the picture.

First consider the problem of inversion of $I - G_2$. Here we shall use the elementary lemma:

Lemma 5.3 *Let $K : V \to W$ and $T : W \to V$ be linear mappings between vector spaces. Then $I - KT : W \to W$ is bijective if and only if $I - TK : V \to V$ is bijective, and*

$$(I - KT)^{-1} = I + K(I - TK)^{-1}T. \tag{5.19}$$

Proof: A straightforward verification. □

The lemma will be applied with $K = K_2$ (going from sections of $E'_1 \oplus E'_1$ to sections of E) and $T = S_0 T_0$ (going the other way). This replaces the construction of the inverse of $I - KT = I - G_2$ by the construction of the inverse of $I - TK = I - S_0T_0K_2$; so that

$$(I - G_2)^{-1} = I + K_2(I - S_1)^{-1}S_0T_0 \quad \text{with} \quad S_1 = S_0T_0K_2 \tag{5.20}$$

holds when $I - S_1$ is invertible. The advantage of this reduction is that S_1 is a ψdo on the boundaryless manifold X'. The factor T_0K_2 is a strongly polyhomogeneous ψdo of order -1 by Lemma 5.2, and it remains to examine the other factor in S_1 and the composition, and to apply this to construct the inverse $(I - S_1)^{-1}$.

Here we go more in details with the symbol classes introduced in Section 3.1. The following class will play a special role:

Definition 5.4 Let r be integer ≥ 0, and let $S = \mathrm{OP}(s(x,\xi,\mu))$ (or let S have the symbol s in local coordinates). S and its symbol will be called **special parameter-dependent of order** $-r$, when

$$\begin{aligned} s(x,\xi,\mu) &\in S^{-r,0}(\mathbf{R}^n,\mathbf{R}^n,\Gamma) \cap S^{0,-r}(\mathbf{R}^n,\mathbf{R}^n,\Gamma) \text{ with} \\ \partial_\mu^m s(x,\xi,\mu) &\in S^{-r-m,0}(\mathbf{R}^n,\mathbf{R}^n,\Gamma) \cap S^{0,-r-m}(\mathbf{R}^n,\mathbf{R}^n,\Gamma) \end{aligned}$$

for any m, all $\partial_\mu^m s$ being weakly polyhomogeneous.

Example 5.5 To give examples, we first note that any *strongly* polyhomogeneous symbol of degree $-r$ satisfies Definition 5.4 by Theorem 3.7. But there are also important weakly polyhomogeneous examples, such as the symbol $(a(x,\xi) + \mu^r)^{-1}$ (μ in a sector Γ), where $a(x,\xi)$ is homogeneous of degree r in ξ for $|\xi| \geq 1$ and $a(x,\xi) + \mu^r$ is invertible when $\mu \in \Gamma$ (by [14], Th. 1.17).

For the operators entering in the APS problem we have:

Proposition 5.6 *The ψdo $\mathcal{S}_{\mathcal{B},\mu}$ on X', with μ running in $\pm\Gamma_0$, is special parameter-dependent of order* 0. *So are $\mathcal{B}$ and the composition $S_0 = \mathcal{S}_{\mathcal{B},\mu}\mathcal{B}$.*

Proof: (Indication.) For the proof we split $\mathcal{S}_{\mathcal{B},\mu}$ in several terms:

$$\begin{aligned} \mathcal{S}_{\mathcal{B},\mu} &\equiv \begin{pmatrix} B+\mu^{-1}(A_\mu+A)B^\perp \\ \mu^{-1}(A_\mu-A)B+B^\perp \end{pmatrix} \\ &= \begin{pmatrix} \mu^{-1}(A_\mu+A)\Pi_< \\ \mu^{-1}(A_\mu-A)\Pi_\geq \end{pmatrix} + \begin{pmatrix} B \\ B^\perp \end{pmatrix} + \begin{pmatrix} -\mu^{-1}(A_\mu+A)B_0 \\ \mu^{-1}(A_\mu-A)B_0 \end{pmatrix}. \end{aligned} \tag{5.21}$$

The second term has a polyhomogeneous symbol in $S^0 \subset S^{0,0}$ (cf. (3.4)) and is independent of μ, hence is special parameter-dependent of order 0. (This proves the statement on $\mathcal{B}$.) The third term is of order $-\infty$, and its boundedness in μ (with improved estimates for derivatives) is seen from considerations on the involved eigenspaces for eigenvalues of modulus $\leq R$.

It is the first term in (5.21) that requires most of the analysis. The crucial fact used here is that

$$\begin{aligned} \mu^{-1}(A_\mu + A)\Pi_< &= \mu^{-1}(A_\mu + A)(A_\mu - A)(A_\mu + |A|)^{-1}\Pi_< \\ &= \mu(A_\mu + |A|)^{-1}\Pi_<, \\ \mu^{-1}(A_\mu - A)\Pi_\geq &= \mu^{-1}(A_\mu - A)(A_\mu + A)(A_\mu + |A|)^{-1}\Pi_\geq \\ &= \mu(A_\mu + |A|)^{-1}\Pi_\geq. \end{aligned} \tag{5.22}$$

Again $\Pi_<$ and $\Pi_\geq$ are in $S^0 \subset S^{0,0}$ and independent of μ, hence special of order 0. In view of the composition rules (cf. (3.7)), it remains to prove the statement for $\mu(A_\mu + |A|)^{-1}$. The advantage of this expression is that A_μ and $|A|$ are both "positive" (strongly elliptic), so that the inverse of $A_\mu + |A|$ can be described by a natural elliptic construction. (Details are given in [14], Proposition 3.5.) The statement on S_0 now follows from the composition rules. □

These operators act on X', of dimension $n-1$ (where the space variable and cotangent variable are denoted x' and ξ'). For $s \in \mathbf{R}$ we define the space $H^{s,\mu}(\mathbf{R}^{n-1})$ as the Sobolev space with norm

$$\|u\|_{s,\mu} = \|\langle(\xi',\mu)\rangle^s \hat{u}(\xi')\|_{L_2(\mathbf{R}^{n-1})}, \tag{5.23}$$

and extend the notion to sections of a Hermitian bundle E'' over X' by use of a finite family of local coordinate systems (the space is then denoted $H^{s,\mu}(E'')$). Note that $H^{0,\mu}(E'') \simeq L_2(E'')$.

We shall need

Proposition 5.7 *Let S be a special parameter-dependent ψdo of order -1 in a bundle E'' over X', with μ running in a sector Γ. Then for $s \in \mathbf{R}$, S is continuous from $H^{s,\mu}(E'')$ to $H^{s+1,\mu}(E'')$, uniformly for μ in closed subsectors Γ' of Γ, $|\mu| \geq 1$; and its norm as an operator in $H^{s,\mu}(E'')$ satisfies*

$$\|S\|_{\mathcal{L}(H^{s,\mu}(E''))} = O(|\mu|^{-1}) \text{ for } |\mu| \to \infty,\ \mu \in \Gamma'. \tag{5.24}$$

For each Γ' there is an $r_{\Gamma'} > 0$ such that $I - S$ is invertible for $\mu \in \Gamma'$ with $|\mu| \geq r_{\Gamma'}$. The inverse equals

$$(I - S)^{-1} = I + S', \quad S' = \sum_{j=1}^{\infty} S^j, \tag{5.25}$$

where the series converges in the norm of operators in $L_2(E'')$.

Moreover, S' is a special parameter-dependent ψdo of order -1.

Proof: (Indication.) By the composition rules, S composed with an invertible ψdo with principal symbol $\langle(\xi',\mu)\rangle$ is special parameter-dependent of order 0; it is not hard to show that such an operator is continuous in $H^{s,\mu}$, uniformly as stated. This implies the asserted continuity from $H^{s,\mu}$ to $H^{s+1,\mu}$; and (5.24) follows since

$$|\mu|\|u\|_{s,\mu} \leq \text{const.}\, \|u\|_{s+1,\mu}. \tag{5.26}$$

For each sector Γ', take $r_{\Gamma'}$ so large that the operator norm of S in $L_2(E'')$ is $\leq \frac{1}{2}$ for $|\mu| \geq r_{\Gamma'}$; then (5.25) holds in operator norm.

The powers S^j are special parameter-dependent ψdo's of order $-j$, by the composition rules. Further efforts are needed to show that the sum S' is indeed a ψdo that is special parameter-dependent of order -1; see the details in [14], proof of Theorem 3.8, as explained for S_2 there. □

Now we use these facts to show:

Theorem 5.8 *The operator S_1 in (5.20) is a special parameter-dependent ψdo of order -1 in the bundle $E_1'' = E_1' \oplus E_1'$ over X'. Hence for each closed subsector Γ of Γ_0 (or $-\Gamma_0$) there is an $r_\Gamma > 0$ such that $I - S_1$ is invertible for $\mu \in \Gamma$ with $|\mu| \geq r_\Gamma$, with inverse*

$$(I - S_1)^{-1} = I + S_2, \quad S_2 = \sum_{j=1}^{\infty} S_1^j, \tag{5.27}$$

S_2 being a special parameter-dependent ψdo of order -1 in E_1''.

Furthermore, for such μ,

$$(I - G_2)^{-1} = I + K_2(I + S_2)S_0T_0, \tag{5.28}$$

and finally

$$\begin{aligned} \mathcal{R}_\mu &= (Q_{\mu,+} - G_1)(I - G_2)^{-1} \\ &= (Q_{\mu,+} - K_1S_0T_0)(I + K_2(I + S_2)S_0T_0) \\ &= Q_{\mu,+} - (K_1 - K_3)(I + S_2)S_0T_0, \ \textit{with } K_3 = Q_{\mu,+}K_2. \end{aligned} \tag{5.29}$$

Proof: In the formula (5.20) for S_1, S_0 is a special parameter-dependent ψdo of order 0 by Proposition 5.6, and T_0K_2 is a special parameter-dependent ψdo of order -1 by Lemma 5.2 and Example 5.5, so it follows from the composition rules (cf. (3.7)) that S_1 is a special parameter-dependent ψdo of order -1. Then Proposition 5.7 applies, showing the assertions for S_2.

Now the formula for $(I - G_2)^{-1}$ follows from (5.20). The first two lines in (5.29) then follow from (5.16) and (5.18). Consequently we have:

$$\begin{aligned} \mathcal{R}_\mu &= (Q_{\mu,+} - K_1 S_0 T_0)(I + K_2(I + S_2)S_0 T_0) \\ &= Q_{\mu,+} + Q_{\mu,+} K_2 (I + S_2) S_0 T_0 \\ &\quad - K_1 S_0 T_0 - K_1 S_0 T_0 K_2 (I + S_2) S_0 T_0 \\ &= Q_{\mu,+} + Q_{\mu,+} K_2 (I + S_2) S_0 T_0 \\ &\quad - K_1 S_0 T_0 - K_1 S_1 (I + S_2) S_0 T_0 \\ &= Q_{\mu,+} - (K_1 - Q_{\mu,+} K_2)(I + S_2) S_0 T_0, \end{aligned} \tag{5.30}$$

using formula (5.20) for S_1 and the fact that $I + S_1(I + S_2) = I + S_2$. This ends the proof. □

Taking the structure of the entering Poisson and trace operators into account, we have obtained:

Corollary 5.9 *For each closed subsector Γ of $\pm\Gamma_0$ one can find $r_\Gamma > 0$ so that the resolvent $\mathcal{R}_\mu = (\mathcal{P}_\mathcal{B} + \mu)^{-1}$ for $\mu \in \Gamma$ with $|\mu| \geq r_\Gamma$ is of the form*

$$\mathcal{R}_\mu = Q_{\mu,+} + KST, \tag{5.31}$$

where K resp. T are a strongly polyhomogeneous Poisson resp. trace operator of degree -1 and S is a special parameter-dependent ψdo on X' of order 0. The detailed structure is given in (5.29).

5.2. TRACE CALCULATIONS

Consider $\mathcal{R}_\mu = (\mathcal{P}_\mathcal{B} + \mu)^{-1}$, as described above. Since the injection of $H^s(X)$ into $L_2(X)$ is trace class for $s > n$, the terms in $\partial_\mu^m \mathcal{R}_\mu$ are trace class when $m \geq n$.

Theorem 5.10 *Let φ be any morphism in $E = E_1 \oplus E_2$, and let $m \geq n = \dim X$. Then*

$$\begin{aligned} \operatorname{Tr}(\varphi \partial_\mu^m (\mathcal{P}_\mathcal{B} + \mu)^{-1}) \sim a_0 \mu^{n-m-1} &+ \sum_{j=1}^{\infty} (a_j + b_j) \mu^{n-m-1-j} \\ &+ \sum_{j=0}^{\infty} (c_j \log \mu + c_j') \mu^{-m-1-j}, \quad \text{as } |\mu| \to \infty, \end{aligned} \tag{5.32}$$

for μ in closed subsectors of $\pm\Gamma_0$. The coefficients a_j, b_j and c_j are integrals, $\int_X a_j(x)\,dx$, $\int_{X'} b_j(x')\,dx'$ and $\int_{X'} c_j(x')dx'$, of densities locally determined by the symbols of P and B, while the c_j' are in general globally determined. The coefficients c_0 and c_0' are the same as for the case where the P_j are zero in (1.2) (the product case).

Proof: We find from (5.29):

$$\varphi\partial_\mu^m(\mathcal{P}_\mathcal{B}+\mu)^{-1} = \varphi\partial_\mu^m Q_{\mu,+} - \varphi\partial_\mu^m[K_1S_0T_0] - \varphi\partial_\mu^m[(K_1S_2 - K_3(I+S_2))S_0T_0]. \quad (5.33)$$

First, $\operatorname{Tr}(\varphi\partial_\mu^m Q_{\mu,+})$ contributes the well-known expansion $\sum_0^\infty a_j\mu^{n-m-1-j}$. For the other terms we can use the invariance of the trace under cyclic permutation of the operators, to reduce to a study of operators on X'. For the middle term we find, by the Leibniz rule:

$$\begin{aligned}
&\operatorname{Tr}_X(\varphi\partial_\mu^m[K_1S_0T_0]) \\
&\quad= \sum_{m_1+m_2+m_3=m} c_{m_1,m_2,m_3}\operatorname{Tr}_X(\varphi\partial_\mu^{m_1}K_1\,\partial_\mu^{m_2}S_0\,\partial_\mu^{m_3}T_0) \\
&\quad= \operatorname{Tr}_{X'}\Big(\sum_{m_1+m_2+m_3=m} c_{m_1,m_2,m_3}\partial_\mu^{m_2}S_0\,\partial_\mu^{m_3}T_0\varphi\partial_\mu^{m_1}K_1\Big) \\
&\quad= \operatorname{Tr}_{X'}\partial_\mu^m(S_0T_0\varphi K_1). \quad (5.34)
\end{aligned}$$

By Lemma 5.2, $T_0\varphi K_1$ is a strongly polyhomogeneous ψdo on X' of order -1, hence special parameter-dependent by Theorem 3.7. Then since S_0 is special parameter-dependent by Proposition 5.6, it follows that $\partial_\mu^m(S_0T_0\varphi K_1)$ is a special parameter-dependent ψdo on X' of order $-m-1$.

To this we can apply our general Theorem 3.8 and its corollary, after a reduction to local trivializations by use of a partition of unity. Since the symbol has degrees $-m-1-j$, $j\ge 0$, and μ-exponent $d=-m-1$, we get an expansion in a series of locally determined terms $b_{k,1}\mu^{-m-1+(n-1)-k}$, $k\ge 0$, together with a series of terms $(c_{j,1}\log\mu + c'_{j,1})\mu^{-m-1-j}$, $j\ge 0$, with $c_{j,1}$ locally determined.

The third term is treated similarly; here the circular permutation of the terms resulting from the Leibniz rule gives a special parameter-dependent ψdo of order $-m-2$, so Corollary 3.9 gives an expansion in a series of locally determined terms $b_{k,2}\mu^{-m-2+(n-1)-k}$, $k\ge 0$, together with a series of terms $(c_{j,2}\log\mu + c'_{j,2})\mu^{-m-1-j}$, $j\ge 1$, with $c_{j,2}$ locally determined.

Taking the contributions together we get the expansion (5.32). One observes moreover that the terms $(c_0\log\mu + c'_0)\mu^{-m-1}$ in (5.32) come only from $\operatorname{Tr}(\varphi\partial_\mu^m[K_1S_0T_0])$, which leads to the last statement in the theorem. For, K_1 and S_0 are the same as for the case where the P_j and P'_j are 0. The third factor $T_0=\gamma_0Q_{\mu,+}$ uses the symbol of $(\mathcal{P}+\mu)^{-1}$ evaluated at $x_n=0$. The leading term of this is the same as for the case where P_j and P'_j are 0, and the lower order terms contribute ultimately with special parameter-dependent ψdo's of order $-m-2$ only; the first possible nonlocal and log contributions from this are the terms with μ^{-m-2} and $\mu^{-m-2}\log\mu$. □

In view of (5.4), it is now easy to draw conclusions from this on asymptotic expansions for traces of λ-derivatives of $\varphi(\Delta_1 - \lambda)^{-1} = \varphi(P_B{}^* P_B - \lambda)^{-1}$ and $\varphi P_B(\Delta_1 - \lambda)^{-1} = \varphi P_B(P_B{}^* P_B - \lambda)^{-1}$, etc.

Corollary 5.11 *Let $\varphi_{kl} : E_l \to E_k$ be morphisms, for $k, l = 1, 2$.*

The traces $\operatorname{Tr}(\varphi_{11}\partial_\lambda^m(\Delta_1-\lambda)^{-1})$ and $\operatorname{Tr}(\varphi_{22}\partial_\lambda^m(\Delta_2-\lambda)^{-1})$ have asymptotic expansions (for $k = 1$ resp. 2):

$$a_{0,kk}(-\lambda)^{\frac{n}{2}-m-1} + \sum_{j=1}^{\infty}(a_{j,kk} + b_{j,kk})(-\lambda)^{\frac{n-j}{2}-m-1} + \sum_{j=0}^{\infty}(c_{j,kk}\log\lambda + c'_{j,kk})(-\lambda)^{\frac{-j}{2}-m-1}; \quad (5.35)$$

and $\operatorname{Tr}(\varphi_{12}\partial_\lambda^m P_B(\Delta_1 - \lambda)^{-1})$ and $\operatorname{Tr}(\varphi_{21}\partial_\lambda^m P_B{}^*(\Delta_2 - \lambda)^{-1})$ have asymptotic expansions (for $\{k, l\} = \{1, 2\}$ resp. $\{2, 1\}$):

$$a_{0,kl}(-\lambda)^{\frac{n-1}{2}-m} + \sum_{j=1}^{\infty}(a_{j,kl} + b_{j,kl})(-\lambda)^{\frac{n-j-1}{2}-m} + \sum_{j=0}^{\infty}(c_{j,kl}\log\lambda + c'_{j,kl})(-\lambda)^{\frac{-j-1}{2}-m}; \quad (5.36)$$

with coefficients described as in Theorem 5.10.

The coefficients $c_{0,kl}$ and $c'_{0,kl}$ are the same as those for the product case.

Proof: Using (5.4), take

$$\varphi = \begin{pmatrix} \varphi_{11} & 0 \\ 0 & 0 \end{pmatrix}, \begin{pmatrix} 0 & 0 \\ 0 & \varphi_{22} \end{pmatrix}, \begin{pmatrix} 0 & \varphi_{12} \\ 0 & 0 \end{pmatrix}, \text{ resp. } \begin{pmatrix} 0 & 0 \\ \varphi_{21} & 0 \end{pmatrix}, \quad (5.37)$$

in Theorem 5.10, and divide by μ in the first two cases. Now replace μ by $(-\lambda)^{\frac{1}{2}}$ and note that $\partial_\lambda = (2\mu)^{-1}\partial_\mu$. □

These results yield asymptotic expansions of the traces of heat operators $\varphi_{11}e^{-t\Delta_1}$, $\varphi_{12}P_B e^{-t\Delta_1}$, etc., and power operators $\varphi_{11}(\Delta_1)^{-s}$, $\varphi_{12}P_B(\Delta_1)^{-s}$, etc., by use of the transition formulas in Section 2:

Theorem 5.12 *There are coefficients $\tilde{a}_{j,kl}$, $\tilde{b}_{j,kl}$, $\tilde{c}_{j,kl}$, $\tilde{c}'_{j,kl}$, related by suitable gamma factors to those in Corollary 5.11 (cf. Theorems 2.1 and 2.3) such that, with $\nu_1 = \operatorname{Tr}(\varphi_{11}\Pi_0(P_B))$, $\nu_2 = \operatorname{Tr}(\varphi_{22}\Pi_0(P_B{}^*))$, the zeta and* *eta functions have singularity structures described by:*

$$\Gamma(s)\operatorname{Tr}(\varphi_{kk}Z(\Delta_1,s)) \sim \frac{-\nu_k}{s} + \frac{\tilde a_{0,kk}}{s-\frac{n}{2}} + \sum_{j=1}^{\infty}\frac{\tilde a_{j,kk}+\tilde b_{j,kk}}{s-\frac{n-j}{2}}$$
$$+\sum_{j=0}^{\infty}\Big(\frac{\tilde c_{j,kk}}{(s+\frac{j}{2})^2}+\frac{\tilde c'_{j,kk}}{s+\frac{j}{2}}\Big);$$
$$\Gamma(s)\operatorname{Tr}(\varphi_{12}P_B Z(\Delta_1,s)) \text{ resp. } \Gamma(s)\operatorname{Tr}(\varphi_{21}{P_B}^* Z(\Delta_2,s)) \tag{5.38}$$
$$\sim \frac{\tilde a_{0,kl}}{s-\frac{n+1}{2}} + \sum_{j=1}^{\infty}\frac{\tilde a_{j,kl}+\tilde b_{j,kl}}{s-\frac{n-j+1}{2}} + \sum_{j=0}^{\infty}\Big(\frac{\tilde c_{j,kl}}{(s+\frac{j-1}{2})^2}+\frac{\tilde c'_{j,kl}}{s+\frac{j-1}{2}}\Big);$$

and the heat traces have the asymptotic behavior for $t\to 0$:

$$\operatorname{Tr}(\varphi_{kk}e^{-t\Delta_1}) \sim \tilde a_{0,kk}t^{-\frac{n}{2}} + \sum_{j=1}^{\infty}(\tilde a_{j,kk}+\tilde b_{j,kk})t^{\frac{j-n}{2}}$$
$$+\sum_{j=0}^{\infty}(-\tilde c_{j,kk}t^{\frac{j}{2}}\log t + \tilde c'_{j,kk}t^{\frac{j}{2}}),$$
$$\operatorname{Tr}(\varphi_{12}P_B e^{-t\Delta_1}) \text{ resp. } \operatorname{Tr}(\varphi_{21}{P_B}^* e^{-t\Delta_2}) \tag{5.39}$$
$$\sim \tilde a_{0,kl}t^{-\frac{n+1}{2}} + \sum_{j=1}^{\infty}(\tilde a_{j,kl}+\tilde b_{j,kl})t^{\frac{j-n-1}{2}} + \sum_{j=0}^{\infty}(-\tilde c_{j,kl}t^{\frac{j-1}{2}}\log t + \tilde c'_{j,kl}t^{\frac{j-1}{2}}),$$

The $\tilde c'_{j,kl}$ and ν_k are in general globally defined, while the other coefficients are local. The coefficients $\tilde c_{0,kl}$ and $\tilde c'_{0,kl}$ are the same as those for the product case.

A detailed account is given in [14]. [14] and [15] also give some information on variations of parameter-dependent situations.

Remark 5.13 Similar considerations allow the calculation of $\operatorname{Tr}(D\partial_\mu^m\mathcal{R}_\mu)$ when D is an arbitrary differential operator on X, for $m\ge n+d$, $d=$ the order of D. One finds that

$$\operatorname{Tr}(D\partial_\mu^m\mathcal{R}_\mu) \sim a_0(D,P)\mu^{n-m+d-1} + \sum_{j=1}^{\infty}(a_j(D,P)+b_j(D,P_B))\mu^{n-m+d-1-j}$$
$$+\sum_{j=0}^{\infty}(c_j(D,P_B)\log\mu + c'_j(D,P_B))\mu^{-m+d-1-j} \tag{5.40}$$

(the primed coefficients global, the others local); and consequences are drawn as above for the corresponding zeta and eta functions and exponential traces.

References

1. M. S. Agranovič: *Some asymptotic formulas for elliptic pseudodifferential operators.* J. Functional Analysis and Appl., **21** (1987), 63–65.
2. M. F. Atiyah, V. K. Patodi and I. M. Singer: *Spectral asymmetry and Riemannian geometry, I.* Math. Proc. Camb. Phil. Soc. **77** (1975), 43–69.
3. N. Bourbaki: " Fonctions d'une variable réelle." **IV**, Ed. Hermann, Paris 1951.
4. L. Boutet de Monvel: *Boundary problems for pseudo-differential operators.* Acta Math. **126** (1971), 11–51.
5. T. Branson and P. B. Gilkey: *Residues of the eta function for an operator of Dirac type.* J. Funct. Analysis **108** (1992), 47–87.
6. J. Brüning and M. Lesch: *On the eta-invariant of certain non-local boundary value problems.* To appear.
7. J. S. Dowker, J. S. Apps, K. Kirsten and M. Bordag: *Spectral invariants for the Dirac equation on the d-ball with various boundary conditions.* Class. Quantum Grav., to appear.
8. J. J. Duistermaat and V. W. Guillemin: *The spectrum of positive elliptic operators and periodic bicharacteristics.* Inventiones Math. **29** (1975), 39–79.
9. A. Friedmann: " Partial Differential Equations." Holt, Rinehart and Winston, New York 1969.
10. P. B. Gilkey: *On the index of geometrical operators for Riemannian manifolds with boundary.* Adv. in Math. **102** (1993), 129–183.
11. P. B. Gilkey and G. Grubb: *Logarithmic terms in asymptotic expansions of heat operator traces.* To appear.
12. G. Grubb: " Functional Calculus of Pseudo-Differential Boundary Problems." Progress in Math., Vol. 65, Birkhäuser, Boston 1986. Second edition 1996.
13. G. Grubb: *Heat operator trace expansions and index for general Atiyah-Patodi-Singer boundary problems.* Comm. P. D. E. **17** (1992), 2031–2077.
14. G. Grubb and R. T. Seeley: *Weakly parametric pseudodifferential operators and Atiyah-Patodi-Singer boundary problems.* Inventiones Math. **121** (1995), 481–529.
15. G. Grubb and R. T. Seeley: *Zeta and eta functions for Atiyah-Patodi-Singer operators.* Journal of Geometric Analysis, to appear. Copenhagen Univ. Math. Dept. Prepr. Ser. **11**, 1993.
16. E. Hille and R. Phillips: "Functional Analysis and Semi-Groups." Amer. Math. Soc. Colloq. Publ. **31**, Providence, Rhode Island 1957.
17. L. Hörmander: *Pseudo-differential operators and non-elliptic boundary problems.* Ann. of Math. **83** (1966), 129–209.
18. L, Hörmander: *Pseudo-differential operators and hypoelliptic equations.* Proc. Symp. Pure Math. **10** (1967), 138–183.
19. L. Hörmander: "The Analysis of Linear Partial Differential Operators, III." Springer Verlag, Heidelberg 1985.
20. T. Kato: "Perturbation Theory for Linear Operators." Springer Verlag, Berlin 1966.
21. R. T. Seeley: *Singular integrals and boundary value problems.* Amer. J. Math. **88** (1966), 781–809.
22. R. T. Seeley: *Complex powers of an elliptic operator.* Amer. Math. Soc. Proc. Symp. Pure Math. **10** (1967), 288–307.
23. R. T. Seeley: *Topics in pseudo-differential operators.* CIME Conference on Pseudo-Differential Operators 1968, Edizioni Cremonese, Roma 1969, pp. 169–305.
24. M. A. Shubin: "Pseudodifferential Operators and Spectral Theory." Nauka, Moscow 1978.

BOUNDARY VALUE PROBLEMS AND EDGE PSEUDO-DIFFERENTIAL OPERATORS

B.-W. SCHULZE
Institut für Mathematik
Universität Potsdam
Postfach 60 15 53
14415 Potsdam
Germany

Introduction

The analysis of pseudo-differential operators on a closed compact C^∞ manifold (in its standard form) allows the construction of parametrices of elliptic operators by inverting local symbols and forming the associated operators. Elliptic regularity and the Fredholm property of elliptic operators in Sobolev spaces are consequences of the basic calculus of pseudo-differential operators. It is well-known how the interplay between symbolic and operator level, together with homotopy and operator algebra aspects, are involved in the index theory in K-theoretic terms, cf. Atiyah, Singer [2], in the program to express the index by analytical formulas, cf. Fedosov [9] or in other strategies for analyzing and interpreting the index, e.g., by the heat kernel asymptotics. For interesting classes of singular or non-compact manifolds, essential problems like adequate operator algebras with symbolic structures, the definition of ellipticity, and index theory, are unsolved.

For pseudo-differential boundary value problems on manifolds with C^∞ boundary, Boutet de Monvel [3] found appropriate operator algebras (for symbols with the transmission property with respect to the boundary), Rempel, Schulze [21] (for general symbols, not necessarily having the transmission property), cf. also the work of Vishik, Eskin [46] and Eskin [8]. Let us mention in this context also the monographs of Rempel, Schulze [21], Grubb [13] and of the author [34]. The present exposition will show how boundary value problems fit into a more general class of pseudo-differential operators on manifolds with edges. As in the (classical) C^∞ situation there are to be expected new interactions to other fields of mathematics, in par-

L. Rodino (ed.), Microlocal Analysis and Spectral Theory, 165–226.

ticular, geometry and topology. Motivations for the analysis on singular manifolds come from (applications in) mathematical physics and engineering. Here, for instance, piece-wise smooth configurations in concrete models are not less classical than smooth ones; however a transparent analysis for higher edge and corner orders (also for non-elliptic and non-linear equations) is still an enormous challenge. It is primarily an analytic problem to invent manageable formalisms in terms of symbolic structures and operator and distribution spaces. The conical and edge singularities are crucial for understanding the hierarchy of polyhedral singularities of growing orders. We will explain essential ideas of this theory, but we will not give the complete calculus. The notions and results of our approach may also be regarded as axiomatic elements for operator algebras on spaces of higher singularity orders, e.g., (warped) polyhedra or their lower-dimensional skeletons. The simplest non-trivial singularity is the conical one. The (infinite) cone over a base space X is the quotient space $X^\Delta = (\overline{\boldsymbol{R}}_+ \times X)/(\{0\} \times X)$, where $\{0\} \times X$ corresponds to the vertex (the conical singularity). In the following, X will be a closed, compact C^∞ manifold. For instance, if X is embedded in the unit sphere S^N of $\boldsymbol{R}^{N+1}$, then

$$X^\Delta \cong \{\tilde{x} \in \boldsymbol{R}^{N+1} : \ \tilde{x} = 0 \ \text{ or } \ \tilde{x}/|\tilde{x}| \in X\}.$$

On the "open stretched" cone $\boldsymbol{R}_+ \times X$, different splittings of coordinates (t, x) and $(\tilde{t}, \tilde{x})$ are said to be equivalent if $(t, x) \to (\tilde{t}(t, x), \tilde{x}(t, x))$ extends to a diffeomorphism $\boldsymbol{R} \times X \to \boldsymbol{R} \times X$; then $\tilde{t}(0, x) = 0$ for all x. For $\dim X = 0$ we have $X^\Delta = \overline{\boldsymbol{R}}_+$ with the conical singularity $t = 0$ of $\overline{\boldsymbol{R}}_+$. Singularities of cusp type as they were studied in Schulze, Shatalov, Sternin [40] and Schulze, Tarkhanov [42] under different aspects, will not be discussed here, though there are many links between the theories of conical singularities and cusps.

A topological Hausdorff space B is called a "manifold with conical singularities" if there is a finite subset $S \subset B$ such that $B \backslash S$ is a paracompact C^∞ manifold, and every $v \in S$ has a neighborhood V which is homeomorphic to the cone X^Δ over some closed compact C^∞ manifold $X = X(v)$, such that $V \setminus \{v\}$ is diffeomorphic to $\boldsymbol{R}_+ \times X$. We define the "stretched" manifold $\boldsymbol{B}$ associated with B by attaching the sets $[0, 1) \times X(v)$, $v \in S$, to $B \setminus S$. Then $\boldsymbol{B}$ is a C^∞ manifold with compact C^∞ boundary $\partial \boldsymbol{B} \cong \times_{v \in S} X(v)$, and we have $\boldsymbol{B} \setminus \partial \boldsymbol{B} \cong B \setminus S$.

For an open set $\Omega \subseteq \boldsymbol{R}^q$, and a cone X^Δ we define the wedge $X^\Delta \times \Omega$. We call a topological Hausdorff space W a manifold with edges Y, $Y \subseteq W$, if $W \backslash Y$ and Y are paracompact C^∞ manifolds of dimensions $1+n+q$ and q, respectively, such that W is locally (near each of the $y \in Y$) homeomorphic to a wedge $X^\Delta(y) \times \Omega(y)$ with n-dimensional cone bases $X(y)$ and open $\Omega(y) \subseteq \boldsymbol{R}^q$, which means that there are "local coordinates" $(t, x, y) \in$

$\boldsymbol{R}_+ \times X \times \Omega$ outside the edge and local coordinates $y \in \Omega$ on the edge. Any pair $(t, x, y) \in \boldsymbol{R}_+ \times X \times \Omega$, $(\tilde{t}, \tilde{x}, \tilde{y}) \in \boldsymbol{R}_+ \times X \times \tilde{\Omega}$ of coordinates has to be compatible, i.e. the diffeomorphism

$$(t, x, y) \to (\tilde{t}(t, x, y), \tilde{x}(t, x, y), \tilde{y}(t, x, y))$$

is the restriction of a diffeomorphism $\boldsymbol{R} \times X \times \Omega \to \boldsymbol{R} \times X \times \tilde{\Omega}$ to $\boldsymbol{R}_+ \times X \times \Omega$, where $\tilde{t}(0, x, y) = 0$ for all x, y, and $\tilde{y}(0, x, y)$ is independent of x. The analysis requires the stretched wedge, which locally looks like $\overline{\boldsymbol{R}}_+ \times X \times \Omega$. The stretched manifold $\boldsymbol{W}$ associated with W, is defined by glueing together the sets $[0, 1) \times X(y) \times \Omega(y)$ and attaching this to $W \setminus Y$. It is a C^∞ manifold with C^∞ boundary $\partial \boldsymbol{W}$ which is a bundle over Y with fiber X, and we have $\boldsymbol{W} \setminus \partial \boldsymbol{W} \cong W \setminus Y$.

A (paracompact) C^∞ manifold with C^∞ boundary can always be regarded as a manifold W with edge Y, setting $\partial W = Y$, $\dim X = 0$, and the model cone of the wedge near every $y \in Y$ is $\overline{\boldsymbol{R}}_+$, the inner normal to the boundary (with respect to some Riemannian metric). For instance, let $G \subset \boldsymbol{R}^n$ be a domain with C^∞ boundary, $W = \overline{G}$. Consider an elliptic differential operator A in $\boldsymbol{R}^n$ with smooth coefficients, say the Laplacian. Clearly A contains no specific information on the geometry of ∂G. The structure of elliptic boundary conditions for A is the consequence of a certain behaviour of A in normal direction to the boundary. We define the so-called boundary symbol of A, which it an operator family acting on $\boldsymbol{R}_+$, parameterized by the points of $T^* \partial G$. The operator A is reformulated as a pseudo-differential operator along ∂G, with the boundary symbol as its (operator-valued) symbol. In this sense A is expressed in anisotropic terms relative to ∂G. This reformulation does not change any properties of A far from ∂G.

In general, on a manifold with edges, we translate the operators near the edges into pseudo-differential operators with specific operator-valued symbols. These symbols take values in a pseudo-differential algebra on the model cone. Far from the edges the operators have to remain "isotropic" pseudo-differential operators. The calculus of pseudo-differential boundary value problems can be formulated as one with operator-valued symbols, cf. Schulze [34]. This permits to read off the basic structures of a calculus also for general edge singularities, cf. Egorov, Schulze [7] and the monograph [37]. This point of view was systematically developed in a sequence of papers, starting with [30] and then continued under various aspects for operator algebras with continuous and variable branching asymptotics in [29], [35], [36], moreover, for corner singularities in [32] and in a joint paper with Dorschfeldt [6] and for non-compact manifolds jointly with Dorschfeldt, Grieme in [5] and in Seiler [43]. In particular, the algebra of pseudo-differential boundary value problems with the transmission property in the sense of Boutet de Monvel [3] found a new interpretation as an edge pseudo-differential calculus. This was elaborated in this form in detail in the joint papers with Schrohe [24], [25] as a tool to treat boundary value problems for conical singularities. As mentioned at the beginning, many questions on elliptic operators in the standard calculus on a C^∞ manifold are also meaningful on a manifold with singularities. This concerns, in particular, an extension of Fedosov's analytical index formula to elliptic operators on manifolds with edges, cf. Fedosov, Schulze, Tarkhanov [11], Schrohe, Seiler [23], or the analysis of asymptotics of solutions, cf. Schulze, Shatalov, Sternin [38], [39].

Let us finally note that a pseudo-differential calculus for manifolds with higher edge and corner singularities requires parameter-dependent variants of the already achieved operator algebras on a given manifold with singularities. The parameter is interpreted as a additional covariable to used either for the Mellin transform along a new corner axis or the Fourier transform along a new edge. This iterative procedure should be based on an axiomaticdescription of the "higher" operator algebras. The elements of the present exposition are chosen to be an ingredient of a future pseudo-differential calculus on manifolds with higher singularities.

The details can be of enormous complexity unless the most efficient strategies are discovered. This program, of course, may also be a challenge for young mathematicians who want to be active in this field. Also in the context of parabolic and hyperbolic operators much work is to be done.

The author thanks M. Gerisch (Max-Planck-Arbeitsgruppe "Partielle Differentialgleichungen und Komplexe Analysis", University of Potsdam) for valuable remarks to the manuscript.

1. Edge Sobolev spaces and operator-valued symbols

1.1. NOTATIONS AND CLASSICAL BACKGROUND

This section recalls some elementary material on Sobolev spaces and pseudo differential operators. For more details we refer to standard monographs such as Hörmander [14], Treves [45], Kumano-go [16]. We will employ Fréchet topologies in symbol and operator spaces. The various statements may be regarded as exercises in pseudo-differential calculus.

If $\Omega \subseteq \boldsymbol{R}^n$ is an open set then $C^\infty(\Omega)$ is the space of all infinitely differentiable functions in Ω, $C_0^\infty(\Omega)$ the subspace of all elements with compact support. $\mathcal{D}'(\Omega) = (C_0^\infty(\Omega))'$ is the space of all distributions in Ω, $\mathcal{E}'(\Omega) = (C^\infty(\Omega))'$ the subspace of all distributions with compact support. If $U \subseteq \boldsymbol{C}^n$ is an open set then $\mathcal{A}(U)$ is the space of all holomorphic functions in U. We will employ the standard locally convex topologies in the spaces. Analogous notations make sense for functions (or distributions) with values in a (say Fréchet) space E, namely $C^\infty(\Omega, E), C_0^\infty(\Omega, E), \ldots, \mathcal{A}'(U, E)$. All occurring Fréchet spaces here can be written as projective limits of Banach spaces

$$\{E^j\}_{j\in \boldsymbol{N}} \text{ with continuous embeddings } E^{j+1} \hookrightarrow E^j \text{ for all } j \in \boldsymbol{N}. \quad (1.1)$$

Here $\boldsymbol{N} = \{0, 1, 2, \ldots\}$. Hilbert spaces in this exposition are assumed to be separable. Given locally convex vector spaces $E, \tilde{E}$, the space of linear continuous operators $E \to \tilde{E}$ will be denoted by $\mathcal{L}(E, \tilde{E})$ or $\mathcal{L}(E)$ for $E = \tilde{E}$. If $E, \tilde{E}$ are Banach spaces then $\mathcal{L}(E, \tilde{E})$ will be considered in the operator norm topology. $\mathcal{S}(\boldsymbol{R}^n)$ will denote the Schwartz space in $\boldsymbol{R}^n \ni x = (x_1, \ldots, x_n)$, defined as the subspace of all $u \in C^\infty(\boldsymbol{R}^n)$ for which the semi-norms

$$u \to \sup_{x\in \boldsymbol{R}^n} |x^\alpha D_x^\beta u(x)|$$

are finite for all $\alpha, \beta \in \boldsymbol{N}^n$. This defines a Fréchet topology in $\mathcal{S}(\boldsymbol{R}^n)$. If $L^2(\boldsymbol{R}^n)$ is the space of all square integrable functions in $\boldsymbol{R}^n$, i.e. the measurable functions u on $\boldsymbol{R}^n$ with $\|u\|_{L^2(\boldsymbol{R}^n)} = \{\int |u(x)|^2 dx\}^{1/2} < \infty$, then

$$u \to \|x^\alpha D_x^\beta u(x)\|_{L^2(\boldsymbol{R}^n)}$$

for all $\alpha, \beta \in \boldsymbol{N}^n$, is an equivalent semi-norm system on $\mathcal{S}(\boldsymbol{R}^n)$. The dual $\mathcal{S}'(\boldsymbol{R}^n)$ is the space of temperate distributions in $\boldsymbol{R}^n$. The Fourier transform

$$\hat{u}(\xi) = (Fu)(\xi) = \int e^{-ix\xi} u(x) dx,$$

with $\xi = (\xi_1, \ldots, \xi_n)$, $x\xi = \sum_{i=1}^n x_i \xi_i$, induces an isomorphism $\mathcal{F} : \mathcal{S}(\boldsymbol{R}^n) \to \mathcal{S}(\boldsymbol{R}^n)$; its inverse is given by the formula

$$(\mathcal{F}^{-1} g)(x) = \int e^{ix\xi} g(\xi) đ\xi,$$

where $đ\xi = (2\pi)^{-n} d\xi$. The Fourier transform extends to an isomorphism $\mathcal{F} : \mathcal{S}'(\boldsymbol{R}^n) \to \mathcal{S}'(\boldsymbol{R}^n)$. We shall also write $\mathcal{F} = \mathcal{F}_{x\to\xi}$ and $\mathcal{F}^{-1} = \mathcal{F}^{-1}_{\xi\to x}$.

The Sobolev space $H^s(\boldsymbol{R}^n)$ of smoothness $s \in \boldsymbol{R}$ is defined as the closure of $\mathcal{S}(\boldsymbol{R}^n)$ with respect to the norm

$$\|u\|_{H^s(\boldsymbol{R}^n)} = \left\{ \int \langle\xi\rangle^{2s} |\hat{u}(\xi)|^2 đ\xi \right\}^{\frac{1}{2}}. \quad (1.2)$$

Here $\langle\xi\rangle = (1 + |\xi|^2)^{\frac{1}{2}}$. The space $H^s(\boldsymbol{R}^n)$ can also be characterized as the subspace of all $u \in \mathcal{S}'(\boldsymbol{R}^n)$ for which $\langle\xi\rangle^s \hat{u}(\xi) \in L^2(\boldsymbol{R}^n_\xi)$. Instead of $\langle\xi\rangle$ we may equivalently use the function $[\xi]$, defined as any element in $C^\infty(\boldsymbol{R}^n)$ for which $[\xi] > 0$ and $[\xi] = |\xi|$ for all $|\xi| > c$ for a constant $c > 0$. Then

$$c_1\langle\xi\rangle \leq [\xi] \leq c_2\langle\xi\rangle \quad \text{for suitable } \ c_1, c_2 \ \text{ for all } \ \xi \in \boldsymbol{R}^n.$$

Note that $[\lambda\xi] = \lambda[\xi]$ for all $\lambda \geq 1$, $|\xi| \geq c$ for a constant $c > 0$. For an open $\Omega \subseteq \boldsymbol{R}^n$ we set

$$H^s_{loc}(\Omega) = \{u \in \mathcal{D}'(\Omega) : \ \varphi u \in H^s(\boldsymbol{R}^n) \quad \text{for all} \quad \varphi \in C_0^\infty(\Omega)\}.$$

The symbols of pseudo-differential operators are defined as follows: For $\mu \in \boldsymbol{R}$ and an open set $U \subseteq \boldsymbol{R}^m$, $S^\mu(U \times \boldsymbol{R}^n)$ is the space of all $a(x,\xi) \in C^\infty(U \times \boldsymbol{R}^n)$ such that

$$\sup_{\substack{x \in K \\ \xi \in \boldsymbol{R}^n}} \langle\xi\rangle^{-\mu+|\beta|} |D_x^\alpha D_\xi^\beta a(x,\xi)| \tag{1.3}$$

is finite for all $\alpha \in \boldsymbol{N}^m$, $\beta \in \boldsymbol{N}^n$, and arbitrary $K \subset\subset U$. The system of semi-norms (1.3) defines a Fréchet topology on the space $S^\mu(U \times \boldsymbol{R}^n)$. Denote by $S^\mu(\boldsymbol{R}^n)$ the subspace of x-independent elements (symbols with "constant coefficients"). This is a closed subspace of $S^\mu(U \times \boldsymbol{R}^n)$, and we have

$$S^\mu(U \times \boldsymbol{R}^n) = C^\infty(U, S^\mu(\boldsymbol{R}^n)).$$

Let $S^{(\mu)}(U \times \boldsymbol{R}^n)$ for $\mu \in \boldsymbol{R}$ be the subspace of all $a_{(\mu)}(x,\xi) \in C^\infty(U \times (\boldsymbol{R}^n \setminus \{0\}))$ satisfying $a_{(\mu)}(x, \lambda\xi) = \lambda^\mu a_{(\mu)}(x,\xi)$ for all $\lambda > 0$ and $x \in U$, $\xi \in \boldsymbol{R}^n \setminus \{0\}$. For every excision function $\chi(\xi)$ in $\boldsymbol{R}^n$ (i.e., $\chi(\xi) \in C^\infty(\boldsymbol{R}^n)$, $\chi(\xi) = 0$ for $|\xi| < c_0$, $\chi(\xi) = 1$ for $|\xi| > c_1$ for certain $0 < c_0 < c_1 < \infty$), we have $\chi(\xi) S^{(\mu)}(U \times \boldsymbol{R}^n) \subset S^\mu(U \times \boldsymbol{R}^n)$.

The subspace of classical symbols $S^\mu_{cl}(U \times \boldsymbol{R}^n) \subset S^\mu(U \times \boldsymbol{R}^n)$ is defined by the following condition: To $a(x,\xi)$ there exists a sequence

$$a_{(\mu-j)}(x,\xi) \in S^{(\mu-j)}(U \times (\boldsymbol{R}^n \setminus \{0\})), \quad j \in \boldsymbol{N},$$

such that for any excision function $\chi(\xi)$

$$a(x,\xi) - \chi(\xi) \sum_{j=0}^{N} a_{(\mu-j)}(x,\xi) \in S^{\mu-(N+1)}(U \times \boldsymbol{R}^n) \tag{1.4}$$

for all $N \in \boldsymbol{N}$. The functions $a_{(\mu-j)}(x,\xi)$, the homogeneous components of $a(x,\xi)$ of order $\mu - j$, are uniquely determined by $a(x,\xi)$. In particular, $a_{(\mu)}(x,\xi)$ is called the homogeneous principal part of $a(x,\xi)$ of order μ.

By requireing continuity of the homogeneous component maps of all orders and of the remainders we get a (nuclear) Fréchet topology in the space $S^{\mu}_{cl}(U \times \boldsymbol{R}^n)$ that is stronger than the one induced by $S^{\mu}(U \times \boldsymbol{R}^n)$. The subspace $S^{\mu}_{cl}(\boldsymbol{R}^n)$ of classical symbols with constant coefficients is closed in $S^{\mu}_{cl}(U \times \boldsymbol{R}^n)$, and we have $S^{\mu}_{cl}(U \times \boldsymbol{R}^n) = C^{\infty}(U, S^{\mu}_{cl}(\boldsymbol{R}^n))$.

Note that

$$\begin{aligned} S^{-\infty}(U \times \boldsymbol{R}^n) &= \cap_{\mu \in \boldsymbol{R}} S^{\mu}(U \times \boldsymbol{R}^n) \\ &= \cap_{j \in \boldsymbol{N}} S^{-j}_{cl}(U \times \boldsymbol{R}^n) \\ &= C^{\infty}(U, \mathcal{S}(\boldsymbol{R}^n)). \end{aligned}$$

In particular, $S^{-\infty}(\boldsymbol{R}^n) = \mathcal{S}(\boldsymbol{R}^n)$. Another obvious relation is $S^{\mu}(\boldsymbol{R}^n_{\xi}) \subset \mathcal{S}'(\boldsymbol{R}^n_{\xi})$ for every $\mu \in \boldsymbol{R}$, which implies $\mathcal{F}^{-1}_{\xi \to \zeta}(S^{\mu}(\boldsymbol{R}^n)) \subset \mathcal{S}'(\boldsymbol{R}^n_{\zeta})$.

Theorem 1.1 *Let $\chi(\zeta)$ be an arbitrary excision function and $\psi(\zeta) = 1 - \chi(\zeta)$. Then $a(\xi) \in S^{\mu}(\boldsymbol{R}^n)$ implies*

$$\chi(\zeta)(\mathcal{F}^{-1}_{\xi \to \zeta} a)(\zeta) \in \mathcal{S}(\boldsymbol{R}^n_{\zeta}),$$

$$h_0(\xi) := (\mathcal{F}_{\zeta \to \xi} \psi(\mathcal{F}^{-1}_{\xi \to \zeta} a))(\xi) \in S^{\mu}(\boldsymbol{R}^n).$$

There is a function $h(\xi + i\eta) \in \mathcal{A}(\boldsymbol{C}^n)$ such that $h_{\eta}(\xi) := h(\xi + i\eta)$ satisfies $h_{\eta}(\xi)|_{\eta=0} = h_0(\xi)$ and

$$h_{\eta}(\xi) \in S^{\mu}(\boldsymbol{R}^n_{\xi}) \quad \text{for every} \quad \eta \in \boldsymbol{R}^n,$$

where $\{h_{\eta} : \eta \in K\}$ is bounded in $S^{\mu}(\boldsymbol{R}^n)$ for every compact subset $K \subset \boldsymbol{R}^n$. Analogous relations hold for classical symbols.

The map $S^{\mu}(\boldsymbol{R}^n) \to S^{\mu}(\boldsymbol{R}^n), a(\xi) \to h_0(\xi)$, which produces a symbol that extends to a holomorphic function in $\xi + i\eta \in \boldsymbol{C}^n$ with the mentioned property will also be called kernel cut-off, with the cut-off function $\psi(\zeta)$. We set

$$h(\xi + i\eta) = (H(\psi)a)(\xi + i\eta),$$

where $H(\psi)$ is the continuous operator

$$H(\psi): \ S^{\mu}(\boldsymbol{R}^n) \to \mathcal{A}(\boldsymbol{C}^n), \ a(\xi) \to h(\xi + i\eta).$$

Since $H(\psi)$ acts only on the ξ-variables, it can be extended to a continuous operator

$$H(\psi): S^{\mu}(U \times \boldsymbol{R}^n) \to C^{\infty}(U, \mathcal{A}(\boldsymbol{C}^n)),$$

and similarly for classical symbols.

Pseudo-differential operators based on the Fourier transform in $\boldsymbol{R}^n$ are defined as

$$\mathrm{Op}(a)u(x) = \int\int e^{i(x-x')\xi} a(x, x', \xi) u(x') dx' d\!\!\bar{}\xi$$

for $a(x,x',\xi) \in S^{\mu}(\Omega \times \Omega \times \mathbf{R}^n)$, $\Omega \subseteq \mathbf{R}^n$ open, in the oscillatory integral sense. If we first assume $u \in C_0^{\infty}(\Omega)$ then

$$\mathrm{Op}(a): \ C_0^{\infty}(\Omega) \to C^{\infty}(\Omega)$$

is continuous. Denote by $L^{\mu}(\Omega)$ ($L^{\mu}_{cl}(\Omega)$) the space of all $\mathrm{Op}(a)$ for arbitrary $a(x,x',\xi) \in S^{\mu}(\Omega \times \Omega \times \mathbf{R}^n)$ ($\in S^{\mu}_{cl}(\Omega \times \Omega \times \mathbf{R}^n)$). Then $L^{-\infty}(\Omega) = \bigcap_{\mu \in \mathbf{R}} L^{\mu}(\Omega)$ is the space of all integral operators with kernels in $C^{\infty}(\Omega \times \Omega)$, called the smoothing operators. In general, every $\mathrm{Op}(a)$ has a distributional kernel

$$k(a)(x,x',x-x') = \int e^{i(x-x')\xi} a(x,x',\xi) đ\xi \in \mathcal{D}'(\Omega \times \Omega)$$

with singular support in $\mathrm{diag}\,(\Omega \times \Omega) = \{(x,x) : \ x \in \Omega\}$, cf. 1.1. A closed subset $K \subset \Omega \times \Omega$ if called proper if $\pi_i^{-1} M \cap K$ is compact for every $M \subset\subset \Omega$, with $\pi_i : \ \Omega \times \Omega \to \Omega$ being the projection to the i th component, $i = 1,2$. Denote by $L^{\mu}(\Omega)_K$ the subspace of all $A \in L^{\mu}(\Omega)$ for which the distributional kernel is supported in a proper set K with $\mathrm{diag}\,(\Omega \times \Omega) \subseteq \mathrm{int}\, K$. Then, from the above statement on the singular support of the distributional kernel of A we obtain that

$$L^{\mu}(\Omega) = L^{\mu}(\Omega)_K + L^{-\infty}(\Omega) \tag{1.5}$$

in the sense of vector spaces. By definition we have $L^{-\infty}(\Omega) \cong C^{\infty}(\Omega \times \Omega)$. Moreover, $A_0 \in L^{\mu}(\Omega)_K$ induces continuous operators

$$A_0: \ C_0^{\infty}(\Omega) \to C_0^{\infty}(\Omega), \quad C^{\infty}(\Omega) \to C^{\infty}(\Omega).$$

Using the fact that $a_0(x,\xi) := \ e_{-\xi} A_0 e_{\xi}$ for $e_{\xi} = e^{ix\xi}$ belongs to $S^{\mu}(\Omega \times \mathbf{R}^n)$ with $\mathrm{Op}(a_0) = A_0$ and that the map $A_0 \to a_0$ is an isomorphism

$$L^{\mu}(\Omega)_K \to S^{\mu}(\Omega \times \mathbf{R}^n)_K := \{a_0(x,\xi) : \ A_0 \in L^{\mu}(\Omega)_K\},$$

where $S^{\mu}(\Omega \times \mathbf{R}^n)_K$ is a closed subspace of $S^{\mu}(\Omega \times \mathbf{R}^n)$, the space $L^{\mu}(\Omega)_K$ can be equipped with a natural Fréchet topology. So $L^{\mu}(\Omega)$ is a Frèchet space since it is a non-direct sum of Fréchet spaces. In an analogous manner we get a Fréchet topology in $L^{\mu}_{cl}(\Omega)$.

Let us recall the general definition of a non-direct sum of two Fréchet spaces E_0, E_1 contained as vector subspaces in a certain topological Hausdorff space. The non-direct sum

$$E_0 + E_1 = \{e_0 + e_1 : \ e_0 \in E_0, e_1 \in E_1\}$$

is isomorphic to $E_0 \oplus E_1 / \Delta$ with $\Delta = \{(e_1 - e) : \ e \in E_0 \cap E_1\}$. Both $E_0 \oplus E_1$ and Δ are Fréchet spaces in a natural way. So $E_0 + E_1$ is Fréchet with the quotient topology. The construction can easily be generalized to finitely many summands.

Another useful notation is the following. If E is a Fréchet space which is a left module over an algebra A, we set

$$[a]E = \text{closure of } \quad \{ae : \ e \in E\} \ \text{ in } \ E,$$

for $a \in A$. In an analogous sense we use the notation $E[b]$ when E is a right module over an algebra B, for $b \in B$, or notation like $[a]E[b]$. In particular, we will use spaces $[\varphi]L^{\mu}(\Omega)[\psi]$ or $[\varphi]L^{\mu}_{cl}(\Omega)[\psi]$ for $\varphi, \psi \in C_0^{\infty}(\Omega)$.

Next we remind of the invariance of pseudo-differential operators under diffeomorphisms $\chi : \ \Omega \to \tilde{\Omega}$ for open sets $\Omega, \tilde{\Omega} \subseteq \boldsymbol{R}^n$. Denoting by $\chi^* : C_0^{\infty}(\tilde{\Omega}) \to C_0^{\infty}(\Omega), \ \ C^{\infty}(\tilde{\Omega}) \to C^{\infty}(\Omega)$ the function pull-backs, to every $A \in L^{\mu}(\Omega)$ we can form the operator push-forward

$$\chi_* A = (\chi^*)^{-1} A \chi^* : \ C_0^{\infty}(\tilde{\Omega}) \to C^{\infty}(\tilde{\Omega}).$$

Then χ_* induces isomorphisms

$$\chi_* : \ L^{\mu}(\Omega) \stackrel{\cong}{\to} L^{\mu}(\tilde{\Omega}), \quad L^{\mu}_{cl}(\Omega) \stackrel{\cong}{\to} L^{\mu}_{cl}(\tilde{\Omega})$$

This gives rise to pseudo-differential operators on C^{∞} manifolds. For instance, let X be a closed compact C^{∞} manifold and $\kappa : \ U \to \Omega$ a chart on X. Then the invariance allows us to define the spaces $L^{\mu}(U) = (\kappa^{-1})_* L^{\mu}(\Omega)$ and $L^{\mu}_{cl}(U) = (\kappa^{-1})_* L^{\mu}_{cl}(\Omega)$. Now let $\{U_1, \ldots, U_N\}$ be an open covering of X by coordinate neighborhoods, $\{\varphi_1, \ldots, \varphi_N\}$ a subordinate partition of unity, and $\{\psi_1, \ldots, \psi_N\}$ another system of functions $\psi_j \in C_0^{\infty}(U_j)$ with $\varphi_j \psi_j = \varphi_j$ for all j. Then we can form

$$L^{\mu}(X) = \sum_{j=1}^{N} [\varphi_j] L^{\mu}(U_j)[\psi_j] + L^{-\infty}(X)$$

as a non-direct sum of Fréchet spaces, where $L^{-\infty}(X)$ is identified with $C^{\infty}(X \times X)$ via some Riemannian metric on X. Analogously we obtain $L^{\mu}_{cl}(X)$. This construction can easily be generalized to any paracompact C^{∞} manifold X. The invariance of Sobolev space distributions under diffeomorphisms allows us the corresponding global definitions. If X is a paracompact C^{∞} manifold we have an evident definition of the space $H^s_{comp}(X)$ of compactly supported distributions of Sobolev smoothness $s \in \boldsymbol{R}$ and the space $H^s_{loc}(X)$ of distributions that are locally of Sobolev smoothness $s \in \boldsymbol{R}$. The latter space is Fréchet in a natural way while $H^s_{comp}(X)$ is an inductive limit of Hilbert spaces. If X is compact then

$$H^s(X) := \ H^s_{comp}(X) = H^s_{loc}(X).$$

Theorem 1.2 *Every $A \in L^{\mu}(X)$, $A: C_0^{\infty}(X) \to C^{\infty}(X)$, extends to a continuous operator*

$$A: H^s_{comp}(X) \to H^s_{loc}(X)$$

for every $s \in \boldsymbol{R}$.

Remark 1.3 *(i) The operator $\mathcal{M}_{\varphi}$ of multiplication by a function $\varphi \in \mathcal{S}(\boldsymbol{R}^n)$ induces a continuous operator $\mathcal{M}_{\varphi}: H^s(\boldsymbol{R}^n) \to H^s(\boldsymbol{R}^n)$, and $\varphi \to \mathcal{M}_{\varphi}$ is continuous as operator $\mathcal{S}(\boldsymbol{R}^n) \to \mathcal{L}(H^s(\boldsymbol{R}^n))$ for every $s \in \boldsymbol{R}$.*

(ii) The pseudo-differential operator $Op(a)$ for a symbol $a(\xi) \in S^{\mu}(\boldsymbol{R}^n)$ induces a continuous operator

$$Op(a): H^s(\boldsymbol{R}^n) \to H^{s-\mu}(\boldsymbol{R}^n),$$

and $\mathrm{Op}: S^{\mu}(\boldsymbol{R}^n) \to \mathcal{L}(H^s(\boldsymbol{R}^n), H^{s-\mu}(\boldsymbol{R}^n))$ *is continuous for every $s \in \boldsymbol{R}$.*

1.2. ABSTRACT WEDGE SOBOLEV SPACES

The abstract wedge Sobolev spaces were introduced in [30] for studying pseudo-differential operators on a manifold with edges. The definition can be motivated by an anisotropic reformulation of $H^s(\boldsymbol{R}^{n+q})$ with respect to the fictitious edge $\boldsymbol{R}^q \ni y$ in $\boldsymbol{R}^n \times \boldsymbol{R}^q \ni (x, y)$. Consider a group of isomorphisms $\kappa_{\lambda}: H^s(\boldsymbol{R}^n) \to H^s(\boldsymbol{R}^n)$, $\lambda \in \boldsymbol{R}_+$, continuous in λ with respect to the strong operator topology, given by

$$(\kappa_{\lambda} u)(x) = \lambda^{\frac{n}{2}} u(\lambda x), \quad \lambda \in \boldsymbol{R}_+, \quad u \in H^s(\boldsymbol{R}^n),$$

$s \in \boldsymbol{R}$. Set

$$\kappa(\eta) := \kappa_{[\eta]} \quad \text{for} \quad \eta \in \boldsymbol{R}^q$$

with the function $\eta \to [\eta]$ defined in the previous section. Then we have the following elementary result:

Proposition 1.4 *The space $H^s(\boldsymbol{R}^{n+q})$, $s \in \boldsymbol{R}$, is the closure of $\mathcal{S}(\boldsymbol{R}^{n+q})$ with respect to the norm*

$$\left\{ \int [\eta]^{2s} \| \kappa^{-1}(\eta) (\mathcal{F}_{y \to \eta} v)(\eta) \|^2_{H^s(\boldsymbol{R}^n)} d\eta \right\}^{\frac{1}{2}},$$

where $v \in \mathcal{S}(\boldsymbol{R}^{n+q})$ is interpreted as an element $v(y) \in \mathcal{S}(\boldsymbol{R}^q_y, \mathcal{S}(\boldsymbol{R}^n))$, $\mathcal{F}_{y \to \eta}$ is the Fourier transform in $\boldsymbol{R}^q$, applied to vector-valued functions, and $\kappa(\eta)$ acts on the values of v in $\mathcal{S}(\boldsymbol{R}^n_x)$ for every $\eta \in \boldsymbol{R}^q$.

Definition 1.5 *Let E be a Hilbert space, and $\{\kappa_\lambda\}_{\lambda\in\mathbf{R}_+}$ be a strongly continuous group of isomorphisms on E, i.e.,*

(i) $\kappa_\lambda : E \to E$ is an isomorphism, $\lambda \in \mathbf{R}_+$,

(ii) $\lambda \to \kappa_\lambda e \in C(\mathbf{R}_+, E)$ for every $e \in E$,

(iii) $\kappa_\lambda\kappa_\rho = \kappa_{\lambda\rho}$ for all $\lambda, \rho \in \mathbf{R}_+$, $\kappa_1 = id$.

Then $\mathcal{W}^s(\mathbf{R}^q, E)$, $s \in \mathbf{R}$, is the closure of $\mathcal{S}(\mathbf{R}^q, E)$ with respect to the norm

$$\|u\|_{\mathcal{W}^s(\mathbf{R}^q,E)} = \left\{\int [\eta]^{2s}\|\kappa^{-1}(\eta)(\mathcal{F}_{y\to\eta}u)(\eta)\|_E^2 d\eta\right\}^{\frac{1}{2}},$$

$\kappa(\eta) = \kappa_{[\eta]}$. This space is called an abstract wedge Sobolev space of smoothness $s \in \mathbf{R}$ with respect to $\{\kappa_\lambda\}_{\lambda\in\mathbf{R}_+}$.

Instead of (ii) we also write $\{\kappa_\lambda\}_{\lambda\in\mathbf{R}_+} \in C(\mathbf{R}_+, \mathcal{L}_\sigma(E))$, σ indicating the strong operator topology.

From the properties (i), (ii), (iii) it follows that there are constants $c > 0$, $M > 0$ such that $\|\kappa(\eta)\|_{\mathcal{L}(E)} \le c[\eta]^M$ for all $\eta \in \mathbf{R}^q$.

Remark 1.6 *(i) Equivalent norms in E give rise to equivalent norms in $\mathcal{W}^s(\mathbf{R}^q, E)$.*

(ii) Replacing $[\eta]$ by $\langle\eta\rangle$ in the norm expression yields an equivalent norm in $\mathcal{W}^s(\mathbf{R}^q, E)$.

(iii) The choice of $\{\kappa_\lambda\}_{\lambda\in\mathbf{R}_+}$ is essential for the space $\mathcal{W}^s(\mathbf{R}^q, E)$; it is fixed in concrete cases and therefore suppressed in the notation.

(iv) Definition 1.5 also makes sense for a Banach space E. Many results on abstract wedge Sobolev spaces remain true in this case.

Example 1.7 *(i) For $E = H^s(\mathbf{R}^n)$, $(\kappa_\lambda u)(x) = \lambda^{\frac{n}{2}}u(\lambda x)$, $\lambda \in \mathbf{R}_+$, we have $\{\kappa_\lambda\}_{\lambda\in\mathbf{R}_+} \in C(\mathbf{R}_+, \mathcal{L}_\sigma(H^s(\mathbf{R}^n))$ and*

$$\mathcal{W}^s(\mathbf{R}^q, H^s(\mathbf{R}^n)) = H^s(\mathbf{R}^q \times \mathbf{R}^n)$$

for every $s \in \mathbf{R}$.

(ii) For $E = H^s(\mathbf{R}_+)$ $(= \{u|_{\mathbf{R}_+} : u \in H^s(\mathbf{R})\})$, $(\kappa_\lambda u)(t) = \lambda^{\frac{1}{2}}u(\lambda t)$, $\lambda \in \mathbf{R}_+$, we have $\{\kappa_\lambda\}_{\lambda\in\mathbf{R}_+} \in C(\mathbf{R}_+, \mathcal{L}_\sigma(H^s(\mathbf{R}_+))$ and

$$\mathcal{W}^s(\mathbf{R}^q, H^s(\mathbf{R}_+)) = H^s(\mathbf{R}^q \times \mathbf{R}_+) = \{v|_{\mathbf{R}^q\times\mathbf{R}_+} : v \in H^s(\mathbf{R}^q \times \mathbf{R})\}$$

for every $s \in \mathbf{R}$.

(iii) Let E be an arbitrary Hilbert space, $\kappa_\lambda = id_E$ for all $\lambda \in \mathbf{R}$. Then Definition 1.5 gives us $H^s(\mathbf{R}^q, E)$ which is the Sobolev space of E-valued distributions of smoothness s in the standard sense, i.e., with the norm

$$\left\{\int [\eta]^{2s}\|(\mathcal{F}u)(\eta)\|_E^2 d\eta\right\}^{\frac{1}{2}}.$$

Setting $T = \mathcal{F}^{-1}\kappa^{-1}(\eta)\mathcal{F}$ *for an arbitrary fixed* $\{\kappa_\lambda\}_{\lambda\in\boldsymbol{R}_+} \in C(\boldsymbol{R}_+, \mathcal{L}_\sigma(E))$ *we obtain for the associated space* $\mathcal{W}^s(\boldsymbol{R}^q, E)$ *an isometric isomorphism*

$$T:\ \mathcal{W}^s(\boldsymbol{R}^q, E) \to H^s(\boldsymbol{R}^q, E)$$

for every $s \in \boldsymbol{R}$. *In particular,* $H^0(\boldsymbol{R}^q, E) = L^2(\boldsymbol{R}^q, E)$ *(= the space of square integrable E-valued functions in* $\boldsymbol{R}^q$*); this is a consequence of Plancherel's theorem in the Hilbert space-valued case. Moreover,*

$$\mathcal{W}^\infty(\boldsymbol{R}^q, E) = \cap_{s\in\boldsymbol{R}}\mathcal{W}^s(\boldsymbol{R}^q, E)$$

is independent of the particular choice of $\{\kappa_\lambda\}_{\lambda\in\boldsymbol{R}_+}$, *i.e.,* $\mathcal{W}^\infty(\boldsymbol{R}^q, E) = H^\infty(\boldsymbol{R}^q, E)$.

(iv) For $E = \boldsymbol{C}^N$ *we always set* $\kappa_\lambda = id_E$ *for all* $\lambda \in \boldsymbol{R}_+$. *Then* $\mathcal{W}^s(\boldsymbol{R}^q, \boldsymbol{C}^N) = H^s(\boldsymbol{R}^q, \boldsymbol{C}^N) = H^s(\boldsymbol{R}^q) \otimes \boldsymbol{C}^N$.

(v) $\mathcal{W}^s(\boldsymbol{R}^q, E)$ *can be endowed with a Hilbert space scalar product that generates the norm. The action*

$$(\chi_\lambda f)(y) = \lambda^{\frac{q}{2}}\kappa_\lambda f(\lambda y) \quad \textit{for} \quad f(y) \in \mathcal{S}(\boldsymbol{R}^q, E),$$

$\lambda \in \boldsymbol{R}_+$, *extends by continuity to a group*

$$\{\chi_\lambda\}_{\lambda\in\boldsymbol{R}_+} \in C(\boldsymbol{R}_+, \mathcal{L}_\sigma(\mathcal{W}^s(\boldsymbol{R}^q, E)))$$

with the properties of Definition 1.5, now with respect to $\mathcal{W}^s(\boldsymbol{R}^q, E)$. *Then*

$$\mathcal{W}^s(\boldsymbol{R}^p, \mathcal{W}^s(\boldsymbol{R}^q, E)) = \mathcal{W}^s(\boldsymbol{R}^{p+q}, E).$$

Analogously to the scalar theory we have the following characterization of $\mathcal{W}^s(\boldsymbol{R}^q, E)$ as a subspace of $\mathcal{S}'(\boldsymbol{R}^q, E) = \mathcal{L}(\mathcal{S}(\boldsymbol{R}^q), E)$:

Proposition 1.8 *For every fixed* $s \in \boldsymbol{R}$, $\mathcal{W}^s(\boldsymbol{R}^q, E)$ *equals the subspace of all* $u \in \mathcal{S}'(\boldsymbol{R}^q, E)$ *for which* $\langle\eta\rangle^s\kappa^{-1}(\eta)(\mathcal{F}_{y\to\eta}u)(\eta) \in L^2(\boldsymbol{R}^q, E)$.

For an open set $\Omega \subseteq \boldsymbol{R}^q$ we define

$$\mathcal{W}^s_{comp}(\Omega, E) = \{u \in \mathcal{W}^s(\boldsymbol{R}^q, E) :\ \text{supp}\ u \subset \Omega\ \ \text{compact}\},$$
$$\mathcal{W}^s_{loc}(\Omega, E) = \{u \in \mathcal{D}'(\Omega, E) :\ \varphi u \in \mathcal{W}^s_{comp}(\Omega, E) \text{ for every } \varphi \in C_0^\infty(\Omega)\},$$

$s \in \boldsymbol{R}$; $\mathcal{D}'(\Omega, E) = \mathcal{L}(C_0^\infty(\Omega), E)$. The space $\mathcal{W}^s_{comp}(\Omega, E)$ is an inductive limit of Hilbert spaces and $\mathcal{W}^s_{loc}(\Omega, E)$ is a Fréchet space. Let $E = \text{proj lim}_{j\in\boldsymbol{N}} E^j$ be the projective limit of Hilbert spaces $\{E^j\}_{j\in\boldsymbol{N}}$ with continuous embeddings $E^{j+1} \hookrightarrow E^j$ for all $j \in \boldsymbol{N}$ and an action

$$\{\kappa_\lambda\}_{\lambda\in\boldsymbol{R}_+} \in C(\boldsymbol{R}_+, \mathcal{L}_\sigma(E^0))$$

with the properties in Definition 1.5 that restricts to

$$\{\kappa_\lambda\}_{\lambda\in\boldsymbol{R}_+} \in C(\boldsymbol{R}_+, \mathcal{L}_\sigma(E^j))$$

with the analogous properties for all j. We then obtain natural embeddings $\mathcal{W}^s(\boldsymbol{R}^q, E^{j+1}) \hookrightarrow \mathcal{W}^s(\boldsymbol{R}^q, E^j)$ for all j, and we set

$$\mathcal{W}^s(\boldsymbol{R}^q, E) = \text{ind } \lim_{j\in\boldsymbol{R}} \mathcal{W}^s(\boldsymbol{R}^q, E^j).$$

Analogously to the scalar theory we have invariance of the wedge Sobolev spaces under diffeomorphisms $\chi: \ \Omega \to \tilde{\Omega}$ for open $\Omega, \tilde{\Omega} \subseteq \boldsymbol{R}^q$:

Theorem 1.9 *Let $\chi: \ \Omega \to \tilde{\Omega}$ be a diffeomorphism. Then the pull-back $\chi^*: \ \mathcal{D}'(\tilde{\Omega}, E) \to \mathcal{D}'(\Omega, E)$ restricts to isomorphisms*

$$\chi^*: \ \mathcal{W}^s_{comp}(\tilde{\Omega}, E) \to \mathcal{W}^s_{comp}(\Omega, E), \quad \mathcal{W}^s_{loc}(\tilde{\Omega}, E) \to \mathcal{W}^s_{loc}(\Omega, E)$$

for all $s \in \boldsymbol{R}$.

For a proof, cf. [34], [37] or [6].

This permits us to define the spaces

$$\mathcal{W}^s_{comp}(Y, E), \quad \mathcal{W}^s_{loc}(Y, E) \tag{1.6}$$

on any paracompact C^∞ manifold Y analogously to the case of $\boldsymbol{C}$-valued Sobolev spaces $H^s_{comp}(Y)$ and $H^s_{loc}(Y)$, respectively.

1.3. PSEUDO-DIFFERENTIAL OPERATORS WITH OPERATOR-VALUED SYMBOLS

Let E and $\tilde{E}$ be Hilbert spaces with strongly continuous groups of isomorphisms $\{\kappa_\lambda\}_{\lambda\in\boldsymbol{R}_+}$ and $\{\tilde{\kappa}_\lambda\}_{\lambda\in\boldsymbol{R}_+}$, respectively, cf. Definition 1.5. For $\mu \in \boldsymbol{R}$ and open $U \subseteq \boldsymbol{R}^p$, we denote $S^{(\mu)}(U \times (\boldsymbol{R}^q \setminus \{0\}); E, \tilde{E})$ as the subspace of all $a_{(\mu)}(y, \eta) \in C^\infty(U \times (\boldsymbol{R}^q \setminus \{0\}), \mathcal{L}(E, \tilde{E}))$ satisfying

$$a_{(\mu)}(y, \lambda\eta) = \lambda^\mu \tilde{\kappa}_\lambda a_{(\mu)}(y, \eta) \kappa_\lambda^{-1} \quad \text{for all} \quad \lambda \in \boldsymbol{R}_+, \quad y \in U, \eta \in \boldsymbol{R}^q \setminus \{0\}$$

Definition 1.10 *Let $U \subseteq \boldsymbol{R}^p$ be open and $\mu \in \boldsymbol{R}$. Then $S^\mu(U \times \boldsymbol{R}^q; E, \tilde{E})$ is the space of all $a(y, \eta) \in C^\infty(U \times \boldsymbol{R}^q, \mathcal{L}(E, \tilde{E}))$ such that the semi-norms*

$$\sup_{\substack{y\in K \\ \eta\in\boldsymbol{R}^q}} \langle\eta\rangle^{-\mu+|\beta|} \|\tilde{\kappa}^{-1}(\eta)\{D_y^\alpha D_\eta^\beta a(y, \eta)\}\kappa(\eta)\|_{\mathcal{L}(E,\tilde{E})} \tag{1.7}$$

are finite for all $\alpha \in \boldsymbol{N}^p$, $\beta \in \boldsymbol{N}^q$, $K \subset\subset U$. Moreover, $S^\mu_{cl}(U \times \boldsymbol{R}^q; E, \tilde{E})$ denotes the subspace of all

$$a(y, \eta) \in S^\mu(U \times \boldsymbol{R}^q; E, \tilde{E})$$

for which there exists a sequence $a_{(\mu-j)}(y,\eta) \in S^{(\mu-j)}(U\times(\mathbf{R}^q\setminus\{0\});E,\tilde{E})$, $j\in\mathbf{N}$, *such that for any excision function* $\chi(\eta)$ *in* $\mathbf{R}^q$

$$a(y,\eta)-\chi(\eta)\sum_{j=0}^{N}a_{(\mu-j)}(y,\eta)\in S^{\mu-(N+1)}(U\times\mathbf{R}^q;E,\tilde{E}) \tag{1.8}$$

for all $N\in\mathbf{N}$. *The elements of* $S^\mu(U\times\mathbf{R}^q;E,\tilde{E})$ *are called operator-valued symbols, those in* $S^\mu_{cl}(U\times\mathbf{R}^q;E,\tilde{E})$ *classical operator-valued symbols of order* μ.

The homogeneous components $a_{(\mu-j)}(y,\eta)$, $j\in\mathbf{N}$, of an $a(y,\eta)\in S^\mu_{cl}(U\times\mathbf{R}^q;E,\tilde{E})$ are uniquely determined. Similarly to the scalar case (see Section 1.1) the spaces $S^\mu(U\times\mathbf{R}^q;E,\tilde{E})$ and $S^\mu_{cl}(U\times\mathbf{R}^q;E,\tilde{E})$ may be endowed with natural Fréchet topologies. The subspaces of y-independent elements $S^\mu(\mathbf{R}^q;E,\tilde{E})$ and $S^\mu_{cl}(\mathbf{R}^q;E,\tilde{E})$ are closed in the topology induced by $S^\mu(U\times\mathbf{R}^q;E,\tilde{E})$ and $S^\mu_{cl}(U\times\mathbf{R}^q;E,\tilde{E})$, respectively. We have

$$S^\mu(U\times\mathbf{R}^q;E,\tilde{E})=C^\infty(U,S^\mu(\mathbf{R}^q;E,\tilde{E})) \tag{1.9}$$

and analogously for classical symbols.

Example 1.11 *Let* $A=\sum_{|\alpha+\beta|\le\mu}a_{\alpha\beta}(x,y)D_x^\alpha D_y^\beta$ *be a differential operator in* $\mathbf{R}^n\times\Omega\ni(x,y)$ *for an open set* $\Omega\subseteq\mathbf{R}^q$, *with* $a_{\alpha\beta}(x,y)\in C^\infty(\mathbf{R}^n\times\Omega)$. *Assume that* $a_{\alpha\beta}(x,y)$ *is independent of* x *for* $|x|>$ *const. Set* $E=H^s(\mathbf{R}^n)$, $\tilde{E}=H^{s-\mu}(\mathbf{R}^n)$, *both endowed with the group actions* $\kappa_\lambda:\ u(x)\to\lambda^{\frac{n}{2}}u(\lambda x)$, $\lambda\in\mathbf{R}_+$. *Then the operator family*

$$a(y,\eta):=\sum_{|\alpha+\beta|\le\mu}a_{\alpha\beta}(x,y)D_x^\alpha\eta^\beta\in C^\infty(\Omega_y\times\mathbf{R}^q_\eta,\mathcal{L}(H^s(\mathbf{R}^n),H^{s-\mu}(\mathbf{R}^n)))$$

is an element of $S^\mu(\Omega\times\mathbf{R}^q;H^s(\mathbf{R}^n),H^{s-\mu}(\mathbf{R}^n))$ *for all* $s\in\mathbf{R}$. *If the coefficients* $a_{\alpha\beta}$ *are independent of* x *then*

$$a(y,\eta)\in S^\mu_{cl}(\Omega\times\mathbf{R}^q;H^s(\mathbf{R}^n),H^{s-\mu}(\mathbf{R}^n))\quad\textit{for all}\quad s\in\mathbf{R},$$

and the homogeneous component of order $\mu-j$ *is*

$$a_{(\mu-j)}(y,\eta)=\sum_{|\alpha+\beta|=\mu-j}a_{\alpha\beta}(x,y)D_x^\alpha\eta^\beta.$$

Remark 1.12 *In the applications the spaces* E *and* $\tilde{E}$ *run over scales* $\{E^s\}_{s\in\mathbf{R}}$ *and* $\{\tilde{E}^t\}_{t\in\mathbf{R}}$, *respectively, and then it is natural to consider symbols of the classes*

$$S^\mu(U\times\mathbf{R}^q;E^s,\tilde{E}^{s-\mu})\quad\textit{for all}\quad s\in\mathbf{R}$$

(or s in some subset of $\boldsymbol{R}$). The above example shows that instead of (1.6) we may expect the more precise property

$$D_y^\alpha D_\eta^\beta S^\mu(U \times \boldsymbol{R}^q; E^s, \tilde{E}^{s-\mu}) \subseteq S^{\mu-|\beta|}(U \times \boldsymbol{R}^q; E^s, \tilde{E}^{s-\mu+|\beta|})$$

for all multi-indices $\alpha \in \boldsymbol{N}^p$, $\beta \in \boldsymbol{N}^q$, and all s. This will be the case in the concrete symbol classes for manifolds with edges.

Example 1.13 *The operator $\mathcal{M}_\varphi$ of multiplication by a function*

$$\varphi(x, y) \in C^\infty(\Omega_y, \mathcal{S}(\boldsymbol{R}_x^n)),$$

which $\Omega \subseteq \boldsymbol{R}^q$ is open, is a symbol in $S^0(\Omega \times \boldsymbol{R}^q; H^s(\boldsymbol{R}^n), H^s(\boldsymbol{R}^n))$, and the map

$$C^\infty(\Omega_y, \mathcal{S}(\boldsymbol{R}_x^n)) \to S^0(\Omega \times \boldsymbol{R}^q; H^s(\boldsymbol{R}^n), H^s(\boldsymbol{R}^n)) : \ \varphi \to \mathcal{M}_\varphi$$

is continuous for all $s \in \boldsymbol{R}$. Note that $\mathcal{M}_\varphi$ is independent of η.

Example 1.14 *Let us set $r'u := u(0)$ for $u \in \mathcal{S}(\boldsymbol{R})$. It is easy to see that $r' : \mathcal{S}(\boldsymbol{R}) \to \boldsymbol{C}$ extends to a continuous operator $r' : H^s(\boldsymbol{R}) \to \boldsymbol{C}$ for all $s > \frac{1}{2}$. Then*

$$r' \in S^{\frac{1}{2}}(\boldsymbol{R}^q; H^s(\boldsymbol{R}), \boldsymbol{C}) \quad \textit{for every} \quad s > \frac{1}{2}.$$

Also here there is no dependence on $\eta \in \boldsymbol{R}^q$. Moreover, the map $k(\eta) : \boldsymbol{C} \to \mathcal{S}(\boldsymbol{R})$ defined by $k(\eta)c = \psi(t[\eta])c$ for any $\psi \in \mathcal{S}(\boldsymbol{R})$ is a symbol

$$k(\eta) \in S^{-\frac{1}{2}}(\boldsymbol{R}^q; \boldsymbol{C}, H^s(\boldsymbol{R})) \quad \textit{for every} \quad s \in \boldsymbol{R}.$$

If $\psi(t) \in C_0^\infty(\boldsymbol{R})$ is identically 1 in a neighborhood of $t = 0$ we have $r'k(\eta) = id_{\boldsymbol{C}}$.

We will obtain below many other non-trivial examples of operator-valued symbols.

Remark 1.15 *The kernel cut-off construction of Theorem 1.1 has an obvious analogue in the operator-valued case.*

To every for $a(y, y', \eta) \in S^\mu(\Omega \times \Omega \times \boldsymbol{R}^q; E, \tilde{E})$ for $\Omega \subseteq \boldsymbol{R}^q$ and $\mu \in \boldsymbol{R}$ we can form the associated pseudo-differential operator

$$\mathrm{Op}(a)u(y) = \int\int e^{i(y-y')\eta} a(y, y', \eta) u(y') dy' đ\eta,$$

interpreted as an operator-valued analogue of an oscillatory integral. Then

$$\mathrm{Op}(a) : \ C_0^\infty(\Omega, E) \to C^\infty(\Omega, \tilde{E})$$

is continuous. This gives rise to the spaces of pseudo-differential operators with operator-valued symbols

$$L^\mu(\Omega; E, \tilde{E}) = \{\mathrm{Op}(a) : \ a(y, y', \eta) \in S^\mu(\Omega \times \Omega \times \boldsymbol{R}^q; E, \tilde{E})\}$$

and analogously to $L^\mu_{cl}(\Omega; E, \tilde{E})$. The space

$$L^{-\infty}(\Omega; E, \tilde{E}) = \cap_{\mu \in \boldsymbol{R}} L^\mu(\Omega; E, \tilde{E})$$

coincides with the space of all integral operators with kernels in $C^\infty(\Omega \times \Omega, \mathcal{L}(E, \tilde{E}))$.

Let $\chi : \ \Omega \to \tilde{\Omega}$ be a diffeomorphism for open sets $\Omega, \tilde{\Omega} \subseteq \boldsymbol{R}^q$. Then the function pull-backs

$$\chi^* : \ C_0^\infty(\tilde{\Omega}, E) \to C_0^\infty(\Omega, E), \quad C^\infty(\tilde{\Omega}, \tilde{E}) \to C^\infty(\Omega, \tilde{E})$$

give rise to the operator push-forward

$$\chi_* A = (\chi^*)^{-1} A \chi^* : \ C_0^\infty(\tilde{\Omega}, E) \to C^\infty(\tilde{\Omega}, \tilde{E}).$$

Analogously to the scalar calculus we obtain an isomorphism

$$\chi_* : \ L^\mu(\Omega; E, \tilde{E}) \to L^\mu(\tilde{\Omega}; E, \tilde{E})$$

for every $\mu \in \boldsymbol{R}$, which restricts to an isomorphism between the corresponding spaces of classical pseudo-differential operators.

On a paracompact C^∞ manifold Y we an define (as in Section 1.1) the global spaces of pseudo-differential operators

$$L^\mu(Y; E, \tilde{E}) \ \text{ and } \ L^\mu_{cl}(Y; E, \tilde{E}), \tag{1.10}$$

respectively. The homogeneous principal symbol of order μ of an operator $A \in L^\mu_{cl}(Y; E, \tilde{E})$ is invariantly defined as a function on the cotangent bundle minus the zero section $T^*Y \setminus 0$ with values in $\mathcal{L}(E, \tilde{E})$. It will be denoted by $\sigma^\mu_\wedge(A)(y, \eta)$, where homogeneity of order μ is defines as

$$\sigma^\mu_\wedge(A)(y, \lambda\eta) = \lambda^\mu \tilde{\kappa}_\lambda \sigma^\mu_\wedge(A)(y, \eta) \kappa_\lambda^{-1} \quad \text{for all} \quad \lambda \in \boldsymbol{R}_+.$$

Theorem 1.16 *Every* $A \in L^\mu(Y; E, \tilde{E})$, $A : \ C_0^\infty(Y, E) \to C^\infty(Y, \tilde{E})$, *extends to a continuous operator*

$$A : \ \mathcal{W}^s_{comp}(Y, E) \to \mathcal{W}^{s-\mu}_{loc}(Y, \tilde{E})$$

for every $s \in \boldsymbol{R}$.

The proof can be reduced to a corresponding result in local coordinates. Let $\Omega \subseteq \boldsymbol{R}^q$ open. Writing $S^\mu(\Omega \times \boldsymbol{R}^q; E, \tilde{E})$ as a projective tensor product of $C^\infty(\Omega)$ and $S^\mu(\boldsymbol{R}^q; E, \tilde{E})$ see (1.8), the assertion follows from the continuity of the operator $\mathcal{M}_\varphi$ of multiplication by $\varphi \in C^\infty(\Omega)$, cf. Example 1.13, and of $\mathrm{Op}(a)$ when $a(\eta)$ has constant coefficients. The tensor product argument is based on the following theorem : Let E, F be Fréchet spaces. Every $g \in E\hat{\otimes}_\pi F$ can be written as a convergent sum

$$g = \sum_{j=0}^{\infty} \lambda_j e_j \otimes f_j \quad \text{for suitable} \quad \lambda_j \in \boldsymbol{C}, e_j \in E, f_j \in F \tag{1.11}$$

with $\sum |\lambda_j| < \infty$ and e_j and f_j tending to zero in E and F, respectively, as $j \to \infty$.

Remark 1.17 *(i) The operator $\mathcal{M}_\varphi$ of multiplication by a function $\varphi \in \mathcal{S}(\boldsymbol{R}^q)$ induces a continuous operator*

$$\mathcal{M}_\varphi : \mathcal{W}^s(\boldsymbol{R}^q, E) \to \mathcal{W}^s(\boldsymbol{R}^q, E),$$

and the operator $\mathcal{S}(\boldsymbol{R}^q) \to \mathcal{L}(\mathcal{W}^s(\boldsymbol{R}^q, E)) : \varphi \to \mathcal{M}_\varphi$ is continuous for every $s \in \boldsymbol{R}$.

(ii) For a symbol $a(\eta) \in S^\mu(\boldsymbol{R}^q; E, \tilde{E})$, the pseudo-differential operator

$$Op(a) : \mathcal{W}^s(\boldsymbol{R}^q, E) \to \mathcal{W}^{s-\mu}(\boldsymbol{R}^q, \tilde{E}),$$

is continuous and

$$\mathrm{Op} : S^\mu(\boldsymbol{R}^q; E, \tilde{E}) \to \mathcal{L}(\mathcal{W}^s(\boldsymbol{R}^q, E), \mathcal{W}^{s-\mu}(\boldsymbol{R}^q, \tilde{E})),$$

$a \to \mathrm{Op}(a)$, *is continuous for every $s \in \boldsymbol{R}$.*

The symbol classes of Definition 1.10 are necessary also in the version of Fréchet spaces $E, \tilde{E}$ that are projective limits of Hilbert spaces $\{E^j\}_{j \in \boldsymbol{N}}$ with continuous embeddings $E^{j+1} \hookrightarrow E^j$, $\{\tilde{E}^j\}_{j \in \boldsymbol{N}}$ with continuous embeddings $\tilde{E}^{j+1} \hookrightarrow \tilde{E}^j$ for all j, and where the corresponding groups $\{\kappa_\lambda\}_{\lambda \in \boldsymbol{R}_+}$ and $\{\tilde{\kappa}_\lambda\}_{\lambda \in \boldsymbol{R}_+}$, first given on E^0 and $\tilde{E}^0$, respectively, have restrictions to

$$\{\kappa_\lambda\}_{\lambda \in \boldsymbol{R}_+} \in C(\boldsymbol{R}_+, \mathcal{L}_\sigma(E^j)), \quad \{\tilde{\kappa}_\lambda\}_{\lambda \in \boldsymbol{R}_+} \in C(\boldsymbol{R}_+, \mathcal{L}_\sigma(\tilde{E}^j))$$

with the mentioned properties, for all j. An operator $a : E \to \tilde{E}$ is continuous if for every $k \in \boldsymbol{N}$ there exists some $j(k) \in \boldsymbol{N}$ such that $a : E^{j(k)} \to \tilde{E}^k$ is continuous (Here a is restricted to $E^{j(k)}$). If $j : \boldsymbol{N} \to \boldsymbol{N}$ is a given function we obtain the space

$$\text{proj lim}_{k \in \boldsymbol{N}} S^\mu(U \times \boldsymbol{R}^q; E^{j(k)}, \tilde{E}^k) \tag{1.12}$$

and then $S^\mu(U \times \boldsymbol{R}^q; E, \tilde{E})$ is defined as the union over all (1.12) where j runs over all $j: \boldsymbol{N} \to \boldsymbol{N}$. In an analogous manner we can proceed for classical symbols. This yields the corresponding classes of classical pseudo-differential operators (1.9) for Fréchet spaces $E, \tilde{E}$. The simpler case when E is a Hilbert space and $\tilde{E}$ a Fréchet space of the mentioned kind, is of particular interest. Then the symbol and operator spaces are Fréchet in a natural way. The elements of the pseudo-differential calculus have straightforward generalizations to the case of Fréchet spaces E, $\tilde{E}$ (up to minor modifications, concerning restrictions to fixed $j: \boldsymbol{N} \to \boldsymbol{N}$).

1.4. EXAMPLES: GREEN, TRACE AND POTENTIAL OPERATORS IN BOUNDARY VALUE PROBLEMS WITH THE TRANSMISSION PROPERTY

The pseudo-differential operators with the transmission property in the sense of the algebra [3] on a manifold with boundary contain an ideal of operators that is responsible for the structure of Green's function and boundary (trace) and potential conditions in elliptic boundary value problems. We shall give a description here in terms of operator-valued symbols in local coordinates $(t, y) \in \overline{\boldsymbol{R}}_+ \times \Omega$ for an open set $\Omega \subseteq \boldsymbol{R}^q$.

Set $\mathcal{S}(\overline{\boldsymbol{R}}_+ \times \overline{\boldsymbol{R}}_+) = \mathcal{S}(\boldsymbol{R} \times \boldsymbol{R})|_{\overline{\boldsymbol{R}}_+ \times \overline{\boldsymbol{R}}_+}$, $\mathcal{S}(\overline{\boldsymbol{R}}_+) = \mathcal{S}(\boldsymbol{R})|_{\overline{\boldsymbol{R}}_+}$; these are (nuclear) Fréchet spaces. Write $\mathcal{S}(\overline{\boldsymbol{R}}_+)$ as projective limit

$$\mathcal{S}(\overline{\boldsymbol{R}}_+) = \text{proj lim}_{k \in \boldsymbol{N}} \langle t \rangle^{-k} H^k(\boldsymbol{R}_+).$$

On the spaces $E^k = \langle t \rangle^{-k} H^k(\boldsymbol{R}_+)$ we define the group of isomorphisms $\{\kappa_\lambda\}_{\lambda \in \boldsymbol{R}_+}$, acting as $(\kappa_\lambda u)(t) = \lambda^{\frac{1}{2}} u(\lambda t)$, $\lambda \in \boldsymbol{R}_+$. Recall that a continuous operator $a: L^2(\boldsymbol{R}_+) \to L^2(\boldsymbol{R}_+)$ is an integral operator with kernel $g_a(t, t') \in \mathcal{S}(\overline{\boldsymbol{R}}_+ \times \overline{\boldsymbol{R}}_+)$ if and only if

$$a: L^2(\boldsymbol{R}_+) \to \mathcal{S}(\overline{\boldsymbol{R}}_+), \quad a^*: L^2(\boldsymbol{R}_+) \to \mathcal{S}(\overline{\boldsymbol{R}}_+)$$

are continuous. Here a^* is the $L^2(\boldsymbol{R}_+)$-adjoint.

Theorem 1.18 *Let*

$$g(y, \eta; t, t') \in C^\infty(\Omega_y, \mathcal{S}(\overline{\boldsymbol{R}}_{+,t} \times \overline{\boldsymbol{R}}_{+,t'}, S^0_{cl}(\boldsymbol{R}^q_\eta)) \tag{1.13}$$

be an arbitrary function and $\mu \in \boldsymbol{R}$. *Then*

$$a(y, \eta) u(t) := [\eta]^{\mu+1} \int_0^\infty g(y, \eta; t[\eta], t'[\eta]) u(t') dt'$$

is a (y, η)*-dependent family of Hilbert-Schmidt operators in* $L^2(\boldsymbol{R}_+)$, *and we have*

$$\begin{aligned} a(y, \eta) &\in S^\mu_{cl}(\Omega \times \boldsymbol{R}^q; L^2(\boldsymbol{R}_+), \mathcal{S}(\overline{\boldsymbol{R}}_+)), \\ a^*(y, \eta) &\in S^\mu_{cl}(\Omega \times \boldsymbol{R}^q; L^2(\boldsymbol{R}_+), \mathcal{S}(\overline{\boldsymbol{R}}_+)), \end{aligned}$$

where $*$ *denotes the point-wise adjoint in* $L^2(\boldsymbol{R}_+)$.

Remark 1.19 *The symbols $a(y,\eta)$ of Theorem 1.18 induce by restriction or extension symbols*

$$a(y,\eta) \in S^{\mu}_{cl}(\Omega \times \boldsymbol{R}^q; H^s(\boldsymbol{R}_+), \mathcal{S}(\overline{\boldsymbol{R}}_+))$$

with

$$a^*(y,\eta) \in S^{\mu}_{cl}(\Omega \times \boldsymbol{R}^q; H^s(\boldsymbol{R}_+), \mathcal{S}(\overline{\boldsymbol{R}}_+)$$

for all $s \in \boldsymbol{R}$, $s > -\frac{1}{2}$.

More generally, for every $d \in \boldsymbol{N}$ we can form

$$a(y,\eta) = \sum_{j=0}^{d} a_j(y,\eta) \frac{\partial^j}{\partial t^j} \tag{1.14}$$

for symbols a_j of order $\mu - j$ in the sense of Theorem 1.18. Then

$$a(y,\eta) \in S^{\mu}_{cl}(\Omega \times \boldsymbol{R}^q; H^s(\boldsymbol{R}_+), \mathcal{S}(\overline{\boldsymbol{R}}_+)) \quad \text{for every} \quad s > d - \frac{1}{2}.$$

An operator $\mathrm{Op}(a) + c$ with an operator-valued symbol, $a(y,\eta)$ of the form (1.13) and $C = \sum_{j=0}^{d} C_j \frac{\partial^j}{\partial t^j}$, where C_j is an integral operator with kernel in $C^\infty(\overline{\boldsymbol{R}}_+ \times \Omega \times \overline{\boldsymbol{R}}_+ \times \Omega)$, is called a Green operator of order μ and type d in the algebra of boundary value problems with the transmission property. For $\mu \in \boldsymbol{N}$ this coincides with the definition in [3], though the equivalence is not completely obvious; it was obtained in [33].

In pseudo-differential boundary value problems it is interesting to generate the trace and potential operators by symbols in an analogous manner. To this end we pass to symbols of block matrix form

$$a(y,\eta) := \begin{pmatrix} a_{11} & a_{12} \\ a_{21} & a_{22} \end{pmatrix} (y,\eta) : \begin{array}{c} H^s(\boldsymbol{R}_+) \\ \oplus \\ \boldsymbol{C}^{N_-} \end{array} \to \begin{array}{c} \mathcal{S}(\overline{\boldsymbol{R}}_+) \\ \oplus \\ \boldsymbol{C}^{N_+} \end{array}, \quad s > d - \frac{1}{2}, \tag{1.15}$$

for certain $N_-, N_+ \in \boldsymbol{N}$. The left upper corner $a_{11}(y,\eta)$ is assumed as in (1.13). For describing the structure of the remaining entries we consider for simplicity $N_- = N_+ = 1$. Then $a_{22}(y,\eta)$ is an element of $S^{\mu}_{cl}(\Omega \times \boldsymbol{R}^q)$.

Theorem 1.20 *Let*

$$g_{12}(y,\eta;t) \in C^\infty(\Omega_y, \mathcal{S}(\overline{\boldsymbol{R}}_{+,t}, S^0_{cl}(\boldsymbol{R}^q_\eta)))$$

be an arbitrary function and $\mu \in \boldsymbol{R}$. Then

$$a_{12}(y,\eta)c := [\eta]^{\mu+\frac{1}{2}} g_{12}(y,\eta;t[\eta])c \tag{1.16}$$

for $c \in \boldsymbol{C}$ *is a symbol in* $S^{\mu}_{cl}(\Omega \times \boldsymbol{R}^q; \boldsymbol{C}, \mathcal{S}(\overline{\boldsymbol{R}}_+))$. *Moreover, let*

$$g_{21}(y,\eta;t') \in C^{\infty}(\Omega_y, \mathcal{S}(\overline{\boldsymbol{R}}_{+,t'}, S^0_{cl}(\boldsymbol{R}^q_\eta))$$

be an arbitrary function and

$$a_{21}(y,\eta)u = [\eta]^{\mu+\frac{1}{2}} \int_0^\infty g_{21}(y,\eta;t'[\eta])u(t')dt' \tag{1.17}$$

for $u \in H^s(\boldsymbol{R}_+)$, $s > -\frac{1}{2}$. *Then*

$$a_{21}(y,\eta) \in S^{\mu}_{cl}(\Omega \times \boldsymbol{R}^q; H^s(\boldsymbol{R}_+), \boldsymbol{C})$$

for every $s > -\frac{1}{2}$.

More generally, if $a_{21,j}(y,\eta)$ is of the form (1.16) with $\mu - j$ instead of μ, we have from the second part of Theorem 1.20

$$a_{21}(y,\eta) = \sum_{j=0}^{d} a_{21,j}(y,\eta)\frac{\partial^j}{\partial t^j} \in S^{\mu}_{cl}(\Omega \times \boldsymbol{R}^q; H^s(\boldsymbol{R}^q), \boldsymbol{C}) \tag{1.18}$$

for every $s > d - \frac{1}{2}$.

If we assume a_{11}, a_{12}, a_{21}, a_{22} in the mentioned form where, in particular, a_{21} is given by (1.17), then we obtain for (1.14) (in the case $N_+ = N_- = 1$)

$$a(y,\eta) \in S^{\mu}_{cl}(\Omega \times \boldsymbol{R}^q; H^s(\boldsymbol{R}_+) \oplus \boldsymbol{C}, \mathcal{S}(\overline{\boldsymbol{R}}_+) \oplus \boldsymbol{C}), \quad s > d - \frac{1}{2}.$$

In particular, for $d = 0$, it follows that

$$a^*(y,\eta) \in S^{\mu}_{cl}(\Omega \times \boldsymbol{R}^q; H^s(\boldsymbol{R}_+) \oplus \boldsymbol{C}, \mathcal{S}(\overline{\boldsymbol{R}}_+) \oplus \boldsymbol{C}), \quad s > -\frac{1}{2}.$$

Here a^* is defined point-wise by

$$(u, a^*v)_{L^2(\boldsymbol{R}_+)\oplus \boldsymbol{C}} = (au, v)_{L^2(\boldsymbol{R}_+)\oplus \boldsymbol{C}}$$

for all $u, v \in C_0^\infty(\boldsymbol{R}_+) \oplus \boldsymbol{C}$.

For arbitrary N_+, N_- the relations are analogous.

Corollary 1.21 *Let* $a(y,\eta)$ *be given by (1.14). Then*

$$\operatorname{Op}(a) : \begin{array}{c} H^s_{comp(y)}(\boldsymbol{R}_+ \times \Omega) \\ \oplus \\ H^s_{comp}(\Omega, \boldsymbol{C}^{N_-}) \end{array} \to \begin{array}{c} H^{s-\mu}_{loc(y)}(\boldsymbol{R}_+ \times \Omega) \\ \oplus \\ H^{s-\mu}_{loc}(\Omega, \boldsymbol{C}^{N_+}) \end{array},$$

is continuous for every $s > d - \frac{1}{2}$. *The subscript comp(y), loc(y) indicate comp, loc with respect to* $y \in \Omega$.

The operators in Corollary 1.21 are, modulo smoothing operators, the Green, trace and potential operators in boundary value problems, including the right lower corners which are classical pseudo-differential operators on Ω. We omit here an explicit definition of the smoothing trace and potential operators. More details, including a new complete description of the algebra of boundary value problems with the transmission property in terms of operator-valued symbols, will be given in [37]. The definitions and results of this section play the role of examples to the general set-up of Sections 1.2, 1.3. The proofs of the theorems are based on tensor product arguments and the observation that, for instance, when we assume instead of (1.12)

$$g(y;t,t') \in C^{\infty}(\Omega_y, \mathcal{S}(\overline{\boldsymbol{R}}_{+,t} \times \overline{\boldsymbol{R}}_{+,t'})), \tag{1.19}$$

the operator-valued symbol

$$a(y,\eta)u(t) = [\eta]^{\mu+1} \int_0^{\infty} g(y;t[\eta],t'[\eta])u(t')dt' \tag{1.20}$$

satisfies

$$a(y,\lambda\eta) = \lambda^{\mu}\kappa_{\lambda}a(y,\eta)\kappa_{\lambda}^{-1} \quad \text{for all} \quad \lambda \geq 1, |\eta| \geq c$$

for a constant $c > 0$. Moreover, the correspondence $g(y;t,t') \to a(y,\eta)$ is continuous in the sense

$$C^{\infty}(\Omega, \mathcal{S}(\overline{\boldsymbol{R}}_{+} \times \overline{R}_{+})) \to S^{\mu}_{cl}(\boldsymbol{R}^{q}; L^{2}(\boldsymbol{R}_{+}), \mathcal{S}(\overline{\boldsymbol{R}}_{+})),$$

and the same for the adjoints. Similarly we can argue for the other entries of the block matrix symbols.

2. Parameter-dependent pseudo-differential operators and cone theory

2.1. PARAMETER-DEPENDENT PSEUDO-DIFFERENTIAL CALCULUS

Pseudo-differential operators on a manifold with conical singularities require the parameter-dependent calculus on the base X of the cone. The parameter-dependent families will be used below as operator-valued symbols for pseudo-differential operators with respect to the Mellin transform along $\boldsymbol{R}_{+}$, the cone axis. Let us stress that symbols of this kind are not operator-valued symbols in the sense of Section 1.2. The calculus here, also called the order reduction approach, contains symbols acting on a space globally without any reference to some group $\{\kappa_{\lambda}\}_{\lambda \in \boldsymbol{R}_{+}}$.

In this section the parameter space will be $\Lambda = \boldsymbol{R}^{l} \ni \lambda$. The dependence of symbols on the parameters will be assumed in a way (which is sufficient for our purposes) that λ is formally involved as an additional covariable. In other words we consider symbols

$$a(x,\xi,\lambda) \in S^{\mu}_{(cl)}(U \times \boldsymbol{R}^{n+l}_{\xi,\lambda}) \quad \text{for open} \quad U \subseteq \boldsymbol{R}^m, \quad \mu \in \boldsymbol{R}.$$

By (cl) we indicate that the notions and results make sense both for classical and non-classical symbols in (ξ,λ). Note that

$$a(x,\xi,\lambda_0) \in S^{\mu}_{(cl)}(U \times \boldsymbol{R}^{n}_{\xi}) \quad \text{for every fixed} \quad \lambda_0.$$

Thus, if $\Omega \subseteq \boldsymbol{R}^n$ is an open set, every $a(x,x',\xi,\lambda) \in S^{\mu}_{(cl)}(\Omega \times \Omega \times \boldsymbol{R}^{n+l}_{\xi,\lambda})$ gives rise to a space of λ-dependent pseudo-differential operators

$$L^{\mu}_{(cl)}(\Omega;\Lambda) = \left\{\mathrm{Op}(a)(\lambda) : \ a \in S^{\mu}_{(cl)}(\Omega \times \Omega \times \boldsymbol{R}^{n+l}_{\xi,\lambda})\right\}.$$

In this calculus the space of smoothing elements is

$$L^{-\infty}(\Omega;\Lambda) = \mathcal{S}(\Lambda, L^{-\infty}(\Omega)).$$

The invariance under diffeomorphisms $\chi: \ \Omega \to \tilde{\Omega}$ holds in the parameter-dependent case as well, here in the sense that the point-wise operator push-forward induces an isomorphism

$$\chi_* : \ L^{\mu}_{(cl)}(\Omega;\Lambda) \to L^{\mu}_{(cl)}(\tilde{\Omega};\Lambda).$$

Thus, if X is a (paracompact) C^∞ manifold and $\kappa : \ U \to \Omega$ a chart, we can introduce $L^{\mu}_{(cl)}(U;\Lambda)$.

The space $L^{-\infty}(X)$ is identified with $C^\infty(X \times X)$ via a given Riemannian metric on X. So we can talk about

$$L^{-\infty}(X;\Lambda) := \ \mathcal{S}(\Lambda, L^{-\infty}(X))$$

which is the space of parameter-dependent smoothing operators on X. If $\{U_j\}_{j\in\boldsymbol{N}}$ is a locally finite open covering of X by coordinate neighborhoods, $\{\varphi_j\}_{j\in\boldsymbol{N}}$ a subordinate partition of unity, further $\{\psi_j\}_{j\in\boldsymbol{N}}$ a system of functions $\psi_j \in C^\infty(U_j)$ with $\varphi_j\psi_j = \varphi_j$ for all j, we denote by $L^{\mu}_{(cl)}(X;\Lambda)$ the space of all operators

$$A(\lambda) = \sum_{j\in\boldsymbol{N}} \varphi_j A_j(\lambda)\psi_j + C(\lambda) \tag{2.1}$$

for arbitrary $A_j(\lambda) \in L^{\mu}_{(cl)}(U_j;\Lambda)$ and $C(\lambda) \in L^{-\infty}(X;\Lambda)$. The space $L^{\mu}_{(cl)}(X;\Lambda)$ has a natural Fréchet topology. In Section 1.1 we have discussed in detail how to introduce adequate Fréchet topologies in symbol and operator spaces. In the present case the arguments are analogous. Also in future, if we introduce some space and speak about its Fréchet topology, it will usually be an immediate consequence of the definition. The easy details will be left to the reader. Every $A(\lambda) \in L^{\mu}_{cl}(X;\Lambda)$ has a well-defined parameter-dependent homogeneous principal symbol of order μ

$$\sigma^{\mu}_{\psi;\lambda}(A)(x,\xi,\lambda) \in C^{\infty}(T^*X \times \Lambda \setminus 0),$$

0 indicates $(\xi, \lambda) = 0$, and the homogeneity means

$$\sigma^{\mu}_{\psi;\lambda}(A)(x,\delta\xi,\delta\lambda) = \delta^{\mu}\sigma^{\mu}_{\psi}(A)(x,\xi,\lambda)$$

for all $\delta \in \boldsymbol{R}_+$.

Definition 2.1 $A(\lambda) \in L^{\mu}_{cl}(X;\Lambda)$ *is called parameter-dependent elliptic of order* μ *if* $\sigma^{\mu}_{\psi;\lambda}(A) \neq 0$ *on* $T^*X \times \Lambda \setminus 0$.

The parameter-dependent ellipticity can also be studied in the non-classical case and all essential consequences hold in analogous form. However we discuss from now on the simpler classical case which is of importance in the applications below.

Theorem 2.2 *Let* $A(\lambda) \in L^{\mu}_{cl}(X;\Lambda)$ *be parameter-dependent elliptic of order* μ. *Then there exists a* $B(\lambda) \in L^{-\mu}_{cl}(X;\Lambda)$ *such that*

$$A(\lambda)B(\lambda) - 1, \quad B(\lambda)A(\lambda) - 1 \in L^{-\infty}(X;\Lambda), \tag{2.2}$$

where $\sigma^{-\mu}_{\psi;\lambda}(B) = (\sigma^{\mu}_{\psi;\lambda}(A))^{-1}$

An operator family $B(\lambda) \in L^{-\mu}_{cl}(X;\Lambda)$ satisfying the relations (2.2) is called a parameter-dependent parametrix of $A(\lambda)$. Note that the existence of the compositions is ensured by a particular choice of $B(\lambda)$ namely to be properly supported. We will not comment on this further, since we are mainly interested in the case that X is compact. This will be assumed in the following.

Recall that we have on X the scale of Sobolev spaces $\{H^s(X)\}_{s \in \boldsymbol{R}}$. In the above notions concerning $\Lambda = \boldsymbol{R}^l$ we may also assume $l = 0$, i.e., that Λ disappears.

We denote the homogeneous principal symbol of $A \in L^{\mu}_{cl}(X)$ of order μ by $\sigma^{\mu}_{\psi}(A)$ which is an element in $C^{\infty}(T^*X \setminus 0)$, homogeneous of order μ in the covariables.

The above parameter-dependent ellipticity induces for $A \in L^{\mu}_{cl}(X)$ the "usual" ellipticity, which requires $\sigma^{\mu}_{\psi}(A) \neq 0$ on $T^*X \setminus 0$, and we want to mention the folloing classical result.

Theorem 2.3 *Let $A \in L^{\mu}_{cl}(X)$ be given. Then the following conditions are equivalent:*

(i) A is elliptic of order μ,

(ii) the operator

$$A: \ H^s(X) \to H^{s-\mu}(X) \tag{2.3}$$

is Fredholm for an $s = s_0 \in \mathbf{R}$.

Moreover the ellipticity of A of order μ implies the existence of a $B \in L^{-\mu}_{cl}(X)$ with $AB - 1$, $BA - 1 \in L^{-\infty}(X)$ and (2.3) is Fredholm for all $s \in \mathbf{R}$. Here $\sigma^{-\mu}_{\psi}(B) = (\sigma^{\mu}_{\psi}(A))^{-1}$.

Remark 2.4 *Another well-known property is the elliptic regularity of solutions of $Au = f \in H^r(X)$, $r \in \mathbf{R}$, when A is elliptic. It says that every solution $u \in H^{-\infty}(X)$ belongs to $H^{r+\mu}(X)$. In particular, $\ker A \subset C^{\infty}(X)$. Since $A^* \in L^{\mu}_{cl}(X)$ is also elliptic, we have $\ker A^* \subset C^{\infty}(X)$ and*

$$\begin{aligned} \text{ind } A &= \dim\ker A - \dim\text{coker } A \\ &= \dim\ker A - \dim\ker A^* \end{aligned}$$

is independent of s.

If $A(\lambda) \in L^{\mu}_{cl}(X;\Lambda)$ is parameter-dependent elliptic of order μ, the operator $A(\lambda_0) \in L^{\mu}_{cl}(X)$ is elliptic in the usual sense for every fixed λ_0. Hence

$$A(\lambda): \ H^s(X) \to H^{s-\mu}(X) \tag{2.4}$$

is a λ-dependent family of Fredholm operators. If we say nothing else we will always assume in the parameter-dependent case that $l \geq 1$.

Example 2.5 *Let us form an operator $A(\lambda)$ by the expression (2.1) with $C(\lambda) = 0$ and $A_j(\lambda)$ defined in local coordinates by* $\mathrm{Op}((c^2 + |\xi|^2 + |\lambda|^2)^{\mu/2})$ *for some $c > 0$. Then $\sigma^{\mu}_{\psi;\lambda}(A) = (|\xi|^2 + |\lambda|^2)^{\mu/2}$, and hence $A(\lambda)$ is parameter-dependent elliptic of order μ.*

Remark 2.6 *Let $C(\lambda) \in L^{-\infty}(X;\Lambda)$ and assume that*

$$1 + C(\lambda): \ H^s(X) \to H^s(X) \tag{2.5}$$

is invertible for a fixed $s \in \mathbf{R}$, for all $\lambda \in \Lambda$. Then (2.5) is invertible for all $s \in \mathbf{R}$, $\lambda \in \Lambda$, and there exists a $G(\lambda) \in L^{-\infty}(X;\Lambda)$ such that

$$(1 + C(\lambda))^{-1} = 1 + G(\lambda).$$

Theorem 2.7 *Let $A(\lambda) \in L^{\mu}_{cl}(X;\Lambda)$ be parameter-dependent elliptic of order μ. Then (2.4) is a family of Fredholm operators (2.4) of index zero, and there exists a constant $c > 0$ such that (2.4) is an isomorphism for all $\lambda \in \Lambda$ with $|\lambda| \geq c$ and all $s \in \mathbf{R}$.*

Theorem 2.8 *To every $\mu \in \boldsymbol{R}$ there exists an $R^{\mu}(\lambda) \in L^{\mu}_{cl}(X;\Lambda)$ which is parameter-dependent elliptic of order μ such that*

$$R^{\mu}(\lambda): \ H^{s}(X) \to H^{s-\mu}(X)$$

is an isomorphism for all $\lambda \in \Lambda$, $s \in \boldsymbol{R}$, and $(R^{\mu}(\lambda))^{-1} \in L^{-\mu}_{cl}(X;\Lambda)$.

We will call any such $R^{\mu}(\lambda)$ a parameter-dependent reduction of orders (of order μ).

2.2. OPERATORS OF FUCHS TYPE

Let B be a manifold with conical singularities and $\boldsymbol{B}$ the associated stretched manifold. According to the notations in the beginning we have a collar neighborhoods V of $\partial\boldsymbol{B} \cong X$ of the form $[0,1) \times X$ with a corresponding splitting of coordinates in (t,x). Since only a neighborhoods of $t=0$ is of interest in the specific assertions concerning the conical singularities we may (and sometimes will) identify the neighborhoods with $\overline{\boldsymbol{R}}_{+} \times X$. If M is a paracompact C^{∞} manifold we denote by Doff$^{\mu}(M)$ the space of all differential operators on M of order μ with C^{∞} coefficients (in local coordinates). Diff$^{\mu}(M)$ is a Fréchet space in a natural way.

An operator $A \in \text{Diff}^{\mu}(\text{int } \boldsymbol{B})$ is said to be of Fuchs type if it has the form

$$A = t^{-\mu}\sum_{j=0}^{\mu} a_j(t)(-t\frac{\partial}{\partial t})^j \tag{2.6}$$

near $\partial\boldsymbol{B}$ in the splitting of coordinates (t,x) over V, where

$$a_j(t) \in C^{\infty}([0,1), \text{Diff}^{\mu-j}(X)), \quad j = 0,\ldots,\mu.$$

Let us consider the Mellin transform

$$(Mu)(z) = \int_0^{\infty} t^{z-1}u(t)dt$$

for $u(t) \in C_0^{\infty}(\boldsymbol{R}_{+}, C^{\infty}(X))$, $z \in \boldsymbol{C}$. Then $(Mu)(z) \in \mathcal{A}(\boldsymbol{C}, C^{\infty}(X))$ and

$$(M^{-1}g)(t) = \frac{1}{2\pi}\int_{\Gamma_\beta} t^{-z}g(z)dz,$$

where $\Gamma_\beta = \{z: \ \text{Re } z = \beta\}$.

In the following we will use notations like

$$\mathcal{S}(\Gamma_\beta), \quad H^{s}(\Gamma_\beta), \quad L^{\mu}_{cl}(X;\Gamma_\beta), \ldots$$

in the sense of the corresponding objects with respect to $\boldsymbol{R}$, identified with Γ_β via $\tau \to \beta + i\tau$.

Observe that

$$M^{-1}zM = -t\frac{\partial}{\partial t}.$$

This suggests to introduce pseudo-differential operators with respect to the Mellin transform with operator-valued symbols

$$h(t,t',z) \in C^\infty(\boldsymbol{R}_+ \times \boldsymbol{R}_+, L^\mu_{cl}(X;\Gamma_{\frac{1}{2}})), \quad \mu \in \boldsymbol{R},$$

where $z \in \Gamma_{\frac{1}{2}}$ is the covariable:

$$(\mathrm{op}_M(h)u)(t) = M^{-1}_{z\to t}\{M_{t'\to z}h(t,t',z)u(t)\},$$

$u \in C_0^\infty(\boldsymbol{R}_+, C^\infty(X))$. The action of the symbol function is taken as a (t,t',z)-dependent pseudo-differential operator on $u(t')$ with respect to the dependence of x globally along X; then the operators follows by applying a Mellin oscillatory integral argument with operator-valued symbols.

We will also consider weighted Mellin pseudo-differential operators

$$(\mathrm{op}_M^\gamma(h)u)(t) = t^\gamma \mathrm{op}_M(T^{-\gamma}h)t^{-\gamma}u$$

for $(T^{-\gamma}h)(t,t',z) = h(t,t',z-\gamma)$, $\gamma \in \boldsymbol{R}$, where here

$$h(t,t',z) \in C^\infty(\boldsymbol{R}_+ \times \boldsymbol{R}_+, L^\mu_{cl}(X;\Gamma_{\frac{1}{2}-\gamma})).$$

Then

$$\mathrm{op}_M^\gamma(h):\ C_0^\infty(\boldsymbol{R}_+, C^\infty(X)) \to C^\infty(\boldsymbol{R}_+, C^\infty(X)) \tag{2.7}$$

is continuous for every γ.

In particular, let A be of Fuchs type on $X^\wedge = \boldsymbol{R}_+ \times X$, i.e., of the form (2.7). Then we have

$$h(t,z) = \sum_{j=0}^{\mu} a_j(t)z^j \in C^\infty(\overline{\boldsymbol{R}}_+, L^\mu_{cl}(X;\Gamma_\rho))$$

for every $\rho \in \boldsymbol{R}$, and $A = t^{-\mu}\mathrm{op}_M^\beta(h)$ for every $\beta \in \boldsymbol{R}$, as an operator on $C_0^\infty(\boldsymbol{R}_+, C^\infty(X))$.

Note that when A is of Fuchs type over $\boldsymbol{B}$ the homogeneous principal symbol of A of order μ

$$\sigma_\psi^\mu \in C^\infty(T^*(\mathrm{int}\ \boldsymbol{B}) \setminus 0) \tag{2.8}$$

is over $T^*(\mathrm{int}\ V) \setminus 0$, $V \cong [0,1) \times X$, of the form

$$t^{-\mu}p_{(\mu)}(t,x,t\tau,\xi) \tag{2.9}$$

where $p_{(\mu)}(t,x,\tilde{\tau},\xi)$ is C^∞ up to $t=0$. With A we also associate a principal conormal symbol

$$\sigma_M^\mu(A)(z) = \sum_{j=0}^{\mu} a_j(0)t^j \tag{2.10}$$

which belongs to $L_{cl}^\mu(X;\Gamma_\beta)$ for every $\beta \in \boldsymbol{R}$.

Definition 2.9 *An operator $A \in \mathit{Diff}^\mu(\mathrm{int}\,\boldsymbol{B})$ of Fuchs type is called elliptic of order μ, with respect to the weight $\gamma \in \boldsymbol{R}$, if*

(i) $\sigma_\psi^\mu(A) \neq 0$ on $T^*(\mathrm{int}\,\boldsymbol{B}) \setminus 0$ and if $t^\mu \sigma_\psi^\mu(A)(t,x,t^{-1}\tau,\xi) \neq 0$ on $T^*V \setminus 0$

(ii)

$$\sigma_M^\mu(A)(z): \ H^s(X) \to H^{s-\mu}(X)$$

is an isomorphism for every $z \in \Gamma_{\frac{n+1}{2}-\gamma}$, $s \in \boldsymbol{R}$, where $n = \dim X$.

In view of the remarks after Theorem 2.3 it suffices to require the condition (ii) only for a particular $s = s_0 \in \boldsymbol{R}$. Then it is satisfied for all $s \in \boldsymbol{R}$.

The ellipticity should allow the parametrix construction within an adapted algebra of pseudo-differential operators. This is just the "cone algebra" which is then closed under parametrix construction for elliptic elements in general. The ellipticity of Fuchs type differential operators was studied by Kondrat'ev [15], who also has established the Fredholm property in weighted Sobolev spaces for the case of compact $\boldsymbol{B}$. In addition there was characterized the elliptic regularity with asymptotics near the conical singularities.

The weighted Sobolev space $\mathcal{H}^{s,\gamma}(X^\wedge)$ of smoothness $s \in \boldsymbol{R}$ and weight $\gamma \in \boldsymbol{R}$ on the stretched cone $X^\wedge$ can be defined as closure of $C_0^\infty(\boldsymbol{R}_+, C^\infty(X))$ with respect to the norm

$$\left\{ \frac{1}{2\pi i} \int_{\Gamma_{\frac{n+1}{2}-\gamma}} \| R^s(\mathrm{Im}\, z)(Mu)(z) \|_{L^2(X)}^2 dz \right\}^{\frac{1}{2}},$$

where $R^s(\tau) \in L_{cl}^s(X;\boldsymbol{R})$ is an order reducing family in the sense of Theorem 2.8. Another choice of $R^s(\tau)$ gives rise to an equivalent norm. There are, of course alternative definitions of $\mathcal{H}^{s,\gamma}(X^\wedge)$. In particular, for $n=0$ and $s \in \boldsymbol{N}$, $\gamma = 0$, we have

$$H^{s,0}(\boldsymbol{R}_+) = \left\{ u \in L^2(\boldsymbol{R}_+) : \ \left(t\frac{\partial}{\partial t} \right)^k u \in L^2(\boldsymbol{R}_+) \quad \text{for} \quad 0 \leq k \leq s \right\}.$$

Then $\mathcal{H}^{s,0}(\boldsymbol{R}_+)$ for arbitrary $s \in \boldsymbol{R}$ can be defined by duality and interpolation.

We have

$$\mathcal{H}^{s,\gamma+\rho}(X^\wedge) = t^\rho \mathcal{H}^{s,\gamma}(X^\wedge) \quad \text{for every} \quad s,\gamma,\rho \in \boldsymbol{R}$$

and

$$\mathcal{H}^{s,\gamma}(X^\wedge) \subset H^s_{loc}(X^\wedge) \quad \text{for every} \quad s,\gamma \in \boldsymbol{R}.$$

It can be proved that the operator of multiplication $\mathcal{M}_\varphi$ by a $\varphi(t,x) \in C_0^\infty(\overline{\boldsymbol{R}}_+ \times X)$ is continuous in the sense

$$\mathcal{M}_\varphi : \mathcal{H}^{s,\gamma}(X^\wedge) \to \mathcal{H}^{s,\gamma}(X^\wedge) \tag{2.11}$$

and that $\varphi \to \mathcal{M}_\varphi$ is continuous as $C_0^\infty(\overline{\boldsymbol{R}}_+ \times X) \to \mathcal{L}(\mathcal{H}^{s,\gamma}(X^\wedge))$ for every$s,\gamma \in \boldsymbol{R}$.

A cut-off function on $\overline{\boldsymbol{R}}_+$ is any $\omega(t) \in C_0^\infty(\overline{\boldsymbol{R}}_+)$ which is real valued and $\omega(t) \equiv 1$ in a neighborhoods of $t = 0$.

Theorem 2.10 *Let $\omega(t)$, $\tilde{\omega}(t)$ be cut-off functions,*

$$h(t,t',z) \in C^\infty(\overline{\boldsymbol{R}}_+ \times \overline{\boldsymbol{R}}_+, L^\mu(X;\Gamma_{\frac{n+1}{2}-\gamma}));$$

then

$$\omega \mathrm{op}_M^{\gamma-\frac{n}{2}}(h)\tilde{\omega} : \mathcal{H}^{s,\gamma}(X^\wedge) \to \mathcal{H}^{s-\mu,\gamma}(X^\wedge)$$

is continuous for every $s \in \boldsymbol{R}$.

This is straightforward when h is independent of t,t'. In general the assertion follows from a tensor product argument, using (2.11). There is also a direct proof, analogously to well-known techniques for standard pseudo-differential operators in Sobolev spaces based on the Fourier transform.

Let us also define a space $H^s_{cone}(X^\wedge) \ni u(t,x)$, first for the case when $u(t,x)$ is supported in a coordinate neighborhoods U of X. Let $\kappa_1 : U \to W$ be a diffeomorphism to an open set $W \subset S^n = \{\tilde{x} \in \boldsymbol{R}^{n+1} : |\tilde{x}| = 1\}$ and $\kappa : \ \boldsymbol{R}_+ \times U \to W^\wedge = \{\tilde{x} \in \boldsymbol{R}^{n+1} \setminus \{0\} : \ \tilde{x}/|\tilde{x}| \in W\}$ defined by $\kappa(t,x) = t\kappa_1(x)$. Then the condition is $(1-\omega(t))u(t(\tilde{x}),x(\tilde{x})) \in H^s(\boldsymbol{R}^{n+1}_{\tilde{x}})$, where $(t(\tilde{x}),x(\tilde{x})) = \kappa^{-1}(\tilde{x})$ for an arbitrary cut-off function $\omega(t)$. The space $H^s_{cone}(X^\wedge)$ in general then follows by a partition of unity argument. Let $H^s_{cone}(X^\wedge)_\varepsilon$ for any $\varepsilon > 0$ be the subspace of all $u \in H^s_{cone}(X^\wedge)$ that vanish for $0 < t < \varepsilon$. Then $H^s_{cone}(X^\wedge)_\varepsilon$ is a Hilberzitable Banach space in a natural way. Set $[1-\omega]H^s_{cone}(X^\wedge) = [1-\omega]H^s_{cone}(X^\wedge)_\varepsilon$ for a cut-off function $\omega(t)$ such that $1-\omega$ vanishes for $0 < t < \varepsilon$ and define

$$\mathcal{K}^{s,\gamma}(X^\wedge) = [\omega]\mathcal{H}^{s,\gamma}(X^\wedge) + [1-\omega]H^s_{cone}(X^\wedge)$$

as non-direct sum of spaces for every $s,\gamma \in \boldsymbol{R}$. Then, also $\mathcal{K}^{s,\gamma}(X^\wedge)$ is a Hilbertizable Banach space contained in $H^s_{loc}(X^\wedge)$, and it is independent of the particular choice of ω. For $n = \dim X = 0$ we simply have

$$\mathcal{K}^{s,\gamma}(\boldsymbol{R}_+) = [\omega]\mathcal{H}^{s,\gamma}(\boldsymbol{R}_+) + [1-\omega]H^s(\boldsymbol{R}_+),$$

$H^s(\boldsymbol{R}_+) = H^s(\boldsymbol{R})|_{\boldsymbol{R}_+}$.

Remark 2.11 *Set* $(\kappa_\lambda u)(t,x) = \lambda^{\frac{n+1}{2}} u(\lambda t, x)$ *for* $\lambda \in \mathbf{R}_+$. *Then* $\{\kappa_\lambda\}_{\lambda \in \mathbf{R}_+}$ *is a group of isomorphisms on the space* $\mathcal{K}^{s,\gamma}(X^\wedge)$, *with the properties required in Definition 1.5. This gives rise to a scale of weighted wedge Sobolev spaces*

$$\mathcal{W}^s(\mathbf{R}^q, \mathcal{K}^{s,\gamma}(X^\wedge)) \quad \textit{for} \quad s, \gamma \in \mathbf{R}.$$

Theorem 2.12 *Let* $A \in \mathit{Diff}^\mu(\mathrm{int}\, \mathbf{B})$ *be of Fuchs type and assume that* $p_{(\mu)}(0, x, \tilde{\tau}, \xi) \neq 0$ *for all* x *and* $(\tilde{\tau}, \xi) \neq 0$ *(cf. the notation (2.9)). Then* $\sigma_M^\mu(A)(z)|_{\Gamma_\beta} \in L_{cl}^\mu(X; \Gamma_\beta)$ *is parameter-dependent elliptic for every* $\beta \in \mathbf{R}$, *uniformly in* $c \leq \beta \leq c'$ *for every* $c \leq c'$. *Moreover, there is a countable set* $D \subset \mathbf{C}$ *with* $D \cap K$ *finite for every* $K \subset\subset \mathbf{C}$ *such that*

$$\sigma_M^\mu(A)(z): \; H^s(X) \to H^{s-\mu}(X)$$

is an isomorphism for all $z \in \mathbf{C} \setminus D$, $s \in \mathbf{R}$.

This will be explained in more detail in Section 2.3 below, where we formulate a refinement of this result.

Corollary 2.13 *Under the conditions of Theorem 2.12 there exists a countable set* $E \subset \mathbf{R}$ *with* $E \cap K$ *finite for every* $K \subset\subset \mathbf{R}$ *such that*

$$\sigma_M^\mu(A)(z): \; H^s(X) \to H^{s-\mu}(X)$$

is an isomorphism for all $z \in \Gamma_\beta$, $\beta \notin E$, $s \in \mathbf{R}$.

Theorem 2.14 *Let* $A \in \mathit{Diff}^\mu(\mathrm{int}\, \mathbf{B})$ *be of Fuchs type. Then the following conditions are equivalent:*
(i) A *is elliptic of order* μ, *with respect to the weight* $\gamma \in \mathbf{R}$,
(ii) the operator

$$A: \; \mathcal{H}^{s,\gamma}(\mathbf{B}) \to \mathcal{H}^{s-\mu,\gamma-\mu}(\mathbf{B}) \tag{2.12}$$

is Fredholm for an $s = s_0 \in \mathbf{R}$.

This result is standard after the work of Kondrat'ev [15]. It is true also for general elliptic operators in the cone algebra below, as well as the elliptic regularity, which says that when A is elliptic and $u \in \mathcal{H}^{-\infty,\gamma}(\mathbf{B})$ is a solution of $Au = f \in \mathcal{H}^{s,\gamma-\mu}(\mathbf{B})$ then $u \in \mathcal{H}^{r+\mu,\gamma}(\mathbf{B})$, for arbitrary $r \in \mathbf{R}$. The elliptic regularity also holds with respect to subspaces with asymptotics as they will be described below.

Remark 2.15 *(i) or (ii) in Theorem 2.3 imply that (2.12) is a Fredholm operator for every* $s \in \mathbf{R}$. *Then* ker A *and* coker A *are s-independent subspaces of* $\mathcal{H}^{\infty,\gamma}(\mathbf{B})$ *and* $\mathcal{H}^{\infty,\gamma-\mu}(\mathbf{B})$, *respectively, and hence* ind A *is independent of* s.

2.3. HOLOMORPHIC OPERATOR FUNCTIONS

The differential operators on int $\boldsymbol{B}$ of order μ of Fuchs type will be elements of the cone algebra on $\boldsymbol{B}$. They have the form

$$A = \omega t^{-\mu} \mathrm{op}_M^{\gamma - \frac{n}{2}}(h)\omega_0 + (1-\omega)A_{int}(1-\omega_1)$$

for an arbitrary $A_{int} \in \mathrm{Diff}^{\mu}(\mathrm{int}\,\boldsymbol{B})$, $h(t,z) = \sum_{j=0}^{\mu} a_j(t)z^j$ for coefficients $a_j(t) \in C^{\infty}(\overline{\boldsymbol{R}}_+, \mathrm{Diff}^{\mu-j}(X))$, and cut-off functions ω, ω_0, ω_1 supported in a small collar neighborhoods $V \cong [0,1) \times X$ of $\partial\boldsymbol{B}$ (notations for pull-backs or push-forwards under $[0,1) \times X \to \boldsymbol{B}$ will often be suppressed for convenience; this should not cause confusions). As mentioned above we have a "hierarchy" of principal symbols

$$(\sigma_{\psi}^{\mu}(A), \sigma_M^{\mu}(A)) = (\text{principal interior symbol, principal conormal symbol})$$

involved in the ellipticity. It is obvious that for two Fuchs-type differential operators A and B of order μ and ν, respectively, AB is of Fuchs type and of order $\mu + \nu$, and

$$\sigma_{\psi}^{\mu+\nu}(AB) = \sigma_{\psi}^{\mu}(A)\sigma_{\psi}^{\nu}(B), \quad \sigma_M^{\mu+\nu}(AB) = (T^{\nu}\sigma_M^{\mu}(A))\,(\sigma_M^{\nu}(B))$$

(recall that, for instance, $\sigma_M^{\mu}(A)(z) = \sum_{j=0}^{\mu} a_j(0)z^j$, and $(T^{\nu}h)(t) = h(z+\nu)$). Now in the parametrix construction for an elliptic operator of Fuchs type we have to invert both symbol components. The discussion of the inverse conormal symbols is particularly interesting. First observe that $h(z) = \sigma_M^{\mu}(A)(z)$ belongs to the space $M_O^{\mu}(X)$ in the sense of the following definition.

Definition 2.16 *$M_O^{\mu}(X)$ for $\mu \in \boldsymbol{R}$ is the subspace of all $h(z) \in \mathcal{A}(\boldsymbol{C}, L_{cl}^{\mu}(X))$ such that*

$$h(z)|_{\Gamma_\beta} \in L_{cl}^{\mu}(X;\Gamma_\beta) \quad \textit{for every} \quad \beta \in \boldsymbol{R},$$

uniformly in $c \le \beta \le c'$ for every $c \le c'$.

The space $M_O^{\mu}(X)$ is Fréchet in a canonical way. $M_O^{-\infty}(X)$ (the intersection over all $M_O^{\mu}(X)$) is a unclear Fréchet space.

Theorem 2.17 *To every $f(z) \in L_{cl}^{\mu}(X;\Gamma_\rho)$ $\rho \in \boldsymbol{R}$ fixed, there exists an $h(z) \in M_O^{\mu}(X)$ such that*

$$h(z)|_{\Gamma_\rho} - f(z) \in L^{-\infty}(X;\Gamma_\rho).$$

The proof follows by a kernel cut-off argument with respect to the parameter.

Note that $h_1(z), h_2(z) \in M_O^{\mu}(X)$ and $(h_1(z) - h_2(z))|_{\Gamma_\rho} \in L^{-\infty}(X;\Gamma_\rho)$ imply $h_1(z) - h_2(z) \in M_O^{-\infty}(X)$.

Remark 2.18 *Let $f(z) \in L^{\mu}_{cl}(X;\Gamma_\rho)$ be parameter-dependent elliptic of order μ. Then $h(z)$ obtained in Theorem 2.17 has the property that $h(z)|_{\Gamma_\beta} \in L^{\mu}_{cl}(X;\Gamma_\beta)$ is parameter-dependent elliptic of order μ, for all $\beta \in \boldsymbol{R}$, uniformly in $c \leq \beta \leq c$ for every $c \leq c'$.*

Let us now mention a well-known result on holomorphic Fredholm families (a proof may be found, for instance, in [31] or in [24]).

Proposition 2.19 *Let H_1, H_2 be Hilbert spaces, $G \subseteq \boldsymbol{C}$ open (arc-wise connected), and let $h(z) \in \mathcal{A}(G, \mathcal{L}(H_1, H_2))$ be an operator function such that $h(z): H_1 \to H_2$ is a Fredholm operator for every $z \in G$. Assume that there is a $z_1 \in G$ such that $h(z): H_1 \to H_2$ is invertible. Then there is a countable subset $D \subset G$, $D \cap K$ finite for every $K \subset\subset G$, such that $h(z): H_1 \to H_2$ is invertible for all $z \in G \setminus D$. In addition $h^{-1}(z)$ extends from $G \setminus D$, $D = \{p_j\}_{j \in \boldsymbol{Z}}$, to a meromorphic operator function with poles in $p_j \in D$ of multiplicities $m_j + 1$ for certain $m_j \in \boldsymbol{N}$, and finite-dimensional Laurent coefficients at $(z - p_j)^{-(k+1)}$, $0 \leq k \leq m_j$, for all j.*

An inspection of the proof of Proposition 2.19 together with Remark 2.18 and Remark 2.4 gives us the following result:

Theorem 2.20 *Let A be of Fuchs type, elliptic of order μ with respect to $\sigma^{\mu}_{\psi}(A)$ (i.e., only (i) of Definition 2.9 is required). Then $h(z) := \sigma^{\mu}_{M}(A)(z) = \sum_{j=0}^{\mu} a_j(0) z^j$,*

$$h(z): H^s(X) \to H^{s-\mu}(X),$$

has the property that $h(z)|_{\Gamma_\beta} \in L^{\mu}_{cl}(X;\Gamma_\beta)$ is parameter-dependent elliptic of order μ, for every $\beta \in \boldsymbol{R}$, uniformly in $c \leq \beta \leq c'$ for every $c \leq c'$, and there is a countable set $D = \{p_j\}_{j \in \boldsymbol{Z}} \subset \boldsymbol{C}$ with $D \cap \{z : c \leq \operatorname{Re} z \leq c'\}$ finite for every $c \leq c'$, such that $h(z)$ is invertible for all $z \in \boldsymbol{C} \setminus D$. Moreover, $h^{-1}(z)$ extends to a meromorphic $L^{-\mu}_{cl}(X)$-valued function, and the Laurent-coefficients at $(z - p_j)^{-(k+1)}$, $0 \leq k \leq m_j$, are finite-dimensional operators in $L^{-\infty}(X)$ for all $0 \leq k \leq m_j$, $j \in \boldsymbol{Z}$. If $A \subseteq \boldsymbol{C}$ is any subset we call a $\chi \in C^{\infty}(\boldsymbol{C})$ an A-excision function if $\chi(z) = 0$ for dist $(z, \overline{A}) < \varepsilon_0$, $\chi(z) = 1$ for dist $(z, \overline{A}) > \varepsilon$, for certain $0 < \varepsilon_0 < \varepsilon_1 < \infty$.

Definition 2.21 *Let $R = \{p_j, m_j, L_j\}_{j \in \boldsymbol{Z}}$ be a sequence, where*

$$(p_j, m_j) \in \boldsymbol{C} \times \boldsymbol{N}, \quad |\operatorname{Re} p_j| \to \infty \quad as \quad |j| \to \infty, \quad and \quad L_j \subset L^{-\infty}(X)$$

is a finite-dimensional subspace of finite-dimensional operators for every j. Such a sequence is called a discrete asymptotic type of Mellin symbols. Moreover, $M^{-\infty}_{R}(X)$ denotes the subspace of all $L^{-\infty}(X)$-valued meromorphic functions $f(z)$ in $\boldsymbol{C}$ such that

(i) $f(z)$ has poles in p_j of multiplicities m_j+1 with Laurent coefficients at $(z-p_j)^{-(k+1)}$ belonging to L_j for $0 \le k \le m_j$, $j \in \mathbf{Z}$,

(ii) if $\chi(z)$ is any $\pi_{\mathbf{C}}R$-excision function, for $\pi_{\mathbf{C}}R = \{p_j\}_{j\in\mathbf{R}}$, then $\chi(z)f(z)|_{\Gamma_\beta} \in L^{-\infty}(X;\Gamma_\beta)$ for every $\beta \in \mathbf{R}$, uniformly in $c \le \beta \le c'$ for all $c \le c'$.

Remark 2.22 *$M_R^{-\infty}(X)$ is a nuclear Fréchet space in a natural way.*

Let us define

$$M_R^\mu(X) = M_O^\mu(X) + M_R^{-\infty}(X)$$

as a non-direct sum of Fréchet spaces.

Remark 2.23 *Under the assumptions of Theorem 2.20 we have*

$$(\sigma_M^\mu(A)(z))^{-1} \in M_R^{-\mu}(X)$$

for a certain discrete Mellin asymptotic type R.

We will call an $h(z) \in M_R^\mu(X)$ for some discrete Mellin asymptotic type elliptic of order μ if for any decomposition $h = h_0 + h_{-\infty}$, $h_0 \in M_O^\mu(X)$, $h_\infty \in M_R^{-\infty}(X)$, $h_0(z)|_{\Gamma_\beta} \in L_{cl}^\mu(X;\Gamma_\beta)$ is parameter-dependent elliptic of order μ (then also $h_0(z)|_{\Gamma_\rho}$ is parameter-dependent elliptic of order ρ for every ρ, uniformly in $c \le \rho \le c'$, for every $c \le c'$). This is a correct definition, i.e., independent of the particular choice of the decomposition.

The following results are easy to verify:

Proposition 2.24 *Let $h(z) \in M_P^\mu(X)$, $g(z) \in M_Q^\mu(X)$ be given, with certain discrete Mellin asymptotic types P, Q. Then the point-wise composition $h(z)g(z)$ belongs to $M_R^{\mu+\nu}(X)$ for some resulting discrete Mellin asymptotic type R. If h and g are elliptic then so is hg.*

Theorem 2.25 *Let $h(z) \in M_R^\mu(X)$ be elliptic of order μ, where R is any discrete Mellin asymptotic type. Then $h^{-1}(z) \in M_Q^{-\mu}(X)$ is elliptic of order $-\mu$, with some resulting discrete Mellin asymptotic type Q.*

2.4. ELEMENTS OF THE CONE ALGEBRA

For studying elliptic regularity of solutions on a manifold with conical singularities it is interesting to look at subspaces of the weighted Sobolev spaces with asymptotics. Let us fix a weight $\gamma \in \mathbf{R}$, consider the associated weight line $\Gamma_{\frac{n+1}{2}-\gamma} \subset \mathbf{C}$, $n = \dim X$, and choose a "weight interval" $\Theta = (\vartheta, 0]$, $-\infty \le \vartheta < 0$, that defines for every weight γ a strip $\Theta_\gamma = \{z \in \mathbf{C} : \frac{n+1}{2} - \gamma + \vartheta < \operatorname{Re} z < \frac{n+1}{2} - \gamma\}$. Let first Θ be finite and call a sequence

$$P = \{(p_j, m_j, L_j)\}_{j=0,\dots,N} \quad \text{for} \quad N = N(P)$$

a discrete asymptotic type associated with the weight data (γ, Θ), if $(p_j, m_j) \in \boldsymbol{C} \times \boldsymbol{N}$, $p_j \in \Theta_\gamma$, and L_j a finite-dimensional subspace of $C^\infty(X)$ for all $j = 0, \ldots, N$.

Definition 2.26 *An element $u(t,x) \in \mathcal{K}^{s,\gamma}(X^\wedge)$ has asymptotics of type P for $t \to 0$ if there are coefficients $c_{jk} = c_{jk}(u) \in L_j$, $0 \le k \le m_j$, such that*

$$u(t,x) - \omega(t) \left\{ \sum_{j=0}^{N} \sum_{k=0}^{m_j} c_{jk} t^{-p_j} \log^k t \right\} \in \mathcal{K}^{s,(\gamma-\vartheta)-}(X^\wedge) \tag{2.13}$$

for any cut-off function $\omega(t)$. Here

$$\mathcal{K}^{s,\rho-}(X^\wedge) = \cap_{\varepsilon>0} \mathcal{K}^{s,\rho-\varepsilon}(X^\wedge), \quad \rho \in \boldsymbol{R},$$

in the Fréchet topology of the projective limit. Denote by $\mathcal{K}_P^{s,\gamma}(X^\wedge)$ the subspace of $\mathcal{K}^{s,\gamma}(X^\wedge)$ consisting of all elements with these asymptotics.

The unique coefficients $c_{jk}(u)$ define linear maps

$$c_{jk} : \mathcal{K}_P^{s,\gamma}(X^\wedge) \to L_j, \quad 0 \le k \le m_j, \quad j = 0, \ldots, N. \tag{2.14}$$

The remainder (2.13) then gives rise to a linear map

$$r : \mathcal{K}_P^{s,\gamma}(X^\wedge) \to \mathcal{K}^{s,(\gamma-\vartheta)-}(X^\wedge) \tag{2.15}$$

for a fixed cut-off function ω. Then $\mathcal{K}_P^{s,\gamma}(X^\wedge)$ becomes a Fréchet space in the topology of the projective limit with respect to (2.14), (2.13), that is independent of ω. For infinite $\Theta = (-\infty, 0]$ we can talk about discrete asymptotic types of the form $P = \{(p_j, m_j, L_j)\}_{j \in \boldsymbol{N}}$, with $\frac{n+1}{2} - \gamma > \operatorname{Re} p_j$ for all j and $\operatorname{Re} p_j \to -\infty$ for $j \to \infty$. Then, to every $k \in \boldsymbol{N} \setminus \{0\}$ we have $P_k = \{(p, m, L) \in P : \frac{n+1}{2} - \gamma - k < \operatorname{Re} p_j < \frac{n+1}{2} - \gamma\}$ in the above sense, where $P_{k+1} \subset P_k$ for every k. It is clear that then $\mathcal{K}_{P_{k+1}}^{s,\gamma}(X^\wedge) \hookrightarrow \mathcal{K}_{P_k}^{s,\gamma}(X^\wedge)$ is continuous for every k, and we can define the space

$$\mathcal{K}_P^{s,\gamma}(X^\wedge) = \lim \operatorname{proj}_{k \in \boldsymbol{R}} \mathcal{K}_{P_k}^{s,\gamma}(X^\wedge)$$

as projective limit.

We will also set

$$\mathcal{S}_P^\gamma(X^\wedge) = [\omega] \mathcal{K}_P^{\infty,\gamma}(X^\wedge) + [1-\omega] \mathcal{S}(\overline{\boldsymbol{R}}_+ \times X),$$

where $\mathcal{S}(\overline{\boldsymbol{R}}_+ \times X) = \mathcal{S}(\boldsymbol{R}, C^\infty(X))|_{\overline{\boldsymbol{R}}_+}$. Then $\mathcal{S}_P^\gamma(X^\wedge)$ is a nuclear Fréchet space.

Analogously to $\mathcal{K}_P^{s,\gamma}(X^\wedge)$ we denote by $\mathcal{H}_P^{s,\gamma}(\boldsymbol{B})$ the subspace of all $u \in H_{loc}^s(\text{int } \boldsymbol{B})$ for which $\omega u \in \mathcal{K}_P^{s,\gamma}(X^\wedge)$ for any cut-off function $\omega(t)$ supported in a collar neighborhoods V of $\partial \boldsymbol{B}$, $v \cong [0,1) \times X \ni (t,x)$.

Remark 2.27 *The space $\mathcal{K}_P^{s,\gamma}(X^\wedge)$ for every fixed $s \in \boldsymbol{R}$ and P, associated with (γ, Θ), can he written as projective limit of Hilbert spaces $\{E^j\}_{j\in\boldsymbol{N}}$ with continuous $E^{j+1} \hookrightarrow E^j \hookrightarrow E^0 = \mathcal{K}^{s,\gamma}(X^\wedge)$, where $\{\kappa_\lambda\}_{\lambda\in\boldsymbol{R}_+}$ on E^0 induces $\{\kappa_\lambda\}_{\lambda\in\boldsymbol{R}_+} \in C(\boldsymbol{R}_+, \mathcal{L}_\sigma(E^j))$ for all j, with the properties required above in Definition 1.5. An analogous remark holds for the spaces $\mathcal{S}_P^\gamma(X^\wedge)$.*

Denote the weighted Mellin transform with the weight $\rho \in \boldsymbol{R}$ by M_ρ, i.e., $(M_\rho u)(t) = M(t^{-\rho}u)(z+\rho)$. Let P be a discrete asymptotic type associated with (γ, Θ), $\Theta = (\vartheta, 0]$ for some $\gamma \in \boldsymbol{R}$ and $-\infty \leq \vartheta < 0$. Let $\widehat{H}^s(\Gamma_\rho \times X)$ for $s \in \boldsymbol{R}$ be the image of $H^s(\Gamma_\rho \times X)$ with respect to the Fourier transform along Γ_ρ.

Theorem 2.28 *Let $\omega(t)$ be an arbitrary cut-off function. Then $M_{\gamma-\frac{n}{2}}(\omega\mathcal{K}_P^{s,\gamma}(X^\wedge))$ is a space of meromorphic $H^s(X)$-valued functions in $\mathrm{Re}\, z > \frac{n+1}{2} + \vartheta - \gamma$, with poles in p_j of multiplicities $m_j + 1$, and Laurent coefficients at $(z-p_i)^{-(k+1)}$ in L_j for $0 \leq k \leq m_j$. If $\chi(z)$ is any $\pi_{\boldsymbol{C}}P$-excision function, $\pi_{\boldsymbol{C}}P = \{p_j\}_{j=0,\ldots,N}$ ($N = N(P)$ or $N = \infty$), then for every $f(z) \in M_{\gamma-\frac{n}{2}}(\omega\mathcal{K}_P^{s,\gamma}(X^\wedge))$ we have*

$$(\chi f)|_{\Gamma_\rho} \in \widehat{H}^s(\Gamma_\rho \times X)$$

for all $\rho > \frac{n+1}{2} + \varphi - \gamma$, uniformly in compact ρ-intervals

Theorem 2.29 *Let $h(t,t',z) \in C^\infty(\overline{\boldsymbol{R}}_+ \times \overline{\boldsymbol{R}}_+, M_R^\mu(X))$, $\mu \in \boldsymbol{R}$, for a certain discrete Mellin asymptotic type R, and let $\omega(t)$, $\tilde{\omega}(t)$ be cut-off functions. Then, if $\pi_{\boldsymbol{C}}R \cap \Gamma_{\frac{n+1}{2}-\gamma} = \emptyset$,*

$$\omega \mathrm{op}_M^\gamma(h)\tilde{\omega} : \; \mathcal{K}_P^{s,\gamma}(X^\wedge) \to \mathcal{K}_Q^{s-\mu,\gamma}(X^\wedge)$$

is continuous for every $s \in \boldsymbol{R}$ and every discrete asymptotic type P with some resulting discrete asymptotic type Q, associated with given weight data (γ, Θ).

We now turn to an analogue of the Green operators from the calculus of boundary value problems, where here the inner normal to the boundary $\boldsymbol{R}_+$ is replaced by $X^\wedge$.

Let us first note that

$$\mathcal{K}^{0,0}(X^\wedge) \cong t^{-\frac{n}{2}} L^2(\boldsymbol{R}_+ \times X)$$

and that the $\mathcal{K}^{0,0}(X^\wedge)$-scalar product

$$(.,.): \; C_0^\infty(X^\wedge) \times C_0^\infty(X^\wedge) \to \boldsymbol{C}$$

extends to a non-degenerate sesquilinear pairing.

$$(.,.): \mathcal{K}^{s,\gamma}(X^\wedge) \times \mathcal{K}^{-s,-\gamma}(X^\wedge) \to \boldsymbol{C}$$

for every $s, \gamma \in \boldsymbol{R}$. The formal adjoint A^* of an operator

$$A \in \cap_{s\in\boldsymbol{R}} \mathcal{L}(\mathcal{K}^{s,\gamma}(X^\wedge), \mathcal{K}^{s-\nu,\delta}(X^\wedge))$$

is an element

$$A^* \in \cap_{s\in\boldsymbol{R}} \mathcal{L}(\mathcal{K}^{s,-\delta}(X^\wedge), \mathcal{K}^{s-\nu,-\gamma}(X^\wedge)).$$

Definition 2.30 *An operator*

$$G \in \cap_{s\in\boldsymbol{R}} \mathcal{L}(\mathcal{K}^{s,\gamma}(X^\wedge), \mathcal{K}^{\infty,\delta}(X^\wedge))$$

is called a Green operator on $X^\wedge$ if for certain (G-dependent) asymptotic types P and Q, associated with (δ, Θ) and $(-\gamma, \Theta)$, respectively, we have

$$\begin{array}{lll} G & \in & \cap_{s\in\boldsymbol{R}} \mathcal{L}(\mathcal{K}^{s,\gamma}(X^\wedge), \mathcal{S}^{\delta}_{P}(X^\wedge)), \\ G^* & \in & \cap_{s\in\boldsymbol{R}} \mathcal{L}(\mathcal{K}^{s,-\delta}(X^\wedge), \mathcal{S}^{-\gamma}_{Q}(X^\wedge)). \end{array}$$

Analogously a $G \in \cap_{s\in\boldsymbol{R}} \mathcal{L}(\mathcal{H}^{s,\gamma}(\boldsymbol{B}), \mathcal{H}^{\infty,\delta}(\boldsymbol{B}))$ is called a Green operator on $\boldsymbol{B}$ if

$$\begin{array}{lll} G & \in & \cap_{s\in\boldsymbol{R}} \mathcal{L}(\mathcal{H}^{s,\gamma}(\boldsymbol{B}), \mathcal{H}^{\infty,\delta}_{P}(\boldsymbol{B})), \\ G & \in & \cap_{s\in\boldsymbol{R}} \mathcal{L}(\mathcal{H}^{s,-\delta}(\boldsymbol{B}), \mathcal{H}^{\infty,-\gamma}_{Q}(\boldsymbol{B})), \end{array}$$

for certain (G-dependent) P, Q.

Let $h_j(z) \in M^{-\infty}_{R_j}(X)$, $j \in N$, for some discrete Mellin asymptotic type R_j, $\gamma, \gamma_j \in \boldsymbol{R}$, $\pi_{\boldsymbol{C}} R \cap \Gamma_{\frac{n+1}{2}-\gamma_j} = \emptyset$, where $-\gamma + j + \gamma_j \geq 0$, $-\gamma_j + \gamma \geq 0$. Then

$$M_j := \omega t^{\delta-\gamma+j} op_M^{\gamma_j - \frac{n}{2}}(h_j)\omega_0 : \ \mathcal{K}^{s,\gamma}(X^\wedge) \to \mathcal{K}^{\infty,\delta}(X^\wedge) \qquad (2.16)$$

is continuous for every $s \in \boldsymbol{R}$ for cut-off functions $\omega, \tilde{\omega}$. According to Theorem 2.29 M_j induces continuous operators between subspaces with asymptotics.

Proposition 2.31 *Let $\tilde{M}_j = \tilde{\omega} t^{\delta-\gamma+j} \mathrm{op}_M^{\tilde{\gamma}_j - \frac{n}{2}}(h_j)\tilde{\omega}_0$ for another pair of cut-off functions $\tilde{\omega}, \tilde{\omega}_0$ and $\tilde{\gamma}_j \in \boldsymbol{R}$, $\pi_{\boldsymbol{C}} R_j \cap \Gamma_{\frac{n+1}{2}-\tilde{\gamma}_j} = \emptyset$, $-\gamma + j + \tilde{\gamma}_j \geq 0$, $-\tilde{\gamma}_j + \gamma \geq 0$. Then $M_j - \tilde{M}_j$ is a Green operator in the sense of Definition 2.30.*

The algebra of cone pseudo-differential operators on $\boldsymbol{B}$ with respect to the weight data (γ, δ, Θ) for $\gamma, \delta \in \boldsymbol{R}$ and $\Theta = (-k, 0]$, $k \in \boldsymbol{N} \setminus \{0\}$, is defined as the subspace of all operators

$$A = \omega t^{\delta-\gamma} \mathrm{op}_M^{\gamma-\frac{n}{2}}(h)\omega_0 + (1-\omega)A_{int}(1-\omega_1) + M + G, \qquad (2.17)$$

for arbitrary $h(t,z) \in C^\infty(\overline{\boldsymbol{R}}_+ \times X)$, $A_{int} \in L^\mu_{cl}(\mathrm{int}\ \boldsymbol{B})$, and

$$M = \sum_{j=0}^{k-1} M_j$$

with M_j being of the form (2.16), and G a Green operator in the sense of Definition 2.30. $\omega, \omega_0, \omega_1$ are arbitrary cut-off functions supported in a collar neighborhood of $\partial\boldsymbol{B}$, with $\omega\omega_0 = \omega$, $\omega\omega_1 = \omega_1$. M is a called smoothing Mellin operator in the cone algebra.

Remark 2.32 *The operators (2.17) belong to*

$$L^\mu_{cl}(int\ \boldsymbol{B}) \cap \cap_{s\in\boldsymbol{R}} \mathcal{L}(\mathcal{H}^{s,\gamma}(\boldsymbol{B}), \mathcal{H}^{s-\mu,\delta}(\boldsymbol{B})).$$

To A we have $\sigma^\mu_\psi(A)$, the homogeneous principal symbol of order μ, and $\sigma^\mu_M(A)$, the principal conormal symbol, defined as $\sigma^\mu_M(A)(z) = h(0,z) + h_0(z)$ in the notations of (2.17) and (2.16). It can be proved that $t^\mu\sigma^\mu_\psi(A)(t,x,t^{-1}\tau,\xi)$ in a collar neighborhoods V of $\partial\boldsymbol{B}$, $V \cong [0,1) \times X$, is C^∞ up to $t = 0$.

A is called elliptic if the conditions of Definition 2.9 (where σ^μ_M has to be replaced by $\sigma^{\gamma-\delta}_M$) are satisfied. Then Theorem 2.14 permits the following refinement:

Theorem 2.33 *A is elliptic of order μ, with respect to the weight γ if and only if*

$$A:\ \mathcal{H}^{s,\gamma}(\boldsymbol{B}) \to \mathcal{H}^{s-\mu,\delta}(\boldsymbol{B})$$

is a Fredholm operator for an $s = s_0 \in \boldsymbol{R}$. The ellipticity implies that A is Fredholm for all $s \in \boldsymbol{R}$. Moreover, there is a parametrix B in the cone algebra of order $-\mu$, to the weight data (δ, γ, Θ) such that $AB - 1$ and $BA - 1$ are Green operators with respect to the corresponding weight data.

Theorem 2.34 *Let A be elliptic of order μ, with respect to the weight γ. Then a solution $u \in \mathcal{H}^{-\infty,\gamma}(\boldsymbol{B})$ of $Au = f \in \mathcal{H}^{s-\mu,\delta}(\boldsymbol{B})$, $s \in \boldsymbol{R}$, belongs to $\mathcal{H}^{s,\gamma}(\boldsymbol{B})$. In particular, for $f \in \mathcal{H}^{s-\mu,\delta}_Q(\boldsymbol{B})$ for some discrete asymptotic type Q we obtain $u \in \mathcal{H}^{s,\gamma}_P(\boldsymbol{B})$ for some resulting discrete asymptotic type R.*

3. Pseudo-differential calculus on manifolds with edges

3.1. EDGE-DEGENERATE DIFFERENTIAL OPERATORS

Let G be a domain in $\boldsymbol{R}^N$ with piece-wise C^∞ geometry, locally being of wedge type. For instance, the wedge may have the form

$$\{(\tilde{x}, y) \in \boldsymbol{R}^{n+1} \times \boldsymbol{R}^q : \tilde{x} = 0 \quad \text{or} \quad \tilde{x}/|\tilde{x}| \in \Sigma, \quad y \in \Omega\},$$

where $\Sigma \subset S^n = \{\tilde{x} : |\tilde{x}| = 1\}$ is an open subset with smooth boundary, and $\Omega \subseteq \boldsymbol{R}^q$ open. If $\tilde{A}(\tilde{x}, y, D_{\tilde{x},y})$ is an elliptic differential operator in $\boldsymbol{R}^{n+1} \times \Omega$ with C^∞ coefficients we may study boundary value problems for $\tilde{A}$ with elliptic conditions along the smooth faces of ∂G. A common method to do this is to introduce polar coordinates into $\tilde{A}$ with respect to $\tilde{x}$. An operator $\tilde{A} \in \mathrm{Diff}^\mu(\boldsymbol{R}^{n+1}_{\tilde{x}} \times \Omega_y)$ in polar coordinates $\tilde{x} \to (t, x)$, $t = |\tilde{x}|$, $x = \tilde{x}/|\tilde{x}| \in S^n$, takes the form

$$A = t^{-\mu} \sum_{j+|\alpha| \leq \mu} a_{j\alpha}(t, y)(-t\frac{\partial}{\partial t})^j (tD_y)^\alpha \tag{3.1}$$

with $a_{j\alpha}(t, y) \in C^\infty(\overline{\boldsymbol{R}}_+ \times \Omega, \mathrm{Diff}^{\mu-(j+|\alpha|)}(S^n))$. A useful model for studying the above problems in to first consider the case when the base of the model cone of the wedge is closed compact. This can be combined with the methods from the standard calculus of boundary problems, say with the transmission property, in a parameter-dependent form, cf. Schrohe, Schulze [26], [27], [28]. We will content ourselves here with the case of closed cone bases X.

A wedge in general has form $X^\Delta \times \Omega$ for $X^\Delta = (\overline{\boldsymbol{R}}_+ \times X)/(\{0\} \times X)$. Operators will be given on the open stretched wedge $X^\wedge \times \Omega$. An operator $A \in \mathrm{Diff}^\mu(X^\wedge \times \Omega)$ is called edge-degenerate if it has the form (3.1) for coefficients $a_{j\alpha}(t, y) \in C^\infty(\overline{\boldsymbol{R}}_+ \times \Omega, \mathrm{Diff}^{\mu-(j+|\alpha|)}(X))$. For studying ellipticity and parametrix constructions it will be necessary to pose additional conditions along Ω of trace and potential type, similarly to the case $n = \dim X$. Note that when $\tilde{A} = \sum_{|\beta| \leq \mu} a_\beta(\tilde{x}) D^\beta_{\tilde{x}}$ is given on $\boldsymbol{R}_+ \times \Omega$ with coefficients $a_\beta(\tilde{x}) \in C^\infty(\overline{\boldsymbol{R}}_+ \times \Omega)$ the operator $\tilde{A}$ in the coordinates $\tilde{x} = (t, y)$, $t \in \boldsymbol{R}_+$, $y \in \Omega$ can also be written

$$A = t^{-\mu} \sum_{j+|\alpha| \leq \mu} a_{j\alpha}(t, y)(-t\frac{\partial}{\partial t})^j (tD_y)^\alpha$$

for coefficients $a_{j\alpha}(t, y) \in C^\infty(\overline{\boldsymbol{R}}_+ \times \Omega)$. This shows once again that differential operators on $\overline{\boldsymbol{R}}_+ \times \Omega$ in usual form are automatically edge-degenerate.

For instance, the Laplace operator takes the form

$$\Delta = \frac{\partial^2}{\partial t^2} + \sum_{k=1}^{q} \frac{\partial^2}{\partial y^2} = t^{-2}\left\{\left(t\frac{\partial}{\partial t}\right)^2 - t\frac{\partial}{\partial t} + \sum_{k=1}^{q}\left(t\frac{\partial}{\partial y_k}\right)^2\right\}.$$

Among the elliptic boundary (or trace) conditions along Ω are the Dirichlet conditions. If $H^s_{loc(y)}(\boldsymbol{R}_+ \times \Omega)$ denotes the subspace of all $u(t,y) \in H^s_{loc}(\boldsymbol{R}_+ \times \Omega)$ such that $\varphi(y)u(t,y) \in H^s(\boldsymbol{R} \times \boldsymbol{R}^q)|_{\boldsymbol{R}_+ \times \Omega}$ for every $\varphi \in C_0^\infty(\Omega)$, the local Dirichlet problem takes the form of an operator column matrix

$$\begin{pmatrix} \Delta \\ T \end{pmatrix} : H^s_{loc(y)}(\boldsymbol{R}_+ \times \Omega) \to \begin{matrix} H^{s-2}_{loc(y)}(\boldsymbol{R}_+ \times \Omega) \\ \oplus \\ H^{s-\frac{1}{2}}_{loc}(\Omega) \end{matrix}, \tag{3.2}$$

$s > \frac{3}{2}$, and a local parametrix will be an operator row matrix (P, K) mapping in the reverse direction (properly supported in y-coordinates). For solving (3.2) we may pass to RT instead of T for an elliptic reduction of orders $R \in L^{\frac{3}{2}}_{cl}(\Omega)$. It is well-known that elliptic boundary value problems for pseudo-differential operators require both trace and potential conditions with respect to the boundary, the latter ones mapping distributions on the boundary into distributions in the domain, like the potential K just mentioned, cf. [3], [20]. Analogously an edge-degenerate differential operator A on a (stretched) wedge $X^\wedge \times \Omega$, which is elliptic in the sense $\sigma^\mu_\psi(A) \neq 0$ on $T^*(X^\wedge \times \Omega) \setminus 0$ and $t^\mu \sigma^\mu_\psi(A)(t,x,y,t^{-1}\tau,\xi,t^{-1}\eta) \neq 0$ up to $t = 0$ for $(\tau,\xi,\eta) \neq 0$, requires for the local parametrix constructions (and globally for the Fredholm property) additional conditions of trace and potential type with respect to Ω, including pseudo-differential operators on Ω. In other words we have to consider block matrices

$$\begin{pmatrix} A & K \\ T & Q \end{pmatrix} : \begin{matrix} \mathcal{W}^{s,\gamma}_{comp(y)}(X^\wedge \times \Omega) \\ \oplus \\ H^s_{comp}(\Omega, \boldsymbol{C}^{N_-}) \end{matrix} \to \begin{matrix} \mathcal{W}^{s-\mu,\gamma-\mu}_{loc(y)}(X^\wedge \times \Omega) \\ \oplus \\ H^{s-\mu}_{loc}(\Omega, \boldsymbol{C}^{N_+}) \end{matrix}$$

Here

$$\mathcal{W}^{s,\gamma}_{comp(y)}(X^\wedge \times \Omega) = \left\{u \in \mathcal{W}^s(\boldsymbol{R}^q, \mathcal{K}^{s,\gamma}(X^\wedge)) : \operatorname{supp}_{(y)} u \subset\subset \Omega \,\text{compact}\right\}$$

and $\mathcal{W}^{s,\gamma}_{loc(y)}(X^\wedge \times \Omega) = \{u \in \mathcal{D}'(\Omega, \mathcal{K}^{s,\gamma}(X^\wedge)) : \varphi u \in \mathcal{W}^{s,\gamma}_{comp(y)}(X^\wedge \times \Omega)$ for every $\varphi \in C_0^\infty(\Omega)\}$, cf. also Remark 2.11. An idea of the wedge pseudo-differential calculus is to treat the problems in terms of pseudo-differential operators along the edge Ω with operator-valued symbols operating along the model cone $X^\wedge$. A starting point are the edge-degenerate symbols.

A symbol $p(t,x,y,\tau,\xi,\eta) \in S^{\mu}_{cl}(\boldsymbol{R}_+ \times \Sigma \times \Omega \times \boldsymbol{R}^{1+n+q})$ for open sets $\Sigma \subseteq \boldsymbol{R}^n$, $\Omega \subseteq \boldsymbol{R}^q$, is called edge-degenerate if

$$p(t,x,y,\tau,\xi,\eta) = \tilde{p}(t,x,y,t\tau,\xi,t\eta)$$

for a $\tilde{p}(t,x,y,\tilde{\tau},\xi,\tilde{\eta}) \in S^{\mu}_{cl}(\overline{\boldsymbol{R}}_+ \times \Sigma \times \Omega \times \boldsymbol{R}^{1+n+q})$.

Proposition 3.1 *Let $p(t,x,y,\tau,\xi,\eta)$ be an edge-degenerate symbol, $\tilde{p}_{(\mu)}(t,x,y,\tilde{\tau},\xi,\tilde{\eta})$ the homogeneous principal part of $\tilde{p}(t,x,y,\tilde{\tau},\xi,\tilde{\eta})$ in $(\tilde{\tau},\xi,\tilde{\eta}) \neq 0$ of order μ and assume that*

$$\tilde{p}_{(\mu)}(t,x,y,\tilde{\tau},\xi,\tilde{\eta}) \neq 0 \quad \text{for all} \quad (t,x,y) \in \overline{\boldsymbol{R}}_+ \times \Sigma \times \Omega, (\tilde{\tau},\xi,\tilde{\eta}) \neq 0.$$

Then there exists an edge-degenerate symbol $r(t,x,y,\tau,\xi,\eta)$ of order $-\mu$ such that

$$(t^{-\mu}p)\#_{t,x,y}(t^{\mu}r) = 1 \ \mathrm{mod}\ S^{-\infty}(\boldsymbol{R}_+ \times \Sigma \times \Omega \times \boldsymbol{R}^{1+n+q}),$$

where # means the Leibniz product between the symbols, taken with respect to the indicated variables. The same is true for the multiplication in reverse order or for the Leibniz product only with respect to t,x.

Similarly to the calculus on a cone it is adequate to formulate operators globally along X and to consider corresponding (t,y,τ,η)-dependent operator families.

Let us fix an open covering of X by coordinate neighborhoods $\{U_1,\ldots,U_N\}$ together with a system of charts $\chi_j : U_j \to \Sigma_j$, $\Sigma_j \subseteq \boldsymbol{R}^n$ open, a subordinate partition of unity $\{\varphi_1,\ldots,\varphi_N\}$ and functions $\{\psi_1,\ldots,\psi_N\}$, $\psi_j \in C^\infty(U_j)$, satisfying $\varphi_j\psi_j = \varphi_j$ for all j.

Given a system of edge-degenerate symbols

$$p_j(t,x,y,\tau,\xi,\eta) \in S^{\mu}_{cl}(\boldsymbol{R}_+ \times \Sigma_j \times \Omega \times \boldsymbol{R}^{1+n+q}), \quad j = 1,\ldots,N, \tag{3.3}$$

with the associated symbols $\tilde{p}_j(t,x,y,\tilde{\tau},\xi,\tilde{\eta})$ which are smooth up to $t=0$ we can form an operator family

$$\tilde{P}(t,y,\tilde{\tau},\tilde{\eta}) = \sum_{j=1}^{N} \varphi_j \left\{(\chi_j^{-1})_* \mathrm{op}_x(\tilde{p}_j)(t,y,\tilde{\tau},\tilde{\eta})\right\} \psi_j$$

where op_x denotes the pseudo-differential action in $\Sigma_j \subseteq \boldsymbol{R}^n$ with respect to x. Then

$$\tilde{P}(t,y,\tilde{\tau},\tilde{\eta}) \in C^\infty(\overline{\boldsymbol{R}}_+ \times \Omega, L^{\mu}_{cl}(X;\boldsymbol{R}^{1+q}_{\tilde{\tau},\tilde{\eta}})). \tag{3.4}$$

For obtaining a pseudo-differential operator on $X^\wedge \times \Omega$ we have to carry out the action also with respect to $t \in \boldsymbol{R}_+$ and to $y \in \Omega$. Concerning the action in t we want to have a control up to $t=0$. For this reason we formulate for the edge-degenerate case a Mellin operator convention that allows us to apply the pseudo-differential calculus along $\boldsymbol{R}_+$ with respect to the Mellin transform (it is needed only for a neighborhoods of $t=0$). In other words we generalize some constructions from the cone calculus in a suitable parameter-dependent form.

Definition 3.2 *$M_O^\mu(X;\boldsymbol{R}^q_\eta)$ for $\mu\in\boldsymbol{R}$ is the subspace of all $h(z,\eta)\in\mathcal{A}(\boldsymbol{C},L^\mu_{cl}(X;\boldsymbol{R}^q_\eta))$ such that*

$$h(z,\eta)|_{\Gamma_\beta}\in L^\mu_{cl}(X;\Gamma_\beta\times\boldsymbol{R}^q_\eta)$$

for every $\beta\in\boldsymbol{R}$, uniformly in $c\le\beta\le c'$ for every $c\le c'$.

Theorem 3.3 *To every $f(z,\eta)\in L^\mu_{cl}(X;\Gamma_\rho\times\boldsymbol{R}^q_\eta)$, $\rho\in\boldsymbol{R}$ fixed, there exists an $h(z,\eta)\in M_O^\mu(X;\boldsymbol{R}^q)$ such that*

$$h(z,\eta)|_{\Gamma_\rho}-f(z,\eta)\in L^{-\infty}(X;\Gamma_\rho\times\boldsymbol{R}^q_\eta).$$

The proof can be obtained analogously to that of Theorem 2.17 by a kernel cut-off argument. Note also that $h_1(z,\eta)$, $h_2(z,\eta)\in M_O^\mu(X;\boldsymbol{R}^q_\eta)$ and $h_1(z,\eta)-h_2(z,\eta)|_{\Gamma_\rho}\in L^{-\infty}(X;\Gamma_\rho\times\boldsymbol{R}^q_\eta)$ imply $h_1(z,\eta)-h_2(z,\eta)\in M_O^{-\infty}(X;\boldsymbol{R}^q_\eta)$. Since the kernel cut-off only acts on covariables, we have analogous results for the case when $f(t,y,z,\eta)$ depends on $(t,y)\in\overline{\boldsymbol{R}}_+\times\Omega$, up to $t=0$. Then the other occurring objects also depend C^∞ on $(t,y)\in\overline{\boldsymbol{R}}_+\times\Omega$.

The Mellin operator convention for the wedge calculus consists of the following result:

Theorem 3.4 *To every $\tilde{P}(t,y,\tilde{\tau},\tilde{\eta})\in C^\infty(\overline{\boldsymbol{R}}_+\times\Omega,L^\mu_{cl}(X;\boldsymbol{R}^{1+q}_{\tilde{\tau},\tilde{\eta}}))$ there exists an $\tilde{h}(t,y,z,\tilde{\eta})\in C^\infty\times\Omega,M_O^\mu(X;\boldsymbol{R}^q_{\tilde{\eta}}))$ such that*

$$\mathrm{op}_t(P)(y,\eta)=\mathrm{op}_M^\beta(h)(y,\eta)\quad\mathrm{mod}\quad C^\infty(\Omega,L^{-\infty}(X^\wedge;\boldsymbol{R}^q_\eta))$$

for $P(t,y,\tau,\eta):=\tilde{P}(t,y,t\tau,t\eta)$, $h(t,y,z,\eta):=\tilde{h}(t,y,z,t\eta)$, and every such $\tilde{h}(t,y,z,\tilde{\eta})$ is unique mod $C^\infty(\overline{\boldsymbol{R}}_+\times\Omega,M_O^{-\infty}(X;\boldsymbol{R}^q_{\tilde{\eta}}))$. The pseudo-differential action op_t along $\boldsymbol{R}_+$ relies on the Fourier transform, and the operator families are understood in the sense

$$C_0^\infty(\boldsymbol{R}_+,C^\infty(X))\to C^\infty(\boldsymbol{R}_+,C^\infty(X)).$$

The proof of Theorem 3.4 is technical, however the idea is easy. In a first step we choose an $\tilde{f}_0(t,y,z,\tilde{\eta})\in C^\infty(\overline{\boldsymbol{R}}_+\times\Omega,L^\mu_{cl}(X;\Gamma_0\times\boldsymbol{R}^q_{\tilde{\eta}}))$, where $\Gamma_0=\{i\tau:\tau\in\boldsymbol{R}\}\cong\boldsymbol{R}$, such that

$$\tilde{f}_0(t,y,-i\tau,\tilde{\eta})=\tilde{P}(t,y,\tau,\tilde{\eta}).$$

Then a calculation shows that there is a

$$\tilde{P}_1(t,y,\tilde{\tau},\tilde{\eta}) \in C^\infty(\overline{\boldsymbol{R}}_+ \times \Omega, L_{cl}^{\mu-1}(X;\Gamma_0 \times \boldsymbol{R}_{\tilde{\tau}\tilde{\eta}}^{1+q}))$$

such that

$$\mathrm{op}_t(P)(y,\eta) = \mathrm{op}_M^\beta(f_0)(y,\eta) + \mathrm{op}_t(P_1)(y,\eta)$$

$C^\infty(\Omega, L^{-\infty}(X^\wedge;\boldsymbol{R}_\eta^q))$, where $f_0(t,y,z,\eta) := \tilde{f}_0(t,y,z,t\eta)$, $P_1(t,y,\tau,\eta) := \tilde{P}_1(t,y,t\tau,t\eta)$. This allows us to start an iteration which yields a sequence

$$\tilde{f}_k(t,y,z,\tilde{\eta}) \in C^\infty(\overline{\boldsymbol{R}}_+ \times \Omega, L_{cl}^{\mu-k}(X;\Gamma_0 \times \boldsymbol{R}_{\tilde{\eta}}^q))$$

for all $k \in \boldsymbol{N}$.

The asymptotic sum

$$\tilde{f}(t,y,z,\tilde{\eta}) \sim \sum_{k=0}^\infty f_k(t,y,z,\tilde{\eta}) \quad \text{in} \quad C^\infty(\overline{\boldsymbol{R}}_+ \times \Omega, L_{cl}^\mu(X;\Gamma_0 \times \boldsymbol{R}_{\tilde{\eta}}^q))$$

then gives an $f(t,y,z,\eta) = \tilde{f}(t,y,z,t\eta)$ such that

$$\mathrm{op}_t(P)(y,\eta) = \mathrm{op}_M^{\frac{1}{2}}(f)(y,\eta) \mod C^\infty(\Omega, L^{-\infty}(X^\wedge;\boldsymbol{R}_\eta^q)).$$

Applying Theorem 3.3 in the (t,y)-dependent form and using the fact that $\mathrm{op}\beta_M(\ldots) = \mathrm{op}_M^\rho(\ldots)$ for arbitrary $\beta,\rho \in \boldsymbol{R}$ for Mellin symbols that are holomorphic in z we obtain the assertion.

Remark 3.5 *Let us set*

$$\tilde{P}_0(y,\tilde{\tau},\tilde{\eta}) = \tilde{P}(0,y,\tilde{\tau},\tilde{\eta}), \quad \tilde{h}_0(y,z,\tilde{\eta}) = \tilde{P}(0,y,z,\tilde{\eta}) \;\; \textit{and}$$
$$\tilde{P}_0(t,y,\tau,\eta) = \tilde{P}_0(y,t\tau,t\eta), \quad h_0(t,y,z,\eta) = \tilde{h}_0(y,z,t\eta).$$

Then, if we insert $\tilde{P}_0(y,\tilde{\tau},\tilde{\eta})$ instead of $\tilde{P}(t,y,\tilde{\tau},\tilde{\eta})$ in Theorem 3.4 it follows that

$$\mathrm{op}_t(P)(y,\eta) = \mathrm{op}_M^\beta(h_0)(y,\eta) \mod C^\infty(\Omega, L^{-\infty}(X^\wedge;\boldsymbol{R}_\eta^q)).$$

Let us now choose cut-off functions $\omega(t)$, $\omega_0(t)$, $\omega_1(t)$ satisfying $\omega\omega_0 = \omega_1$, $\omega\omega_1 = \omega_1$. Then, using Theorem 3.4 together with the pseudo-locality of pseudo-differential operators in parameter-dependent form, we obtain

$$\begin{aligned} \mathrm{op}_t(P)(y,\eta) &= \omega(t[\eta])\mathrm{op}_M^\beta(h)(y,\eta)\omega_0(t[\eta]) \\ &+ (1-\omega(t[\eta]))\mathrm{op}_t(P)(y,\eta)(1-\omega_1(t[\eta])) \end{aligned} \tag{3.5}$$

mod $C^\infty(\Omega, L^{-\infty}(X^\wedge;\boldsymbol{R}_\eta^q))$, for every $\beta \in \boldsymbol{R}$.

Theorem 3.6 *Set*

$$\begin{aligned} a_0(y,\eta) &= t^{-\mu}\omega(t[\eta])\mathrm{op}_M^{\gamma-\frac{n}{2}}(h)(y,\eta)\omega_0(t[\eta]),\\ a_1(y,\eta) &= t^{-\mu}(1-\omega(t[\eta]))\mathrm{op}_t(P)(y,\eta)(1-\omega_1(t[\eta])\end{aligned}$$

for P and h of Theorem 3.4, and let $\underline{\omega}(t)$, $\underline{\omega}_0(t)$ be arbitrary cut-off functions. Then

$$\begin{aligned} a(y,\eta) &= \underline{\omega}(t)\{a_0(y,\eta)+a_1(y,\eta)\}\underline{\omega}_0(t)\\ &\in S^{\mu}(\Omega\times \boldsymbol{R}^q;\mathcal{K}^{s,\gamma}(X^\wedge),\mathcal{K}^{s-\mu,\gamma-\mu}(X^\wedge)) \end{aligned} \tag{3.6}$$

for every $s\in \boldsymbol{R}$. Moreover,

$$\mathrm{Op}_y(a)=\mathrm{Op}_y(\underline{\omega}\mathrm{op}_t(P)\underline{\omega}_0) \quad \mathrm{mod} \quad L^{-\infty}(X^\wedge\times\Omega), \tag{3.7}$$

where $Op_y(\ldots)=\mathcal{F}^{-1}_{\eta\to y}(\ldots)\mathcal{F}_{y'\to\eta}$.

The proof of (3.6) in Theorem 3.6 is based on estimates of the norms of pseudo-differential operators by the symbols. (3.7) is a consequence of (3.5).

Remark 3.7 *Let us set with the notation of Remark 3.5*

$$\begin{aligned} \sigma_\wedge^\mu(a)(y,\eta) &= t^{-\mu}\omega(t|\eta|)\mathrm{op}_M^{\gamma-\frac{n}{2}}(h_0)(y,\eta)\omega_0(t|\eta|)\\ &+ t^{-\mu}(1-\omega(t|\eta|))\mathrm{op}_t(P_0)(y,\eta)(1-\omega_1(t|\eta|)) \end{aligned}$$

for $(y,\eta)\in\Omega\times(\boldsymbol{R}^q\setminus\{0\})$. Then

$$\sigma_\wedge^\mu(a)(y,\eta):\ \mathcal{K}^{s,\gamma}(X^\wedge)\to\mathcal{K}^{s-\mu,\gamma-\mu}(X^\wedge)$$

is a family of continuous operators and

$$\sigma_\wedge^\mu(a)(y,\lambda\eta)=\lambda^\mu\kappa_\lambda\sigma_\wedge^\mu(a)(y,\eta)\kappa_\lambda^{-1} \quad \text{for all} \quad \lambda\in\boldsymbol{R}_+$$

and all $(y,\eta)\in\Omega\times(\boldsymbol{R}^q\setminus\{0\})$, $s\in\boldsymbol{R}$.

Remark 3.8 *$a(y,\eta)$ in the notation of Theorem 3.6 is a parameter-dependent family of cone pseudo-differential operators on the infinite (open stretched) cone $X^\wedge$.*

Remark 3.9 *The relation (3.7) of Theorem 3.6 shows that for $a(y,\eta)$ given by (3.6) we have*

$$\mathrm{Op}(a)\in L^\mu_{cl}(X^\wedge\times\Omega).$$

Since (3.4) that is involved in the definition can be generated by a system p_j of local edge-degenerate symbols (3.3) via (3.4), the correspondence

$$\{t^{-\mu}p_1,\ldots,t^{-\mu}p_N\}\to\mathrm{Op}(a)$$

may be interpreted as an operator convention for such degenerate symbols. If we assume in addition that the symbol p_j *over* $\Sigma_j \cap \chi_j(U_j \cap U_k)$ *is compatible with* p_k *over* $\Sigma_k \cap \chi_k(U_k \cap U_j)$ *modulo symbols of order* $-\infty$, *in the sense of the rule for the symbol push-forward, (associated with the operator push-forward to the transition diffeomorphisms* $\{\Sigma_j \cap \chi_j(U_j \cap U_k)\} \to \{\Sigma_k \cap \chi_k(U_k \cap U_j)\}$*) for all* k, j, *then* $\mathrm{Op}(a)$ *has* $\{t^{-\mu}p_1, \ldots, t^{-\mu}p_N\}$ *as the local symbols over* $\boldsymbol{R}_+ \times \Sigma_j \times \Omega$, $j = 1, \ldots, N$.

Theorem 3.10 *Let* $a(y, \eta)$ *be given by (3.6). Then* $Op(a)$ *induces continuous operators*

$$\mathrm{Op}(a): \mathcal{W}^s_{comp(y)}(\Omega, \mathcal{K}^{s,\gamma}(X^\wedge)) \to \mathcal{W}^{s-\mu}_{loc(y)}(\Omega, \mathcal{K}^{s-\mu,\gamma-\mu}(X^\wedge))$$

for all $s \in \boldsymbol{R}$.

3.2. GREEN, TRACE AND POTENTIAL EDGE SYMBOLS

Definition 3.11 *An operator-valued symbol*

$$g(y, y', \eta) \in \cap_{s\in\boldsymbol{R}} S^\mu_{cl}(\Omega \times \Omega \times \boldsymbol{R}^q; \mathcal{K}^{s,\gamma}(X^\wedge) \oplus \boldsymbol{C}^{N_-}, \mathcal{K}^{\infty,\delta}(X^\wedge) \oplus \boldsymbol{C}^{N_+})$$

for $\mu, \gamma, \delta \in \boldsymbol{R}$ *and open* $\Omega \subseteq \boldsymbol{R}^q$ *is called a Green edge symbol (with discrete asymptotics) if*

$$\begin{aligned} g(y, y', \eta) &\in \cap_{s\in\boldsymbol{R}} S^\mu_{cl}(\Omega \times \Omega \times \boldsymbol{R}^q; \mathcal{K}^{s,\gamma}(X^\wedge) \oplus \boldsymbol{C}^{N_-}, \mathcal{S}^\delta_P(X^\wedge) \oplus \boldsymbol{C}^{N_+}), \\ g(y, y', \eta) &\in \cap_{s\in\boldsymbol{R}} S^\mu_{cl}(\Omega \times \Omega \times \boldsymbol{R}^q; \mathcal{K}^{s,-\delta}(X^\wedge) \oplus \boldsymbol{C}^{N_+}, \mathcal{S}^{-\gamma}_Q(X^\wedge) \oplus \boldsymbol{C}^{N_-}) \end{aligned}$$

for g*-dependent discrete asymptotic types* P *and* Q, *associated with the corresponding weight data. Here* $*$ *indicates the point-wise formal adjoint in the sense*

$$(gu, v)_{\mathcal{K}^{0,0}(X^\wedge)\oplus\boldsymbol{C}^{N_+}} = (u, g^*v)_{\mathcal{K}^{0,0}(X^\wedge)\oplus\boldsymbol{C}^{N_-}}$$

for all $u \in C_0^\infty(X^\wedge) \oplus \boldsymbol{C}^{N_-}$, $v \in C_0^\infty(X^\wedge) \oplus \boldsymbol{C}^{N_+}$.

Recall that the group action on a space $E \oplus \boldsymbol{C}^N$ equals $\kappa_\lambda \oplus \mathrm{id}_{\boldsymbol{C}^N}$, $\lambda \in \boldsymbol{R}_+$, when $\{\kappa_\lambda\}_{\lambda\in\boldsymbol{R}_+}$ is the given group action on E. Here, for E we have $\mathcal{K}^{s,\gamma}(X^\wedge)$ with $(\kappa_\lambda u)(t, x) = \lambda^{\frac{n+1}{2}} u(\lambda t, x)$, or E^j in the meaning of Remark 2.18. By definition we have $g(y, y', \eta) = (g_{ij})(y, y', \eta)_{i,j=1,2}$.

The entry $g_{21}(y, y', \eta)$ has the meaning of a vector of N_+ trace symbols, $g_{12}(y, y', \eta)$ of N_- potential symbols, whereas $g_{22}(y, y', \eta)$ is an $N_- \times N_+$-matrix of symbols in $S^\mu_{cl}(\Omega \times \Omega \times \boldsymbol{R}^q)$.

Remark 3.12 *Note that the operator families in Definition 3.11 may be regarded as a generalization of those in Section 1.4 (for the case $d = 0$). The asymptotic types in Section 1.4 are the Taylor asymptotics in the image (i.e., smoothness up to $t = 0$). This belongs to the motivation for our notation to call the symbols of Definition 3.11 Green symbols.*

As classical symbols the Green symbols $g(y, y', \eta)$ have a unique sequence $g_{(\mu-j)}(y, y', \eta)$ of components of homogeneity $\mu - j$ for all $j \in \boldsymbol{N}$. We set

$$\sigma_\wedge^\mu(g)(y,\eta) = g_{(\mu)}(y, y', \eta)|_{y'=y}. \tag{3.8}$$

This will be regarded as an operator family

$$\sigma_\wedge^\mu(g)(y,\eta) : \begin{array}{c} \mathcal{K}^{s,\gamma}(X^\wedge) \\ \oplus \\ \boldsymbol{C}^{N_-} \end{array} \rightarrow \begin{array}{c} \mathcal{K}^{s-\mu,\delta}(X^\wedge) \\ \oplus \\ \boldsymbol{C}^{N_+} \end{array}$$

for every s, where

$$\sigma_\wedge^\mu(g)(y,\lambda\eta) = \lambda^\mu \begin{pmatrix} \kappa_\lambda & 0 \\ 0 & 1 \end{pmatrix} \sigma_\wedge^\mu(g)(y,\eta) \begin{pmatrix} \kappa_\lambda & 0 \\ 0 & 1 \end{pmatrix}^{-1}$$

for all $\lambda \in \boldsymbol{R}_+$ and $(y,\eta) \in \Omega \times (\boldsymbol{R}^q \setminus \{0\})$.

Theorem 3.13 *Let $g(y, y', \eta)$ be a Green symbol of order μ. Then* $\mathrm{Op}(g)$ *induces continuous operators*

$$\mathrm{Op}(g) : \begin{array}{c} \mathcal{W}^s_{comp(y)}(\Omega, \mathcal{K}^{s,\gamma}(X^\wedge)) \\ \oplus \\ H^s_{comp}(\Omega, \boldsymbol{C}^{N_-}) \end{array} \rightarrow \begin{array}{c} \mathcal{W}^{s-\mu}_{loc(y)}(\Omega, \mathcal{K}_P^{\infty,\delta}(X^\wedge)) \\ \oplus \\ H^{s-\mu}_{loc}(\Omega, \boldsymbol{C}^{N_+}) \end{array}$$

for all $s \in \boldsymbol{R}$, with some asymptotic type P dependent on g.

For the proof it suffices to apply Theorem 1.16.

The pseudo-differential calculus on a manifold with edges gives rise to another interesting class of operator-valued symbols, namely those of smoothing Mellin type. These are (y, y', η)-dependent families of operators of analogous structure as in the cone theory, cf. Section 2.4. Let us fix a discrete Mellin asymptotic type $R_{j\alpha}$ and let $h_{j\alpha}(y, y') \in C^\infty(\Omega \times \Omega, M^{-\infty}_{R_{j\alpha}}(X))$ for $j \in \boldsymbol{N}$, $\alpha \in \boldsymbol{N}^q$, $|\alpha| \le j$. Assume that for given $\gamma, \gamma_{j\alpha} \in \boldsymbol{R}$ we have $\pi_{\boldsymbol{C}} R_{j\alpha} \cap \Gamma_{\frac{n+1}{2}-\gamma_{j\alpha}} = \emptyset$, and $-\gamma + j + \gamma_{j\alpha} \ge 0$, $-\gamma_{j\alpha} + \gamma \ge 0$. Then, for arbitrary cut-off functions $\omega(t)$, $\omega_0(t)$

$$m_{j\alpha}(y, y', \eta) = \omega(t[\eta]) t^{-\mu+j} \mathrm{op}_M^{\gamma_{j\alpha}-\frac{n}{2}}(h_j)(y, y') \eta^\alpha \omega_0(t[\eta])$$

is a family of continuous operators $\mathcal{K}^{s,\gamma}(X^\wedge) \to \mathcal{K}^{\infty,\gamma-\mu}(X^\wedge)$ for every $s \in \boldsymbol{R}$, C^∞ dependent on y, y', η. From

$$m_{j\alpha}(y, y', \lambda\eta) = \lambda^{\mu-j+|\alpha|}\kappa_\lambda m_{j\alpha}(y, y', \eta)\kappa_\lambda^{-1}$$

for all $\lambda \geq 1$, $|\eta| \geq$ const for a constant > 0 we see that

$$m_{j\alpha}(y, y', \eta) \in S_{cl}^{\mu-j+|\alpha|}(\Omega \times \Omega \times \boldsymbol{R}^q; \mathcal{K}^{s,\gamma}(X^\wedge), \mathcal{K}^{\infty,\gamma-\mu}(X^\wedge)) \tag{3.9}$$

for all $s \in \boldsymbol{R}$. Analogously we have

$$m_{j\alpha}(y, y', \eta) \in S_{cl}^{\mu-j+|\alpha|}(\Omega \times \Omega \times \boldsymbol{R}^q; \mathcal{K}_P^{s,\gamma}(X^\wedge), \mathcal{K}_Q^{\infty,\gamma-\mu}(X^\wedge))$$

for every discrete asymptotic type P with some resulting discrete asymptotic type Q, associated with the corresponding weight data. A smoothing Mellin edge symbol in the edge symbolic calculus is defined as

$$m(y, y', \eta) = \sum_{j=0}^{k-1} \sum_{|\alpha| \leq j} m_{j\alpha}(y, y', \eta) \tag{3.10}$$

for arbitrary $m_{j\alpha}$ of the form (3.9) and $k \in \boldsymbol{N} \setminus \{0\}$. Here $(-k, 0]$ is interpreted as weight strip for the discussion of asymptotics. As classical elements the symbols we have unique components $m_{(\mu-l)}(y, y', \eta)$ of order $\mu - l$, $l \in \boldsymbol{N}$, and we set

$$\sigma_\wedge^\mu(m)(y, \eta) = m_{(\mu)}(y, y', \eta)|_{y'=y}. \tag{3.11}$$

Note that only the summands with $j = |\alpha|$ contribute to the term of order μ and that in this case

$$m_{j\alpha,(\mu)}(y, y', \eta) = \omega(t|\eta|)t^{-\mu+j}\mathrm{op}_M^{\gamma_{j\alpha}-\frac{n}{2}}(h_j)(y, y')\eta^\alpha\omega_0(t|\eta|).$$

Remark 3.14 *Analogously to Proposition 2.31 the symbol $m(y, y', \eta)$ remains unchanged modulo some Green symbol $g(y, y', \eta)$ (of type of a left upper corner in the block matrices) when we change the cut-off functions or the weight $\gamma_{j\alpha}$ (under the mentioned assumptions).*

Theorem 3.15 *Let $m(y, y', \eta)$ be a smoothing Mellin symbol of order μ in the sense of (3.10). Then $\mathrm{Op}(m)$ induces a continuous operator*

$$\mathrm{Op}(m): \ \mathcal{W}^s_{comp(y)}(\Omega, \mathcal{K}^{s,\gamma}(X^\wedge)) \to \mathcal{W}^{s-\mu}_{loc(y)}(\Omega, \mathcal{K}^{\infty,\gamma-\mu}(X^\wedge))$$

and

$$\mathrm{Op}(m): \ \mathcal{W}^s_{comp(y)}(\Omega, \mathcal{K}_P^{s,\gamma}(X^\wedge)) \to \mathcal{W}^{s-\mu}_{loc(y)}(\Omega, \mathcal{K}_Q^{\infty,\gamma-\mu}(X^\wedge))$$

for every $s \in \boldsymbol{R}$ and every (discrete) asymptotic type P with some resulting (discrete) asymptotic type Q.

3.3. CONTINUOUS ASYMPTOTICS

For higher-dimensional singularities, e.g., of edge type, the analysis of solutions of elliptic equations with asymptotics near the singular set (say in terms of the distance variable $t \in \boldsymbol{R}_+$, $t \to 0$) leads to exponents of t and of $\log t$ that may depend on the edge variable y. The Mellin transform $M_{t\to z}$ of such a y-dependent distribution with asymptotics then consists of a family of meromorphic functions in $\boldsymbol{C} \ni z$, where in general the poles have not constant multiplicity under varying y. For instance, if $a(y), b(y), c(y) \in C^\infty(\Omega)$ for open $\Omega \subseteq \boldsymbol{R}^q$, $\operatorname{Re} a(y), \operatorname{Re} b(y), \operatorname{Re} c(y) < \frac{1}{2}$ for all $y \in \Omega$, then

$$u(t,y) = \omega(t) M_{z\to t}^{-1} (z - c(y))\{(z - a(y))(z - b(y))\}^{-1}$$

belongs to $C^\infty(\Omega, L^2(\boldsymbol{R}_+))$ and we have for every fixed $y \in \Omega$

$$u(t,y) \sim \gamma(y) t^{a(y)} \log t \quad \text{for} \quad t \to 0$$

when $a(y) = b(y)$ and $a(y) \neq c(y)$,

$$u(t,y) \sim \alpha(y) t^{a(y)} + \beta(y) t^{b(y)} \quad \text{for} \quad t \to 0$$

for $a(y) \neq b(y)$ and $a(y) \neq c(y)$, $b(y) \neq c(y)$, whereas

$$u(t,y) \sim \delta(y) t^{b(y)} \quad \text{for} \quad t \to 0$$

for $a(y) = c(y)$, $b(y) \neq c(y)$, and similarly for $a(y) \neq c(y)$, $b(y) = c(y)$. Here $\gamma(y), \alpha(y), \beta(y), \delta(y)$ are certain complex coefficients. In other words we obtain branching discrete asymptotics, and the y-dependent behavior can be extremely complicated.

For the systematic calculus that enables us to characterize such effects in general, also in the vector-valued situation for weighted wedge distributions of the classes $\mathcal{W}^s_{loc}(\Omega, \mathcal{K}^{s,\gamma}(X^\wedge))$, it is useful to extend the concept of discrete asymptotics and to pass to the continuous asymptotics. This was first mentioned in Rempel, Schulze [22] and then intensely studied by the author in [29], [31], [35], [36]. We shall describe here some basic ideas.

If $K \subset\subset \boldsymbol{C}$ is a compact set we have the space $\mathcal{A}'(K)$ of analytic functionals carried by K in its nuclear Fréchet topology, cf. Hörmander [14, Vol. 1], (for the Fréchet topology cf. [31], [34]) To every $p \in \boldsymbol{C}$, $m \in \boldsymbol{N}$, we can define an element $\zeta \in \mathcal{A}'(\{p\})$ by $\langle \zeta, h \rangle = \sum_{k=0}^m C_k \left(\frac{d}{dz}\right)^k h(z)|_{z=p}$, $h \in \mathcal{A}(\boldsymbol{C})$. Then, in particular,

$$\langle \zeta, t^{-z} \rangle = t^{-p} \sum_{k=0}^{m} (-1)^k c_k \log^k t.$$

This shows that the discrete asymptotics for $t \to 0$ can also be written as follows. There are compact sets $K_j \subset\subset \boldsymbol{C}$, $K_j \subset \{z: \ \operatorname{Re} z < \frac{1}{2} - \delta\}$ for some $\delta \in \boldsymbol{R}$, $j \in \boldsymbol{N}$, with $\sup\{\operatorname{Re} z: \ z \in K_j\} \to -\infty$ as $j \to \infty$, such that there are analytic functionals $\zeta_j \in \mathcal{A}'(K_j)$ with

$$u(t) \sim \sum_{j=0}^{\infty} \langle \zeta_j, t^{-z} \rangle \quad \text{for} \quad t \to 0.$$

The discrete asymptotics as a special case correspond to $K_j = \{p_j\}$, where ζ_j is a linear combination of derivatives of the Dirac distribution in p_j. Up to now we have considered the scalar case, i.e., asymptotics for distributions on $\boldsymbol{R}_+$, say in $L^2(\boldsymbol{R}_+)$, where $\delta = 0$. In general, for $\mathcal{K}^{s,\gamma}(X^\wedge)$ with a non-trivial cone base we take $C^\infty(X)$-valued analytic functionals. If $K \subset\subset \boldsymbol{C}$ is given then $\mathcal{A}'(K, E) = \mathcal{A}'(K) \hat{\otimes}_\pi E$ is the space of E-valued analytic functions, carried on K, for any Fréchet space E.

If a set $V \subset \boldsymbol{C}$ is given we denote by V^c the complement of the union of all unbounded connected components of $\boldsymbol{C} \setminus \overline{V}$.

Definition 3.16 *Let* $V \subset \{z: \ \frac{n+1}{2} - \gamma + \vartheta < \operatorname{Re} z < \frac{n+1}{2} - \gamma\}$ *be a compact set* $V^c = V$, $-\infty < \vartheta < 0$, $\gamma \in \boldsymbol{R}$. *Then* $u(t) \in \mathcal{K}^{s,\gamma}(X^\wedge)$ *belongs to* $\mathcal{K}_V^{s,\gamma}(X^\wedge)$ *for the continuous asymptotic type* V *if for some cut-off function* $\omega(t)$ *and an element* $\zeta \in \mathcal{A}'(V, C^\infty(X))$

$$\omega(u - \langle \zeta, t^{-z} \rangle) \in \mathcal{K}^{s,(\gamma-\vartheta)-}(X^\wedge).$$

The space $\mathcal{K}_V^{s,\gamma}(X^\wedge)$ can be endowed with a natural Fréchet topology.

More generally if $V \subset \boldsymbol{C}$ is a closed set, $V^c = V$, $V \subset \{z: \ \operatorname{Re} z < \frac{n+1}{2} - \gamma\}$, and $V \cap \{z: \ c \leq \operatorname{Re} z \leq c'\}$ compact for every $c \leq c'$, then we set $V_k = V \cap \{z: \ \operatorname{Re} z \geq \frac{n+1}{2} - \gamma - (k+1)\}$, $k \in \boldsymbol{N}$, and define

$$\mathcal{K}_V^{s,\gamma}(X^\wedge) = \cap_{k \in \boldsymbol{N}} \mathcal{K}_{V_k}^{s,\gamma}(X^\wedge),$$

endowed with the projective limit topology.

Theorem 3.17 *For arbitrary* V *with the required properties,* $-\infty \leq \vartheta < 0$, *and every* $s \in \boldsymbol{R}$ *we have*

$$\mathcal{K}_V^{s,\gamma}(X^\wedge) = \mathcal{K}^{s,(\gamma+\vartheta)-}(X^\wedge) + \mathcal{K}_V^{\infty,\gamma}(X^\wedge). \tag{3.12}$$

Moreover, for $V_1 + V_2 := (V_1 \cup V_2)^c$

$$\mathcal{K}_V^{s,\gamma}(X^\wedge) = \mathcal{K}_{V_1}^{s,\gamma}(X^\wedge) + \mathcal{K}_{V_2}^{s,\gamma}(X^\wedge), \tag{3.13}$$

in the sense of non-direct sums of Fréchet spaces. In (3.12), for $\vartheta = -\infty$, $\mathcal{K}^{s,(\gamma+\vartheta)-}(X^\wedge)$ *is defined as* $\mathcal{K}^{s,\infty}(X^\wedge)$.

Theorem 3.17 follows from a decomposition result for (vector-valued) analytic functionals with respect to decompositions of the carrier sets. This employs, in particular, a Cousin problem argument.

Also operator-valued Mellin symbols with continuous asymptotic data can be introduced:

Definition 3.18 *Let $V \subset \boldsymbol{C}$ be a closed set, $V^c = V$, $V \cap \{z : \ c \leq Re\, z \leq c'\}$ compact for every $c \leq c'$. Then $M_V^{-\infty}(X)$ denotes the space of all $h(z) \in \mathcal{A}(\boldsymbol{C} \setminus V, L^{-\infty}(X))$ such that for every $V_k := \{z : \ -(k+1) \leq Re\, z \leq k+1\}$, $k \in \boldsymbol{N}$, there exists a $\zeta \in \mathcal{A}'(V_k, L^{-\infty}(X))$ with*

$$h(z) - (M_{\delta, t \to z}\omega\langle\zeta, t^{-z}\rangle)(z) \in \mathcal{A}(\{|Re\, z| < k+1, L^{-\infty}(X)\})$$

for some cut-off function $\omega(t)$, and $\delta \in \boldsymbol{R}$ with $\left|\frac{1}{2} - \delta\right| < k+1$, and moreover, for every V-excision function $\chi(z)$ we have

$$\chi(z)h(z)|_{\Gamma_\beta} \in L^{-\infty}(X; \Gamma_\beta)$$

for every $\beta \in \boldsymbol{R}$, uniformly in $c \leq \beta \leq c'$ for every $c \leq c'$.

$M_V^{-\infty}(X)$ is a (nuclear) Fréchet space in a natural way. We then define

$$M_V^\mu(X) = M_O^\mu(X) + M_V^{-\infty}(X)$$

for $\mu \in \boldsymbol{R}$ as a non-direct sum of Fréchet spaces. Similarly to Theorem 3.17 we have

$$M_V^\mu(X) = M_{V_1}^\mu(X) + M_{V_2}^\mu(X)$$

for $V = V_1 + V_2$.

The pseudo-differential calculus on a manifold with conical singularities is possible also in the framework of the continuous asymptotics, cf. [34]. Let us mention here only some few facts.

Theorem 3.19 *Let $h(t, t', z) \in C^\infty(\overline{\boldsymbol{R}}_+ \times \overline{\boldsymbol{R}}_+, M_V^\mu(X))$ be given with a $V \subset \boldsymbol{C}$ of the mentioned kind, where $V \cap \Gamma_{\frac{n+1}{2}-\gamma} = \emptyset$. Then, if $\omega(t), \tilde{\omega}(t)$ are cut-off functions, $\omega \mathrm{op}_M^{\gamma-\frac{n}{2}}(h)\tilde{\omega}$ induces a continuous operator*

$$\omega \mathrm{op}_M^{\gamma-\frac{n}{2}}(h)\tilde{\omega} : \ \mathcal{K}_B^{s,\gamma}(X^\wedge) \to \mathcal{K}_C^{s-\mu,\gamma}(X^\wedge)$$

for every continuous asymptotic type B with some resulting continuous asymptotic type C, for all $s \in \boldsymbol{R}$.

The weighted Sobolev spaces $\mathcal{H}^{s,\gamma}(\boldsymbol{B})$ on a stretched manifold $\boldsymbol{B}$ with conical singularities also contain subspaces with continuous asymptotics in a natural way.

Theorem 3.20 *The elliptic regularity of Theorem 2.34 holds in analogous form for continuous asymptotic types.*

Remark 3.21 *The y-dependent discrete (branching) asymptotics of elements in $\mathcal{W}^s_{loc(y)}(\Omega, \mathcal{K}^{s,\gamma}(X^\wedge))$ can be formulated in terms of C^∞ functions of y with values in $\mathcal{A}'(K, C^\infty(X))$ for certain $K \subset\subset \boldsymbol{C}$, where y-wise the analytic functionals are finite linear combinations of derivatives of Dirac measures in points $p_j \in \boldsymbol{C}$, with coefficients in $C^\infty(X)$. This concept for the elliptic regularity for edge problems was applied in [33]. The general calculus for variable branching asymptotics in the case of boundary value problems may be found in [35], [36]. The edge case with non-trivial cone base is similar.*

Similarly as in Section 2.4 we define for a continuous asymptotic type P

$$\mathcal{S}^\gamma_P(X^\wedge) = [\omega]\mathcal{K}^{\infty,\gamma}_P(X^\wedge) + [1-\omega]\mathcal{S}(\overline{\boldsymbol{R}}_+ \times X).$$

For the spaces $\mathcal{K}^{s,\gamma}_P(X^\wedge)$ and $\mathcal{S}^\gamma_P(X^\wedge)$ we have an analogue of Remark 2.18. In particular, we can also form the wedge spaces

$$\mathcal{W}^s_{comp(y)}(\Omega, \mathcal{K}^{\infty,\gamma}_P(X^\wedge)) \quad \text{and} \quad \mathcal{W}^s_{loc(y)}(\Omega, \mathcal{K}^{\infty,\gamma}_P(X^\wedge))$$

in the framework of the continuous asymptotics.

By replacing the discrete asymptotic types of Definition 3.11 by continuous ones we define the Green symbols with continuous asymptotics. In an analogous manner we introduce smoothing Mellin symbols of the form (3.10) with continuous asymptotics in the involved $h_{j\alpha}(y, y') \in C^\infty(\Omega \times \Omega, M^{-\infty}_{R_{j\alpha}}(X))$. This will tacitly be used below in Section 3.4.

3.4. ELLIPTICITY OF PSEUDO-DIFFERENTIAL OPERATORS ON MANIFOLDS WITH EDGES

For discussing ellipticity of operators on a manifold with edges we want to consider once again the edge-degenerate differential operators on a (stretched) wedge $X^\wedge \times \Omega \ni (t, x, y)$:

$$A = t^{-\mu} \sum_{j+|\alpha| \le \mu} a_{j\alpha}(t, y) \left(-t\frac{\partial}{\partial t}\right)^j (tD_y)^\alpha$$

for coefficients $a_{j\alpha}(t, y) \in C^\infty(\overline{\boldsymbol{R}}_+ \times \Omega, \mathrm{Diff}^{\mu-(j+|\alpha|)}(X))$.

Let us assume for simplicity that $a_{j\alpha}(t, y)$ is independent of t for $|t| >$ const for a constant > 0. If we set

$$a(y, \eta) := t^{-\mu} \sum_{j+|\alpha| \le \mu} a_{j\alpha}(t, y) \left(-t\frac{\partial}{\partial t}\right)^j (t\eta)^\alpha : \mathcal{K}^{s,\gamma}(X^\wedge) \to \mathcal{K}^{s-\mu,\gamma-\mu}(X^\wedge),$$

we have $a(y,\eta) \in S^\mu(\Omega \times \boldsymbol{R}^q; \mathcal{K}^{s,\gamma}(X^\wedge), \mathcal{K}^{s-\mu,\gamma-\mu}(X^\wedge))$ and $A = \mathrm{Op}(a)$. Recall that $a(y,\eta)$ is classical when $a_{j\alpha}(t,y)$ is independent of t. According to the notation in Section 3.1 the operator A has two principal symbols, the homogeneous principal interior symbol of order μ, denoted by $\sigma_\psi^\mu(A)$, and the homogeneous principal edge symbol of order μ which is of the form

$$\sigma_\wedge^\mu(A)(y,\eta) = t^{-\mu} \sum_{j+|\alpha|\le\mu} a_{j\alpha}(0,y)\left(-t\frac{\partial}{\partial t}\right)^j (t\eta)^\alpha. \tag{3.14}$$

The homogeneity means $\sigma_\wedge^\mu(A)(y,\lambda\eta) = \lambda^\mu \kappa_\lambda \sigma_\wedge^\mu(A)(y,\eta)\kappa_\lambda^{-1}$ for all $\lambda \in \boldsymbol{R}_+$. The operators (3.14) are of Fuchs type on $X^\wedge$ for every fixed (y,η). Hence we have from the cone operator calculus a principal conormal symbol derived from (3.14), namely

$$\sigma_M^\mu \sigma_\wedge^\mu(A)(y,z) = \sum_{j=0}^{\mu} a_{j0}(0,y) z^j : \; H^s(X) \to H^{s-\mu}(X),$$

$s \in \boldsymbol{R}$. Here the dependence on η disappears. Of course, there is also a subordinate homogeneous principal interior symbol of $\sigma_\wedge^\mu(A)$ of order μ in the cone sense, namely $(\sigma_\psi^\mu \sigma_\wedge^\mu(A))(t,x,y,\tau,\xi)$ that we obtain from $\sigma_\psi^\mu(A)$ by inserting $\eta = 0$ and freezing the coefficients in $t = 0$, where the weight factor $t^{-\mu}$ remains untouched. This allows us to apply various results from Section 2 in the present case, here in a corresponding y-dependent form. Let us write

$$\sigma_\psi^\mu(A)(t,x,y,\tau,\xi,\eta) = t^{-\mu}\tilde{p}_{(\mu)}(t,x,y,t\tau,\xi,t\eta)$$

for a corresponding $\tilde{p}_{(\mu)}(t,x,y,\tilde{\tau},\xi,\tilde{\eta})$ that is C^∞ in t up to $t = 0$ and homogeneous of order μ in $(\tilde{\tau},\xi,\tilde{\eta}) \neq 0$.

Theorem 3.22 *Let A be an edge-degenerate differential operator on $X^\wedge \times \Omega$ and assume that $\tilde{p}_{(\mu)}(0,x,y,\tilde{\tau},\xi,\tilde{\eta}) \neq 0$ for all x,y and $(\tilde{\tau},\xi,\tilde{\eta}) \neq 0$. Then for every $y \in \Omega$ there exists a countable set $E(y) \subset \boldsymbol{R}$ with $E(y) \cap K$ finite for every $K \subset\subset \boldsymbol{R}$ such that*

$$\sigma_\wedge^\mu(A)(y,\eta) : \; \mathcal{K}^{s,\gamma}(X^\wedge) \to \mathcal{K}^{s-\mu,\gamma-\mu}(X^\wedge) \tag{3.15}$$

is a Fredholm operator for every $\gamma \in \boldsymbol{R} \setminus E(y)$, $\eta \neq 0$, for all $s \in \boldsymbol{R}$.

The theorem is an analogue of Theorem 2.14, where (y,η) are fixed. However $\overline{\boldsymbol{R}}_+ \times X$ is not compact with respect to $t \to \infty$. For the Fredholm property of operators on $X^\wedge$ we need the conditions of Definition 2.9 that are satisfied here for $\sigma_\wedge^\mu(A)(y,\eta)$. In addition there is required an "exit ellipticity" due to the exit $t \to \infty$, cf. [7], [34]. This is automatically satisfied for $\eta \neq 0$, so we drop further details here. The countable set $E(y)$ that determines the exceptional weights $\gamma = \gamma(y)$ comes from Corollary 2.13. The allowed weights for the ellipticity of A in the wedge sense will be those for which $\gamma \notin E(y)$ for all $a \in \Omega$. Note that the index of the Fredholm operator (3.15) may change when we choose another admissible weight γ.

The operator family (3.15) plays an analogous role as the homogeneous principal boundary symbol of an elliptic operator in the context of boundary value problems, say in a bounded domain in Euclidean space, with C^∞ boundary. This corresponds to the case $\dim X = 0$. The Fredholm property of the boundary symbol $\sigma_\wedge^\mu(A)$ as a (y, η)-dependent operator family, between Sobolev spaces on the half axis (the inner normal to the boundary) is the reason for the boundary conditions (which are of trace and potential type in the pseudo-differential case). They complete $\sigma_\wedge^\mu(A)$ to a (y, η)-dependent family of isomorphisms. The latter property is an ellipticity in the sense of operator-valued symbols. In boundary value problems it is called the Shapiro-Lopatinskii condition. Together with the interior ellipticity this finally implies the Fredholm property of a corresponding operator block matrix on the level of global operators between Sobolev spaces.

An analogous strategy can be applied to operators on a manifold with edges, where the contribution to the ellipticity from the edges may be described on the wedge $X^\wedge \times \Omega$. For the ellipticity of edge symbols (the analogues of the boundary symbols) both trace and potential conditions are necessary at the same time, even for differential operators (in contrast to the case of boundary value problems). The minimal number of trace and potential conditions will be $\dim \ker \sigma_\wedge^\mu(A)(y, \eta)$ and $\dim \operatorname{coker} \sigma_\wedge^\mu(A)(y, \eta)$, respectively. All this is possible under some topological condition on $\sigma_\psi^\mu(A)$. By virtue of the role of the open set $\Omega \subseteq \boldsymbol{R}^q$ as local coordinates on an edge Y which is a q-dimensional C^∞ manifold we may choose a smaller $\Omega_0 \subset \Omega$ with $\overline{\Omega}_0 \subset\subset \Omega$ and formulate everything for $y \in \overline{\Omega}_0$. Then for simplifying notations we speak about $\overline{\Omega}$, again. In addition we may assume Ω to be simply connected, with C^∞ boundary. Let $\gamma \notin E(y)$ for all $y \in \overline{\Omega}$. Then (3.15) is a family of Fredholm operators on $T^*\overline{\Omega} \setminus 0$ that is uniquely determined by its restriction to the cosphere bundle $S^*\overline{\Omega}$ induced by $T^*\overline{\Omega}$ which is compact. There is then an index element in the K theoretic sense

$$\operatorname{ind}_{S^*\overline{\Omega}} \sigma_\wedge^\mu(A) \in K(S^*\overline{\Omega}).$$

The condition is now that $\operatorname{ind}_{S^*\overline{\Omega}} \sigma_\wedge^\mu(A) \in \pi^* K(\overline{\Omega})$, where $\pi: S^*\overline{\Omega} \to \overline{\Omega}$ is the canonical projection. This will be assumed whenever we talk about ellipticity.

Now $\sigma^\mu_\wedge(A)(y,\eta)$ can be completed to a block matrix

$$\begin{pmatrix} \sigma^\mu_\wedge(A) & \sigma^\mu_\wedge(K) \\ \sigma^\mu_\wedge(T) & \sigma^\mu_\wedge(Q) \end{pmatrix}(y,\eta): \begin{array}{c} \mathcal{K}^{s,\gamma}(X^\wedge) \\ \oplus \\ J^-_y \end{array} \to \begin{array}{c} \mathcal{K}^{s-\mu,\gamma-\mu}(X^\wedge) \\ \oplus \\ J^+_y \end{array} \tag{3.16}$$

for vector bundles $J^-, J^+ \in \text{Vect}\,(Y)$, subscript y indicates fibres over y (in local coordinates), where (3.16) is an isomorphism for all $(y,\eta) \in T^*\Omega \setminus 0$. The choice of $\sigma^\mu_\wedge(T), \sigma^\mu_\wedge(K)$ and $\sigma^\mu_\wedge(Q)$ is possible in such a way that when we denote (3.16) also by $\sigma^\mu_\wedge(\mathcal{A})(y,\eta)$, we have homogeneity in the sense

$$\sigma^\mu_\wedge(\mathcal{A})(y,\lambda\eta) = \lambda^\mu \begin{pmatrix} \kappa_\lambda & 0 \\ 0 & 1 \end{pmatrix} \sigma^\mu_\wedge(\mathcal{A})(y,\eta) \begin{pmatrix} \kappa_\lambda & 0 \\ 0 & 1 \end{pmatrix}^{-1} \tag{3.17}$$

for all $\lambda \in \boldsymbol{R}_+$, $(y,\eta) \in T^*\Omega \setminus 0$. Finally the additional entries in (3.16) can be chosen to be the form of (3.8) for a suitable operator-valued symbol $g(y,\eta)$ in the sense of Definition 3.11, where $N_\pm$ is the fibre dimension of $J^\pm$.

From $(\sigma^\mu_\psi(A), \sigma^\mu_\wedge(A))$ we can pass to an operator block matrix $\mathcal{A} = (A_{ij})_{i,j=1,2}$ with $A = A_{11}$,

$$A_{21} = \text{Op}(\chi\sigma^\mu_\wedge(T)), \quad A_{12} = \text{Op}(\chi\sigma^\mu_\wedge(K)), \quad A_{22} = \text{Op}(\chi\sigma^\mu_\wedge(Q)),$$

where Op is the pseudo-differential action in y and $\chi(\eta)$ an excision function in $\boldsymbol{R}^q$. We then also set $T = A_{21}$, $K = A_{12}$, $Q = A_{22}$ which gives us a continuous operator

$$\mathcal{A} = \begin{pmatrix} A & K \\ T & Q \end{pmatrix}: \begin{array}{c} \mathcal{W}^s_{comp(y)}(\Omega, \mathcal{K}^{s,\gamma}(X^\wedge)) \\ \oplus \\ H^s_{comp}(\Omega, J^-) \end{array} \to \begin{array}{c} \mathcal{W}^{s-\mu}_{loc(y)}(\Omega, \mathcal{K}^{s-\mu,\gamma-\mu}(X^\wedge)) \\ \oplus \\ H^{s-\mu}_{loc}(\Omega, J^+) \end{array},$$

for every $s \in \boldsymbol{R}$, where $H^s_{comp,loc}(\Omega, J^\pm)$ are the corresponding spaces of distributional sections of $J^\pm$ of Sobolev smoothness $s \in \boldsymbol{R}$. Now the parametrix construction for $\mathcal{A}$ requires to invert both $\sigma^\mu_\psi(A)$ and $\sigma^\mu_\wedge(\mathcal{A})$ and to pass to a corresponding operator in reverse direction. This is possible within the full pseudo-differential calculus for manifolds with edges as it may be found in [7], [30], cf. also [34] for the analogue in boundary value problems. It turns out that the inversion of $\sigma^\mu_\wedge(\mathcal{A})$ also leads to contributions of the form (3.11) in the left upper corners, such that it is necessary to consider such smoothing Mellin operators in the algebra from the very beginning.

Let $\boldsymbol{W}$ be a (stretched) manifold with edges Y, cf. the notation in the introduction. Assume for simplicity that $\boldsymbol{W}$ is compact. Then we have the global weighted wedge Sobolev spaces

$$\mathcal{W}^{s,\gamma}(\boldsymbol{W}) \subset H^s_{loc}(\text{int}\,\boldsymbol{W})$$

that are locally near the edge defined by $\mathcal{W}^s_{loc(y)}(\Omega, \mathcal{K}^{s,\gamma}(X^\wedge))$, and the subspaces with asymptotics of type P (say continuous) $\mathcal{W}^{s,\gamma}_P(\boldsymbol{W})$ locally being defined by $\mathcal{W}^s_{loc(y)}(\Omega, \mathcal{K}^{s,\gamma}_P(X^\wedge))$. The coordinate changes are supposed to be (t,x) independent for small t. Then we have invariance and the definitions are correct. Here and from now on we assume that $\partial\boldsymbol{W}$ has a collar neighborhood V in which a global splitting of variables $(t,x,y) \in [0,1) \times X \times Y \cong V$ is possible. In addition for every $J \in \text{Vect}\,(Y)$ we denote by $H^s(Y,J)$ the space of distributional sections in J of Sobolev smoothness s.

An operator $\mathcal{G}$ is called a global smoothing operator if it induces continuous operators

$$\mathcal{G}: \begin{array}{c} \mathcal{W}^{s,\gamma}(\boldsymbol{W}) \\ \oplus \\ H^s(Y,J^-) \end{array} \to \begin{array}{c} \mathcal{W}^{\infty,\delta}_P(\boldsymbol{W}) \\ \oplus \\ H^\infty(Y,J^+) \end{array}, \quad \mathcal{G}^*: \begin{array}{c} \mathcal{W}^{s,-\delta}(\boldsymbol{W}) \\ \oplus \\ H^s(Y,J^+) \end{array} \to \begin{array}{c} \mathcal{W}^{\infty,-\gamma}_Q(\boldsymbol{W}) \\ \oplus \\ H^\infty(Y,J^-) \end{array}$$

for all $s \in \boldsymbol{R}$ with $\mathcal{G}$-dependent asymptotic types P and Q associated to weight data (δ,Θ) and $(-\gamma,\Theta)$, respectively, $\Theta = (\vartheta, 0]$, $-\infty \le \vartheta < 0$ fixed. The formal adjoint refers to the pairing with respect to the scalar product in $\mathcal{W}^{0,0}(\boldsymbol{W})$.

The pseudo-differential operators on $\boldsymbol{W}$ will be given near $\partial\boldsymbol{W}$ in terms of the local forms on the wedges $X^\wedge \times \Omega$, where $\Omega \subseteq \boldsymbol{R}^q$ corresponds to local coordinates on Y.

The "algebra" of pseudo-differential operators on $\boldsymbol{W}$ with respect to the weight data $(\gamma, \gamma-\mu, \Theta)$ for $\gamma \in \boldsymbol{R}$, $\Theta = (-k, 0]$, $k \in \boldsymbol{N}\setminus\{0\}$, and with continuous asymptotics, is defined as the space of all operators

$$\mathcal{A} = \begin{pmatrix} A + M + A_{int} & 0 \\ 0 & 0 \end{pmatrix} + \mathcal{G}_1 + \mathcal{G}. \tag{3.18}$$

Here A is locally in $X^\wedge \times \Omega$ of the form $\text{Op}(a)$ for an operator-valued symbol $a(y,\eta)$ like (3.6), M is locally $\text{Op}(m)$ for $m(y,\eta)$ like (3.10) (with continuous asymptotics), $A_{int} \in L^\mu_{cl}(\text{int}\,\boldsymbol{W})$, $\mathcal{G}_1 = \text{Op}(g)$, where $g(y,y',\eta)$ is a symbol as in Definition 3.11, also with continuous asymptotics, $N_\pm$ being the fibre dimension of $J^\pm$, and $\mathcal{G}$ is a global smoothing operator with respect to the weight data $(\gamma, \gamma-\mu, \Theta)$.

The operators of the form $\mathcal{G}_1 + \mathcal{G}$ will be called the Green operators in the algebra on $\boldsymbol{W}$.

Theorem 3.23 *An operator of the form (3.18) induces continuous operators*

$$\mathcal{A}: \begin{array}{c} \mathcal{W}^{s,\gamma}(\boldsymbol{W}) \\ \oplus \\ H^s(Y,J^-) \end{array} \to \begin{array}{c} \mathcal{W}^{s-\mu,\gamma-\mu}(\boldsymbol{W}) \\ \oplus \\ H^{s-\mu}(Y,J^+) \end{array} \tag{3.19}$$

and

$$\mathcal{A}: \begin{array}{c} \mathcal{W}_P^{s,\gamma}(\boldsymbol{W}) \\ \oplus \\ H^s(Y, J^-) \end{array} \rightarrow \begin{array}{c} \mathcal{W}_Q^{s-\mu,\gamma-\mu}(\boldsymbol{W}) \\ \oplus \\ H^{s-\mu}(Y, J^+) \end{array}$$

for every $s \in \boldsymbol{R}$ and arbitrary continuous asymptotic type P with some resulting continuous asymptotic type Q.

The operators (3.18) have two principal symbols, namely $\sigma_\psi^\mu(\mathcal{A})$, the homogeneous principal interior symbol of order μ which is defined by the left upper corner $\in L_{cl}^\mu(\text{int}\,\boldsymbol{W})$, and $\sigma_\wedge^\mu(\mathcal{A})$, the homogeneous principal edge symbol of order μ. This is an operator family

$$\sigma_\wedge^\mu(\mathcal{A})(y,\eta): \begin{array}{c} \mathcal{K}^{s,\gamma}(X^\wedge) \\ \oplus \\ J_y^- \end{array} \rightarrow \begin{array}{c} \mathcal{K}^{s-\mu,\gamma-\mu}(X^\wedge) \\ \oplus \\ J_y^+ \end{array} \tag{3.20}$$

parametrized by $T^*Y \setminus 0$, with the homogeneity (3.16). The entries were defined in Remark 3.7, (3.8) and (3.11). Thus we have a "hierarchy" of principal symbols

$$(\sigma_\psi^\mu(\mathcal{A}), \sigma_\wedge^\mu(\mathcal{A})) = (\text{principal interior symbol, principal edge symbol}).$$

The composition of two operators $\mathcal{A}$ and $\mathcal{B}$ of orders μ and ν, respectively of the mentioned kind is possible when the bundles over Y and the weights fit together. Then $\mathcal{AB}$ is of analogous structure, and

$$\sigma_\psi^{\mu+\nu}(\mathcal{AB}) = \sigma_\psi^\mu(\mathcal{A})\sigma_\psi^\nu(\mathcal{B}), \quad \sigma_\wedge^{\mu+\nu}(\mathcal{AB}) = \sigma_\wedge^\mu(\mathcal{A})\sigma_\wedge^\nu(\mathcal{B}).$$

Definition 3.24 *An operator (3.18) is called elliptic of order μ if*
(i) $\sigma_\psi^\mu(\mathcal{A}) \neq 0$ on $T^*(\text{int}\,\boldsymbol{W}) \setminus 0$ and

$$t^\mu \sigma_\psi^\mu(\mathcal{A})(t, x, y, t^{-1}\tau, \xi, t^{-1}\eta) \neq 0 \quad \text{on} \quad T^*V \setminus 0,$$

where $V \cong [0,1) \times X \times Y$ is a collar neighborhood of $\partial \boldsymbol{W}$ (in the corresponding splitting of coordinates)
(ii) $\sigma_\wedge^\mu(\mathcal{A})$ is an isomorphism (3.20) for some $s \in \boldsymbol{R}$ and all $(y,\eta) \in T^*Y \setminus 0$.

Theorem 3.25 *An operator (3.18) is elliptic if and only if (3.19) is a Fredholm operator for an $s = s_0 \in \boldsymbol{R}$. The ellipticity implies that (3.19) is a Fredholm operator for all $s \in \boldsymbol{R}$. Moreover, there is a parametrix $\mathcal{B}$ of analogous structure, now of order $-\mu$ and belonging to the weight data $(\gamma - \mu, \gamma, \Theta)$, such that $\mathcal{AB} - 1$ and $\mathcal{BA} - 1$ are Green operators with respect to the corresponding weight data.*

Theorem 3.26 *Let $\mathcal{A}$ be elliptic. Then a solution*

$$u \in \mathcal{W}^{-\infty,\gamma}(\boldsymbol{W}) \oplus H^{-\infty}(Y, J^-) \quad \text{of } \mathcal{A}u = f \in \mathcal{W}^{s-\mu,\gamma-\mu}(\boldsymbol{W}) \oplus H^{s-\mu}(Y, J^+),$$

$s \in \boldsymbol{R}$, belongs to $\mathcal{W}^{s,\gamma}(\boldsymbol{W}) \oplus H^s(Y, J^-)$. In particular, for

$$f \in \mathcal{W}_Q^{s-\mu,\gamma-\mu}(\boldsymbol{W}) \oplus H^{s-\mu}(Y, J^+)$$

for some continuous asymptotic type Q we obtain

$$u \in \mathcal{W}_R^{s,\gamma}(\boldsymbol{W}) \oplus H^s(Y, J^-)$$

for some resulting continuous asymptotic type R.

4. Boundary value problems in the framework of the edge calculus

As we have seen in the previous section the calculus of pseudo-differential operators on a manifold with edges may be regarded as a generalization of the calculus on a manifold with boundary. In the case of boundary value problems we have certain interesting subalgebras and modifications. One point is that the interior symbols may be the usual ones (i.e., not necessarily edge-degenerate), with smooth dependence in the independent variables up to the boundary. It is not evident at first glance that the corresponding operators have some relation to the edge formalism. The following sections will show that this is the case. Move details many be found in [34].

4.1. BOUNDARY SYMBOLS

We want to obtain operator conventions to symbols $a(t, y, \tau, \eta) \in S^\mu_{cl}(\overline{\boldsymbol{R}}_+ \times \Omega \times \boldsymbol{R}^{1+q}_{\tau,\eta})$, $\mu \in \boldsymbol{R}$, for the "half space" $\boldsymbol{R}_+ \times \Omega$, $\Omega \subseteq \boldsymbol{R}^q$ open.

Let us first consider $a(t, y, \tau, \eta) \in S^\mu(\boldsymbol{R} \times \Omega \times \boldsymbol{R}^{1+q})$, and set

$$\mathrm{op}(a)(y, \eta)u(t) = \int\int e^{i(t-t')\tau} a(t, y, \tau, \eta) u(t') dt' đ\tau,$$

interpreted as an operator family $C_0^\infty(\boldsymbol{R}) \to C^\infty(\boldsymbol{R})$. By restricting the action to $C_0^\infty(\boldsymbol{R}_+)$ we can form

$$\mathrm{op}^+(a)(y, \eta) = r^+ \mathrm{op}(a)(y, \eta) e^+ : \; C_0^\infty(\boldsymbol{R}_+) \to C^\infty(\boldsymbol{R}_+) \qquad (4.1)$$

with $e^+ : \; C_0^\infty(\boldsymbol{R}_+) \to C^\infty(\boldsymbol{R})$ (extension by zero) and $r^+ : \; C^\infty(\boldsymbol{R}) \to C^\infty(\boldsymbol{R}_+)$ (restriction to $\boldsymbol{R}_+$). Then, for $\mathrm{Op}(\ldots) = \mathcal{F}^{-1}_{\eta \to y}(\ldots)\mathcal{F}_{y' \to \eta}$, we have

$$\mathrm{Op}(\mathrm{op}^+(a))v(y) = \int\int e^{i(y-y')\eta} \mathrm{op}^+(a)(y, \eta) v(y') dy' đ\eta.$$

Here $v(y)$ is regarded as a $C_0^\infty(\boldsymbol{R}_+)$-valued function of $y \in \Omega$, identified with a $w(t, y) \in C_0^\infty(\boldsymbol{R}_+ \times \Omega)$.

Remark 4.1 *If $a(t,y,\tau,\eta) \in S^{\mu}_{cl}(\boldsymbol{R} \times \Omega \times \boldsymbol{R}^{1+q})$ is a symbol with the transmission property with respect to $t = 0$, $\mu \in \boldsymbol{Z}$, and if $a(t,y,\tau,\eta)$ is independent of t for $|t| > const$, then $\mathrm{op}^+(a)(y,\eta)$ extends to an operator family*

$$\mathrm{op}^+(a)(y,\eta): \ H^s(\boldsymbol{R}_+) \to H^{s-\mu}(\boldsymbol{R}_+)$$

for $s > -\frac{1}{2}$, and we have

$$\mathrm{op}^+(a)(y,\eta) \in S^{\mu}(\Omega \times \boldsymbol{R}^q; H^s(\boldsymbol{R}_+), H^{s-\mu}(\boldsymbol{R}_+)), \tag{4.2}$$

for $s > -\frac{1}{2}$, where the operator-valued symbols refer to $(\kappa_\lambda u)(t) = \lambda^{\frac{1}{2}}\kappa(\lambda t)$, $\lambda \in \boldsymbol{R}_+$.

A proof of this result may be found in [24], cf. also [37]. From (4.2) we obtain an operator convention $a \to \mathrm{Op}(\mathrm{op}^+(a))$ in the half space.

Many interesting symbols have not the transmission property, e.g., $(1+|\tau|^2+|\eta|^2)^{\frac{1}{2}}$. Such symbols appear when we reduce mixed elliptic problems to the boundary, for instance, the Zaremba problem for the Laplacian. General operator conventions can be obtained in terms of edge-degenerate symbols. We will often assume that the symbols are given over $\overline{\boldsymbol{R}}_+ \times \Omega$. In connection with the operation e^+ they are to be regarded then as symbols for $t \in \boldsymbol{R}$, however the subsequent restriction to $\boldsymbol{R}_+$, cf. (4.1), makes the operator conventions independent of the particular choice of the smooth extension of the symbol in $t \in \boldsymbol{R}$.

The only specific step in operator conventions for boundary value problems, compared with the general constructions from Section 3, is the following observation:

Proposition 4.2 *To every $a(t,y,\tau,\eta) \in S^{\mu}_{cl}(\overline{\boldsymbol{R}}_+ \times \Omega \times \boldsymbol{R}^{1+q}_{\tau,\eta})$, $\mu \in \boldsymbol{R}$, there exists a $\tilde{b}(t,y,\tilde{\tau},\tilde{\eta}) \in S^{\mu}_{cl}(\overline{\boldsymbol{R}}_+ \times \Omega \times \boldsymbol{R}^{1+q}_{\tilde{\tau},\tilde{\eta}})$ such that $b(t,y,\tau,\eta) := \tilde{b}(t,y,t\tau,t\eta)$ satisfies*

$$\mathrm{op}^+(a)(y,\eta) = t^{-\mu}\mathrm{op}^+(b)(y,\eta) \mod C^\infty(\Omega, L^{-\infty}(\boldsymbol{R}_+;\boldsymbol{R}^q)),$$

(regarded as an operator family $C_0^\infty(\boldsymbol{R}_+) \to C^\infty(\boldsymbol{R}_+)$).

In fact, if especially $a(t,y,\tau,\eta)$ satisfies $a(t,y,\lambda\tau,\lambda\eta) = \lambda^{\mu-j}a(t,y,\tau,\eta)$ for $|\tau,\eta| \geq const$, $\lambda \geq 1$, for any $j \in \boldsymbol{N}$, then we may set

$$\tilde{b}(t,y,\tilde{\tau},\tilde{\eta}) = t^j a(t,y,\tilde{\tau},\tilde{\eta}).$$

In general we obtain $\tilde{b}$ by an asymptotic summation, using the homogeneous components of a.

Note that the $\tilde{b}(t,y,t\tau,t\eta)$ are particular edge-degenerate symbols with the boundary $\{t=0\}$ as the edge.

The version of Definition 3.2 for $\dim X = 0$ is the following. $M_O^\mu(\boldsymbol{R}^q)$ for $\mu \in \boldsymbol{R}$ is the subspace of all $h(z,\eta) \in \mathcal{A}(\boldsymbol{C}, S_{cl}^\mu(\boldsymbol{R}_\eta^q))$ such that $h(z,\eta)|_{\Gamma_\beta} \in S_{cl}^\mu(\Gamma_\beta \times \boldsymbol{R}^q)$ for every $\beta \in \boldsymbol{R}$, uniformly in $c \leq \beta \leq c'$ for every $c < c'$.

Theorem 4.3 *To every $b(t,y,\tau,\eta)$ in the notations of Proposition 4.2 there exists an $\tilde{h}(t,y,z,\tilde{\eta}) \in C^\infty(\overline{\boldsymbol{R}}_+ \times \Omega, M_O^\mu(\boldsymbol{R}_{\tilde{\eta}}^q))$ such that $h(t,y,z,\eta) = \tilde{h}(t,y,z,t\eta)$ satisfies*

$$\mathrm{op}^+(b)(y,\eta) = \mathrm{op}_M^\beta(h)(y,\eta) \mod C^\infty(\Omega, L^{-\infty}(\boldsymbol{R}_+;\boldsymbol{R}^q)),$$

for every $\beta \in \boldsymbol{R}$.

This is a special case of Theorem 3.4.

Corollary 4.4 *To every $a(t,y,\tau,\eta) \in S_{cl}^\mu(\overline{\boldsymbol{R}}_+ \times \Omega \times \boldsymbol{R}^{1+q})$, $\mu \in \boldsymbol{R}$, there exists an $\tilde{h}(t,y,z,\tilde{\eta}) \in C^\infty(\overline{\boldsymbol{R}}_+ \times \Omega, M_O^\mu(\boldsymbol{R}_{\tilde{\eta}}^q))$ such that for $h(t,y,z,\eta) := \tilde{h}(t,y,z,t\eta)$*

$$\mathrm{op}^+(a)(y,\eta) = t^{-\mu}\mathrm{op}_M^\beta(h)(y,\eta) + g(y,\eta)$$

for every $\beta \in \boldsymbol{R}$, where $\mathrm{op}^+(a)(y,\eta)$ is regarded as an operator family $C_0^\infty(\boldsymbol{R}_+) \to C^\infty(\boldsymbol{R}_+)$, and $g(y,\eta) \in C^\infty(\Omega, \mathcal{S}(\boldsymbol{R}^q, L^{-\infty}(\boldsymbol{R}_+))$.

Note that $\mathrm{Op}(g) : C_0^\infty(\boldsymbol{R}_+ \times \Omega) \to C^\infty(\boldsymbol{R}_+ \times \Omega)$ belongs to $L^{-\infty}(\boldsymbol{R}_+ \times \Omega)$.

Recall that $\mathcal{H}^{s,\gamma}(\boldsymbol{R}_+)$ is the weighted Mellin Sobolev space of smoothness $s \in \boldsymbol{R}$ and weight $\gamma \in \boldsymbol{R}$ on $\boldsymbol{R}_+$. For $s \in \boldsymbol{N}$ we have

$$\mathcal{H}^{s,\gamma}(\boldsymbol{R}_+) = \{u(t) \in t^\gamma L^2(\boldsymbol{R}_+) : \left(t\frac{d}{dt}\right)^j u(t) \in t^\gamma L^2(\boldsymbol{R}_+) \text{ for } j = 0,\dots,s\}.$$

The spaces $\mathcal{H}^{s,\gamma}(\boldsymbol{R}_+)$ for $s \in \boldsymbol{R}$ can be defined by duality and interpolation. The weighted Mellin transform

$$(M_\gamma u)(z) = M(t^{-\gamma}u)(z+\gamma)$$

induces an isomorphism

$$M_\gamma : t^\gamma L^2(\boldsymbol{R}_+) \to L^2(\Gamma_{\frac{1}{2}-\gamma})$$

and isomophisms

$$M_\gamma : \mathcal{H}^{s,\gamma}(\boldsymbol{R}_+) \to \hat{H}^s(\Gamma_{\frac{1}{2}-\gamma}),$$

for all $s \in \boldsymbol{R}$, where $\hat{H}^s(\Gamma_{\frac{1}{2}-\gamma}) = \{u \in \mathcal{S}'(\Gamma_{\frac{1}{2}-\gamma}) : \langle \mathrm{Im}\, z\rangle^s u(z) \in L^2(\Gamma_{\frac{1}{2}-\gamma})\}$. Recall also that $\mathcal{H}^{s,\gamma}(\boldsymbol{R}_+) \subset H_{loc}^s(\boldsymbol{R}_+)$ and

$$\mathcal{K}^{s,\gamma}(\boldsymbol{R}_+) = \omega\mathcal{H}^{s,\gamma}(\boldsymbol{R}_+) + (1-\omega)H^s(\boldsymbol{R}_+)$$

for any cut-off function $\omega(t)$. Then $\mathcal{K}^{s,\gamma}(\boldsymbol{R}_+)$ is a Hilbert space, $\mathcal{K}^{0,0}(\boldsymbol{R}_+) = L^2(\boldsymbol{R}_+)$, and $(\kappa_\lambda u)(t) = \lambda^{\frac{1}{2}}u(\lambda t)$, $\lambda \in \boldsymbol{R}_+$, is a strongly continuous group of isomorphisms on $\mathcal{K}^{s,\gamma}(\boldsymbol{R}_+)$.

Now the wedge Sobolev spaces specialize for the set $\boldsymbol{R}_+ \times \Omega$ to

$$\mathcal{W}^s_{loc(y)}(\Omega, \mathcal{K}^{s,\gamma}(\boldsymbol{R}_+)), \quad s, \gamma \in \boldsymbol{R}. \tag{4.3}$$

and we have

$$H^s_{comp}(\boldsymbol{R}_+ \times \Omega) \subset \mathcal{W}^s_{loc(y)}(\Omega, \mathcal{K}^{s,\gamma}(\boldsymbol{R}_+)) \subset H^s_{loc}(\boldsymbol{R}_+ \times \Omega).$$

The spaces of the type (4.3) are adequate for boundary value problems without the transmission property.

For every pair of cut-off functions $\omega(t)$, $\omega_0(t)$ the operator

$$\omega t^{-\mu} \mathrm{op}^\gamma_M(h)(y,\eta)\omega_0 : \ \mathcal{K}^{s,\gamma}(\boldsymbol{R}_+) \to \mathcal{K}^{s-\mu,\gamma-\mu}(\boldsymbol{R}_+)$$

is continuous for every $s \in \boldsymbol{R}$. Instead of $\omega(t)$, $\omega_0(t)$ we also employ $\omega(t[\eta])$, $\omega_0(t[\eta])$.

Proposition 4.5 *We have*

$$\omega(t[\eta])t^{-\mu}\mathrm{op}^\gamma_M(h)(y,\eta)\omega_0(t[\eta]) \in S^\mu(\Omega \times \boldsymbol{R}^q; \mathcal{K}^{s,\gamma}(\boldsymbol{R}_+), \mathcal{K}^{s-\mu,\gamma-\mu}(\boldsymbol{R}_+))$$

for all $s \in \boldsymbol{R}$.

Proposition 4.6 *Let* $a(t,y,\tau,\eta) \in S^\mu(\overline{\boldsymbol{R}}_+ \times \Omega \times \boldsymbol{R}^{1+q})$ *be independent of* t *for* $t > const$, *and* $\omega(t), \omega_1(t)$ *arbitrary cut-off functions. Then*

$$(1-\omega(t[\eta]))\mathrm{op}^+(a)(y,\eta)(1-\omega_1(t[\eta])) \in S^\mu(\Omega \times \boldsymbol{R}^q; \mathcal{K}^{s,\gamma}(\boldsymbol{R}_+), \mathcal{K}^{s-\mu,\delta}(\boldsymbol{R}_+))$$

for every $s \in \boldsymbol{R}$ *and* $\gamma, \delta \in \boldsymbol{R}$.

Now our operator convention for the half space for symbols $a(t,y,\tau,\eta) \in S^\mu_{cl}(\overline{\boldsymbol{R}}_+ \times \Omega \times \boldsymbol{R}^{1+q})$ that are independent of t for large t is

$$a(t,y,\tau,\eta) \to \mathrm{Op}(p) : \ \mathcal{W}^s_{comp(y)}(\Omega, \mathcal{K}^{s,\gamma}(\boldsymbol{R}_+)) \to \mathcal{W}^{s-\mu}_{loc(y)}(\Omega, \mathcal{K}^{s-\mu,\gamma-\mu}(\boldsymbol{R}_+))$$

where

$$\begin{aligned} p(y,\eta) &= \omega(t[\eta])t^{-\mu}\mathrm{op}^\gamma_M(h)(y,\eta)\omega_0(t[\eta]) \\ &+ (1-\omega(t[\eta]))\mathrm{op}^+(a)(y,\eta)(1-\omega_1(t[\eta])) \end{aligned}$$

with arbitrary cut-off functions $\omega, \omega_0, \omega_1$ satisfying $\omega\omega_0 = \omega$, $\omega\omega_1 = \omega_1$, and h is obtained from a by Corollary 4.4. $p(y,\eta)$ plays the role of a (complete) boundary symbol to $a(t,y,\eta,\tau)$. We shall not go into further details how to specify the material of Section 4 to boundary value problems. Let us only mention here that the concept of ellipticity including boundary symbols applies and that we have the corresponding versions Theorem 3.25 and Theorem 3.26.

4.2. CONORMAL SYMBOLS AND MELLIN EXPANSIONS

As noted at the beginning, the half axis $\overline{\boldsymbol{R}}_+$ may be regarded as a manifold with conical singularities. Also here we may ask to what extent the pseudo-differential operators $\mathrm{op}^+(a)$ for $a(t,\tau) \in S^{\mu}_{cl}(\overline{\boldsymbol{R}}_+ \times \boldsymbol{R})$ fit into the cone operator calculus. The answer was practically given in the preceding section in terms of a complete boundary symbol to a, if we keep the parameters y, η fixed. Let us mention some further details for the special case $a(\tau) \in S^0_{cl}(\boldsymbol{R})$ and illustrate the role of conormal symbols as it was done on the principal symbolic level in Eskin's book [8].

We consider the operators $\mathrm{op}^+(a) = r^+\mathrm{op}(a)e^+$ in $L^2(\boldsymbol{R}_+)$, where $e^+ : L^2(\boldsymbol{R}_+) \to L^2(\boldsymbol{R})$ is the extension by zero, $r^+ : L^2(\boldsymbol{R}) \to L^2(\boldsymbol{R}_+)$ the restriction to $\boldsymbol{R}_+$. A question is how to characterize the subalgebra of $\mathcal{L}(L^2(\boldsymbol{R}_+))$ that is generated by $\{\mathrm{op}^+(a) : a(\tau) \in S^0_{cl}(\boldsymbol{R})\}$. The answer is a consequence of the following theorem and of the general results on the cone algebra:

Theorem 4.7 *Let $a(\tau) \in S^0_{cl}(\boldsymbol{R})$. Then $\mathrm{op}^+(a)$ can be written*

$$\mathrm{op}^+(a) = \omega \mathrm{op}_M(h)\omega_0 + (1-\omega)\mathrm{op}^+(a)(1-\omega_1) + m + g$$

for arbitrary cut-off functions $\omega(t)$, $\omega_0(t)$, $\omega_1(t)$, some $h(t,z) \in C^\infty(\overline{\boldsymbol{R}}_+, M^0_O)$ and a smoothing Mellin + Green operator with discrete asymptotics, with respect to the infinite weight integral $(-\infty, 0]$.

Note that the smoothing Green operators in this case are continuous operators

$$g : L^2(\boldsymbol{R}_+) \to \mathcal{S}_P(\boldsymbol{R}_+) \quad \text{with} \quad g^* : L^2(\boldsymbol{R}_+) \to \mathcal{S}_Q(\boldsymbol{R}_+),$$

for certain discrete asymptotic types P and Q ($*$ indicates the $L^2(\boldsymbol{R}_+)$-adjoint). The smoothing Mellin operators are asymptotic expansions of the form

$$m = \sum_{j=0}^{\infty} \omega(c_j t) t^j \mathrm{op}_M^{\gamma_j}(h_j)\tilde{\omega}(c_j t)$$

for suitable $h_j(z) \in M^{-\infty}_{R_j}$, $R_j = \{(r_j, n_j)\}_{j\in\boldsymbol{Z}}$, and $\gamma_j \in \boldsymbol{R}$, $\gamma_j \geq 0$, $j+\gamma_j \geq 0$, $\pi_{\mathbb{C}} R_j \cap \Gamma_{\frac{1}{2}-\gamma_j} = \emptyset$ for all j. The constants c_j in the cut-off functions $\omega, \tilde{\omega}$ are increasing sufficiently fast as $j \to \infty$, and $\gamma_j \to \infty$, $j+\gamma_j \to \infty$ as $j \to \infty$.

Corollary 4.8 *The subalgebra of* $\mathcal{L}(L^2(\boldsymbol{R}_+))$ *generated by* $\mathrm{op}^+(a)$, $a(\tau) \in S^0_{cl}(\boldsymbol{R})$, *is contained in the algebra of all operators of the form* $\mathrm{op}^+(a) + m + g$ *for arbitrary* $a(\tau) \in S^0_{cl}(\boldsymbol{R})$ *and smoothing Mellin + Green operators as in Theorem 4.7, which is a subalgebra of the cone algebra on* $\boldsymbol{R}_+$.

Let us set

$$g^+(z) = (1 - e^{-2\pi i z})^{-1}, \quad g^-(z) = 1 - g^+(z).$$

Moreover, $f_j(z) = 1$ for $j = 0$, $f_j(z) = \Pi_{k=1}^{j}(k - z)^{-1}$ for $j \geq 1$. Write the asymptotic expansion of $a(\tau) \in S^0_{cl}(\boldsymbol{R})$ in the form

$$a(\tau) \sim \sum_{j=0}^{\infty} a_j^{\pm}(i\tau)^{-j} \quad \text{for} \quad \tau \to \pm\infty,$$

$i = \sqrt{-1}$. Then

$$\sigma_M^{-j}(\mathrm{op}^+(a)(z) = \{a_j^+ g^+(z) + a_j^- g^-(z)\} f_j(z) \quad \text{for all} \quad j \in \boldsymbol{N}.$$

This expansion for $j = 0$ is due to Eskin [8]. For $j \in \boldsymbol{N} \setminus \{0\}$ it was obtained by Rempel, Schulze , cf. the monograph [20].

Let us finally note that there is an analogue of Theorem 4.7 to symbols $a(t, \tau) \in S^{\mu}_{cl}(\overline{\boldsymbol{R}}_+ \times \boldsymbol{R})$, $a(t, \tau)$ independent of t for large t. The analogue of Corollary 4.8 is to be modified by subtracting finite-dimensional Green operators on $\boldsymbol{R}_+$ with discrete asymptotics. Also the explicit formulas for the conormal symbols can be generalized to arbitrary $a(t, \tau)$.

References

1. M.S. Agranovich and M.I. Vishik. Elliptic problems with parameter and parabolic problems of general type. *Uspekhi. Mat. Nauk*, 19: 3, 53–161, 1964.
2. M.F. Atiyah and I.M. Singer. The index of elliptic operators. I. *Ann. Math.*, 87: 484–530, 1968.
3. L. Boutet de Monvel. Boundary problems for pseudo-differential operators. *Acta Math.*, 126 (1–2): 11–51, 1971.
4. T. Buchholz and B.-W. Schulze. Anisotropic edge pseudo-differential operators with discrete asymptotics (In preparation).
5. Ch. Dorschfeldt, U. Grieme, and B.-W. Schulze. Pseudo-differential calculus in the Fourier-edge approach on non-compact manifolds. Preprint MPI/96-79., Max-Planck-Inst. für Math., Bonn, 1996.
6. Ch. Dorschfeldt and B.-W. Schulze. Pseudo-differential operators with operator-valued symbols in the Mellin-edge-approach. *Ann. of Global Anal. Geometry*, 12: 2, 135–171, 1994.
7. Ju. V. Egorov and B.-W. Schulze. *Pseudo-Differential Operators, Singularities, Applications*. Birkhäuser Verlag, Basel et al., 1997.
8. G. I. Eskin. *Boundary Problems for Elliptic Pseudo-Differential Equations*. Nauka, Moscow, 1973. (Transl. of Math. Monographs 52, Amer. Math. Soc. Providence, Rhode Island 1980)

9. B. V. Fedosov. Analytical index formulas for elliptic operators. *Trudy Mosk. Mat. Obshch.*, 30:159–241, 1974 (Russian).
10. B. V. Fedosov and B.-W. Schulze. On the index of elliptic operators on a cone. In: Advances in Partial Differential Equations (Schrödinger Operators, Markov Semigroups, Wavelet Analysis, Operator Algebras) Akademie Verlag, Berlin, 348–372, 1996.
11. B. V. Fedosov and B.-W. Schulze and N.N. Tarkhanov. On the index of elliptic operators on a wedge (In preparation)
12. J. Gil and B.-W. Schulze and J. Seiler. *Holomorphic operator-valued symbols for edge-degenerate pseudo-differential operators.* (to appear in Proceedings Conference "Partial Differential Equations" Potsdam, 1996. Math. Research 100, Akademie Verlag, 1997).
13. G. Grubb. *Functional Calculus of Pseudo-Differential Boundary Problems.* Birkhäuser Verlag, Basel et al., 1986.
14. L. Hörmander. *The Analysis of Linear Partial Differential Operators.* Vols. 1-4. Springer-Verlag, New York, 1983/85.
15. V. A. Kondrat'ev. Boundary value problems for elliptic equations in domains with conical points. *Trudy Mosk. Mat. Obshch*, 16, 209–292, 1967.
16. H. Kumano-go. *Pseudo-Differential Operators.* MIT Press, Cambridge, Mass., 1981.
17. G. Luke. Pseudo-differential operators on Hilbert bundles. *J. of Diff. Equ.*, 12, 566–589, 1972.
18. F. Mantlik. Norm closure and extension of the symbolic calculus for the cone algebra. *Ann. of Global Anal. Geometry*, 13: 4, 339–376, 1995.
19. R.B. Melrose and G.A. Mendoza. Elliptic operators of totally characteristic type. MSRI, Preprint. 1983.
20. S. Rempel and B.-W. Schulze. *Index Theory of Elliptic Boundary Problems.* Akademie-Verlag, Berlin, 1982.
21. S. Rempel and B.-W. Schulze. Parametrices and boundary symbolic calculus for elliptic boundary problems without transmission property. *Math. Nachr.* 105: 45–149, 1982.
22. S. Rempel and B.-W. Schulze. Complete Mellin symbols and the conormal asymptotics in boundary value problems. Proc. Journees *Equ. aux Dériv. Part.* Conf. No. V, St.- Jean de Monts. 1984.
23. E. Schrohe and J. Seiler. An analytical index formula of Fedosov type for pseudo-differential operators on non-compact wedges (In preparation).
24. E. Schrohe and B.-W. Schulze. Boundary value problems in Boutet de Monvel's algebra for manifolds with conical singularities. I. In *Advances in Partial Differential Equations (Pseudo-Differential Calculus and Mathematical Physics)*, pages 97–209. Akademie Verlag, Berlin, 1994.
25. E. Schrohe and B.-W. Schulze. Boundary value problems in Boutet de Monvel's algebra for manifolds with conical singularities. II. In *Advances in Partial Differential Equations (Boundary Value Problems, Schrödinger Operators, Deformation Quantization)*. Akademie Verlag, Berlin.
26. E. Schrohe and B.-W. Schulze. Mellin operators in a pseudodifferential calculus for boundary value problems on manifolds with edges. Preprint MPI/96-74., Max-Planck-Inst. für Math., Bonn, 1996.
27. E. Schrohe and B.-W. Schulze. Smoothing Mellin and Green symbols for boundary value problems on manifolds with edges (To appear).
28. E. Schrohe and B.-W. Schulze. A symbol algebra for pseudo-differential boundary value problems on manifolds with edges. (to appear in Proceedings Conference "Partial Differential Equations" Potsdam, 1996. Math. Research 100, Akademie Verlag, 1997).
29. B.-W. Schulze. Regularity with continuous and branching asymptotics for elliptic operators on manifolds with edges. *Integral Equ. and Operator Theory*, 11: 557–602, 1988.

30. B.-W. Schulze. Pseudo-differential operators on manifolds with edges. In: *Symposium "Partial Differential Equation"*, Holzhau 1988. Teubner-Texte zur Mathematik, Leipzig, 112: 259–287, 1989.
31. B.-W. Schulze. *Pseudo-Differential Operators on Manifolds with Singularities.* North-Holland, Amsterdam, 1991.
32. B.-W. Schulze. The Mellin pseudo-differential calculus on manifolds with corners. In: *Symposium "Analysis in Domains and on Manifolds with Singularities"*, Breitenbrum 1990. Teubner-Texte zur Mathematik, Leipzig, 131: 208–289, 1992.
33. B.-W. Schulze. The variable discrete asymptotics of solutions of singular boundary value problems. In: *Symposium "Operator Calculus and Spectral Theory"*, Lambrecht 1991. Operator Theory: Advances and Applications Birkhäuser Verlag, Basel, 271–289, 1992.
34. B.-W. Schulze. *Pseudo-Differential Boundary Value Problems, Conical Singularities, and Asymptotics.* Akademie Verlag, 1994.
35. B.-W. Schulze. The variable discrete asymptotics in pseudo-differential boundary value problems. I. In *Advances in Partial Differential Equations (Pseudo-Differential Calculus and Mathematical Physics)*, pages 9–96. Akademie Verlag, Berlin, 1994.
36. B.-W. Schulze. The variable discrete asymptotics in pseudo-differential boundary value problems. II. In *Advances in Partial Differential Equations (Boundary Value Problems, Schrödinger Operators, Deformation Quantization)*, pages 9–96. Akademie Verlag, Berlin, 1995.
37. B.-W. Schulze. Boundary value problems and singular pseudo-differential operators. J. Wiley, Chichester. 1997 (To appear)
38. B.-W. Schulze and B. Sternin and V. Shatalov. Differential equations on manifolds with singularities in classes of resurgent functions. Preprint MPI/95-88., Max-Planck-Inst. für Math., Bonn, 1995.
39. B.-W. Schulze and B. Sternin and V. Shatalov. On some global aspects of the theory of partial differential equations on manifolds with singularities. Preprint MPI/96-28., Max-Planck-Inst. für Math., Bonn, 1996.
40. B.-W. Schulze and B. Sternin and V. Shatalov. An operator algebra on manifolds with cups-type singularities. Preprint MPI/96-111., Max-Planck-Inst. für Math., Bonn, 1996.
41. B.-W. Schulze and N.N. Tarkhanov. Green pseudodifferential operators on manifolds with edges. Comm. Partial Differential Equations (To appear)
42. B.-W. Schulze and N.N. Tarkhanov. The index of elliptic operators on manifolds with cusps (To appear in Proceedings Conference "Partial Differential Equations" Potsdam, 1996. Math. Research, Akademie Verlag, 1997).
43. J. Seiler. Continuity of edge and corner pseudo-differential operators. *Math. Nachr* (To appear).
44. B. Sternin and V. Shatalov. *Borel-Laplace Transform and Asymptotic Theory.* CRC Press, Boca Raton, New York, London, Tokyo 1996.
45. F. Treves. *Introduction to Pseudodifferential and Fourier Integral Operators.* Vols. 1,2. New Jork, Plenum, 1985.
46. M.I. Vishik and G.I. Eskin. Convolution equations in a bounded region. *Uspechi Mat. Nauk*, 20: 3, 89–152, 1965.

WODZICKI'S NONCOMMUTATIVE RESIDUE AND TRACES FOR OPERATOR ALGEBRAS ON MANIFOLDS WITH CONICAI SINGULARITIES

ELMAR SCHROHE
Max-Planck-Arbeitsgruppe
"Partielle Differentialgleichungen und Komplexe Analysis"
Universität Potsdam
14415 Potsdam
Germany

Introduction

In 1984 M. Wodzicki found a trace on the algebra $\Psi_{cl}(M)$ of all classical pseudodifferential operators on a closed compact manifold M; he called it the *noncommutative residue*. This trace vanishes on the ideal $\Psi^{-\infty}(M)$ of smoothing operators; it even is the unique trace (up to constant multiples) on $\Psi_{cl}(M)/\Psi^{-\infty}(M)$, provided M is connected and $\dim M > 1$.

Although it first seems a rather exotic object, this trace has found a wide range of applications both in mathematics and in mathematical physics. In appreciation of Wodzicki's accomplishment the name *Wodzicki residue* has become generally accepted.

Also various extensions and analogs of the noncommutative residue have been established, e.g. for certain algebras of Fourier integral operators (Guillemin [11]), manifolds with boundary (Fedosov, Golse, Leichtnam, and Schrohe [7, 8]), manifolds with conical singularities (Schrohe [26]), or cusp pseudodifferential operators (Melrose and Nistor [21]).

In these four lectures I shall first give a short review of Wodzicki's residue and some of its applications. Next I will explain the idea of B.-W. Schulze's 'cone algebra', a pseudodifferential calculus for manifolds with conical singularities. For every conical singularity we shall obtain a trace on this algebra. These traces vanish on operators supported in the interior and are therefore different from Wodzicki's. On the other hand, there is a natural ideal in the cone algebra having a trace which extends the classical noncommutative residue. All these traces vanish on smoothing operators.

L. Rodino (ed.), Microlocal Analysis and Spectral Theory, 227–250.

They are moreover seen to be the unique traces with this property on a slightly extended version of the cone algebra. In view of the fact that this ASI focuses on microlocal analysis and spectral theory, I shall finally sketch Connes theorem linking Wodzicki's residue to Dixmier's trace. For one thing this makes the noncommutative residue an important tool for explicit computations in noncommutative geometry, see Connes [3]; it also shows Weyl's law on the asymptotics of the eigenvalue of the Laplacian.

Lecture 1: Wodzicki's Noncommutative Residue for Pseudodifferential Operators

1.1 Definition. Let $\mathcal{A}$ be an algebra over $\mathbb{C}$. A linear map $\tau : \mathcal{A} \to \mathbb{C}$ is called a *trace* if it vanishes on commutators, i.e., if

$$\tau(P, Q) = \tau(PQ - QP) = 0 \text{ for all } P, Q \in \mathcal{A}.$$

Clearly, if τ is a trace, then $\lambda\tau$ is a trace for each λ in $\mathbb{C}$; moreover, the zero map is always a trace. When we speak of a unique trace, we shall mean that it is non-zero and the only one up to multiples.

1.2 Example. On $M_r(\mathbb{C})$, the algebra of $r \times r$ matrices over $\mathbb{C}$, there is a unique trace, namely the standard one, $Tr : A \mapsto \sum_j^r A_{jj}$. Indeed, let $E_{j,k}$ denote the matrix having a single 1 at position j, k (and zeros else). Then the statement is immediate from the observation that $[E_{j,k}, E_{k,k}] = E_{j,k}$ for $j \neq k$ and $[E_{j,k}, E_{k,j}] = E_{j,j} - E_{k,k}$.

In this lecture we shall be concerned with the following theorem, proven by M. Wodzicki in 1984, as well as with several of its applications.

1.3 Theorem. Let M be closed, compact, connected, $\dim M > 1$. Let $\mathcal{A} = \Psi_{cl}(M)/\Psi^{-\infty}(M)$ be the algebra of all classical pseudodifferential operators on M modulo the ideal of the regularizing elements. Then there is a unique trace on $\mathcal{A}$, the so-called noncommutative residue or Wodzicki residue.

1.4 Applications. (a) As mentioned before, the noncommutative residue plays an important role in Connes' noncommutative geometry due to Connes observation that it coincides with Dixmier's trace on pseudodifferential operators of order $-\dim M$, cf. [2].
(b) As Wodzicki observed, it also is closely related to the residues of zeta functions of elliptic pseudodifferential operators that were computed by Seeley [30] as well as to the coefficients in heat kernel expansions.
(c) Wodzicki's trace is the multi-dimensional analog of the residue Manin [19] and Adler [1] had found in 1978/79 in connection with their work on algebraic aspects of Korteweg-de Vries equations in dimension one.

(d) Guillemin [10] had discovered the noncommutative residue independently as an essential ingredient in his 'soft' proof of Weyl's formula on the asymptotic distribution of eigenvalues. Under rather general axiomatic conditions linking 'classical observables', i.e. functions p on a symplectic manifold, with their 'quantum mechanical counterparts', namely self-adjoint operators on a suitable Hilbert space, he showed that the counting function $N_P(\lambda)$ of the eigenvalues of P satisfies the relation $N_P(\lambda) = c\,\mathrm{vol}\{p \leq \lambda\}$ with a constant c independent of p or P.
(e) The noncommutative residue has been used in conformal field theory in order to construct central extensions of the algebra of pseudodifferential symbols on the circle, cf. Khesin and Kravchenko [16].
(f) It has been applied to derive the Einstein-Hilbert action in the theory of gravitation (Kalau and Walze [13], Kastler [15]).

We shall now go more into the details. We first recall a few facts about pseudodifferential operators:

1.5 Classical pseudodifferential operators on manifolds. Let $m \in \mathbb{Z}$ and let a be as symbol in Hörmander's class $S^m = S^m_{1,0}(\mathbb{R}^n \times \mathbb{R}^n)$. It defines the linear operator $A : \mathcal{S}(\mathbb{R}^n) \to \mathcal{S}(\mathbb{R}^n)$ by

$$Au(x) = \int e^{ix\xi} a(x,\xi)\hat{u}(\xi)\,đ\xi, \quad u \in \mathcal{S}(\mathbb{R}^n)$$

We say that A is a pseudodifferential operator of order m on $\mathbb{R}^n$ and refer to a as its symbol; it is uniquely determined by A, see [18]. We call a classical if it has an asymptotic expansion $a \sim \sum_{j=0}^{\infty} a_{m-j}$ with $a_j \in S^j$ homogeneous of degree j in ξ for large $|\xi|$, i.e., $a_j(x,\lambda\xi) = \lambda^j a_j(x,\xi)$ for $\lambda \geq 1$ and $|\xi| \geq R$. The $\sim$ indicates that upon subtracting of the first N summands from a we obtain an element in S^{m-N}.

In the following we let M be a compact manifold of dimension n, E a vector bundle over M.

We say that a linear operator $A : C^\infty(M,E) \to C^\infty(M,E)$ is classical pseudodifferential operator and write $A \in \Psi_{cl}(M)$ if, in each coordinate neighborhood, the action of A is given by a pseudodifferential operator with a classical symbol, modulo an operator with smooth integral kernel, a so-called *smoothing operator*. We denote those by $\Psi^{-\infty}(M)$. Note that an operator will be smoothing whenever its symbol in in $S^{-\infty} \bigcap_m S^m$ and that $\Psi^{-\infty}(M)$ is an ideal in $\Psi_{cl}(M)$. In the following we let $\mathcal{A} = \Psi_{cl}/\Psi^{-\infty}$.

Any smooth change of the symbols a_j on $\{|\xi| \leq R\}$ modifies a_j by an element in $S^{-\infty}$. Over each coordinate neighborhood U, the equivalence class of a pseudodifferential operator of order m in $\mathcal{A}$ can be therefore be identified with a formal sum of homogeneous functions (taking values in square matrices), $\sum_{j=0}^{\infty} a_{m-j}(x,\xi)$, with $a_j(x,\xi) \in C^\infty(U \times (\mathbb{R}^n \setminus \{0\}))$

homogeneous in ξ of degree j. There are well-known rules for the behavior of $\sum a_j$ under changes of coordinates.

1.6 Definition and Lemma. On $\mathbb{R}^n$, $n \geq 2$, define the $(n-1)$-form

$$\sigma(\xi) = \sum_{j=1}^{n} (-1)^{j+1} \xi_j d\xi_1 \wedge \ldots \wedge \widehat{d\xi_j} \wedge \ldots \wedge d\xi_n.$$

The hat indicates that this differential is omitted. Let p be a smooth function on $\mathbb{R}^n \setminus \{0\}$ which is homogeneous of degree $-n$. Euler's identity $\sum \xi_j \partial_{\xi_j} p = -np$ implies that the form $p\sigma$ is closed:

$$d(p\sigma) = (dp) \wedge \sigma + p d\sigma = -npd\xi_1 \wedge \ldots d\xi_n + pnd\xi_1 \wedge \ldots d\xi_n = 0.$$

The restriction of σ to the unit sphere S^{n-1} is the surface measure.

We can now define the Wodzicki residue $\operatorname{res} A$ of an operator A:

1.7 Theorem. *Let $A \in \Psi_{cl}(M)$, $x \in M$. Suppose that in a neighborhood U of x, the symbol A has the asymptotic expansion $\sum a_j$ with a_j homogeneous of degree j for $|\xi| \geq 1$. Denote by Tr the trace on $\mathcal{L}(E)$ and define*

$$\operatorname{res}_x A = \left(\int_{S^{n-1}} Tr\, a_{-n}(x,\xi) \sigma(\xi) \right) dx_1 \wedge \ldots \wedge dx_n.$$

This is a density on M. It therefore makes sense to set

$$\operatorname{res} A = \int_M \operatorname{res}_x A. \tag{1}$$

Then res *is only depends on the equivalence class of A in $\mathcal{A}$. It is a trace:* $\operatorname{res}[A, B] = 0$ *for all $A, B \in \mathcal{A}$. If M is connected, then any other trace on $\mathcal{A}$ is a multiple of* res.

Note. The local density $a_{-n}(x,\xi)\sigma(\xi) \wedge dx_1 \wedge \cdots \wedge dx_n$ can be patched to a global density Ω_A with $\operatorname{res} A = \int_{S^*M} \Omega_A$: Denoting by ω the canonical symplectic form on T^*M and by ρ the radial vector field one has

$$a_{-n}\sigma \wedge dx_1 \wedge \cdots \wedge dx_n = (-1)^{n(n-1)/2} \frac{1}{n!} (a\, \rho \rfloor \omega^n)_0,$$

where $(\ldots)_0$ is the homogeneous component of degree 0 in an asymptotic expansion of $a\,\rho \rfloor \omega^n$ into homogeneous forms ($\rfloor$ stands for the contraction of forms with vector fields).

The proof relies on the following simple lemma. For a proof see e.g. [8].

1.8 Lemma. (a) *Let the function p be a derivative of a smooth homogeneous function q of degree $-(n-1)$ on $\mathbb{R}^n \setminus \{0\}$, say $p = \frac{\partial}{\partial \xi_k} q$. Then $\int_{S^{n-1}} p\sigma = 0$.*

(b) *Let p be a homogeneous function on $\mathbb{R}^n \setminus \{0\}$. Each of the following conditions is sufficient for p to be a sum of derivatives:*

(i) $\deg p \neq -n$.
(ii) $\deg p = -n$ *and* $\int_S p\sigma = 0$.
(iii) $p = \xi^\alpha \partial^\beta q$ *where q is a homogeneous function and* $|\beta| > |\alpha|$.

Proof of Theorem 1.7. Under a change of variables χ the symbol a transforms to a symbol b with

$$b(y, {}^t\chi'(y)\xi) \sim \sum_{|\alpha| \geq 0} \partial_\xi^\alpha a(\chi(y), \xi) \varphi_\alpha(y, \xi), \tag{1}$$

where the $\varphi_\alpha(y, \xi)$ are polynomials in ξ of degree $\leq |\alpha|/2$ and $\varphi_0 = 1$ (see Hörmander [12, (18.1.30)]. Changing the variable in the integral, and applying first (1), then Lemma 1.8(b.iii) we get

$$\begin{aligned} \int_S b_{-n}(y, \eta)\sigma(\eta) &= |\det \chi'(y)| \int_S b_{-n}(y, {}^t\chi'(y)\xi)\sigma(\xi) \qquad (2) \\ &= |\det \chi'(y)| \sum_{|\alpha|\geq 0} \int_S (\partial_\xi^\alpha a(\chi(y), \xi)\varphi_\alpha(y, \xi))_{-n}\sigma(\xi) \\ &= |\det \chi'(y)| \int_S a_{-n}(\chi(y), \xi)\sigma(\xi). \end{aligned}$$

Hence 1.7(1) is well-defined.

For the proof of the trace property we may employ the linearity of res to confine ourselves to the case of two operators A, B with symbols a and b supported in the same chart U. Also we may assume that we are in the scalar case, since everything commutes under Tr. The symbol of $[A, B]$ is given by

$$\sum_{|\alpha|\geq 0} \frac{(-i)^{|\alpha|}}{\alpha!} (\partial_\xi^\alpha a \, \partial_x^\alpha b - \partial_\xi^\alpha b \, \partial_x^\alpha a). \tag{3}$$

We may rewrite this expression as $\sum_{j=1}^n \partial_{\xi_j} A_j + \partial_{x_j} B_j$, where A_j and B_j are bilinear expressions in a and b and their derivatives; they vanish for $x \notin U$. Thus, the integrals over S of $(\partial_{\xi_j} A_j)_{-n}\sigma$ are zero by Lemma 1.8(a). The same holds for the integrals of $(\partial_{x_j} B_j)_{-n}$ over U, since all B_j have compact x-support in U.

To prove uniqueness, suppose τ is another trace on $\mathcal{A}$, and consider an operator A with symbol $a \sim \sum a_j$ supported in U. Let $\widehat{x}_j$ and $\widehat{\xi}_j$ denote any symbols with x-supports in U coinciding with x_j and ξ_j on the support of a. The symbols of the commutators $[A, \mathrm{op}\,\widehat{x}_j]$ and $[A, \mathrm{op}\,\widehat{\xi}_j]$ then are $-D_{\xi_j} a$ and $D_{x_j} a$, respectively. Since the trace τ vanishes on commutators, it vanishes on all symbols that are derivatives with respect to either x or ξ.

Define $\overline{a}(x) = \frac{1}{\mathrm{vol}\, S} \int_S a_{-n}(x,\xi)\sigma_\xi$. Applying Lemma 1.8(b) to a_j for all $j \neq -n$, there exist n functions $b_{kj}(x,\xi)$, $k = 1, \ldots, n$, homogeneous of degree $j+1$ in ξ such that $a_j = \sum_{k=1}^n \partial_{\xi_k} b_{kj}$. Let $b_k(x,\xi) \sim \sum_{j \le m, j \neq -n} b_{kj}$. Then

$$a(x,\xi) - \overline{a}(x)|\xi|^{-n} = \sum_{k=1}^{n} \partial_{\xi_k} b_k(x,\xi) + (a_{-n}(x,\xi) - \overline{a}(x)|\xi|^{-n}) .$$

Clearly, $\int_S (a_{-n}(x,\xi) - \overline{a}(x)|\xi|^{-n})\sigma(\xi) = 0$. So Lemma 1.8(b.ii) shows that $a_{-n}(x,\xi) - \overline{a}(x)|\xi|^{-n}$ is a finite sum of derivatives with respect to ξ. Hence $\tau(a) = \tau(\overline{a}(x)|\xi|^{-n})$, so $T : f \mapsto \tau(f(x)|\xi|^{-n})$ defines a functional on $C_0^\infty(U)$ which vanishes on derivatives. Now it is no restriction to assume that U is diffeomorphic to an open ball. We easily deduce from Schwartz [29, II.4] that $Tf = c \int f(x)dx$ for a suitable constant c. A priori, the constant might depend on U, but on the intersection of two coordinate neighborhoods the constants must agree. If M is connected, then all are equal, and the proof is complete. ◁

Note that no continuity condition is required for the uniqueness of the noncommutative residue.

1.9 Examples and remarks. (a) Let $A = (I - \Delta)^{-n/2}$. Then $a_{-n}(x,\xi) = |\xi|^{-n}$ and $\mathrm{res}\, A = \int_M \int_{S^{n-1}} |\xi|^{-n}\sigma(\xi)dx = \mathrm{vol}\, S^{n-1} \cdot \mathrm{vol}\, M$. So the volume of M can be found as a noncommutative residue.
(b) If A is a differential operator, then $\mathrm{res}\, A = 0$.
(c) If the order of A is $< -n$, then $\mathrm{res}\, A = 0$, so res is not an extension of the usual operator trace. In fact, as we shall see in Lecture 4, Wodzicki's residue coincides with Dixmier's trace on pseudodifferential operators of order $-n$ and therefore *vanishes* on trace class operators.

1.10 Seeley's results on complex powers. We additionally assume A to be invertible of order $m > 0$. In particular, a is elliptic, but we impose a slightly stronger condition: There exists a ray $R_\theta = \{z = re^{i\theta}, r \ge 0\}$ in $\mathbb{C}$ with no eigenvalue of of $a_m(x,\xi)$ on R_θ for $\xi \neq 0$. Then Seeley [30] showed that the spectrum of A is discrete with only finitely many eigenvalues on R_θ. Shifting θ slightly, R_θ will not intersect the spectrum. Moreover:

(i) There exists a family of complex powers $\{A^s : s \in \mathbb{C}\}$, defined by

$$\begin{aligned} A^s &= \frac{i}{2\pi} \int_C \lambda^s (A - \lambda)^{-1} d\lambda, \quad \mathrm{Re}\, s < 0; \\ A^{s+k} &= A^s A^k, \quad \mathrm{Re}\, s < 0, k \in \mathbb{N}. \end{aligned}$$

Here C is the path in $\mathbb{C}$ going from infinity along R_θ to a small circle around 0, clockwise about the circle, and back along R_θ.

(ii) A^s is a pseudodifferential operator of order $m \operatorname{Re} s$, $s \mapsto A^s$ is analytic.

(iii) For $\operatorname{Re} s < -n/m$, A^s is an integral operator with a continuous integral kernel $k_s(x,y)$. For each $x \in M$, $s \mapsto k_s(x,x)$ extends to a meromorphic map with at most simple poles in $s_j = \frac{j-n}{m}$, $j = 0,1,\dots$. There is no pole in $s = 0$; the residue in s_j is given by an explicit formula. If A is a *differential* operator, then also the residues at the positive integers vanish.

1.11 The noncommutative residue and zeta functions. We use the notation of 1.10. Since the spectrum $\{\lambda_j\}$ of A is discrete and A^s is trace class for $\operatorname{Re} s < -n/m$ we may define the zeta function

$$\zeta_A(s) = \operatorname{trace} A^{-s} = \sum \lambda_j^{-s}, \quad \operatorname{Re} s > n/m.$$

This is a holomorphic function. It coincides with $\int_M k_{-s}(x,x)dx$ hence has a meromorphic extension to $\mathbb{C}$ with at most simple poles in the points s_j. Wodzicki used Seeley's explicit formulas to show that

$$\operatorname{Res}_{s=-1}\zeta_A = (2\pi)^n \operatorname{res} A/\operatorname{ord} A; \tag{1}$$

here $\operatorname{ord} A$ is the order of A; more generally

$$\operatorname{Res}_{s=s_j}\zeta_A = (2\pi)^n \operatorname{res} A^{-s_j}/\operatorname{ord} A. \tag{2}$$

We can use this relation to define res via zeta functions: Let P be an arbitrary pseudodifferential operator. Choose A satisfying the assumptions of 1.10 with $\operatorname{ord} A > \operatorname{ord} P$. Then also $A + uP$, $u \in \mathbb{R}$, will meet the requirements of 1.10, provided $|u|$ small, and (1) shows that

$$\operatorname{res} P = \frac{d}{du} \operatorname{res}(A+uP)|_{u=0} = (2\pi)^{-n} \operatorname{ord} A \operatorname{Res}_{s=-1} \zeta_{A+uP}.$$

1.12 Heat kernels. Starting from the assumptions in 1.10 we additionally ask that$A$ is a positive operator and that the eigenvalues of the principal symbol a_m lie in the right half-plane. Then one can define

$$e^{-tA} = \int_C e^{-t\lambda}(A-\lambda)^{-1}d\lambda,$$

where C is a suitable contour around the spectrum. The operator e^{-tA} is trace class, and $\operatorname{trace} e^{-tA} = \sum_{j=1}^\infty e^{-\lambda_j t}$. The identity

$$\int_0^\infty t^{s-1}e^{-\lambda t}dt = \lambda^{-s}\int_0^\infty (\lambda t)^{s-1}e^{-\lambda t}d(\lambda t) = \lambda^{-s}\Gamma(s).$$

shows that $\Gamma(s)\zeta_A(s) = \int_0^\infty t^{s-1}\mathrm{trace}\,(e^{-tA})dt$ is the Mellin transform of $\mathrm{trace}\, e^{-tA}$. It is a well-known property of the Mellin transform that the asymptotic behavior $\sim t^{-s_j}\ln^k t$ near $t = 0$ produces a pole in s_j of order $k+1$ and vice versa. From the above results for the zeta function one immediately deduces the asymptotic expansion near zero:

$$\mathrm{trace}\, e^{-tA} \sim \sum_{j=0}^{\infty} \alpha_j(A)t^{\frac{j-n}{m}} + \sum_{k=1}^{\infty} \beta_k(A)t^k \ln t$$

Note that there is no term $t^0 \ln t$, since ζ_A regular in 0 while the Gamma function has a simple pole; for the same reason there are no terms $t^k \ln t$ if A is *differential.*

So we get $\mathrm{res}\, A = \mathrm{ord}\, A \cdot \beta_1(A)$. Moreover, we can define the noncommutative residue for a general pseudodifferential operator by choosing an operator A with the above properties and $\mathrm{ord}\, A > \mathrm{ord}\, P$ and letting

$$\mathrm{res}\, P = -\mathrm{ord}\, A \frac{d}{du}\beta_1(A + uP)|_{u=0}.$$

Classically, A is the Laplace–Beltrami operator Δ associated with a Riemannian metric on M, so that one really deals with the heat equation. It is well-known that the coefficients $a_j(\Delta)$ carry important geometric information, see e.g. Gilkey [9].

1.13 Notes and Remarks. The original reference for Wodzicki's residue is [32]; a much more elaborate presentation was given in [33]. Kassel's paper [14] gives a good survey. The proof of Theorem 1.7 here follows [7].

In Theorem 1.7 we asked for simplicity that $n \geq 2$. For $n = 1$ the cosphere bundle has two components. A simpler version of the above arguments then shows that one gets two residues when restricting to orientation preserving changes of coordinates otherwise one residue as before.

Lecture 2: The Cone Algebra

In this lecture we shall review the cone calculus for manifolds with conical singularities introduced by B.-W. Schulze. In the next lecture we shall deal with noncommutative residues for these objects.

Following the general idea of noncommutative geometry, the information about the underlying space is encoded in a suitable algebra of linear operators. From the analysis of the classical case presented in Section 1, we know that Wodzicki's residue recovers the geometric invariants detected by the heat kernel expansion methods. One might therefore hope that a similar result holds for the singular case.

In this context the choice of the operator algebra is very important. Consider for example a manifold M with boundary. One possible operator algebra is, of course, the algebra of classical pseudodifferential operators on the open interior. Yet it is not difficult to see from the proof of Theorem 1.7 that there is *no* trace on this algebra.

On a manifold with boundary, it seems more natural to consider boundary value problems. The canonical analog of the algebra of pseudodifferential operators then is Boutet de Monvel's algebra. As it turns out we then get the desired result [7, 8]:

2.1 Theorem. *There is a trace on the algebra $\mathcal{B}_{cl}$ of classical elements in Boutet de Monvel's calculus on M. It extends Wodzicki's residue, vanishes on the ideal $\mathcal{B}^{-\infty}(M)$ of smoothing elements, and it is the unique trace on the quotient algebra $\mathcal{B}_{cl}(M)/\mathcal{B}^{-\infty}(M)$, provided M is connected and* $\dim M > 1$.

We now introduce the basic elements of Schulze's cone calculus.

2.2 Manifolds with conical singularities. A manifold with conical singularities, B, is a second countable Hausdorff space which is, outside a finite number of points $v \in B$, a smooth manifold.

In a neighborhood of each of the so-called *singularities* or *singular points* v, the manifold is diffeomorphic to a cone $X \times [0, \infty)/X \times \{0\}$, whose cross-section, X, is a closed compact manifold.

In the following we shall confine ourselves to the case of one singularity v. We blow up at v and obtain a manifold with boundary, and a neighborhood of the boundary can be identified with the collar $X \times [0, 1)$. We denote the resulting object by $\mathbb{B}$, while $X^\wedge$ is the cylinder $X^\wedge = X \times \mathbb{R}_+$.

2.3 Idea of the calculus. Apart from the technical complications the basic concept is the following :

- On the smooth part of B use the pseudifferential calculus in its standard form.
- Near singularities use Mellin calculus on $X \times \mathbb{R}_+$ working with smooth families of meromorphicMellin symbols taking values in the algebra of pseudodifferential operators on X.

2.4 Mellin transform. For $u \in C_0^\infty(\mathbb{R}_+)$ we define the *Mellin transform* Mu by

$$(Mu)(z) = \int_0^\infty t^{z-1} u(t) dt, \quad z \in \mathbb{C}$$

This furnishes an entire function which is rapidly decreasing along each line $\Gamma_\beta = \{z \in \mathbb{C} : \operatorname{Re} z = \beta\}$. Plancherel's theorem for the Fourier transform shows that M extends to an isomorphism $L^2(\mathbb{R}_+) \to L^2(\Gamma_{1/2})$. The identity

$(Mu)|_{\Gamma_{1/2-\gamma}}(z) = M_{t\to z}(t^{-\gamma}u)(z+\gamma)$ motivates the following definition of the *weighted* Mellin transform:

$$M_g gu(z) = M(t^{-\gamma}u)(z+\gamma).$$

The inverse of M_γ is given by

$$(M_\gamma^{-1}h)(t) = \frac{1}{2\pi i}\int_{\Gamma_{1/2-\gamma}} t^{-z}h(z)dz.$$

For $v = -t\partial_t u$ one has $Mv(z) = zMu(z)$ in particular $-t\partial_t u = M^{-1}zMu$.

2.5 Cut-off functions. Whenever we speak of a *cut-off function* or use the notation $\omega, \tilde{\omega}, \omega_1, \omega_2, \ldots$ withour further specification we mean a function $\omega \in C_0^\infty(\overline{\mathbb{R}}_+)$ with $\omega(t) = 1$ near $t = 0$. We will also speak of cut-off functions on $\mathbb{B}$, asking that they vanish on the part of $\mathbb{B}$ not identified with the collar.

2.6 Mellin Sobolev spaces. For $s \in \mathbb{N}, \gamma \in \mathbb{R}$ the Mellin Sobolev space $\mathcal{H}^{s,\gamma}(X^\wedge)$ is the set of all $u \in \mathcal{D}'(X^\wedge)$ for which $t^{n/2-\gamma}(t\partial_t)^k Du(x,t) \in L^2(X^\wedge)$ whenever $k \le s$ and D is a differential operator of order $\le s-k$ on X. Interpolation and duality furnish $\mathcal{H}^{s,\gamma}(X^\wedge)$ for all $s, \gamma \in \mathbb{R}$. Note that duality is with respect to the pairing

$$(u,v) = \frac{1}{2\pi i}\int_{\Gamma_{\frac{n+1}{2}}} (Mu(z), Mv(z))_{L^2(X)}dz$$

and that $\mathcal{H}^{0,n/2}(X^\wedge) = L^2(X^\wedge)$.

These spaces make sense on $\mathbb{B}$, too: We pick a cut-off function ω on $\mathbb{B}$ and let $\mathcal{H}^{s,\gamma}(\mathbb{B}) = \{u : \omega u \in \mathcal{H}^{s,\gamma}(X^\wedge), (1-\omega)u \in H^s_{loc}(\operatorname{int}\mathbb{B})\}$.

2.7 Mellin Symbols and Mellin Operators. Let $\mu \in \mathbb{Z}, \gamma \in \mathbb{R}$. By $L^\mu(X;\mathbb{R})$ denote the space of parameter-dependent pseudodifferential operators of order μ on X with parameter space $\mathbb{R}$. $L^\mu(X, \Gamma_{1/2-\gamma})$ is the corresponding space with $\Gamma_{1/2-\gamma}$ identified with $\mathbb{R}$.

Given $f \in C^\infty(\overline{\mathbb{R}}_+, L^\mu(X;\Gamma_{1/2-\gamma}))$ define the Mellin operator with (Mellin) symbol f and weight γ by

$$[\mathrm{op}_M^\gamma f]u(t) = \frac{1}{2\pi i}\int_{\Gamma_{1/2-\gamma}} t^{-z}f(t,z)[M_\gamma u](z)dz$$

for $u \in C_0^\infty(X^\wedge) = C_0^\infty(R_+, C^\infty(X))$. It is easy to see that $\mathrm{op}_M^\gamma f : C_0^\infty(X^\wedge) \to C^\infty(X^\wedge)$ is continuous. Moreover,

$$\omega_1[\mathrm{op}_M^\gamma f]\omega_2 : \mathcal{H}^{s,\gamma+n/2}(X^\wedge) \to \mathcal{H}^{s-\mu,\gamma+n/2}(X^\wedge)$$

is bounded for all s.

We shall now turn to the analysis of asymptotics.

2.8 Example. Let ω be a cut-off function.

(a) Write $M(\omega) = z^{-1}M(-t\partial_t\omega)(z)$. Since $t\partial_t\omega \in C_0^\infty(\mathbb{R}_+)$, we obtain a meromorphic function with a single simple pole in $z = 0$; it is rapidly decreasing along each Γ_β, uniformly for β in compact intervals, including $\beta = 0$, provided we remove a neighborhood of $z = 0$ (by multiplication with a function which vanishes there and is 1 near infinity).

(b) Let $\operatorname{Re} p < 1/2, k \in \mathbb{N}$. Then $M(t^{-p}\ln^k t\omega(t))(z) = \frac{d^k}{dz^k}(M\omega)(z-p)$. This again is a meromorphic function with a single pole in $z = p$ of order $k+1$, it also is rapidly decreasing along each Γ_β, uniformly for β in compact intervals provided we remove a neighborhood of the pole itself.

2.9 Asymptotic types and Mellin Sobolev spaces with asymptotics. Fix $\gamma \in \mathbb{R}$. Recall that a weight datum $\mathbf{g}$ is a triple $\mathbf{g} = (\gamma + n/2, \gamma + n/2, (-1, 0])$ consisting of two reals and an interval.

(a) An asymptotic type associated with $\mathbf{g}$ is a finite set $P = \{(p_j, m_j, C_j) : j = 1, \ldots, J\}$ with $J \in \mathbb{N}$ (possibly $J = 0$, then P is the empty set), $p_j \in \mathbb{C}$ with $-1/2 - \gamma < \operatorname{Re} p_j < 1/2 - \gamma$, $m_j \in \mathbb{N}$, and C_j finite-dimensional subspaces of $C^\infty(X)$. We denote by $\pi_{\mathbb{C}}P$ the set $\{p_j : j = 1, \ldots, J\}$.

(b) A Mellin asymptotic type is a sequence $P = \{(p_j, m_j, L_j) : j \in \mathbb{Z}\}$ with $p_j \in \mathbb{C}, \operatorname{Re} p_j \to \mp\infty$ as $j \to \pm\infty$, $m_j \in \mathbb{N}$, and L_j finite dimensional subspaces of finite rank operators in $L^{-\infty}(X)$. As before we write $\pi_{\mathbb{C}} = \{p_j\}$.

(c) Given an asymptotic type P, and $s, \gamma \in \mathbb{R}$ we let $\mathcal{H}_P^{s,\gamma+n/2}(\mathbb{B})$ be the space of all $u \in \mathcal{H}^{s,\gamma+n/2}(\mathbb{B})$ for which there exist $c_{jk} \in C_j, j = 1, \ldots, J, k = 0, \ldots, m_j$, such that, for all $\varepsilon > 0$,

$$u - \sum_{j=1}^{J}\sum_{k=0}^{m_j} c_{jk}t^{-p_j}\ln^k t\,\omega(t) \in \mathcal{H}^{s,\gamma+n/2+1-\varepsilon}(\mathbb{B})$$

2.10 Meromorphic Mellin symbols.

(a) $M_O^\mu(X)$ is the space of all entire functions $h : \mathbb{C} \to L^\mu(X)$ such that $h|_{\Gamma_\beta} \in L^\mu(X; \Gamma_\beta)$ uniformly for β in compact intervals.

(b) Let P be a Mellin type. $M_P(X)$ is the space of all holomorphic $h : \mathbb{C} \setminus \pi_{\mathbb{C}}P \to L^\mu(X)$ with the following properties:

   (i) In a neighborhood of p_j we have $h(z) = \sum_{k=0}^{m_j} \nu_{jk}(z-p_j)^{-k-1} + h_0(z)$ with $\nu_{jk} \in L_j$ and h_0 analytic near p_j;

(ii) for each interval $[c_1, c_2]$ we find elements ν_{jk} in L_j such that

$$h(\beta + i\tau) = \sum_{\{j:\mathrm{Re}\, p_j \in [c_1,c_2]\}} \sum_{k=0}^{m_j} \nu_{jk} M_{t\to z}(t^{-p_j} \ln^k t\omega(t))(\beta + i\tau) \in L^{\mu}(X, \mathbb{R}_\tau)$$

uniformly for $\beta \in [c_1, c_2]$. We set $M_P^{-\infty}(X) = \bigcap M_P^{\mu}(X)$.

2.11 Theorem. *$M_P^{\mu}(X) = M_O^{\mu}(X) + M_P^{-\infty}(X)$ as a non-direct sum of Fréchet spaces.*

With these notions at hand we are ready to define the full algebra. Fix μ and γ and recall that $\mathbf{g}$ is the weight datum $(\gamma + n/2, \gamma + n/2, (-1, 0])$.

2.12 The residual elements: Green operators. $C_G(\mathbb{B}, \mathbf{g})$ is the space of all operators $G : C_0^{\infty}(\mathrm{int}\, \mathbb{B}) \to \mathcal{D}'(\mathrm{int}\, \mathbb{B})$ with continuous extensions

$$G : \mathcal{H}^{-s,\gamma+n/2}(\mathbb{B}) \to \mathcal{H}_{Q_1}^{\infty,\gamma+n/2}(\mathbb{B}) \text{ and}$$
$$G^* : \mathcal{H}^{s,-\gamma-n/2}(\mathbb{B}) \to \mathcal{H}_{Q_2}^{\infty,-\gamma-n/2}(\mathbb{B})$$

for suitable asymptotic types Q_1, Q_2 and all s. Here, G^* is the adjoint with respect to the pairing $\mathcal{H}^{s,\gamma}, \mathcal{H}^{-s,-\gamma}$.

Note: $\mathcal{H}_{Q_1}^{\infty,\gamma+n/2} \hookrightarrow \mathcal{H}^{N,\gamma+n/2}(\mathbb{B})$ is compact for each N, hence $C_G(\mathbb{B}, \mathbf{g})$ consists of compact operators.

2.13 An ideal: The algebra $C_{M+G}(\mathbb{B}, \mathbf{g})$. $C_{M+G}(\mathbb{B}, \mathbf{g})$ is the space of all operators $R : C_0^{\infty}(\mathrm{int}\, \mathbb{B}) \to \mathcal{D}'(\mathrm{int}\, \mathbb{B})$ that can be written

$$R = \omega_1[\mathrm{op}_M^{\gamma} h]\omega_2 + G,$$

where

(i) $h_0 \in M_{P_0}^{-\infty}(X)$ for some Mellin asymptotic type P_0,
(ii) $\pi_{\mathbb{C}} P_0 \cap \Gamma_{1/2-\gamma} = \emptyset$,
(iii) ω_1, ω_2 cut-off functions, and
(iv) $G \in C_G(\mathbb{B}, \mathbf{g})$

Note: These operators form an algebra which turns out to be an ideal in the final algebra. The Green operators form an ideal in $C_{M+G}(\mathbb{B}, \mathbf{g})$. A change in the choice of the cut-off functions results in a Green operator.

2.14 The full algebra. $C^{\mu}(\mathbb{B}, \mathbf{g})$ is the space of all operators

$$A_M + A_\psi + R,$$

where $A_M = \omega_1 \mathrm{op}_M^{\gamma} h\omega_2$, with $h \in C^{\infty}(\overline{\mathbb{R}}_+, M_O^{\mu}(X))$, is a Mellin operator supported close to the singularity, A_ψ is a pseudodifferential operator of order μ supported in the interior and $R \in C_{M+G}(\mathbb{B}, \mathbf{g})$.

Note: $C^{\mu}(\mathbb{B}, \mathbf{g})$ is a Fréchet space with the natural topology.

2.15 Theorem. *The composition of operators yields a continuous map*

$$C^{\mu}(\mathbb{B}, \mathrm{g}) \times C^{\mu'}(\mathbb{B}, \mathrm{g}) \to C^{\mu+\mu'}(\mathbb{B}, \mathrm{g}).$$

We have the ideal structure:

$$C_G(\mathbb{B}\,\mathrm{g}) \trianglelefteq C_{M+G}(\mathbb{B}, \mathrm{g}) \trianglelefteq C^{\mu}(\mathbb{B}, \mathrm{g}).$$

2.16 Mellin quantization. For $h \in C^\infty(\overline{\mathbb{R}}_+, M_0^\mu(X))$ there is a $p \in C^\infty(\overline{\mathbb{R}}_+, \tilde{L}^\mu(X;\mathbb{R}))$ such that

$$\operatorname{op} p \equiv \operatorname{op}_M^\gamma h \bmod L^{-\infty}(X^\wedge).$$

Here $C^\infty(\mathbb{R}_+, \tilde{L}^\mu(X, \mathbb{R}))$ denotes the space of *totally characteristic symbols* (also Fuchs type symbols), i.e. the elements of $C^\infty(\overline{\mathbb{R}_+} L^\mu(X;\mathbb{R}))$ that can be written $p(t,\tau) = q(t, t\tau)$ for some $q \in C^\infty(\overline{\mathbb{R}}_+ L^\mu(X;\mathbb{R}))$. The symbol p has the asymptotic expansion

$$p(t,\tau) \sim \sum_{k=0}^{\infty} \frac{1}{k!} \partial_{t'}^k D_\tau^k \{h(t, -iT(t,t')\tau)\frac{T(t,t')}{t'}\}|_{t'=t}$$

with $T(t,t') = \frac{t-t'}{\ln t - \ln t'}$. Note that $T(t,t') = t$.

2.17 Symbols. To an operator in the cone algebra we can therefore associate to important symbols, namely

(i) the interior pseudodifferential symbol which is in fact defined up to the boundary with a totally characteristic degeneracy, and

(ii) the operator family $\{h(0,z) + h_0(z) : H^s(X) \to H^{s-\mu}(X) : z \in \Gamma_{1/2-\gamma}\}$, the so-called *conormal symbol* .

The conormal symbol plays an important role in the Fredholm theory on manifolds with conical singularities. The Fredholm property for an operator is equivalent to the invertibility of the interior principal symbol and the invertibility of the conormal symbol on $\Gamma_{1/2-\gamma}$.

2.18 Notes and Remarks. This is a simplified and comprehensive version of the cone calculus. I used the material in the joint work [23, 24]. Other good sources are Egorov and Schulze [6] and Schulze [28].

Lecture 3: Noncommutative Residues on Manifolds with Conical Singularities

We start with a negative result:

3.1 Example. Wodzicki's residue does not extend to cone algebra. In order to see this recall first that $\mathbb{B}$ is $(n+1)$-dimensional. Suppose $h \in C^\infty(\overline{\mathbb{R}}_+, M_O^{-n-1}(X))$, and h vanishes for $t \geq 1$. For $\gamma = 1/2$ we consider the operator $\mathrm{op}_M^{1/2} h$. According to 2.16 we can find a pseudodifferential symbol: $\mathrm{op}_M^{1/2} h \equiv \mathrm{op}\, p \mod L^{-\infty}(X^\wedge)$ with $p_{-n-1}(x,t,\xi,\tau) = h(t)_{-n-1}(x,\xi,-it\tau)$. In order to distinguish it from the densities we shall analyze below, we now write W-res for the Wodzicki density introduced in Theorem 1.7. We then have

$$\begin{aligned}
&\text{W-res}_{(x,t)}\mathrm{op}\, p \\
&= \left(\int_{S^n} p_{-n-1}(x,t,\xi,\tau)\sigma(\xi,\tau)\right) dx\, dt \\
&= \left(\int_{S^n} h(t)_{-n-1}(x,\xi,-it\tau)\sigma(\xi,\tau)\right) dx\, d\tau \\
&= \left(\int_{S^{n-1}} \int_{-\infty}^{\infty} h(t)_{-n-1}(x,\xi,-it\tau) d\tau\, \sigma(\xi)\right) dx\, d\tau \\
&= t^{-1}\left(\int_{S^{n-1}} \int_{-\infty}^{\infty} h(t)_{-n-1}(x,\xi,-is) ds\, \sigma(\xi)\right) dx\, d\tau;
\end{aligned}$$

here $\sigma(\xi,\tau)$ is the n form corresponding to the $n-1$-form σ used in Section 1. For the third equality we have used that the integrand is a closed form, hence we can shift the contour.

In order to compute the noncommutative residue we would have to integrate the density over the collar $X \times [0,1)$. This, however, is not possible, unless $h(t)_{-n-1}$ vanishes for $t = 0$.

We shall now define a different density:

3.2 Definition. Let A be as in 2.14. Near $x \in X$ let $h(0)(x,\xi,i\tau)$ be the local symbol of $h(0,i\tau)$. The subscript $-n-1$ in the notation $h(0)_{-n-1}(x,\xi,i\tau)$, below, indicates the term of homogeneity $-n-1$ with respect to (ξ,τ). Define

$$\mathrm{res}_x A = \left(\int_{S^{n-1}} \int_{-\infty}^{\infty} \mathrm{Tr}\, h(0)_{-n-1}(x,\xi,i\tau) d\tau\, \sigma(\xi)\right) dx_1 \wedge \ldots \wedge dx_n.$$

Since the operators may take values in a vector bundle E, we also introduced a trace Tr on $\mathcal{L}(E)$ in the integral above. For $n = 1$ replace integration over S^{n-1} by $h(0)_{-2}(x,1,i\tau) + h(0)_{-2}(x,-1,i\tau)$.

3.3 Remark.

(a) The decomposition $h + h_O$ is not unique, but h_O is of order $-\infty$ and therefore gives no contribution.

(b) $\operatorname{res}_x A = (\int_{S^n} \operatorname{Tr} h(0)_{-n-1}(x,\xi,i\tau)\sigma(\xi,\tau))\, dx_1\wedge\ldots\wedge dx_n$ in view of the fact that $h(0)_{-n-1}(x,\xi,i\tau)\sigma(\xi,\tau)$ is a closed form of degree $-n-1$.

3.4 Lemma. $\operatorname{res}_x A$ *defines a density on* X.

Proof. We fixed t as a global coordinate. So changes of coordinates are of the form $(x,t) \mapsto (\chi(x),t)$. Hence Lemma follows as in the standard case. ◁

3.5 Definition. For $A \in C^\mu(\mathbb{B},\mathbf{g})$ let

$$\operatorname{res} A = \int_X \operatorname{res}_x A = \int_X \int_{S^{n-1}} \int_{-\infty}^{\infty} \operatorname{Tr} h(0)_{-n-1}(x,\xi,i\tau) d\tau\, \sigma(\xi) dx_1\wedge\ldots\wedge dx_n.$$

May write

$$\operatorname{res} A = W - \operatorname{res} \int_{-\infty}^{\infty} h(0)_{-n-1}(\cdot,\cdot,i\tau) d\tau$$

with Wodzicki's residue of the $(-n)$-homogeneous $\int_{-\infty}^{\infty} h(0)_{-n-1} d\tau$.

3.6 Example. Let Δ be the Laplacian for a two-dimensional manifold with a conical singularity. Close to the singularity a computation shows that $t^2\Delta = c^2\partial_x^2 + (t\partial_t)^2$ has the Mellin symbol $g(t,x,\xi,z) = -c^2|\xi|^2 + t^2$. Her c is a suitable constant depending on the opening angle of the cone. A parametrix A to $t^2\Delta$ therefore has the Mellin symbol $(-c^2|\xi|^2 + z^2)^{-1}$ modulo lower order terms. This is the desired component of order -2, 'integration' over S^*S^1 gives $2(-c^2+z^2)^{-1}$. Thus

$$\operatorname{res} A = -s \int_{S^1} \int_{-\infty}^{\infty} (c^2+\tau^2)^{-1} d\tau\, dx = -4\pi^2/c.$$

We shall now produce an extension of Wodzicki's residue. On the collar we consider the algebra of all Mellin operators with vanishing Mellin symbol in $t = 0$.

3.7 Operators on the collar. Consider the operators in the cone algebra that can be written in the form

$$A = \omega_1[\operatorname{op}_M^\gamma h]\omega_2 + R$$

with

(i) $h \in C^\infty(\overline{\mathbb{R}}_+, M_O^\mu(X)), h(0) = 0,$

(ii) ω_1,ω_2 cut-off functions with $\omega_1\omega_2 = \omega_1$,

(iii) $R \in C_{M+G}(\mathbb{B},\mathbf{g})$.

For convenience assume that $\operatorname{supp}\omega \subseteq [0,1)$. What is important is that we may choose $h(0)=0$ and that $\omega_1\omega_2 = \omega_1$. The latter condition normalizes the representation in a certain sense. For $(x,t) \in X \times (0,1)$ define

$$\operatorname{res}^0_{x,t} A = \left(\int_{S^{n-1}} \int_{-\infty}^{\infty} \omega_1(t) h(t)_{-n-1}(x,\xi,i\tau) d\tau\, \sigma(\xi)\right) dx_1 \wedge \ldots \wedge dx_n \wedge \frac{dt}{t}.$$

3.8 Lemma.

$$\operatorname{res}^0_{x,t} A \int_{S^n_{\xi,\tau}} Tr\, \omega_1(t) h(t)_{-n-1}(x,\xi,i\tau)\sigma(\xi,\tau) dx_1 \wedge \ldots \wedge \frac{dt}{t}.$$

It is a density on $X \times \mathbb{R}_+$.

Proof. The asserted identity follows from the fact that $h(t)_{-n-1}(x,\xi,i\tau)\sigma(\xi,\tau)$ is closed. That it is a density can then be proven as before. ◁

3.9 Theorem. *For A as above let*

$$\operatorname{res}^0 A = \int_0^1 \int_X \operatorname{res}^0_{x,t} A.$$

This makes sense in view of the condition $h(0) = 0$. Moreover, res^0 is a trace on operators of this form.

Proof. This can be shown just like in Theorem 1.7. ◁

3.10 Lemma. *Let p be the totally characteristic pseudodifferential symbol associated with A mod $L^{-\infty}(X \times \mathbb{R}_+)$, cf. 2.16.*

$$p(t,\tau) \sim \omega_1(t) \sum_{k=0}^{\infty} \frac{1}{k!} \partial^k_{t'} D^k_\tau \{h(t, -iT(t,t')\tau)\frac{T(t,t')}{t'}\}|_{t'=t}$$

with $T(t,t') = (t-t')/(\ln t - \ln t')$. Then

$$\operatorname{res}^0 A = W - \operatorname{res} p.$$

Proof.

$$\begin{aligned}
&\int_{S^n_{\xi,\tau}} Tr\, p(t)_{-n-1}(x,\xi,\tau) dx \wedge dt \\
&= \int_{S^n_{\xi,\tau}} Tr \left(\sum_{k=0}^{\infty} \frac{1}{k!} \partial^k_{t'} D^k_\tau \{h(t)(x,\xi,-iT(t,t')\tau)\frac{T(t,t')}{t'}\}|_{t'=t}\right)_{-n-1} \\
&= \int_{S^n_{\xi,\tau}} Tr\, h(t)_{-n-1}(x,\xi,-it\tau)\sigma(\xi,\tau) dx \wedge dt
\end{aligned}$$

for the integrand can be rewritten as $D^l_\tau \ldots \tau^j$ with $j \leq l/2$. Then

$$\begin{aligned}
\text{W-res}\, p &= \int_0^1 \int_X \int_{S^n_{\xi,\tau}} Tr\, h(t)_{-n-1}(x,\xi,-it\tau)\sigma(\xi,\tau) dx \wedge dt \\
&= \int_0^1 \int_X \int_{S^{n-1}_\xi} \int_{-\infty}^{\infty} Tr\, h(t)_{-n-1}(x,\xi,-it\tau) d\tau\, \sigma(\xi) dx \wedge dt \\
&= \operatorname{res}^0 A.
\end{aligned}$$

◁

3.11 The extended cone algebra. On $\mathbb{B}$ choose a smooth function $\underline{t}$ which coincides with the geodesic distance to the boundary near the boundary of $\mathbb{B}$ and which is strictly positive in the interior. Let $C(\mathbb{B},\mathbf{g})^+$ be the space of finite sums of operators of the form $\underline{t}^m B$ with $\operatorname{Re} m \geq 0$ and $B \in C(\mathbb{B},\mathbf{g})$. Similarly define $C_{M+G}(\mathbb{B},\mathbf{g})^+$.

$$\begin{aligned}
C(\mathbb{B},\mathbf{g})^+_0 &= \langle t^m B : B \in C(\mathbb{B},\mathbf{g}), \operatorname{Re} m > 0\rangle \\
C_{M+G}(\mathbb{B},\mathbf{g})^+_0 &= \langle t^m B : B \in C_{M+G}(\mathbb{B},\mathbf{g}), \operatorname{Re} m > 0\rangle
\end{aligned}$$

Why is this an algebra? For the Green operators multiplication by $\underline{t}$ is no problem, neither is it for the pseudodifferential operators in the interior. For the Mellin operators we use the following computation. Note that we may assume $\underline{t} = t$, since we are close to the boundary.

$$\begin{aligned}
\operatorname{op}^\gamma_M h(t^m u)(t) &= \frac{1}{2\pi i}\int_{\Gamma_{1/2-\gamma}} \int_0^\infty (t/t')^{-z} h(t,z)(t'^m u(t'))\frac{dt'}{t'} dz \\
&= \frac{t^m}{2\pi i}\int_{\Gamma_{1/2-\gamma}} \int_0^\infty (t/t')^{-(z+m)} h(t,z) u(t)\frac{dt'}{t'} dz \\
&= \frac{t^m}{2\pi i}\int_{\Gamma_{1/2-\gamma}} \int_0^\infty (t/t')^{-\zeta} h(t,\zeta - m) u(t'))\frac{dt'}{t'} d\zeta \\
&= t^m [\operatorname{op}^\gamma_M T^{-m} h] u(t).
\end{aligned}$$

There is a minor problem if h is a meromorphic smoothing Mellin symbol and $T^{-m}h$ has a singularity on $\Gamma_{1/2-\gamma}$. Then we use the fact that we may write

$$t^m \operatorname{op}^\gamma_M h = t^m \operatorname{op}^{\gamma_0}_M h + G$$

with $\gamma - \mu \leq \gamma_0 \leq \gamma$ and a Green operator G.

With this notation we can state the theorem on uniqueness. A full proof is given in [26].

3.12 Theorem.

(a) *For each conical point we get precisely one continuous trace on the quotient* $C(\mathbb{B},\mathbf{g})^+/C_{M+G}(\mathbb{B},\mathbf{g})^+$.

(b) *On* $C(\mathbb{B},\mathbf{g})_0^+/C_{M+G}(\mathbb{B},\mathbf{g})_0^+$ *there is a unique trace, namely the extension of Wodzicki's residue.*

We understand (a) in the sense that near this conical point the operator algebra is the cone algebra, i.e. there is not a fictitious conical point.

Lecture 4: The Noncommutative Residue and Dixmier's Trace

Dixmier's paper [4] settled a longstanding question: Is every completely additive trace proportional to the standard operator trace on the set where it is finite? Dixmier showed that the answer is 'no' by explicitly constructing counter-examples. We start this section by reviewing his result, following Connes [3] in presentation and terminology.

4.1 The spaces $\mathcal{L}^{(1,\infty)}(H)$ and $\mathcal{L}_0^{(1,\infty)}(H)$. Let H be an (infinite-dimensional) Hilbert space, $T \in \mathcal{K}(H)$, and $|T| = (T^*T)^{1/2}$. Let $\mu_0(T) \geq \mu_1(T) \geq \ldots$ be the sequence of eigenvalues of $|T|$, repeated according to their multiplicity. It is well-known that

$$\mu_j(T) = \inf\{\|T - F\| : \operatorname{rank} F = j\} = \min\{\|T|_{E^\perp}\| : \dim E = j\}. \quad (1)$$

We define $\sigma_N(T) = \sum_{j=0}^N \mu_j(T)$ and let $\mathcal{L}^{(1,\infty)}(H) = \{T \in \mathcal{K}(H) : \sigma_N(T) = O(\log N)\}$, endowed with the norm

$$\|T\|_{1,\infty} = \sup_{N\geq 2} \frac{\sigma_N(T)}{\log N}.$$

We have a natural subspace $\mathcal{L}_0^{(1,\infty)}(H) = \{T \in \mathcal{K}(H) : \sigma_N(T) = o(\log N)\}$.

4.2 Lemma. *Let σ_N be as in 4.1, and let T, T_1, and T_2 be compact.*

(a) $\sigma_N(T) = \max\{\|TP_E\|_1 : \dim E = N\}$, *with the $\mathcal{L}^1$-norm $\|\cdot\|$ and P_E denoting the projection on E.*

(b) $\sigma_n(T) = \max\{\operatorname{trace}(TP_E) : \dim E = N\}$ *for $T \geq 0$.*

(c) $\sigma_N(T_1 + T_2) \leq \sigma_N(T_1) + \sigma_N(T_2)$.

(d) $\sigma_N(T_1) + \sigma_N(T_2) \leq \sigma_{2N}(T_1 + T_2)$ *if $T_1, T_2 \geq 0$.*

(e) $\mathcal{L}_0^{(1,\infty)}, \mathcal{L}^{(1,\infty)}$ *are two-sided ideals in $\mathcal{L}(H)$.*

Proof. For (a) use 4.1(1); the maximum is attained by choosing E the eigenspace with respect to the first N eigenvalues of $|T|$. (a) implies (b); the maximum is attained for the same E. (c) is immediate from (b). Since the dimension is subadditive one gets (d). Finally (e) is a consequence of the estimate $\mu_j(TA) \le \mu_j(T)\|A\|$ valid for bounded A. ◁

4.3 Cesàro Mean. We define the Cesàro mean Mf for $f \in L^\infty(\mathbb{R}_+)$ by

$$(Mf)(t) = \frac{1}{\log t}\int_1^t f(s)\frac{ds}{s}.$$

The function Mf is continuous and bounded; $M : L^\infty(\mathbb{R}_+) \to C_b(1,\infty)$ is continuous. Moreover, $M1 = 1$ and $M(f(\lambda\cdot)) - Mf \in C_{b(0)}(1,\infty)$. Here, $\lambda \in \mathbb{C}$, and the subscript 0 indicates that the function vanishes at infinity.

4.4 The 'limit' $\lim_\omega$. Let $T_1, T_2 \in \mathcal{L}^{(1,\infty)}$ be positive and

$$\alpha_N = \frac{\sigma_N(T_1)}{\log N},\quad \beta_N = \frac{\sigma_N(T_2)}{\log N},\quad \gamma_N = \frac{\sigma_N(T_1+T_2)}{\log N}.$$

Then $\{\alpha_N\}, \{\beta_N\}$, and $\{\gamma_N\}$ are bounded sequences. By 4.2 we have

$$\gamma_N \le \alpha_N + \beta_N \le (\log 2N/\log N)\,\gamma_{2N}, \tag{1}$$

but in general no convergence. We embed $\mathcal{L}^\infty$ into $L^\infty(\mathbb{R}_+)$ in the canonical way by associating to the sequence $\{a_N\}$ the function $f_{\{a_N\}}$ which has the value a_j on the interval $[j-1, j[$, $j = 1, 2, \ldots$. Next we choose a linear form ω on $C_b(1,\infty)$ with (i) $\omega \ge 0$, (ii) $\omega(1) = 1$, and (iii) $\omega(f) = 0$ for $f \in C_{b(0)}$. Then we define $\lim_\omega\{a_N\} = \omega(Mf_{\{a\}})$ with the help of Cesàro's mean.

Note that $\lim_\omega$ coincides with the usual limit on convergent sequences by (ii) and (iii).

4.5 Dixmiers trace. For a positive operator $T \in \mathcal{L}^{(1,\infty)}$ let

$$\operatorname{Tr}_\omega(T) = \lim_\omega \frac{1}{\log N}\sum_{n=0}^{N}\mu_n(T).$$

As Proposition 4.6(a) shows, Tr_ω is additive. We can therefore extend it uniquely to a linear map on $\mathcal{L}^{(1,\infty)}$, also denoted Tr_ω.

4.6 Proposition. *Let* $T, T_1, T_2 \in \mathcal{L}^{(1,\infty)}(H), S \in \mathcal{L}(H)$

(a) $\operatorname{Tr}_\omega(T_1+T_2) = \operatorname{Tr}_\omega(T_1) + \operatorname{Tr}_\omega(T_2)$ *for positive* T_1, T_2.

(b) $\operatorname{Tr}_\omega(T) \ge 0$ *if* $T \ge 0$.

(c) *If* S *is invertible, then* $\operatorname{Tr}_\omega(STS^{-1}) = \operatorname{Tr}_\omega(T)$. *In particular,* Tr_ω *is independent of the inner product in* H.

(d) $\operatorname{Tr}_\omega(ST) = \operatorname{Tr}_\omega(TS)$.

(e) $\operatorname{Tr}_\omega = 0$ *on* $\mathcal{L}_0^{(1,\infty)}$, *so it vanishes on trace class operators.*

Proof. (a) follows from 4.4(1) together with property (iii) of ω. We only have to check (c) for positive T. Then use 4.2(b). Finally (c) implies (d). ◁

4.7 Example. Consider the operator $(1-\Delta)^{-n/2} : L^2(\mathbf{T}^n) \to L^2(\mathbf{T}^n)$, where Δ is the Laplacian. The eigenvalues of Δ are known to be the lengths $\|k\|^2$ as k varies over $\mathbb{Z}^n$, so the eigenvalues of $(1-\Delta)^{-n/2}$ are $(1-\|k\|^2)^{-n/2}$.

Let us show that $(1-\Delta)^{-n/2} \in \mathcal{L}^{1,\infty}$ and $Tr_\omega(1-\Delta)^{-n/2} = \Omega_n/n$, independent of ω. Here, $\Omega_n = \mathrm{vol}\,(S^{n-1})$: We let N_R denote the number of lattice points in B_R, the ball of radius R. Clearly, $N_R \sim \mathrm{vol}\, B_R$, hence $\ln N_R \sim n \ln R$. Moreover,

$$\begin{aligned}\sum_{|k|\le R} (1+|k|)^{n/2} &\sim \Omega_n \int_0^R (1+r^2)^{-n/2} r^{n-1} dr \\ &\sim \Omega_n \int_0^R r^{-1} dr = \Omega_n \ln R.\end{aligned}$$

We conclude that

$$(\ln N_R)^{-1} \sum_{|k|\le R} (1+|k|)^{n/2} \sim \frac{\Omega_n \ln R}{n \ln R} = \frac{\Omega_n}{n}.$$

Recall that for $(1-\Delta)^{-n/2} : L^2(\mathbf{T}^n) \to L^2(\mathbf{T}^n)$ we had computed in 1.9 that

$$\mathrm{res}\,(1-\Delta)^{-n/2} = \mathrm{vol}\, S^{n-1} \mathrm{vol}\, \mathbf{T}^n = \Omega_n (2\pi)^n.$$

In one special case we therefore have proven the following result:

4.8 Theorem (Connes 1988). Let M be closed compact, n-dimensional, E a vector bundle over M; $P : L^2(M,E) \to L^2(M,E)$ a pseudodifferential operator of order $-n$. Then

(a) $P \in \mathcal{L}^{(1,\infty)}(L^2(M,E))$

(b) $\mathrm{res}\, P = (2\pi)^n n \mathrm{Tr}_\omega P$, independent of ω.

Proof. We start with the observation that both res and Tr_ω are local: If $\{\varphi_1, \dots, \varphi_J\}$ is a partition of unity on M and if $\{\psi_1, \dots, \psi_J\}$ are smooth functions with $\varphi_j \psi_j = \varphi_j$, then

$$\mathrm{res}\, P = \sum \mathrm{res}\, \varphi_j P \psi_j \text{ and } \mathrm{Tr}_\omega P = \sum \mathrm{Tr}_\omega \varphi_j P \psi_j,$$

since $\mathrm{res}\, \varphi_j P(1-\psi_j) = \mathrm{res}\,(1-\psi_j)\varphi_j P) = 0$, similarly for Tr_ω. Thus we may assume that $M = \mathbf{T}^n$.

Part (a) now follows from writing $P = (P(1-\Delta)^{n/2})(1-\Delta)^{-n/2}$: The first factor on the right hand side is bounded, $(1-\Delta)^{-n/2} \in \mathcal{L}^{1,\infty}$, and $\mathcal{L}^{1,\infty}$ is an ideal.

In order to see (b) we first note that we proved it for $(1-\Delta)^{-n/2}$, see 4.7. By linearity it is enough to consider $T = P + \lambda(1-\Delta)^{-n/2}$ for $P \geq 0$ and large positive λ. In that case, $T : L^2(M,E) \to H^n(M,E)$ is invertible, and $T = A^{-1}$ for a pseudodifferential operator A of order n satisfying the assumptions of Seeley's theorem. By Wodzicki's formula 1.11(2),

$$\frac{\operatorname{res} T}{n(2\pi)^n} = \frac{\operatorname{res} A^{-1}}{(2\pi)^n \operatorname{ord} A} = \operatorname{Res}_{s=1}\zeta_A = -\operatorname{Res}_{s=-1}\zeta_T = \operatorname{Res}_{s=1}\sum_{j=0}^{\infty}\lambda_j^s,$$

where $\lambda_0 \geq \lambda_1 \geq \ldots$ are the eigenvalues of T, so the λ_j^{-1} are the eigenvalues of A.

Let $\lambda_0 \geq \ldots \lambda_{k_0-1} \geq 1 > \lambda_{k_0} \geq \lambda_{k_0+1}$, denote by θ the characteristic function of $\mathbb{R}_+$, and define $\mu(x) = \sum_{k=0}^{\infty}\theta(x + \log\lambda_{k+k_0})$. This is a positive measure. Its Laplace transform is

$$\int_0^\infty e^{-sx}d\mu(x) = \sum_{k=k_0}^{\infty}\lambda_k^s = \zeta_A(s) - \sum_{k=0}^{k_0-1}\lambda_k^s.$$

According to Seeley's result, this function is analytic for $\operatorname{Re} s > 1$ and extends to $\{\operatorname{Re} s < 1-\varepsilon\}$ with a simple pole in $s = 1$. We can therefore apply Ikehara's Tauberian theorem [5, Section 47] and conclude that $\operatorname{Res}_{s=1}\zeta_A(s) = \lim e^{-x}\mu(x) =: c$.

Now $\mu(x) = \sum_{\{k: x \geq -\log\lambda_{k+k_0}\}} 1 = \sum_{\{k: e^{-x} \leq \lambda_{k+k_0}\}} 1$ so that $\mu(x) = j$ iff $\lambda_{k_0+j+1} < e^{-x} \leq \lambda_{k_0+j}$. From this we derive that $j\lambda_{k_0+j+1} < \mu(x)e^{-x} \leq j\lambda_{k_0+j}$, hence $\lambda_{k_0+j} \sim c/j$ with above c. We conclude that $T \in \mathcal{L}^{(1,\infty)}$ and $\operatorname{Tr}_\omega(T) = \lim_{N\to\infty}\sigma_N/\log N = c$. Note that the limit exists and therefore is independent of ω. ◁

4.9 Corollary: Weyl's theorem. Let M be closed, compact, n-dimensional, let Δ be the Laplace–Beltrami operator on M with respect to some Riemannian metric. For the eigenvalues λ_j of $-\Delta$ we then get the asymptotics

$$\lambda_j \sim 4\pi^2\left(\frac{n}{\operatorname{vol} M}\right)^{2/n}\left(\frac{j}{\Omega_n}\right)^{2/n}$$

Proof. We deduce this from the last part of the proof of the previous theorem rather than from the assertion. Consider $(1-\Delta)^{-n/2}$. Its inverse A satisfies the assumptions of Seeley; ζ_A is analytic on $\{\operatorname{Re} s > 1\}$ and extends to a larger half-plane with a simple pole. We know from 1.9 that res $(I - \Delta)^{-n/2} = \Omega_n \operatorname{vol} M$, hence $\operatorname{Tr}_\omega(I-\Delta)^{-n/2} = (2\pi)^{-n}\operatorname{res}(I-\Delta)^{-n/2}/n = (2\pi)^{-n}\Omega_n \operatorname{vol} M/n =: c_n$. An application of Ikehara's Tauberian theorem as above implies that the eigenvalues μ_j of $(I-\Delta)^{-n/2}$ satisfy $\mu_j \sim c_n/j$. The identity $\lambda_j = \mu_j^{-2/n}$ then proves the result. ◁

4.10 Notes and Remarks. The idea of the proof of Theorem 4.8 was adapted from Várilly and Gracia-Bondia [31].

In the article [2], Connes used the coincidence of the noncommutative residue and Dixmier's trace in the following way:

For an algebra $\mathcal{A}$, a p-summable Fredholm module $(\mathcal{H}, F)$ over $\mathcal{A}$, and a finite projective module $\mathcal{E}$ over $\mathcal{A}$ with an $\mathcal{A}$-valued inner product, one can introduce the notion of connections ∇ and curvature θ.

He then considers the case of a 4-dimensional smooth compact Riemannian Spin^c manifold. The Fredholm module $(\mathcal{H}, F)$ consists of the Hilbert space $\mathcal{H}$ of L^2-spinors and $F = D|D|^{-1}$, where D is the Dirac operator. Under a compatibility assumption he can show that the (abstractly defined) curvature θ is an element of $\mathcal{L}^{2,\infty}$ so that the value of the Dixmier trace $\operatorname{Tr}_\omega(\theta^2) = I(\theta)$ defines a positive functional independent of ω.

Moreover, given a classical connection A, the classical Yang-Mills action $YM(A)$ of A can be recovered by

$$YM(A) = 16\pi^2 \inf I(\theta)$$

with the infimum taken over a suitable class of connections related to A.

References

1. M. Adler. On a trace functional for formal pseudo-differential operators and the symplectic structure of Korteweg–de Vries type equations. *Inventiones Math.*, 50:219–248, 1979.
2. A. Connes. The action functional in non-commutative geometry. *Comm. Math. Physics*, 117:673–683, 1988.
3. A. Connes. *Noncommutative Geometry.* Academic Press, New York, London, Tokyo, 1994.
4. J. Dixmier. Existence de traces non normales. *C.R. Acad. Sc. Paris, Série A*, 262:1107–1108, 1966.
5. W. Donoghue. *Distributions and Fourier Transforms.* Academic Press, New York and London 1969.
6. Yu. Egorov and B.-W. Schulze. *Pseudo-Differential Operators, Singularities, Applications.* Birkhäuser, Boston, Basel, Berlin 1997.
7. B.V. Fedosov, F. Golse, E. Leichtnam, and E. Schrohe. Le résidu non commutatif pour les variétés à bord. *C.R. Acad. Sc. Paris, Série I*, 320:669–674, 1995.
8. B.V. Fedosov, F. Golse, E. Leichtnam, and E. Schrohe. The noncommutative residue for manifolds with boundary. *J. Functional Analysis* (to appear).

9. P. Gilkey. *Invariance Theory, the Heat Equation and the Atiyah–Singer Index Theorem*. CRC Press, Boca Raton 1995
10. V. Guillemin. A new proof of Weyl's formula on the asymptotic distribution of eigenvalues. *Advances Math.*, 55:131 – 160, 1985.
11. V. Guillemin. Residue traces for cerain algebras of Fourier integral operators. *J. Functional Analysis*, 115:391–417, 1993.
12. L. Hörmander. *The Analysis of Linear Partial Differential Operators III*. Grundlehren der mathematischen Wissenschaften 274, Springer Verlag, Berlin, Heidelberg, 1985.
13. W. Kalau and M. Walze. Gravity, non-commutative geometry and the Wodzicki residue, *J. Geometry Physics* 16:327–344, 1995.
14. C. Kassel. Le résidu non commutatif [d'apres M. Wodzicki]. *Astérisque*, 177-178:199–229, 1989. Séminaire Bourbaki, 41ème année, Exposé no. 708, 1988-89.
15. D. Kastler. The Dirac operator and gravitation. *Comm. Math. Phys.* 166:633-643, 1995.
16. B.A. Khesin and O.S. Kravchenko. A central extension of the algebra of pseudodifferential symbols. *Functional Analysis and Appl.*, 25:152–154, 1991.
17. M. Kontsevich and V.I. Vishik. Determinants of elliptic pseudo-differential operators. *GAFA*, 1995.
18. H. Kumano-go. *Pseudo-Differential Operators.* The MIT Press, Cambridge, MA, and London 1981.
19. Yu.I. Manin. Algebraic aspects of nonlinear differential equations. J. Sov. Math., 11:1–122, 1979.
20. R. Melrose. *The Atiyah-Patodi-Singer Index Theorem.* A K Peters, Wellesley, MA 1993.
21. R. Melrose and V. Nistor. Homology of pseudodifferential operators I. Manifolds with boundary. Preprint, MIT 1996.
22. B.A. Plamenevskij. *Algebras of Pseudodifferential Operators* (Russ.). Nauka, Moscow 1986.
23. E. Schrohe and B.-W. Schulze. Boundary value problems in Boutet de Monvel's algebra for manifolds with conical singularities I. In: *Pseudodifferential Operators and Mathematical Physics. Advances in Partial Differential Equations* **1**. Akademie Verlag, Berlin, 1994, 97 – 209.
24. E. Schrohe and B.-W. Schulze. Boundary value problems in Boutet de Monvel's algebra for manifolds with conical singularities II. *Boundary Value Problems, Deformation Quantization, Schrödinger Operators. Advances in Partial Differential Equations* **2**. Akademie Verlag, Berlin 1995, 70 - 205.
25. E. Schrohe. Traces on the cone algebra with asymptotics. *Actes des Journées de Saint Jean de Monts, Journées Equations aux Dérivées Partielles* 1996. Ecole Polytechnique, Palaiseau 1996.
26. E. Schrohe. Noncommutative Residues and Manifolds with Conical Singularities. *J. Functional Analysis* (to appear)
27. B.-W. Schulze. *Pseudo-Differential Operators on Manifolds with Singularities.* North-Holland, Amsterdam 1991.
28. B.-W. Schulze. *Pseudo-Differential Boundary Value Problems, Conical Singularities and Asymptotics.* Akademie Verlag, Berlin 1994.
29. L. Schwartz. *Théorie des Distributions.* Hermann, Paris 1966.
30. R. Seeley. Complex powers of an elliptic operator. *Poc. Symp. Pure Math.* 10:288–307, 1967.
31. J.C. Várilly and J.M. Gracia-Bondía. Connes' noncommutative geometry and the Standard model. *J. Geometry Physics* 12:223–301, 1993.

32. M. Wodzicki. *Spectral Asymmetry and Noncommutative Residue.* Thesis, Stekhlov Institute of Mathematics, Moscow, 1984.
33. M. Wodzicki. Noncommutative residue, Chapter I. Fundamentals. In Yu. I. Manin, editor, *K-theory, Arithmetic and Geometry*, volume 1289 of *Springer LN Math.*, pages 320–399. Springer, Berlin, Heidelberg, New York, 1987.

LOWER BOUNDS FOR PSEUDODIFFERENTIAL OPERATORS

CESARE PARENTI*, ALBERTO PARMEGGIANI**
*Dipartimento di Informatica, **Dipartimento di Matematica
Università di Bologna
Piazza di Porta S.Donato 5, 40127 Bologna, ITALIA

1. Introduction and setting of the problem.

We start off by fixing some notation (see Sjöstrand [6]). Let X be an open subset of $\mathbf{R}^n$ (more generally, X can be a C^∞ n-dimensional manifold without boundary) and let $\Sigma \subset T^*(X)\setminus 0 \simeq X\times(\mathbf{R}^n\setminus\{0\})$ be a C^∞ conic submanifold. With $\mu \in \mathbf{R}$ and $h \in \mathbf{Z}_+ = \{0,1,2,\ldots\}$, we denote by $N^{\mu,h}(X,\Sigma)$ *the set of all classical symbols of order* μ, $p(x,\xi) \sim \sum_{j\geq 0} p_{\mu-j}(x,\xi)$, such that for any $j \geq 0$ one has

$$|p_{\mu-j}(x,\xi)| \lesssim |\xi|^{\mu-j}\mathrm{dist}_\Sigma(x,\xi)^{(h-2j)_+},$$

where $t_+ := \max\{t,0\}$ and $\mathrm{dist}_\Sigma(x,\xi)$ denotes the distance of $(x,\xi/|\xi|)$ to $\{(y,\eta) \in \Sigma;\ |\eta| = 1\}$. $\mathrm{OPN}^{\mu,h}(X,\Sigma)$ will then denote the corresponding class of (*properly-supported*) pseudodifferential operators.
Recall that the notation $f \lesssim g$, stands for: *for any conic subset* U *of* $T^*(X)\setminus 0$ *with compact base, there exists a constant* $C_U > 0$, *for which* $f(x,\xi) \leq C_U g(x,\xi)$, $\forall (x,\xi) \in U$.

We say that p (or the corresponding operator P) is **transversally elliptic** (with respect to Σ) iff

$$|\xi|^\mu \mathrm{dist}_\Sigma(x,\xi)^h \lesssim |p_\mu(x,\xi)|.$$

It is useful to bear in mind that

$$A \in \mathrm{OPN}^{\mu,h}(X,\Sigma),\ B \in \mathrm{OPN}^{\mu',h'}(X,\Sigma) \Longrightarrow AB \in \mathrm{OPN}^{\mu+\mu',h+h'}(X,\Sigma),$$

$$A \in \mathrm{OPN}^{\mu,h}(X,\Sigma) \Longrightarrow A^* \in \mathrm{OPN}^{\mu,h}(X,\Sigma).$$

There are two basic objects related to an operator $P \in \mathrm{OPN}^{\mu,h}(X,\Sigma)$.

L. Rodino (ed.), Microlocal Analysis and Spectral Theory, 251–262.

1. The localized polynomial

Let $P \in \mathrm{OPN}^{\mu,h}(X,\Sigma)$ with principal symbol p_μ; let $\rho \in \Sigma$ and $v = (v_x, v_\xi) \in T_\rho(T^*X)$. One defines the *localized polynomial of* P at $\rho \in \Sigma$ by:

$$p_\rho(v) := \sum_{|\alpha|+|\beta|=h} \frac{1}{\alpha!\beta!}(\partial_x^\alpha \partial_\xi^\beta p_\mu)(\rho)(v_x)^\alpha (v_\xi)^\beta.$$

Notice that for $h = 0$, $p_\rho(v) = p_\mu(\rho)$. Furthermore, for $h \geq 1$, p is transversally elliptic iff for every $\rho \in \Sigma$ one has

$$p_\rho(v) \neq 0, \ \forall v \in N_\rho\Sigma = T_\rho(T^*X)/T_\rho\Sigma, \ v \neq 0.$$

The polynomial p_ρ has an *invariant* meaning (see Sjöstrand [6]).

Examples.

For $h = 1$:

$$p_\rho(v) = \langle dp_\mu(\rho), v\rangle = \sigma(v, H_{p_\mu}(\rho)v),$$

where $\sigma = \sum_j d\xi_j \wedge dx_j$ is the canonical symplectic form of T^*X and $H_{p_\mu}(\rho) = (\partial_\xi p_\mu(\rho), -\partial_x p_\mu(\rho))$ is the Hamiltonian vector-field associated to p_μ.

For $h = 2$:

$$p_\rho(v) = \frac{1}{2}\langle \text{Hess } p_\mu(\rho)v, v\rangle = \sigma(v, F(\rho)v),$$

where Hess $p_\mu(\rho)$ is the Hessian-matrix of p_μ at ρ, and

$$F(\rho) = \frac{1}{2} J \text{ Hess } p_\mu(\rho), \ J = \begin{pmatrix} 0 & I_n \\ -I_n & 0 \end{pmatrix},$$

is the *fundamental matrix* of p_μ at ρ.

2. The localized differential operator

The second main object is a differential operator with polynomial coefficients attached at $\rho \in \Sigma$, for every $P \in \mathrm{OPN}^{\mu,h}(X,\Sigma)$. More precisely, define

$$P_\rho(y, D_y) := \sum_{|\alpha|+|\beta|+2j=h} \frac{1}{\alpha!\beta!}(\partial_x^\alpha \partial_\xi^\beta p_{\mu-j})(\rho) y^\alpha D_y^\beta, \ y \in \mathbf{R}^n$$

($D_y = \partial_y/\sqrt{-1}$). It is well known that the operator P_ρ has an "invariant meaning" (see Sjöstrand [6], Boutet-Grigis-Helffer [1]). Notice that the localized polynomial p_ρ is the "principal symbol" of P_ρ.

We can now state the problem we are dealing with.

Let $A = A^* \in \mathrm{OPN}^{2m,2k}(X,\Sigma)$. Find necessary and/or sufficient conditions on A (and Σ) in order to have the following lower bound: for every compact $K \subset X$, there exists $C_K > 0$ such that

$$(Au,u) \geq -C_K \|u\|^2_{m-\frac{k+1}{2}},\ \forall u \in C_0^\infty(K), \tag{1}$$

where $(\ ,\ )$ denotes the L^2-scalar product and $\|\cdot\|_s$ ($s \in \mathbf{R}$) is the usual H^s-Sobolev norm.

Some remarks are in order to justify the interest in studying the above inequality.

1. The validity of inequality (1) depends **only** on the first $k+1$ terms $a_{2m}, a_{2m-1}, \ldots, a_{2m-k}$ of the symbol of A. In other words, inequality (1) is stable under perturbations of A with operators in $\mathrm{OPS}^{2m-(k+1)}(X)$. In this sense the exponent $m - \frac{k+1}{2}$ is *sharp*.

2. We point out that possible applications are about the maximal hypoellipticity, the well-posedness of the Cauchy problem for weakly hyperbolic operators, just to mention a few of them (see, e.g., Helffer-Nourrigat [2], Hörmander [3], Parmeggiani [5]).

2. Necessary conditions.

We have the following necessary conditions

Proposition 2.1 *Suppose that (1) holds. Then:*

(i) $$a_{2m}(x,\xi) \geq 0,\ \forall (x,\xi) \in T^*(X)\setminus 0;$$

when $k \geq 1$, *for any* $\rho \in \Sigma$:

(ii) $$(A_\rho(y,D_y)v,v) \geq 0,\ \forall v \in C_0^\infty(\mathbf{R}^n).$$

Proof. We will sketch only the proof of *(ii)*. Let $\Sigma \ni \rho = (x_0,\xi_0)$, $|\xi_0| = 1$, and fix any compact neighborhood K of x_0. Let $K' \subset\subset K$ be such that $\mathrm{supp}\,(Au) \subset K'$ if $u \in C_0^\infty(K)$. Take $\chi \in C_0^\infty(X)$ with $\chi \equiv 1$ near $K \cup K'$, so that $Au = \chi A(\chi u) = a(x,D)$, for $u \in C_0^\infty(K)$. For $v \in C_0^\infty(\mathbf{R}^n)$ and $t \geq 1$, put

$$u_t(x) = e^{it^2\langle x,\xi_0\rangle} v(t(x-x_0)).$$

For t large, we have $u_t \in C_0^\infty(K)$ and

$$\hat{u}_t(\xi) = t^{-n} e^{i\langle x_0, t^2\xi_0 - \xi\rangle} \hat{v}(\frac{\xi}{t} - t\xi_0).$$

Hence, as $t \to +\infty$, we have

$$\|u_t\|^2_{m-\frac{k+1}{2}} = t^{4(m-\frac{k+1}{2})-n}\Big(\|v\|_0^2 + o(1)\Big).$$

On the other hand,

$$Au_t(x) = e^{it^2\langle x,\xi_0\rangle}\phi_t(t(x-x_0)),$$

with

$$\phi_t(x) = (2\pi)^{-n}\int e^{i\langle x,\eta\rangle}a(x_0+x/t, t\eta+t^2\xi_0)\hat{v}(\eta)d\eta.$$

Taylor expanding then yields

$$a(x_0+\frac{x}{t}, t\eta+t^2\xi_0) = t^{4(m-\frac{k}{2})}\sum_{|\alpha|+|\beta|+2j=2k}\frac{1}{\alpha!\beta!}(\partial_x^\alpha\partial_\xi^\beta a_{2m-j})(\rho)x^\alpha\eta^\beta + O(t^{4(m-\frac{k+1}{2})}).$$

Hence, by putting $t(x-x_0)=y$, we get

$$(Au_t, u_t) = t^{4(m-\frac{k}{2})-n}(A_\rho(y,D_y)v,v) + O(t^{4(m-\frac{k+1}{2})-n}) \geq$$

$$\geq -C_K t^{4(m-\frac{k+1}{2})-n}(\|v\|_0^2 + o(1)).$$

Upon dividing by $t^{4(m-k/2)-n}$, and letting $t\to+\infty$, we obtain (ii). •

Remark 2.2 *Conditions (i) and (ii) in the above proposition, being only pointwise, are clearly holding true* **without** *supposing that Σ is a C^∞-manifold.*

Sufficiency is in general not true under the sole condition (ii). (From now on we will take $k\geq 1$.)

In the double characteristics case (i.e. $k=1$), the situation has been greatly clarified by Hörmander [3], who proved the following theorem:

Theorem 2.3 *(Hörmander [3]) Let $A = A^* \in \mathrm{OPN}^{2m,2}(X,\Sigma)$, and suppose:*

(a) A is transversally elliptic,

(b) $\mathrm{rank}\,\sigma\big|_\Sigma$ is constant (i.e. $\dim\left(T_\rho\Sigma\cap(T_\rho\Sigma)^\sigma\right)$ is constant as ρ varies in the connected components of Σ).

Then the lower bound (1) (with $k=1$) holds iff

(h1) $a_{2m}\geq 0$ on $T^*(X)\setminus 0$;

(h2) $\mathrm{sub}(A)(\rho) + \mathrm{Tr}^+F_A(\rho)\geq 0,\ \forall\rho\in\Sigma$,

where $\mathrm{sub}(A) = a_{2m-1} + i\langle\partial_x,\partial_\xi\rangle a_{2m}/2$ is the **subprincipal symbol** *of A, F_A is the fundamental matrix of a_{2m} and $\mathrm{Tr}^+F_A(\rho) = \sum_{\mu>0;i\mu\in\mathrm{spec}(F_A(\rho))}\mu$ is the* **positive trace** *of $F_A(\rho)$.*

Using Weyl-calculus, it can be easily shown that condition $(h2)$ is equivalent to the non-negativity of the localized operator $A_\rho(y,D_y)$.

We now briefly recall the proof of sufficiency in Hörmander's theorem, since we shall need it later on (for the details, see [3]).

By a standard argument, we can suppose $A \in \mathrm{OPN}^{2,2}(X,\Sigma)$. At any fixed $\rho \in \Sigma$, we can find a symplectic basis in $T_\rho(T^*X)$ so that, with v having coordinates (x,ξ),

$$a_\rho(v) = \sigma(v, F_A(\rho)v) = \sum_{j=1}^{d} \mu_j(x_j^2+\xi_j^2) + \sum_{k=d+1}^{d+l} x_k^2 \quad (\mu_j > 0, j = 1,\dots,d).$$

The integers d, l are uniquely determined (because of the assumptions (a) and (b)) by

$$2d + l = \operatorname{codim} \Sigma, \; 2n - 2(d+l) = \operatorname{rank} \sigma\Big|_{\Sigma}.$$

Unfortunately, we cannot expect the above normal form to hold smoothly in ρ varying in Σ. However, we have the following (smooth) fiber-bundle decomposition of the normal bundle to Σ :

$$N\Sigma = \operatorname{Im}(F_A^2) \oplus \Big(\operatorname{Ker}(F_A^2)/T\Sigma\Big).$$

We write (as smooth fiber bundles on Σ) :

$$\mathbf{C} \otimes \Big(\operatorname{Im}(F_A^2)\Big) = V \oplus \bar{V},$$

with

$$V(\rho) = \bigoplus_{\mu(\rho)>0} \operatorname{Ker}\Big(F_A(\rho) - i\mu(\rho)\Big), \; \rho \in \Sigma.$$

We can choose smooth vector-fields on Σ :

$$\begin{cases} v_j(\rho) \in V(\rho), & j = 1,\dots,d \\ v_{d+k}(\rho) \in \operatorname{Ker}\Big(F_A(\rho)^2\Big)/T_\rho\Sigma, & k = 1,\dots,l, \end{cases}$$

so that for any $\rho \in \Sigma$:

$$\begin{cases} \sigma\Big(\overline{v_j(\rho)}, F_A(\rho)v_k(\rho)\Big) = 2\delta_{jk}, & j,k = 1,\dots,d \\ \sigma\Big(v_{d+j}(\rho), F_A(\rho)v_{d+k}(\rho)\Big) = \delta_{jk}, & j,k = 1,\dots,l. \end{cases}$$

Now, using Morse Lemma, one can find, in a conic neighborhood of a point $\rho \in \Sigma$, $2d + l$ smooth real-valued functions $f_1,\dots,f_{2d+l}$, homogeneous of degree 1 in ξ, with independent differentials, in such a way that

$$a_2(x,\xi) = \sum_{j=1}^{2d+l} f_j(x,\xi)^2.$$

Moreover, upon defining

$$\begin{cases} X_j = f_{2j-1} + \sqrt{-1} f_{2j}, & j = 1, \dots, d \\ X_{j+d} = f_{2d+j}, & j = 1, \dots, l, \end{cases}$$

we have

$$H_{X_j}(\rho) = F_A(\rho) v_j(\rho), \ j = 1, \dots, d + l.$$

As a consequence (microlocally),

$$A - \sum_{j=1}^{d+l} X_j(x, D)^* X_j(x, D) = B \in \mathrm{OPS}^1.$$

Furthermore, the principal symbol of B is given by

$$b_1(\rho) = \mathrm{sub}(A)(\rho) - \frac{1}{2\sqrt{-1}} \sum_{j=1}^{d} \sigma(H_{\bar{X}_j}(\rho), H_{X_j}(\rho)), \ \rho \in \Sigma.$$

To compute the Poisson brackets at a **fixed** ρ, we use the above normal form of $F_A(\rho)$. Define

$$w_j = \begin{pmatrix} -\frac{\sqrt{-1}}{\sqrt{\mu_j}} e_j \\ \\ \frac{1}{\sqrt{\mu_j}} e_j \end{pmatrix} \in V(\rho), \ j = 1, \dots, d$$

($e_j = (0, \dots, \overset{j}{1}, \dots, 0) \in \mathbf{R}^d$). Since $F_A(\rho) w_j = \sqrt{-1} \mu_j w_j$, we get

$$\sigma(\bar{w}_j, F_A(\rho) w_k) = \sqrt{-1} \mu_k \sigma(\bar{w}_j, w_k) = 2\delta_{jk}, \ j, k = 1 \dots, d.$$

Hence, $v_j(\rho) = \sum_{k=1}^{d} U_{jk} w_k$, $j = 1, \dots, d$, for some unitary $d \times d$ matrix $U = (U_{jk})_{j,k}$. Therefore,

$$\sum_{j=1}^{d} \sigma(H_{\bar{X}_j}(\rho), H_{X_j}(\rho)) = \sum_{j=1}^{d} \sigma\Big(F_A(\rho)\overline{v_j(\rho)}, F_A(\rho) v_j(\rho)\Big) =$$

$$= \sum_{j=1}^{d} \sum_{k,h=1}^{d} \bar{U}_{jk} U_{jh} \mu_k \mu_h \sigma(\bar{w}_k, w_h) =$$

$$= -2\sqrt{-1} \sum_{j=1}^{d} \sum_{k=1}^{d} \bar{U}_{jk} U_{jk} \mu_k = -2\sqrt{-1} \mathrm{Tr}^+ F_A(\rho).$$

In conclusion,

$$b_1(\rho) = \mathrm{sub}(A)(\rho) + \mathrm{Tr}^+ F_A(\rho),$$

which, by hypothesis, is non-negative on Σ. By a suitable 0th-order modification of $X_j(x, D)$, we can suppose $b_1 \geq 0$ in a conic neighborhood of $\rho \in \Sigma$. The Sharp Gårding Estimate finally yields (microlocally)

$$(Au, u) = \sum_{j=1}^{d+l} \|X_j u\|_0^2 + (Bu, u) \geq (Bu, u) \geq -C\|u\|_0^2.$$

One can then patch together the above microlocal estimates . •

From now on, we will suppose $A = A^* \in \mathrm{OPN}^{2m,2k}(X, \Sigma)$, *transversally elliptic* with respect to Σ, and $\mathrm{rank}\sigma\big|_{\Sigma}$ =constant. As far as we know, for $k \geq 2$, no result as complete as the above Hörmander's theorem is available. In particular, we do not know how to get the lower bound (1), even from a strengthened form of the necessary condition, namely

$$\forall \rho \in \Sigma: \ (A_\rho(y, D_y)v, v) > 0, \ \forall v \in C_0^\infty(\mathbf{R}^n), v \neq 0.$$

Note that the above condition implies (see Boutet-Grigis-Helffer [1]) the microlocal hypoellipticity with loss of k derivatives for A. As the following example shows, we need some extra-information on the "spectrum" of the localized operator A_ρ in order to get the lower bound.

3. An example.

Suppose $2m = 2k = 4$, and that the localized polynomial $a_\rho(v)$ is the product of two non-negative quadratic forms:

$$a_\rho(v) = \sigma(v, F_1(\rho)v)\ \sigma(v, F_2(\rho)v),$$

for which

$$\mathrm{Ker}F_1(\rho) = \mathrm{Ker}F_2(\rho) = T_\rho\Sigma, \ \forall \rho \in \Sigma,$$

and

$$[F_1(\rho), F_2(\rho)] = 0, \ \forall \rho \in \Sigma.$$

Hence, without loss of generality, we can suppose

$$A = PQ + R,$$

with $P = P^*$, $Q = Q^* \in \mathrm{OPN}^{2,2}(X, \Sigma)$, $p_2, q_2 \geq 0$ on $T^*(X) \setminus 0$ and transversally elliptic with respect to Σ, $F_1 = F_P$, $F_2 = F_Q$, and $R \in \mathrm{OPN}^{3,2}(X, \Sigma)$. Since $A = A^*$, we get

$$R - R^* = [Q, P], \ \text{ and } \ F_{R-R^*} = \frac{1}{\sqrt{-1}}[F_P, F_Q] = 0.$$

Therefore $R - R^* \in \mathrm{OPN}^{3,3}(X, \Sigma)$ and, in particular, F_R is *real*. We next suppose

$$[F_R(\rho), F_P(\rho)] = [F_R(\rho), F_Q(\rho)] = 0, \ \forall \rho \in \Sigma.$$

Our last assumption will be

$$\mathrm{sub}(P)\Big|_\Sigma = \mathrm{sub}(Q)\Big|_\Sigma = 0,$$

which is no retriction since it can be readily achieved by suitably modifying R, leaving all the commutation relations unchanged.

A standard argument shows that, for any fixed $\rho \in \Sigma$, in suitable symplectic coordinates, one can write

$$\sigma(v, F_P(\rho)v) = \sum_{j=1}^{d} \mu_j(\rho)(x_j^2 + \xi_j^2) + \sum_{k=d+1}^{d+l} x_k^2,$$

$$\sigma(v, F_Q(\rho)v) = \sum_{j=1}^{d} \lambda_j(\rho)(x_j^2 + \xi_j^2) + \sum_{k=d+1}^{d+l} \alpha_k(\rho)x_k^2,$$

$$\sigma(v, F_R(\rho)v) = \sum_{j=1}^{d} \omega_j(\rho)(x_j^2 + \xi_j^2) + \sum_{k=d+1}^{d+l} \beta_k(\rho)x_k^2,$$

with $\mu_j, \lambda_j, \alpha_k > 0$, and $\omega_j, \beta_k \in \mathbf{R}$.

Using Weyl-calculus, it is not difficult to show that the necessary condition $(A_\rho(y, D_y)v, v) \geq 0$, for any $\rho \in \Sigma$, for any $v \in C_0^\infty(\mathbf{R}^n)$, is equivalent to the following algebraic requirement:

$$F_\rho(\zeta, y) := \Big(\langle\mu, \zeta\rangle + \langle y, \vec{1}\rangle + \mathrm{Tr}^+ F_P(\rho)\Big)\Big(\langle\lambda, \zeta\rangle + \langle\alpha, y\rangle + \mathrm{Tr}^+ F_Q(\rho)\Big) + \langle\omega, \zeta\rangle + \tag{2}$$

$$+\langle\beta, y\rangle + \sum_{j=1}^{d} \omega_j + \mathrm{Re}\,\mathrm{sub}(R)(\rho) \geq 0,$$

$\forall \rho \in \Sigma$, $\forall \zeta \in (2\mathbf{Z}_+)^d$, $\forall y \in \bar{\mathbf{R}}^l_+$, where $\langle\mu, \zeta\rangle = \sum_{j=1}^{d} \mu_j \zeta_j$, etc., $\vec{1} = (1, 1, \ldots, 1) \in \bar{\mathbf{R}}^l_+$.

We now microlocally work near any fixed point of Σ. By using the symbols X_j, $j = 1, \ldots, d + l$ constructed before, we can write

$$P = \sum_{j=1}^{d+l} X_j^* X_j + \mathrm{OPS}^1,$$

$$Q = \sum_{j,k=1}^{d+l} X_j^* A_{jk} X_k + \mathrm{OPS}^1,$$

where $A_{jk} = A^*_{kj} \in \mathrm{OPS}^0$, and the matrix $(\sigma_0(A_{jk}))_{j,k}$ is *positive*. Let $(C_{jk})_{j,k}$ be a positive square root of $(A_{jk})_{j,k}$, and define

$$Y_j = \sum_{j,k=1}^{d+l} C_{jk}X_k,\ j = 1,\ldots,d+l,$$

so that $Q = \sum_{j=1}^{d+l} Y_j^*Y_j + \mathrm{OPS}^1$. We now define

$$B = \sum_{j=1}^{d+l} Y_j^* X_j.$$

It turns out that $B \in \mathrm{OPN}^{2,2}(X,\Sigma)$ is *transversally elliptic* and in some coordinates as above

$$\sigma(v, F_B(\rho)v) = \sum_{j=1}^{d} \sqrt{\lambda_j(\rho)\mu_j(\rho)}(x_j^2 + \xi_j^2) + \sum_{k=d+1}^{d+l} \sqrt{\alpha_k(\rho)}x_k^2,$$

so that B looks like a "square root" of PQ. More precisely, let us compute

$$\tilde{A} := \Big(\sum_{j=1}^{d+l} X_j^*X_j\Big)\Big(\sum_{k=1}^{d+l} Y_k^*Y_k\Big) - B^*B.$$

It is not difficult to show that

$$\tilde{A} = L + \sum_{j,k=1}^{d} \Big(X_j^*[X_j, Y_k^*]Y_k - X_j^*[Y_j, Y_k^*]X_k\Big) + \mathrm{OPN}^{3,3}(X,\Sigma),$$

where $L = L^* \in \mathrm{OPN}^{4,4}(X,\Sigma)$ and $(Lu, u) \geq 0$, for any $u \in C_0^\infty$. Thanks to the hypothesis $[F_P, F_Q] = 0$, one can show that

$$\sum_{j,k=1}^{d} \Big(X_j^*[X_j, Y_k^*]Y_k - X_j^*[Y_j, Y_k^*]X_k\Big) \in \mathrm{OPN}^{3,3}(X,\Sigma).$$

In conclusion,

$$A = PQ + R = L + B^*B + \tilde{R},$$

with $L \geq 0$ as above, $\tilde{R} \in \mathrm{OPN}^{3,2}(X,\Sigma)$ and

$$F_{\tilde{R}} = F_R + (\mathrm{Tr}^+ F_P)F_Q + (\mathrm{Tr}^+ F_Q)F_P,$$

$$\mathrm{sub}(\tilde{R}) = \mathrm{sub}(R) - (\mathrm{Tr}^+ F_P)(\mathrm{Tr}^+ F_Q),$$

on Σ. We now disregard L and work with $B^*B+\tilde{R}$ only. To get a lower bound for $B^*B+\tilde{R}$, we use a deformation argument analogous to Hörmander's one (see [3]). Namely, we look for $C=C^*\in \mathrm{OPS}^1$ such that, writing

$$B^*B+\tilde{R}=(B-C)^*(B-C)+\tilde{R}+B^*C+CB-C^2,$$

and observing that $(B-C)^*(B-C)\geq 0$, we can apply Hörmander's theorem to the operator

$$\hat{A}_C:=\tilde{R}+B^*C+CB-C^2\in \mathrm{OPN}^{3,2}(X,\Sigma).$$

Of course, we need only construct $c=\sigma_1(C)$ on Σ. To this purpose, we need (on Σ) the following:

$$\sigma(v,F_{\hat{A}_C}v)=\sum_{j=1}^{d}\Big(\omega_j+2c\sqrt{\lambda_j\mu_j}+(\mathrm{Tr}^+F_P)\lambda_j+(\mathrm{Tr}^+F_Q)\mu_j\Big)(x_j^2+\xi_j^2)+$$
$$+\sum_{k=d+1}^{d+l}\Big(\beta_k+2c\sqrt{\alpha_k}+(\mathrm{Tr}^+F_P)\alpha_k+\mathrm{Tr}^+F_Q\Big)x_k^2>0,\ \forall v\in N_\rho\Sigma\setminus\{0\}, \tag{3}$$

and

$$\mathrm{Re\ sub}(\hat{A}_C)+\mathrm{Tr}^+F_{\hat{A}_C}=\mathrm{Re\ sub}(R)-2c\sum_{j=1}^{d}\sqrt{\lambda_j\mu_j}-(\mathrm{Tr}^+F_P)(\mathrm{Tr}^+F_Q)-c^2+$$
$$+\sum_{j=1}^{d}\Big(\omega_j+2c\sqrt{\lambda_j\mu_j}+(\mathrm{Tr}^+F_P)\lambda_j+(\mathrm{Tr}^+F_Q)\mu_j\Big)\geq 0. \tag{4}$$

It is now convenient to introduce the following quantity:

$$\gamma(\rho):=\max\Big\{\max_{1\leq j\leq d}\Big(-\frac{\omega_j(\rho)+(\mathrm{Tr}^+F_P(\rho))\lambda_j(\rho)+(\mathrm{Tr}^+F_Q(\rho))\mu_j(\rho)}{2\sqrt{\lambda_j(\rho)\mu_j(\rho)}}\Big),$$
$$\max_{1\leq k\leq l}\Big(-\frac{\beta_k(\rho)+(\mathrm{Tr}^+F_P(\rho))\alpha_k(\rho)+\mathrm{Tr}^+F_Q(\rho)}{2\sqrt{\alpha_k(\rho)}}\Big),0\Big\}.$$

Observe that the function $\Sigma\ni\rho\mapsto\gamma(\rho)\geq 0$ is (in general) just continuous. Moreover, define

$$J(\rho)=\mathrm{Re\ sub}(R)(\rho)+\sum_{j=1}^{d}\omega_j(\rho)+(\mathrm{Tr}^+F_P(\rho))(\mathrm{Tr}^+F_Q(\rho))-\gamma(\rho)^2.$$

It can be readily seen that (with obvious notations)

$$F_\rho(\zeta,y) = J(\rho)+$$
$$+\Big\langle \omega(\rho) + (\mathrm{Tr}^+ F_P(\rho))\lambda(\rho) + (\mathrm{Tr}^+ F_Q(\rho))\mu(\rho) + 2\gamma(\rho)\sqrt{\lambda(\rho)\mu(\rho)}, \zeta\Big\rangle+$$
$$+\Big\langle \beta(\rho) + (\mathrm{Tr}^+ F_P(\rho))\alpha(\rho) + (\mathrm{Tr}^+ F_Q(\rho))\vec{1} + 2\gamma(\rho)\sqrt{\alpha(\rho)}, y\Big\rangle+$$
$$+\Big(\gamma(\rho) - \langle\sqrt{\lambda(\rho)\mu(\rho)},\zeta\rangle - \langle\sqrt{\alpha(\rho)},y\rangle\Big)^2+$$
$$+\Big\langle\begin{pmatrix}\mu(\rho)\\ \vec{1}\end{pmatrix},\begin{pmatrix}\zeta\\ y\end{pmatrix}\Big\rangle\Big\langle\begin{pmatrix}\lambda(\rho)\\ \alpha(\rho)\end{pmatrix},\begin{pmatrix}\zeta\\ y\end{pmatrix}\Big\rangle - \Big\langle\begin{pmatrix}\sqrt{\lambda(\rho)\mu(\rho)}\\ \sqrt{\alpha(\rho)}\end{pmatrix},\begin{pmatrix}\zeta\\ y\end{pmatrix}\Big\rangle^2.$$

By virtue of the Cauchy-Schwarz inequality and the definition of γ, one has

$$J(\rho) = \min_{(\zeta,y)\in\bar{\mathbf{R}}_+^d\times\bar{\mathbf{R}}_+^l} F_\rho(\zeta,y).$$

If we have that $J(\rho) > 0$, $\forall\rho\in\Sigma$, then we can obviously construct a **smooth** symbol $c(\rho) > \gamma(\rho)$ for which conditions (3) and (4) (**in strict form**) are satisfied. To get $J(\rho) > 0$ we require:

$(H1)$ $$F_\rho(\zeta,y) > 0,\ \forall\zeta\in(2\mathbf{Z}_+)^d,\ \forall y\in\bar{\mathbf{R}}_+^l,$$

(necessary condition (2) in strict form) and

$(H2)$ $$\min_{(\zeta,y)\in(2\mathbf{Z}_+)^d\times\bar{\mathbf{R}}_+^l} F_\rho(\zeta,y) = \min_{(\zeta,y)\times\bar{\mathbf{R}}_+^d\times\bar{\mathbf{R}}_+^l} F_\rho(\zeta,y).$$

Let us clarify the meaning of condition $(H2)$ above. There are two cases in which $(H2)$ is a trivial consequence of $(H1)$, namely when either $\gamma(\rho) = 0$, in which case $J(\rho) = F_\rho(0,0)$, or

$$0 < \gamma(\rho) = -\frac{\beta_k(\rho) + (\mathrm{Tr}^+ F_P(\rho))\alpha_k(\rho) + \mathrm{Tr}^+ F_Q(\rho)}{2\sqrt{\alpha_k(\rho)}},$$

for some $k\in\{1,\ldots,l\}$, in which case

$$J(\rho) = F_\rho(\zeta = 0, y = \frac{\gamma(\rho)}{\sqrt{\alpha_k(\rho)}}e_k).$$

The troublesome case is when

$$0 < \gamma(\rho) = -\frac{\omega_j(\rho) + (\mathrm{Tr}^+ F_P(\rho))\lambda_j(\rho) + (\mathrm{Tr}^+ F_Q(\rho))\mu_j(\rho)}{2\sqrt{\lambda_j(\rho)\mu_j(\rho)}},$$

for some $j \in \{1, \ldots, d\}$, and

$$\gamma(\rho) > \max_{1 \leq k \leq l} \Big(- \frac{\beta_k(\rho) + (\mathrm{Tr}^+ F_P(\rho))\alpha_k(\rho) + \mathrm{Tr}^+ F_Q(\rho)}{2\sqrt{\alpha_k(\rho)}} \Big).$$

If this case occurs, the subset of $\bar{\mathbf{R}}_+^d \times \{0\} \subset \bar{\mathbf{R}}_+^d \times \bar{\mathbf{R}}_+^l$ where F_ρ attains its minimum $J(\rho)$ may have empty intersection with $(2\mathbf{Z}_+)^d \times \{0\}$, as it can be seen by trivial examples showing that $J(\rho) < 0$, while condition $(H1)$ is still satisfied.
Hence condition $(H2)$ is highly non-trivial exactly in this case. Notice that it can be spelt out as a "lattice" relation among the ω_j, λ_j and μ_j's.
In conclusion, we have the sought for lower bound for the operator A considered above under conditions $(H1)$ and $(H2)$.

Remark 3.1 *1) We are still unable to fill in the gap between the necessary condition*

$$F_\rho(\zeta, y) \geq 0, \ \ (\rho \in \Sigma, \zeta \in (2\mathbf{Z}_+)^d, y \in \bar{\mathbf{R}}_+^l)$$

and the sufficient conditions $(H1)$ and $(H2)$.

2) There are at least two kinds of objections to the example we have considered here.
First of all, the heavy requirements on the product form of the localized polynomial and the commutativity of the involved fundamental matrices. All this was somehow forced by the need of very precise informations on the "spectrum" of $A_\rho(y, D_y)$, required by our approach.
Secondly, in our proof of sufficiency, we completely threw away the "possible" contributions of the 4th-order non-negative terms L and $(B-C)^*(B-C)$. However, it is not clear (al least to us) how to take advantage of them.

References

1. L.Boutet de Monvel-A.Grigis-B.Helffer. Paramétrixes D'Opérateurs Pseudo-Différentiels a Charactéristiques Multiples. *Astérisque* **34-35**, 1976.
2. B.Helffer and J.Nourrigat. *Hypoellipticité Maximale pour des Opérateurs Polynomes de Champs de Vecteurs.* Birkhäuser, 1985.
3. L.Hörmander. The Cauchy Problem for Differential Equations with Double Characteristics. *Journal D'Analyse Mathématique,* **Vol.32**, 1977.
4. C.Parenti and A.Parmeggiani. A Necessary and Sufficient Condition fo a Lower Bound for 4th-Order Pseudodifferential Operators. To appear in *Journal D'Analyse Mathématique.*
5. A.Parmeggiani. An Application of the Almost-Positivity of a Class of 4th-Order Pseudodifferential Operators. Preprint (1995).
6. J.Sjöstrand. Parametrices for Pseudodifferential Operators with Multiple Characteristics. *Arkiv för Matematik* **12**, 1974.

WEYL FORMULA FOR GLOBALLY HYPOELLIPTIC OPERATORS IN $\mathbf{R}^n$

ERNESTO BUZANO
Dipartimento di Matematica, Università di Torino
Via Carlo Alberto 10, 10123 Torino, Italy

1. Introduction

It is well-known that the spectrum of the *harmonic oscillator* in $\mathbf{R}^n$:

$$-\Delta + \|x\|^2 = \sum_{j=1}^{n} D_j^2 + \sum_{j=1}^{n} x_j^2,$$

is given by a sequence of eigenvalues

$$\lambda_\alpha = \sum_{j=1}^{n}(2\alpha_j + 1), \qquad \alpha \in \mathbf{N}^n,$$

($\mathbf{N} = \{0, 1, \ldots\}$), to each one of which there corresponds a single eigenfunction so that the eigenvalues are given by $2k + n$, $k \in \mathbf{N}$. Each $2k + n$ is semi-simple with multiplicity $\binom{n-1+k}{n-1}$.

The asymptotic behavior of λ_α as $|\alpha| \to \infty$, i.e. of the eigenvalues $2k+n$ repeated according to their multiplicities, can be easily deduced from the one of the counting function:

$$N(\lambda) = \sum_{\lambda_\alpha \le \lambda} 1.$$

Because $\lambda_\beta \le \lambda_\alpha$ whenever $\beta \le \alpha$ (i.e. $\beta_i \le \alpha_i$ for $i = 1, \ldots, n$), we have that $N(\lambda)$ is the number of points of $\mathbf{Z}^n$ with odd positive co-ordinates which belong to the n-simplex of side λ. This means that $N(\lambda)$ is asymptotically equivalent to the volume of the n-simplex of side $\lambda/2$. Thus we obtain

$$N(\lambda) = \frac{(\lambda/2)^n}{n!} + o(\lambda^n) = \frac{\sigma_n}{n(2\pi)^n} \int \left[\lambda - \|x\|^2\right]_+^{n/2} dx + o(\lambda^n), \qquad (1)$$

L. Rodino (ed.), Microlocal Analysis and Spectral Theory, 263–306.

as $\lambda \to \infty$, where σ_n is the area of the unit sphere in $\mathbf{R}^n$ and $[\lambda - \|x\|^2]_+$ is the positive part of $\lambda - \|x\|^2$.

In this paper all the asymptotic formulas concerning λ are always for $\lambda \to +\infty$. From now on we shall omit sistematically the sentence "as $\lambda \to +\infty$". Moreover, in order to simplify the notation we employ the following conventions. Given two functions $f, g : \mathbf{R}^n \to \mathbf{R}$, we say that

$$f(x) \prec g(x)$$

if there exists a positive constant C such that

$$f(x) \leq Cg(x), \qquad \text{for all } x.$$

Furthermore, we write

$$R \gg 0$$

to mean that R is a positive, *conveniently large* constant. Thus by

$$f(x) \prec g(x), \qquad \text{for } \|x\| \geq R \gg 0,$$

we mean that we can choose R large enough so that there exists a positive constant C such that

$$f(x) \leq Cg(x), \qquad \text{for all } \|x\| \geq R.$$

Of course, one should pay attention to the order of the logical quantifiers. For example, if we write that for all $\alpha \in \mathbf{N}^n$ we have

$$|\partial^\alpha f(x)| \prec \|x\|^{-|\alpha|}, \qquad \text{for } \|x\| \geq R \gg 0,$$

we mean that we can choose R large enough so that for each $\alpha \in \mathbf{N}^n$ there exists a positive constant C_α such that

$$|\partial^\alpha f(x)| \leq C_\alpha \|x\|^{-|\alpha|}, \qquad \text{for } \|x\| \geq R.$$

Equation (1) is a special case of the Weyl formula, we now explain. Let us consider the Schrödinger operator with real potential $W(x)$ in $\mathbf{R}^n$:

$$-\Delta + W(x) \tag{2}$$

and assume that $W(x) \to \infty$ as $\|x\| \to \infty$. Under these conditions the spectrum of (2) consist of a sequence of real semi-simple eigenvalues λ_j diverging to $+\infty$. Let us assume that the sequence $\{\lambda_j\}$ is arranged in increasing order and that the eigenvalues are repeated according to their multiplicity. Then we can define the *counting function*

$$N(\lambda) = \sum_{\lambda_j \leq \lambda} 1.$$

The Weyl formula for the counting function of the operator (2) is

$$N(\lambda) = V(\lambda)\left(1 + \mathcal{O}(\lambda^{-\epsilon})\right), \tag{3}$$

where ϵ is a suitable positive constant and

$$V(\lambda) = \frac{\sigma_n}{n(2\pi)^n} \int [\lambda - W(x)]_+^{n/2}\, dx$$

is called *Weyl term.*

The first result about (3) we want to mention is due to Tulovskiĭ and Shubin [14] in 1973, which proved the Weyl formula with $\epsilon < 1/2$ for potentials W which are elliptic polynomials of second order, i.e. such that

$$\|x\|^2 \prec W(x) \prec \|x\|^2, \qquad \text{for } \|x\| \geq R \gg 0.$$

This result has been improved by Hörmander [8] in 1979, which obtained $\epsilon < 2/3$, and by Helffer and Robert [6] in 1981, which obtained the optimal error estimate $\epsilon = 1$. The Weyl formula has been extended by Tamura [12] to elliptic potentials of order $m \geq 1$ and by Helffer and Robert [7] and by Mohamed [10] to *quasi-elliptic potentials*, which satisfy some further hypothesis. A smooth function W is quasi-elliptic if there exist n positive constants $m_1, \ldots, m_n$ and $0 < \rho \leq m_j^{-1}$, $j = 1, \ldots, n$ such that for each $\alpha \in \mathbf{N}^n$ we have

$$\partial^\alpha W(x) \prec \left(1 + \sum_{j=1}^{n} |x_j|^{m_j}\right)^{1-\rho|\alpha|},$$

and

$$W(x) \succ 1 + \sum_{j=1}^{n} |x_j|^{m_j}, \qquad \text{for } \|x\| \geq R \gg 0.$$

Under the assumption that W has a *principal part*, i.e. there exists

$$W_0(x) = \lim_{t \to +\infty} t^{-1} W\left(t^{m_1^{-1}} x_1, \ldots, t^{m_n^{-1}} x_n\right),$$

such that for a suitable $\epsilon > 0$ we have

$$\|W(x) - W_0(x)\| \prec \left(1 + \sum_{j=1}^{n} |x_j|^{m_j}\right)^{1-\epsilon},$$

they obtained the Weyl formula

$$N(\lambda) = V(\lambda)\left(1 + \mathcal{O}(\lambda^{-\epsilon})\right) = V_0\lambda^p\left(1 + \mathcal{O}(\lambda^{-\epsilon})\right),$$

where

$$\begin{aligned} p &= \frac{1}{m_1} + \cdots + \frac{1}{m_n} + \frac{n}{2}, \\ V_0 &= \frac{\sigma_n}{n(2\pi)^n}\int [\lambda - W_0(x)]_+^{n/2}\,dx. \end{aligned}$$

In the next two sections we want to see how to extend the Weyl formula to more general hypoelliptic potentials.

2. Newton polyhedra and hypoelliptic polynomials

Let us first consider a polynomial potential

$$W(x) = \sum_{\alpha\in\mathcal{A}} c_\alpha x^\alpha$$

and let associate with W its *Newton polyhedron*, i.e. the convex hull $\mathcal{Q}$ of $\{0\}\cup\mathcal{A}$. The Newton polyhedron $\mathcal{Q}$ is contained in[1] $(\mathbf{R}_0^+)^n$ and it is the convex hull of a finite subset $V(\mathcal{Q}) \subset \mathbf{N}^n$ of convex-linearly independent points called the *vertices of* $\mathcal{Q}$ and univocally determined by $\mathcal{Q}$.

Moreover there exists a finite set

$$N(\mathcal{Q}) = N_0(\mathcal{Q}) \cup N_1(\mathcal{Q}) \subset \mathbf{R}^n,$$

such that

$$\|\nu\| = 1, \qquad \text{for all } \nu \in N_0(\mathcal{Q})$$

and

$$\mathcal{Q} = \{x\in\mathbf{R}^n : \nu\cdot x \geq 0, \forall\nu\in N_0(\mathcal{Q})\} \cap \{x\in\mathbf{R}^n : \nu\cdot x \leq 1, \forall\nu\in N_1(\mathcal{Q})\}.$$

$N_0(\mathcal{Q})$ and $N_1(\mathcal{Q})$ are univocally determined by $\mathcal{Q}$ and the boundary of $\mathcal{Q}$ is made up of faces F_ν which are the convex hull of the vertices of $\mathcal{Q}$ lying on the hyperplane orthogonal to $\nu \in N(\mathcal{Q})$ and of equation

$$\begin{aligned} \nu\cdot x &= 0, \qquad \text{if } \nu\in N_0(\mathcal{Q}), \\ \nu\cdot x &= 1, \qquad \text{if } \nu\in N_1(\mathcal{Q}). \end{aligned}$$

The following definition is due to Volevič and Gindikin:

[1] We adopt the following notations:

$$\mathbf{R}^+ = \{r\in\mathbf{R} : r > 0\}, \qquad \mathbf{R}_0^+ = \{r\in\mathbf{R} : r \geq 0\}.$$

Definition 1 *A convex polyhedron $\mathcal{Q}$ is* complete *if*

1. $V(\mathcal{Q}) \subset \mathbf{N}^n$,
2. $0 \in V(\mathcal{Q}) \neq \{0\}$,
3. $N_0(\mathcal{Q}) = \{e_1, \dots, e_n\}$,
4. $N_1(\mathcal{Q}) \subset (\mathbf{R}^+)^n$,

where

$$e_j = (0, \dots, 0, 1, 0, \dots, 0), \qquad \textit{with } 1 \textit{ in } j\textit{-position.}$$

Given a complete polyhedron $\mathcal{Q}$, for each $\alpha \in \mathbf{N}^n$ we have the estimate (see [2], Ch. 1, Lemma 8.1)

$$|x^\alpha| \prec \Lambda_{\mathcal{Q}}(x)^{k(\mathcal{Q},\alpha)}, \tag{4}$$

with

$$k(\mathcal{Q}, \alpha) = \min\left\{t \in \mathbf{R}^+ : t^{-1}\alpha \in \mathcal{Q}\right\} = \max_{\nu \in N_1(\mathcal{Q})} \nu \cdot \alpha. \tag{5}$$

Newton polyhedra of quasi-elliptic polynomials are simplexes, in particular they are complete. The following proposition shows that the class of polynomials with complete Newton polyhedron is very wide:

Proposition 1 *The Newton polyhedron of a hypoelliptic polynomial is complete.*

Proof: See [2], Ch. 1, Thm. 1.1 □

With a complete polyhedron $\mathcal{Q}$ we can associate a *weight function*

$$\Lambda_{\mathcal{Q}}(x) = \left(\sum_{\alpha \in V(\mathcal{Q})} x^{2\alpha}\right)^{1/2}.$$

We have the estimates:

$$\langle x \rangle^{\mu_0} \prec \Lambda_{\mathcal{Q}}(x) \prec \langle x \rangle^{\mu_1},$$

where

$$\langle x \rangle = \left(1 + \|x\|^2\right)^{1/2},$$

as standard, and

$$\mu_0 = \min_{\alpha \in V(\mathcal{Q}) \setminus 0} |\alpha|, \qquad \mu_1 = \max_{\alpha \in V(\mathcal{Q})} |\alpha|.$$

Moreover, for each $\alpha \in \mathbf{N}^n$ we have

$$\partial_x^\alpha \Lambda_{\mathcal{Q}}(x) \prec \Lambda_{\mathcal{Q}}(x)^{1-|\alpha|/\mu}, \tag{6}$$

where

$$\mu = \max\left\{\frac{1}{\nu_j} : j = 1, \ldots, n,\, \nu \in N_1(\mathcal{Q})\right\}$$

is the *formal order of $\mathcal{Q}$.*

To prove the estimate (6), we observe that if $\beta \leq \alpha \in \mathcal{Q} \cap \mathbf{N}^n$, we have

$$\partial^\beta x^\alpha = \beta! \binom{\alpha}{\beta} x^{\alpha-\beta}$$

and

$$\alpha - \beta \in \mathcal{Q}.$$

Therefore

$$\nu \cdot (\alpha - \beta) \leq 1 - \frac{|\beta|}{\mu}, \qquad \text{for all } \nu \in N_1(\mathcal{Q}).$$

Hence (see (5))

$$k(\mathcal{Q}, \alpha) \leq 1 - \frac{|\beta|}{\mu},$$

thus from (4) we obtain for each $\alpha \in \mathcal{Q} \cap \mathbf{N}^n$ and $\beta \leq \alpha$

$$\left|\partial^\beta x^\alpha\right| \prec \Lambda_{\mathcal{Q}}(x)^{1-|\beta|/\mu}.$$

In general, we can prove the following

Proposition 2 *If W is a real-valued, bounded from below hypoelliptic polynomial, with Newton polyhedron $\mathcal{Q}$ of weight $\Lambda_{\mathcal{Q}}$, then there exist*

$$l \leq 1, \qquad -1/\mu \leq \tau < 0,$$

such that for each $\alpha \in \mathbf{N}^n$ we have

$$\begin{aligned} |\partial^\alpha W(x)| &\prec W(x)\Lambda_{\mathcal{Q}}(x)^{\tau|\alpha|}, & (7)\\ \Lambda_{\mathcal{Q}}(x)^l &\prec W(x) \prec \Lambda_{\mathcal{Q}}(x), & (8) \end{aligned}$$

for $\|x\| \geq R \gg 0$. □

If $l = 1$ we say that W is *multi-quasi-elliptic or $\mathcal{Q}$-elliptic* . Multi-quasi-elliptic polynomials have a long history in the theory of linear partial differential equations. They have been introduced by Volevič and Gindikin [15] and have been extensively studied by several authors among which Friberg [5] and Cattabriga [4].

Of course it may well happen that $l < 1$: for example

$$W(x, y) = \left(x^{2n-1} - y^{2n}\right)^2 + x^{2n-2}y^{2n}$$

fails to be Q-elliptic along the curve $y^{2n} = x^{2n-1}$. Indeed its Newton polyhedron Q has vertices

$$(0,0),\ (0,4n),\ (4n-2,0),$$

the associated weight is

$$\Lambda_Q(x,y) = \left(1 + x^{8n-4} + y^{8n}\right)^{1/2},$$

and we have

$$W(x,y) \succ \Lambda_Q(x,y)^{\frac{4n-3}{4n-2}}, \qquad \text{for } \|x\| + \|y\| \geq R \gg 0.$$

3. The Weyl formula for the Schrödinger operator with hypoelliptic potential

Definition 2 *A complete polyhedron $Q \subset (\mathbf{R}_0^+)^n$ is* non degenerate *if the intersection of its boundary with the diagonal of $(\mathbf{R}^+)^n$ is an* internal point *to a face F_ν.*

This means that there exists $\nu \in N_1(Q)$ such that

$$s = \max\left\{t \in [0,1] : \frac{t}{\nu_j} e_j + \frac{1-t}{|\nu|} e \in F_\nu,\ j = 1, \dots, n\right\} > 0,$$

where

$$e = (1, \dots, 1) \in \mathbf{R}^n.$$

This inequality is equivalent to

$$|x_1 \cdots x_n|^{\frac{1-s}{|\nu|}} \left(|x_1|^{s/\nu_1} + \cdots + |x_n|^{s/\nu_n}\right) \prec \Lambda_Q(x).$$

Now we can state the Weyl formula for non degenerate hypoelliptic potentials. See [2] for the degenerate case in dimension 2 and [3] for fully degenerate polynomial potentials in dimension n but without error estimate.

Theorem 1 *Consider a non degenerate, complete polyhedron $Q \subset (\mathbf{R}_0^+)^n$ with weight Λ_Q. Let W be a real-valued C^∞ potential satisfying (7) and (8), with*

$$l > 0, \qquad -\frac{1}{\mu} \leq \tau < \frac{1}{\mu}.$$

Assume that $W = W_0 + \tilde{W}$, where

1. *W_0 is quasi-homogeneous:*

$$W_0(x)\left(\lambda^{\nu_1}x_1,\ldots,\lambda^{\nu_n}x_n\right)=\lambda W_0(x),\quad \textit{for all } x\neq 0 \textit{ and } \lambda>0,$$

2. *W_0 is non degenerate:*

$$|x_1\cdots x_n|^{\frac{1-s}{|\nu|}}\left(|x_1|^{s/\nu_1}+\cdots+|x_n|^{s/\nu_n}\right)\prec W_0(x),$$

for all $\|x\|\geq R\gg 0$,

3. *there exists $\tilde{s}\in[0,1)$ such that*

$$|\tilde{W}(x)|\prec 1+|x_1|^{\tilde{s}/\nu_1}+\cdots+|x_n|^{\tilde{s}/\nu_n}.$$

Then

$$N(\lambda)=\left(V_0+\mathcal{O}(h(\lambda))\right)\lambda^{|\nu|+n/2}, \tag{9}$$

with

$$V_0 = \frac{\sigma_n}{n(2\pi)^n}\int\left[1-W_0(x)\right]_+^{n/2}\,dx, \tag{10}$$

$$h(\lambda) = \begin{cases}\lambda^{-\epsilon}(\log\lambda)^{2n-2}, & \textit{if } \epsilon_1>\epsilon_3 \textit{ and } \epsilon_2>\epsilon_3,\\ \lambda^{-\epsilon}(\log\lambda)^{2n-1}, & \textit{if } \epsilon_1>\epsilon_2=\epsilon_3,\\ \lambda^{-\epsilon}, & \textit{otherwise,}\end{cases} \tag{11}$$

where

$$\epsilon<\epsilon_1,\qquad \epsilon\leq\epsilon_2,\qquad \epsilon\leq\epsilon_3, \tag{12}$$

$$\epsilon_2=1-\tilde{s},\qquad \epsilon_3=\frac{|\nu|^2}{|\nu|+n/2}\frac{(1-\tilde{s})s}{(1-s)\tilde{s}}.$$

and

$$\epsilon_1=\frac{1}{2}(\rho-\delta),$$

with

$$-\frac{1}{\mu}\leq\rho<\delta\leq\frac{1}{\mu}$$

such that for each α, $\beta\in\mathrm{N}^n$ we have

$$\left|\partial_\xi^\alpha\partial_x^\beta\left(\|\xi\|^2+W(x)\right)\right| \prec\left(\|\xi\|^2+W(x)\right)\left(\|x\|^2+\Lambda_Q(x)^2\right)^{(-\rho|\alpha|+\delta|\beta|)/2}, \tag{13}$$

for $\|x\|+\|\xi\|\geq R\gg 0$.

Remark 1: W_0 is called *principal part of W*.

Remark 2: We can always take $\rho = \frac{1}{\mu}$ and $\delta = \tau_+$, but this choice can be improved in several cases. For example, if W is a polynomial and $l > 1/3$ we have

$$\rho = \frac{1}{\mu} \quad \text{and} \quad \delta = \frac{1-2l}{\mu l},$$

with $\delta < 0$ for $l > 1/2$.

We shall prove this theorem in Section 6.

From Theorem 1 one easily obtains the asymptotic behavior of the eigenvalues λ_k of $-\Delta + W$:

Corollary 1 *Under the same assumption of Theorem 1 we have*

$$\lambda_k = \left(\frac{k}{V_0}\right)^{\frac{1}{|\nu|+n/2}} (1 + \mathcal{O}(h(k))), \qquad \text{as } k \to \infty.$$

□

EXAMPLES

I. Consider

$$\begin{aligned} W(x,y) &= \sum_{j=1}^{4} x^{2h_j} y^{2k_j} + \sum_{j=1}^{J} x^{h'_j} y^{k'_j} \\ &+ \frac{1}{2} \sin\left(\left(\sum_{j=1}^{4} x^{2h_j} y^{2k_j}\right)^{\frac{1}{\mu}+\delta}\right) \end{aligned}$$

where h_j, k_j, h'_j and k'_j are non-negative integers such that

$$\begin{gathered} 0 = h_1 < h_2 < h_3 < h_4, \qquad k_1 > k_2 > k_3 > k_4 = 0, \\ h_2 < k_2, \qquad h_3 > k_3, \\ -\frac{k_3}{h_4 - h_3} > \frac{k_3 - k_2}{h_3 - h_2} > \frac{k_2 - k_1}{h_2} \\ \frac{1}{2}\frac{h_{j+1} - h_j}{h_{j+1}k_j - h_j k_{j+1}} k'_l - \frac{1}{2}\frac{k_{j+1} - k_j}{h_{j+1}k_j - h_j k_{j+1}} h'_l < 1 \end{gathered}$$

for $j = 1, 2, 3$ and $1 \le l \le J$. Moreover

$$\mu = \max\left\{-2\frac{h_2 k_1}{k_2 - k_1}, 2\frac{h_4 k_3}{h_4 - h_3}\right\}$$

and

$$-\frac{1}{\mu} \le \delta < \frac{1}{\mu}.$$

μ is the formal order of the Newton polygon $\mathcal{Q}$ associated with W, i.e. the polygon with vertices

$$(0,0),\ (0,2k_1),\ (2h_2,2k_2),\ (2h_3,2k_3),\ (2h_4,0).$$

The face that intersect the diagonal has normal ν with components

$$\nu_1 = -\frac{k_3-k_2}{2(h_3k_2-h_2k_3)}, \qquad \nu_2 = \frac{h_3-h_2}{2(h_3k_2-h_2k_3)},$$

the principal part is

$$W_0(x,y) = x^{2h_2}y^{2k_2} + x^{2h_3}y^{2k_3}.$$

Finally

$$\begin{aligned} s &= \min\{1-2h_2|\nu|, 1-2k_3|\nu|\}, \\ \tilde{s} &= \max\Big\{2\nu_2k_1,\ 2\nu_1h_4,\ \nu_1h'_j+\nu_2k'_j : j=1,\ldots,J\Big\}. \end{aligned}$$

Thus we have

$$N(\lambda) = (V_0 + \mathcal{O}(\lambda^{-\epsilon}))\,\lambda^{1+|\nu|},$$

where ϵ satisfies (12) and

$$\begin{aligned} V_0 &= \frac{1}{4\pi}\int\Big[1-\Big(x^{2h_2}y^{2k_2}+x^{2h_3}y^{2k_3}\Big)\Big]_+\,dx\,dy \\ &= \frac{\nu_1\nu_2B(\nu_1,\nu_2)}{\pi|\nu|(|\nu|+1)} \end{aligned}$$

(B is the Euler Beta function).

II. The second example is a potential which is not multi-quasi-elliptic, but still satisfies the hypotheses of the theorem.

$$W(x,y) = \Big[\Big(x^{2n-1}-y^{2n}\Big)^2 + x^{2n-2}y^{2n}\Big]\Big(x^{2k}+y^{2m}\Big),$$

with

$$n \geq 1, \qquad k < \frac{2n-1}{2n}m, \qquad m > 2n-1.$$

The Newton polygon $\mathcal{Q}$ has vertices

$$(0,0),\ (0,4n+2m),\ (4n-2,2m),\ (4n-2+2k,0).$$

W fails to be multi-quasi-elliptic along the curve $y^{2n} = x^{2n-1}$, but it is hypoelliptic because it is the product of two hypoelliptic polynomials.

On the other side $\mathcal{Q}$ is non degenerate because the diagonal of $(\mathbf{R}^+)^2$ meets the boundary of $\mathcal{Q}$ in a point internal to the side of vertices $(4n-2, 2m)$, $(4n-2+2k, 0)$. Eventually we obtain

$$N(\lambda) = V_0 \lambda^{\frac{m+k}{2m(2n-1+k)}+1} \left(1 + \mathcal{O}(\lambda^{-\epsilon})\right),$$

with $\epsilon > 0$ satisfying (12) and

$$\begin{aligned} V_0 &= \frac{\nu_1 \nu_2 B(\nu_1, \nu_2)}{\pi |\nu| (|\nu| + 1)} \\ \nu_1 &= \frac{1}{2(2m-1+k)}, \\ \nu_2 &= \frac{k}{2(2m-1+k)m}. \end{aligned}$$

4. Symbolic calculus

In this section we develop a calculus for a class of symbols which contains $\|\xi\|^2 + W(x)$, where W satisfies the estimates (7) and (8).

4.1. SYMBOL CLASSES

Given a complete polyhedron $\mathcal{P} \subset (\mathbf{R}_0^+)^{2n}$ with formal order μ and weight

$$\Lambda_{\mathcal{P}}(x, \xi) = \left(\sum_{(\alpha,\beta) \in V(\mathcal{P})} \xi^\alpha x^\beta \right)^{1/2},$$

we denote by $S^m_{\mathcal{P},\rho,\delta}$ the class of symbols satisfying for each α, $\beta \in \mathbf{N}^n$ the following estimate:

$$\left| \partial_\xi^\alpha \partial_x^\beta a(x, \xi) \right| \prec \Lambda_{\mathcal{P}}(x, \xi)^{m - \rho|\alpha| + \delta|\beta|},$$

with

$$-\rho \le \delta < \rho \le \frac{1}{\mu}, \qquad \text{and} \quad m \in \mathbf{R}.$$

If, for each α, $\beta \in \mathbf{N}^n$, a satisfies the estimates

$$\left| \partial_\xi^\alpha \partial_x^\beta a(x, \xi) \right| \prec |a(x, \xi)| \, \Lambda_{\mathcal{P}}(x, \xi)^{-\rho|\alpha| + \delta|\beta|} \tag{14}$$

$$\Lambda_{\mathcal{P}}(x, \xi)^l \prec |a(x, \xi)| \, \Lambda_{\mathcal{P}}(x, \xi)^m, \tag{15}$$

for $\|x\| + \|\xi\| \ge R \gg 0$ with

$$l \le m,$$

then a is called *globally hypoelliptic.* We denote by $HS^{m,l}_{\mathcal{P},\rho,\delta}$ the class of globally hypoelliptic symbols. If $l = m$, a is called *globally multi-quasi-elliptic or globally $\mathcal{Q}$-elliptic.*

4.2. PROPERTIES OF THE WEIGHT FUNCTION

Now we examinate some properties of the weight function.

1. For each $m \in \mathbf{R}$ we have

$$\Lambda^m_{\mathcal{P}} \in S^m_{\mathcal{P},1/\mu,-1/\mu}. \tag{16}$$

Proof: We already know the result for $m = 1$ (see estimate (6)). For $m = -1$, (16) follows by induction and differentiation from the identity $\Lambda^{-1}_{\mathcal{P}}\Lambda_{\mathcal{P}} = 1$.
For general m (16) follows from the identity

$$\partial^\alpha_\xi \partial^\beta_x \left(\Lambda_{\mathcal{Q}}(x,\xi)^m\right) = \Lambda_{\mathcal{P}}(x,\xi)^{m-|\alpha+\beta|} a_{m,\alpha+\beta}(x,\xi), \tag{17}$$

with

$$a_{m,\gamma} \in S^{(1-1/\mu)|\gamma|}_{\mathcal{P},1/\mu,-1/\mu}.$$

(17) is proven by induction. □

2. $\Lambda_{\mathcal{P}}$ is *slowly varying* in the sense that there exists $\epsilon > 0$ such that

$$\Lambda_{\mathcal{P}}((x,\xi) \prec \Lambda_{\mathcal{P}}(x+y,\xi+\eta) \prec \Lambda_{\mathcal{P}}(x,\xi),$$

for

$$\|y\|^2 + \|\eta\|^2 \leq \epsilon^2 \Lambda_{\mathcal{P}}(x,\xi)^{2/\mu}.$$

Proof: Let Taylor expand $\Lambda_{\mathcal{P}}(x+y,\xi+\eta)^{1/\mu}$:

$$\Lambda_{\mathcal{P}}(x+y,\xi+\eta)^{1/\mu} = \Lambda_{\mathcal{P}}(x,\xi)^{1/\mu} + \int_0^1 (1-t) \sum_{|\alpha+\beta|=1} \eta^\alpha y^\beta \partial^\alpha_\xi \partial^\beta_x \left(\Lambda_{\mathcal{P}}(x+ty,\xi+t\eta)^{1/\mu}\right) dt.$$

Because $\Lambda^{1/\mu}_{\mathcal{P}} \in S^{1/\mu}_{\mathcal{P},1/\mu,-1/\mu}$, the derivatives in the integral are bounded and we have

$$\left|\Lambda_{\mathcal{P}}(x+y,\xi+\eta)^{1/\mu} - \Lambda_{\mathcal{P}}(x,\xi)^{1/\mu}\right| \prec \left(\|y\|^2 + \|\eta\|^2\right)^{1/2} \leq \epsilon \Lambda_{\mathcal{P}}(x,\xi)^{1/\mu}.$$

If we choose $\epsilon > 0$ small enough we obtain the result. □

3. $\Lambda_{\mathcal{P}}$ is *temperate* in the following strong sense:

$$\Lambda_{\mathcal{P}}(x+y,\xi+\eta) \prec \Lambda_{\mathcal{P}}(x,\xi)\Lambda_{\mathcal{P}}(y,\eta),$$

as one easily verifies.

4. We have

$$\Lambda_{\mathcal{P}}(x,\xi) \prec \langle x,\xi\rangle^{\mu},$$

where

$$\langle x,\xi\rangle = \left(1+\|x\|^2+\|\xi\|^2\right)^{1/2},$$

as standard.

It is possible to define a metric in the sense of Hörmander

$$G_{(x,\xi)}(y,\eta) = \Lambda_{\mathcal{P}}(x,\xi)^{2\delta}\|y\|^2 + \Lambda_{\mathcal{P}}(x,\xi)^{-2\rho}\|\eta\|^2,$$

so that the uncertainty principle is satisfied and our classes are the same as Hörmander's:

$$S^m_{\mathcal{P},\rho,\delta} = S\left(g,\Lambda^m_{\mathcal{P}}\right).$$

However for didactic reasons we prefer to recall the basic properties of the symbolic calculus in this particular setting.

4.3. ASYMPTOTIC EXPANSION

Let $\{m_j\}$ be a sequence of real numbers tending to $-\infty$ and consider

$$a_j \in S^{m_j}_{\mathcal{P},\rho,\delta}, \qquad \text{for each } j;$$

then for $a \in S^m_{\mathcal{P},\rho,\delta}$ we write

$$a \sim \sum a_j,$$

if for each N we have

$$a - \sum_{j<N} a_j \in S^{m_N}_{\mathcal{P},\rho,\delta}.$$

The following two propositions are standard:

Proposition 3 *Given a sequence $a_j \in S^{m_j}_{\mathcal{P},\rho,\delta}$ with $m_j \to -\infty$, there exists $a \in S^{m_1}_{\mathcal{P},\rho,\delta}$ such that $a \sim \sum a_j$. Moreover if $b \in S^m_{\mathcal{P},\rho,\delta}$ is such that $b \sim \sum a_j$, then $a - b \in \mathcal{S}(\mathbf{R}^n)$.* □

Proposition 4 *Given $a \in S^m_{\mathcal{P},\rho,\delta}$ and a sequence $a_j \in S^{m_j}_{\mathcal{P},\rho,\delta}$ with $m_j \to -\infty$ assume that*

1. *for each* $\alpha, \beta \in \mathbf{N}^n$ *there exists* $k_{\alpha,\beta}$ *such that*

$$\left|\partial_\xi^\alpha \partial_x^\beta a(x,\xi)\right| \prec \Lambda_{\mathcal{P}}(x,\xi)^{k_{\alpha,\beta}},$$

2. *there exists a sequence* $l_j \to -\infty$, *such that for each* N *we have*

$$\left|a(x,\xi) - \sum_{j<N} a_j(x,\xi)\right| \prec \Lambda_{\mathcal{P}}(x,\xi)^{l_N}.$$

Then

$$a \sim \sum a_j. \square$$

4.4. QUANTIZATION

The *t-quantization* associates with a symbol $a \in S^m_{\mathcal{P},\rho,\delta}$ and a real number t the pseudodifferential operator

$$A_t u(x) = \int e^{i(x-y)\cdot\xi} a(tx + (1-t)y, \xi) u(y)\, dy\, đ\xi,$$

where

$$đ\xi = (2\pi)^{-n}\, d\xi,$$

as standard. This procedure yields a family of continuous operators on the Schwartz class:

$$A_t : \mathcal{S}(\mathbf{R}^n) \to \mathcal{S}(\mathbf{R}^n),$$

which extend by double transposition to temperate distributions

$$A_t : \mathcal{S}'(\mathbf{R}^n) \to \mathcal{S}'(\mathbf{R}^n).$$

Beside the *Weyl quantization*

$$\mathrm{Op}_W(a)u(x) = A_{1/2}u(x) = \int e^{i(x-y)\cdot\xi} a\left(\frac{1}{2}(x+y), \xi\right) u(y)\, dy\, đ\xi,$$

we are also interested in the *left* (or classical) *quantization:*

$$\mathrm{Op}_L(a)u(x) = A_0 u(x) = \int e^{i(x-y)\cdot\xi} a\,(x,\xi)\, u(y)\, dy\, đ\xi.$$

We say that a is the *left symbol* of $\mathrm{Op}_L(a)$ and the *Weyl symbol* of $\mathrm{Op}_W(a)$. Given an operator A we denote by $\mathrm{symb}_L(A)$ and by $\mathrm{symb}_W(a)$ the left and the Weyl symbol of A. We have

$$\mathrm{symb}_L(A) = e^{-ix\cdot\xi} A e^{ix\cdot\xi}.$$

The Weyl symbol a_W can be obtained by the left symbol a_L in terms of the oscillatory integral

$$a_W(x,\xi) = \int e^{-iy\cdot\eta} a_L\left(x + \frac{1}{2}y, \xi - \eta\right) dy\, đ\eta.$$

By Taylor expanding with respect to η and integrating by parts one easily obtains

$$a_W \sim \sum_\alpha \frac{(-1)^{|\alpha|}}{\alpha! 2^{|\alpha|}} \partial_\xi^\alpha D_x^\alpha a_L.$$

In particular

$$a_W - a_L \in S^{m-(\rho-\delta)}_{\mathcal{P},\rho,\delta}.$$

Let us denote by $L^m_{\mathcal{P},\rho,\delta}$ and $HL^{m,l}_{\mathcal{P},\rho,\delta}$ the classes of operators with Weyl symbol belonging respectively to $S^m_{\mathcal{P},\rho,\delta}$ and $HS^{m,l}_{\mathcal{P},\rho,\delta}$.

Also the Schwartz kernel K_A of A can be computed easily from its Weyl symbol a in terms of inverse Fourier transform:

$$K_A(x,y)) = \mathcal{F}^{-1}_{\xi\to x-y}\left\{a\left(\frac{1}{2}(x+y),\xi\right)\right\}. \tag{18}$$

We say that A is *regularizing* if it sends $\mathcal{S}'(\mathbf{R}^n)$ into $\mathcal{S}(\mathbf{R}^n)$. One can prove the following

Proposition 5 *The following statements are equivalent:*

1. *A is regularizing,*
2. *$K_A \in \mathcal{S}(\mathbf{R}^n_x \times \mathbf{R}^n_y)$,*
3. *$a \in \mathcal{S}(\mathbf{R}^n_x \times \mathbf{R}^n_\xi) = \bigcap_{m\in\mathbf{R}} S^m_{\mathcal{P},\rho,\delta}$.*

□

We set

$$L^{-\infty} = \bigcap_{m\in\mathbf{R}} L^m_{\mathcal{P},\rho,\delta}.$$

The advantage of the Weyl quantization over the left quantization is that it makes easy to compute the symbol a^* of the formal adjoint A^*. By definition

$$(A^*u, v)_{L^2} = (u, Av)_{L^2}, \qquad \text{for all } u, v \in \mathcal{S}(\mathbf{R}^n),$$

but

$$\int u(x)\overline{\int e^{i(x-y)\cdot\xi} a\left(\frac{1}{2}(x+y),\xi\right) v(y)\, dy\, đ\xi}\, dx$$
$$= \int\left\{\int e^{i(y-x)\cdot\xi}\overline{a\left(\frac{1}{2}(x+y),\xi\right)} u(x)\, dx\, đ\xi\right\}\overline{v(y)}\, dy,$$

so

$$a^*(x,\xi) = \overline{a(x,\xi)}.$$

Thus A *is formally self-adjoint if and only if a is real-valued.*

4.5. COMPOSITION FORMULA

More complicated is the composition formula of two operators. If $A = \mathrm{Op}_W(a)$ and $B = \mathrm{Op}_W(b)$, with $a \in S^m_{\mathcal{P},\rho,\delta}$ and $b \in S^l_{\mathcal{P},\rho,\delta}$, we denote by $b\#a$ the Weyl symbol of BA. After some computations we obtain

$$\begin{aligned} & b\#a(x,\xi) \\ & \quad = \int e^{i(y\cdot\eta - w\cdot\zeta)} b\left(x+\frac{1}{2}y, \xi+\zeta\right) a\left(x+\frac{1}{2}w,\xi+\eta\right)\, dy\, đ\eta\, dw\, đ\zeta. \end{aligned}$$

By Taylor expanding and integrating by parts we obtain:

$$\begin{aligned} & b\#a(x,\xi) \\ & \sim \sum_{\alpha,\beta} \frac{1}{\alpha!\beta!} \int e^{i(y\cdot\eta - w\cdot\zeta)} \zeta^\alpha \eta^\beta \partial_\xi^\alpha b\left(x+\frac{1}{2}y,\xi\right) \partial_\xi^\beta a\left(x+\frac{1}{2}w,\xi\right) dy\, đ\eta\, dw\, đ\zeta \\ & = \sum_{\alpha,\beta} \frac{(-1)^{|\beta|}}{\alpha!\beta!2^{|\alpha+\beta|}} \cdot \\ & \quad \cdot \int e^{i(y\cdot\eta - w\cdot\zeta)} \partial_\xi^\alpha D_x^\beta b\left(x+\frac{1}{2}y,\xi\right) \partial_\xi^\beta D_x^\alpha a\left(x+\frac{1}{2}w,\xi\right) dy\, đ\eta\, dw\, đ\zeta. \end{aligned}$$

In conclusion, by computing the integrals we obtain

$$b\#a \sim \sum_{\alpha,\beta} \frac{(-1)^{|\beta|}}{\alpha!\beta!2^{|\alpha+\beta|}} \partial_\xi^\alpha D_x^\beta b \partial_\xi^\beta D_x^\alpha a. \tag{19}$$

From this asymptotic expansion we obtain in particular that

$$b\#a \in S^{m+l}_{\mathcal{P},\rho,\delta},$$

and

$$b\#a - ba \in S^{m+l-(\rho-\delta)}_{\mathcal{P},\rho,\delta}.$$

4.6. FURTHER RESULTS

Using symbolic calculus one can prove the following standard results:

Proposition 6 *If $m \leq 0$ every $A \in L^m_{\mathcal{P},\rho,\delta}$ extends to a bounded operator in $L^2(\mathbf{R}^n)$, wich is compact if $m < 0$.* □

Proposition 7 *For each $A \in HL^{m,l}_{\mathcal{P},\rho,\delta}$, there exists a* parametrix, *i.e. an operator $B \in HL^{-l,-m}_{\mathcal{P},\rho,\delta}$ such that $BA - I$ and $AB - I$ are regularizing.* □

Proposition 8 *If A is globally hypoelliptic, then $Au \in \mathcal{S}(\mathbf{R}^n) \implies u \in \mathcal{S}(\mathbf{R}^n)$.* □

The last proposition explains in which sense the operators are globally hypoelliptic.

5. Friedrichs symmetrization and non negative operators

5.1. FRIEDRICHS SYMMETRIZATION

The Friedrichs symmetrization is a technical tool which allows to obtain semi-bounds for operators with semi-bounded symbols. The idea is the following. Given $a \in S^m_{\mathcal{P},\rho,\delta}$, we define a new symbol $a_F \in S^m_{\mathcal{P},\rho,\delta}$ such that

$$\mathrm{Op}_W(a) - \mathrm{Op}_L(a_F) \in L^{m-(\rho-\delta)}_{\mathcal{P},\rho,\delta},$$

and $A_F = \mathrm{Op}_L(a_F)$ is non negative if a is non negative i.e.:

$$a(x,\xi) \geq 0, \quad \forall (x,\xi) \quad \implies \quad (A_F u, u)_{L^2} \geq 0, \quad \forall u \in \mathcal{S}(\mathbf{R}^n).$$

Let us show how to define the *Friedrichs symmetrization* a_F of a. Choose a real-valued $q \in C^\infty(\mathbf{R}^n)$ such that

$$\begin{aligned} q(\sigma) &\geq 0, \quad \text{for all } \sigma, \\ q(\sigma) &= 0, \quad \text{for } \|\sigma\| \geq 1, \\ q(\sigma_1, \ldots, -\sigma_j, \ldots, \sigma_n) &= q(\sigma), \qquad \text{for all } \sigma \text{ and } j = 1, \ldots, n, \end{aligned}$$

and

$$\int q(\sigma)^2 \, d\sigma = 1.$$

Let

$$\tau = \frac{1}{2}(\rho + \delta)$$

and define

$$F(x,\xi,\zeta) = \Lambda_{\mathcal{P}}(x,\xi)^{-n\tau/2} q\left(\Lambda_{\mathcal{P}}(x,\xi)^{-\tau}(\zeta - \xi)\right).$$

Then we set

$$b(\eta,x,\xi) = \int F(x,\eta,\zeta) a(x,\zeta) F(x,\xi,\zeta) \, đ\zeta,$$

and

$$a_F(x,\xi) = \int e^{-i(x-y)\cdot(\xi-\eta)} b(\eta, y, \xi)\, dy\, đ\eta.$$

Following Taylor [13], Ch. VII, §2 or Kumano-go [9] Ch. 3, §4, one can prove the following

Theorem 2 *We have*

$$a_F \in S^m_{\mathcal{P},\rho,\delta}$$

and

$$a_F(x,\xi) \sim a(x,\xi) + \sum_{|\alpha+\beta|>0} \sum_{\gamma\leq\alpha} f_{\alpha,\beta,\gamma}(x,\xi)\partial_\xi^\beta \partial_x^\gamma a(x,\xi), \tag{20}$$

where each $f_{\alpha,\beta,\gamma}$ is independent from a *and*

$$\begin{aligned} f_{0,\beta,0} &= 0, \qquad \textit{if} \quad |\beta| = 1, \\ f_{\alpha,\beta,\gamma} &\in S^{-\rho+\tau|\beta|}_{\mathcal{P},1/\mu,-1/\mu}, \qquad \textit{if} \quad |\alpha| = 1, \\ f_{\alpha,\beta,\gamma} &\in S^{-\tau|2\alpha-\gamma|+\tau|\beta|}_{\mathcal{P},1/\mu,-1/\mu}, \qquad \textit{if} \quad |\alpha| > 1, \end{aligned}$$

so that

$$a - a_F \in S^{m-(\rho-\delta)}_{\mathcal{P},\rho,\delta},$$

and

$$\begin{aligned} \sum_{\gamma\leq\alpha} f_{\alpha,\beta,\gamma}\partial_\xi^\beta\partial_x^\gamma a &\in S^{m-(\rho-\delta)}_{\mathcal{P},\rho,\delta}, \qquad \textit{for } |\alpha+\beta| = 1, \\ \sum_{\gamma\leq\alpha} f_{\alpha,\beta,\gamma}\partial_\xi^\beta\partial_x^\gamma a &\in S^{m-(\rho-\delta)|\alpha+\beta|/2}_{\mathcal{P},\rho,\delta}, \qquad \textit{for } |\alpha+\beta| > 1. \end{aligned}$$

□

5.2. NON NEGATIVE OPERATORS

Now we apply the Friedrichs symmetrization to the non negativity of operators.

Theorem 3 *If a is real-valued, then $A_F = \mathrm{Op}_L(a_F)$ is formally self-adjoint. If $a \geq 0$ then $A_F \geq 0$, i.e.*

$$(A_F u, u)_{L^2} \geq 0, \qquad \textit{for all } u \in \mathcal{S}(\mathbf{R}^n).$$

Proof: We have

$$\begin{aligned} (A_F u, v)_{L^2} &= \int\int e^{ix\cdot\xi} a_F(x,\xi)\hat{u}(\xi)\, đ\xi\, \overline{v(x)}\, dx \\ &= \int e^{ix\cdot\xi} \int e^{-i(x-y)\cdot(\xi-\eta)} b(\eta,y,\xi)\hat{u}(\xi)\overline{v(x)}\, dy\, đ\eta\, dx\, đ\xi \\ &= \int e^{iy\cdot(\xi-\eta)} b(\eta,y,\xi)\hat{u}(\xi)\overline{\hat{v}(\eta)}\, dy\, đ\xi\, đ\eta. \end{aligned}$$

But

$$b(\eta, y, \xi) = b(\xi, y, \eta),$$

and b is real-valued if a is real-valued, thus we have that

$$(A_F u, v)_{L^2} = (u, A_F v)_{L^2} .$$

Let now assume that $a \geq 0$ and let prove that

$$\int e^{iy\cdot(\xi-\eta)} b(\eta, y, \xi) \hat{u}(\xi) \overline{\hat{u}(\eta)} \, d\!\!\!{}^-\xi \, d\!\!\!{}^-\eta$$

$$= \int e^{iy\cdot(\xi-\eta)} \left\{ \int F(y, \eta, \zeta) a(y, \zeta) F(y, \xi, \zeta) \, d\!\!\!{}^-\zeta \right\} \hat{u}(\xi) \overline{\hat{u}(\eta)} \, d\!\!\!{}^-\xi \, d\!\!\!{}^-\eta \geq 0,$$

for all y. By regularizing the integral we may assume that $a(y, \zeta)$ has compact support in ζ. Then we can exchange the order of integration and obtain

$$\int \left| \int e^{iy\cdot\xi} F(y, \xi, \zeta) \hat{u}(\xi) \, d\!\!\!{}^-\xi \right|^2 a(y, \zeta) \, d\!\!\!{}^-\zeta \geq 0.$$

□

Now we can deal with lower bounds.

Theorem 4 *Given $A = \mathrm{Op}_W(a) \in L^m_{\mathcal{P},\rho,\delta}$ there exists $b \in S^m_{\mathcal{P},\rho,\delta}$ such that*

$$\mathrm{Op}_W(a) - \mathrm{Op}_L(b_F) \in L^{-\infty}.$$

Moreover

$$b(x, \xi) \sim \sum_{\alpha,\beta} \sum_{\gamma \leq \alpha} g_{\alpha,\beta,\gamma}(x, \xi) \partial_\xi^\beta \partial_x^\gamma a(x, \xi),$$

where each $g_{\alpha,\beta,\gamma}$ is idependent from a and b and

$$\begin{aligned} g_{0,0,0} &= 1, \\ g_{0,\beta,0} &= 0, \qquad \text{if } |\beta| = 1, \\ g_{\alpha,\beta,\gamma} &\in S^{-\rho+\tau|\beta|}_{\mathcal{P},1/\mu,-1/\mu}, \qquad \text{if } |\alpha| = 1, \\ g_{\alpha,\beta,\gamma} &\in S^{-\tau|2\alpha-\gamma|+\tau|\beta|}_{\mathcal{P},1/\mu,-1/\mu}, \qquad \text{if } |\alpha| > 1. \end{aligned}$$

Proof: Let consider the sequence of symbols

$$\begin{aligned} b_0 &= a \in S^m_{\mathcal{P},\rho,\delta}, \\ b_j &= \mathrm{symb}_W \{\mathrm{Op}_W(b_{j-1}) - \mathrm{Op}_L((b_{j-1})_F)\} \in S^{m-j(\rho-\delta)}_{\mathcal{P},\rho,\delta}, \qquad \text{for } j \geq 1, \end{aligned}$$

and define

$$b \sim \sum b_j.$$

We have

$$\begin{aligned}\mathrm{Op}_W(b_{N+1}) &= \mathrm{Op}_W(b_N) - \mathrm{Op}_L\left((b_N)_F\right)\\ &= \mathrm{Op}_W(b_{N-1}) - \mathrm{Op}_L\left((b_{N-1})_F\right) - \mathrm{Op}_L\left((b_N)_F\right) = \cdots\\ &= \mathrm{Op}_W(b_0) - \mathrm{Op}_L\left((b_0)_F\right) - \cdots - \mathrm{Op}_L\left((b_N)_F\right)\\ &= \mathrm{Op}_W(a) - \mathrm{Op}_L\left((b_0 + \cdots + b_N)_F\right).\end{aligned}$$

This implies that

$$\mathrm{Op}_W(a) - \mathrm{Op}_L(b_F) \in L^{-\infty}.$$

The asymptotic expansion of b follows quite easily from (20). □

Corollary 2 *Let $A = \mathrm{Op}_W(a) \in HL^{m,l}_{\mathcal{P},\rho,\delta}$ and assume that a is real-valued and bounded from below, then also A is bounded from below, i.e.*

$$(Au, u)_{L^2} \succ \|u\|_{L^2}, \qquad \textit{for all } u \in \mathcal{S}(\mathbf{R}^n).$$

Proof: By Theorem 4 there exists b such that $A - \mathrm{Op}_L(b_F)$ is bounded, because regularizing. So it suffices to show that b is bounded from below. But from hypoellipticity

$$\left|\frac{\partial_\xi^\alpha \partial_x^\beta a(x,\xi)}{a(x,\xi)}\right| \prec \Lambda_{\mathcal{P}}(x,\xi)^{-\rho|\alpha|+\delta|\beta|}, \qquad \text{for } \|x\| + \|x\| \geq R \gg 0,$$

thus from the asymptotic expansion of b we have

$$b(x,\xi) = a(x,\xi)\left(1 + \mathcal{O}\left(\Lambda_{\mathcal{P}}(x,\xi)^{-1}\right)\right), \qquad \text{as } \|x\| + \|\xi\| \to \infty.$$

It follows that b is bounded from below. □

5.3. ANTI-WICK QUANTIZATION

We end this section by the construction of an isomorphism of given order $m \in \mathbf{R}$.

Given $a \in S^m_{\mathcal{P},\rho,-\rho}$, with $0 < \rho \leq 1/\mu$, we define the ***anti-Wick quantization*** of a as the operator

$$\mathrm{Op}_{AW}(a) = \mathrm{Op}_W(\sigma * a)$$

where

$$\sigma(x,\xi) = \pi^{-n} e^{-\|x\|^2 - \|\xi\|^2}.$$

The following properties are easily verified:

$\sigma * a - a \in S^{m-2\rho}_{\mathcal{P},\rho,\rho}$,

$\sigma * a$ is real-valued if and only if a is real-valued,

$\sigma * a \geq c$ if $a \geq c$,
$\sigma * a \in HS^{m,l}_{\mathcal{P},\rho,-\rho}$ if and only if $a \in HS^{m,l}_{\mathcal{P},\rho,-\rho}$.

Theorem 5 *If $a \in HS^{m,l}_{\mathcal{P},\rho,-\rho}$ and $a > 0$, then $A = \mathrm{Op}_{AW}(a)$ is an isomorphism of $\mathcal{S}(\mathbf{R}^n)$ with inverse belonging to $HS^{-l,-m}_{\mathcal{P},\rho,-\rho}$.*

Proof: A simple computation yields

$$(Au, u)_{L^2} = 2^n \pi^{n/2} \int \left| \int e^{ix\cdot\xi} e^{-\|x-y\|^2/4} u(x)\, dx \right|^2 a(y,\xi)\, dy\, đ\xi, \tag{21}$$

thus, if $Au = 0$, we obtain $u = 0$ and A is one-to-one.

To show that A is onto, we have to solve in $\mathcal{S}(\mathbf{R}^n)$ the equation

$$Au = f,$$

for $f \in \mathcal{S}(\mathbf{R}^n)$. By hypoellipticity we can solve it in $L^2(\mathbf{R}^n)$. Let $b = a^{-1}$, and $B = \mathrm{Op}_{AW}(b)$, then

$$BA = I + R$$

with $R \in L^{-2\rho}_{\mathcal{P},\rho,-\rho}$ compact, because of negative order, and B one-to-one, because $b > 0$. Then also BA is one-to-one, and so it is onto by Fredholm Theory. But then also A is onto because B is one-to-one.

Finally from the existence of the parametrix we obtain that $A^{-1} \in HS^{-l,-m}_{\mathcal{P},\rho,-\rho}$. □

A simple, but useful example is given by

$$\mathrm{Op}_{AW}(\Lambda^m_{\mathcal{P}}) \in HL^{m,m}_{\mathcal{P},1/\mu,-1/\mu}.$$

6. The Weyl formula for globally hypoelliptic operators

An operator $A \in L^m_{\mathcal{P},\rho,\delta}$ is always closable in $L^2(\mathbf{R}^n)$. Let us denote by $\overline{A}$ its closure. Then one easily proves that $\overline{A}$ is self-adjoint if and only if A is formally self-adjoint i.e. its Weyl symbol is real-valued (see [2], Section 2.1).

Let us consider a hypoelliptic operator $A \in HL^{m,l}_{\mathcal{P},\rho,\delta}$ with real Weyl symbol a. If $l > 0$, then A has a parametrix of negative order $-l$, and therefore $\overline{A}$ has compact resolvent. Thus we have that $\overline{A}$ has a real spectrum which consists of an unbounded sequence of semi-simple eigenvalues λ_j. By hypoellipticity, the corresponding eigenfunctions ϕ_j belong to $\mathcal{S}(\mathbf{R}^n)$, thus A and its closure $\overline{A}$ have the same spectrum. From hypoellipticity and Theorem 4 one proves that a is either bounded from above or from below. By a change of sign we can always assume that a is bounded from below, so that the sequence of the eigenvalues diverges to $+\infty$. Then it make sense to consider the *counting function*

$$N(\lambda) = \sum_{\lambda_j \leq \lambda} 1,$$

where the eigenvalues λ_j are in increasing order and repeated according to their multiplicity.

The following result has been proven by Boggiatto and Buzano in [1] for $\delta = -\rho$, see also [2] for a more throughout exposition of this case.

Theorem 6 *Let $A = \mathrm{Op}_W(a) \in HL^{m,l}_{\mathcal{P},\rho,\delta}$ with a real-valued, semi-bounded from below and*

$$0 < l \leq m, \qquad -\rho \leq \delta < \rho \leq \frac{1}{\mu}.$$

Assume that

1. *$\mathcal{P}$ is non degenerate, so that there exist*

$$0 < s \leq 1, \qquad \textit{and} \qquad \nu = (\nu', \nu'') \in (\mathbf{R}^+)^n \times (\mathbf{R}^+)^n$$

such that

$$\begin{aligned} &|x_1 \cdots x_n \cdot \xi_1 \cdots \xi_n|^{\frac{1-s}{|\nu|}} \cdot \\ &\quad \cdot \left(|x_1|^{s/\nu_1'} + \cdots + |x_n|^{s/\nu_n'} + |\xi_1|^{s/\nu_1''} + \cdots + |\xi_n|^{s/\nu_n''}\right) \\ &\prec \Lambda_{\mathcal{P}}(x, \xi), \end{aligned}$$

2. *$a = a_0 + \tilde{a}$ where*

   (a) *a_0 is quasi-homogeneous:*

$$a_0\left(\lambda^{\nu_1'} x_1, \ldots, \lambda^{\nu_n'} x_n, \lambda^{\nu_1''} \xi_1, \ldots, \lambda^{\nu_n''} \xi_n\right) = \lambda^m a_0(x, \xi),$$

   for $\lambda > 0$ and $(x, \xi) \neq (0, 0)$,

   (b) *a_0 is non degenerate:*

$$\begin{aligned} &|x_1 \cdots x_n \cdot \xi_1 \cdots \xi_n|^{\frac{1-s}{|\nu|}} \cdot \\ &\quad \cdot \left(|x_1|^{s/\nu_1'} + \cdots + |x_n|^{s/\nu_n'} + |\xi_1|^{s/\nu_1''} + \cdots + |\xi_n|^{s/\nu_n''}\right) \\ &\prec a_0(x, \xi), \qquad \textit{for } \|x\| + \|\xi\| \geq R \gg 0, \end{aligned}$$

   (c) *there exists $\tilde{s} \in [0, 1)$ such that*

$$|\tilde{a}(x, \xi)| \prec 1 + |x_1|^{\tilde{s}/\nu_1'} + \cdots + |x_n|^{\tilde{s}/\nu_n'} + |\xi_1|^{\tilde{s}/\nu_1''} + \cdots + |\xi_n|^{\tilde{s}/\nu_n''}.$$

Then we have

$$N(\lambda) = [V_0 + \mathcal{O}(h(\lambda))]\, \lambda^{|\nu|/m},$$

where

$$\begin{aligned} V_0 &= \int \chi(a_0(x,\xi),1)\, dx\, đ\xi, \\ \chi(t,\lambda) &= \begin{cases} 1, & \text{if } t \le \lambda, \\ 0, & \text{if } t \ge \lambda, \end{cases} \\ h(\lambda) &= \begin{cases} \lambda^{-\epsilon}(\log\lambda)^{2n-2}, & \text{if } \epsilon_1 > \epsilon_3 \text{ and } \epsilon_2 > \epsilon_3. \\ \lambda^{-\epsilon}(\log\lambda)^{2n-1}, & \text{if } \epsilon_1 > \epsilon_2 = \epsilon_3, \\ \lambda^{-\epsilon}, & \text{otherwise} \end{cases} \end{aligned} \tag{22}$$

with

$$\epsilon < \epsilon_1, \qquad \epsilon \le \epsilon_2, \qquad \epsilon \le \epsilon_3,$$

and

$$\epsilon_1 = \frac{1}{2}(\rho - \delta), \qquad \epsilon_2 = 1 - \tilde{s}, \qquad \epsilon_3 = \frac{(1-\tilde{s})s}{(1-s)\tilde{s}} \frac{|\nu|}{m}.$$

a_0 is the *principal part of* a (however, in general a_0 is not the principal symbol of A).

We shall prove this theorem in Section 9.

We end this section by showing how to derive Theorem 1 from Theorem 6. Consider the convex polyhedron $\mathcal{P} \subset \mathbf{R}^n_x \times \mathbf{R}^n_\xi$ of vertices

$$\{(0,2e_1),\ldots,(0,2e_n)\} \cup \{(\beta,0) : \beta \in V(\mathcal{Q})\}.$$

$\mathcal{Q}$ is non degenerate, hence the diagonal of $(\mathbf{R}^+)^n$ intersects the boundary of $\mathcal{Q}$ in a point interior to a face F_ν and we have the estimate

$$|x_1 \cdots x_n|^{\frac{1-s}{|\nu|}} \left(|x_1|^{s/\nu_1} + \cdots + |x_n|^{s/\nu_n}\right) \prec \Lambda_{\mathcal{Q}}(x), \qquad \text{for all } x,$$

for a suitable

$$0 < s \le 1.$$

Then the diagonal of $(\mathbf{R}^+)^{2n}$ meets the face

$$F_{(\nu,\frac{1}{2}e)},$$

which is the convex hull of

$$F_\nu \times \{0\} \qquad \text{and} \qquad \left\{\left(0,\frac{1}{2}e_1\right),\ldots,\left(0,\frac{1}{2}e_n\right)\right\},$$

in an interior point and we have the estimate

$$|x_1 \cdots x_n \cdot \xi_1 \cdots \xi_n|^{\frac{1-ts}{|\nu|+n/2}} \cdot \Big(|x_1|^{ts/\nu_1} + \cdots + |x_n|^{ts/\nu_n} + |\xi_1|^{2ts} + \cdots + |\xi_n|^{2ts}\Big) \prec \Lambda_{\mathcal{P}}(x,\xi),$$

with

$$\Lambda_{\mathcal{P}}(x,\xi) = \left(\sum_{(\alpha,\beta)\in V(\mathcal{P})} \xi^{2\alpha} x^{2\beta}\right)^{1/2} = \left(\xi_1^4 + \cdots + \xi_n^4 + \Lambda_{\mathcal{Q}}(x)^2\right)^{1/2},$$

and

$$t = \frac{|\nu|}{|\nu| + n(1-s)/2}.$$

Then we can apply Theorem 6 to $-\Delta + W(x) = \mathrm{Op}_W\left(\|\xi\|^2 + W(x)\right)$ with ρ and δ satisfying (13) and st in place of s. We obtain (9), (10) and (11).

7. The method of approximate spectral projections

In order to evaluate the counting function $N(\lambda)$ we consider the *spectral projection:*

$$E(\lambda) = \sum_{\lambda_j \le \lambda} (u, \phi_j)_{L^2}\, \phi_j,$$

where λ_j and ϕ_j are respectively the eigenvalues and the eigenfunctions of A. Then

$$N(\lambda) = \operatorname{Tr} E(\lambda).$$

Recall that the *trace of a self-adjoint compact operator* T on a Hilbert space H is given by the series of its eigenvalues μ_j, repeated according to their multiplicity:

$$\operatorname{Tr} T = \sum \mu_j.$$

If the series is absolutely convergent, we say that T is a *trace-class operator* and we define its *trace-class norm* as

$$\|T\|_{\mathcal{B}_1(H)} = \sum |\mu_j|.$$

The space of trace-class operators is denoted be $\mathcal{B}_1(H)$. We shall come back on trace-class operators in the next section.

Now we compute the trace of the projection operator. Let

$$K_\lambda(x,y) = \sum_{\lambda_j \le \lambda} \phi_j(x)\overline{\phi_j(y)},$$

be the Schwartz kernel of $E(\lambda)$. Since $\phi_j \in \mathcal{S}(\mathbf{R}^n)$ and the sum is finite, $K_\lambda \in \mathcal{S}(\mathbf{R}^{2n})$, so $E(\lambda)$ is a regularizing operator with Weyl symbol:

$$e_\lambda(x,\xi) = \mathcal{F}^{-1}_{y\to\xi}\left\{K_\lambda\left(x+\frac{1}{2}y+, x-\frac{1}{2}y\right)\right\}.$$

Therefore

$$\begin{aligned}\operatorname{Tr} E(\lambda) &= \sum_{\lambda_j \leq \lambda} \|\phi_j\|^2_{L^2} \\ &= \int K_\lambda(x,x)\,dx \\ &= \int e_\lambda(x,\xi)\,dx\,d\!\!\!{}^-\xi.\end{aligned}$$

So we have to compute e_λ in terms of the Weyl symbol a of A.

Let us first perform a formal computation we justify afterwards. Let

$$\chi(z,\lambda) = \begin{cases} 1, & \text{if } \operatorname{Re} z \leq \lambda, \\ 0, & \text{if } \operatorname{Re} z \geq \lambda, \end{cases}$$

and assume that λ is not an eigenvalue of $\overline{A}$. Then we can express $E(\lambda)$ as a Dunford-Taylor integral

$$E(\lambda) = \frac{1}{2\pi i}\oint \chi(z,\lambda)(\overline{A}-z)^{-1}\,dz \tag{23}$$

where the integral is computed along a simple Jordan curve in $\operatorname{Re} z < \lambda$ and enclosing all $\lambda_j < \lambda$. We know that $(\overline{A}-z)^{-1}$ is a pseudodifferential operator with Weyl symbol $b_z(x,\xi)$ such that

$$b_z(x,\xi) = \frac{1}{a(x,\xi)-z} + \cdots.$$

Let us substitute

$$(\overline{A}-z)^{-1}u(x) = \int e^{i(x-y)\cdot\xi} b_z\left(\frac{1}{2}(x+y),\xi\right) u(y)\,dy\,d\!\!\!{}^-\xi$$

into (23). We obtain

$$\begin{aligned}E(\lambda)u(x) &= \frac{1}{2\pi i}\oint \chi(z,\lambda)\int e^{i(x-y)\cdot\xi} b_z\left(\frac{1}{2}(x+y),\xi\right)\,dy\,d\!\!\!{}^-\xi\,dz \\ &= \int e^{i(x-y)\cdot\xi}\frac{1}{2\pi i}\oint \chi(z,\lambda) b_z\left(\frac{1}{2}(x+y),\xi\right)\,dz\,u(x)\,dy\,d\!\!\!{}^-\xi.\end{aligned}$$

By Cauchy Theorem

$$\begin{aligned}\frac{1}{2\pi i}\oint \chi(z,\lambda)b_z\left(\frac{1}{2}(x+y),\xi\right)dz \\ &= \frac{1}{2\pi i}\oint \frac{\chi(z,\lambda)}{a(x,\xi)-z}\,dz+\cdots \\ &= \chi(a(x,\xi),\lambda)+\cdots.\end{aligned}$$

Thus

$$e_\lambda(x,\xi)=\chi(a(x,\xi),\lambda)+\cdots,$$

and

$$\begin{aligned}N(\lambda) &= \operatorname{Tr}E(\lambda) \\ &= \int \chi(a(x,\xi),\lambda)\,dx\,đ\xi+\cdots.\end{aligned}$$

What we want to do now is to make such an argument rigorous and to estimate the remainder. Let us first remark that $\chi(a(x,\xi),\lambda)$ is not smooth. So the first step is to regularize χ. Consider a positive real number ϵ and a smooth function

$$\psi:\mathbf{R}\to\mathbf{R},$$

such that

$$\int \psi\,dt=1,$$

and

$$\begin{aligned}\psi(t) &\geq 0, &&\text{for all } t,\\ \psi(t) &= 0, &&\text{for } |t|\geq 1.\end{aligned}$$

Define

$$\chi_\epsilon(t,\lambda)=\lambda^{-(1-\epsilon)}\int \chi\left(s,\lambda+\lambda^{1-\epsilon}/2\right)\psi\left(2\lambda^{-(1-\epsilon)}(t-s)\right)ds,$$

then

$$\chi_\epsilon(t,\lambda)=\begin{cases}1, & \text{if } t\leq\lambda,\\ 0, & \text{if } t\geq\lambda+\lambda^{1-\epsilon},\end{cases}$$

and for each $k\in\mathbf{N}$ we have

$$\left|\partial_t^k\chi_\epsilon(t,\lambda)\right|\prec\lambda^{-(1-\epsilon)k},\qquad \text{for } \lambda>0 \text{ and all } t.$$

Now we set

$$e_{\lambda,\epsilon}(x,\xi)=\chi_\epsilon(a(x,\xi),\lambda).$$

We have

$$e_{\lambda,\epsilon}(x,\xi) = \begin{cases} 1, & \text{if } a(x,\xi) \le \lambda, \\ 0, & \text{if } a(x,\xi) \ge \lambda + \lambda^{1-\epsilon}. \end{cases} \tag{24}$$

Thus $e_{\lambda,\epsilon}$ has compact support and defines a regularizing operator

$$E_\epsilon(\lambda) = \mathrm{Op}_W(e_{\lambda,\epsilon}).$$

Of course

$$e_{\lambda,\epsilon} - e_\lambda \to 0, \qquad \text{as } \lambda^{-\epsilon} \to 0,$$

so we can try to approximate $\operatorname{Tr} E(\lambda)$ by $\operatorname{Tr} E_\epsilon(\lambda)$ for λ large.

Observe that $e_{\lambda,\epsilon}$ has compact support, but not uniformly in λ. Actually we have the following estimate

$$\left|\partial_\xi^\alpha \partial_x^\beta e_{\lambda,\epsilon}(x,\xi)\right| \prec \lambda^{\epsilon|\alpha+\beta|} \Lambda_{\mathcal{P}}(x,\xi)^{-\rho|\alpha|+\delta|\beta|}, \tag{25}$$

for $\lambda \ge 1$ and $\{x\| + \|\xi\| \ge R \gg 0$ This is a consequence

1. of the estimate of $\chi_\epsilon(t,\lambda)$,
2. of the fact that
$$\lambda \le a(x,\xi) \le \lambda + \lambda^{1-\epsilon},$$
on the support of the derivatives of $e_{\lambda,\epsilon}$,
3. of the following Faà di Bruno-type estimate for two smooth functions $g : \mathbf{R} \to \mathbf{R}$ and $f : \mathbf{R}^n \to \mathbf{R}$:

$$|\partial_x^\alpha (g(f(x))| \prec \sum_{0<k\le|\alpha|} \left|g^{(k)}(f(x))\right| \cdot \sum_{\beta^1+\cdots+\beta^k=\alpha} \left|\partial_x^{\beta^1} f(x) \cdots \partial_x^{\beta^n} f(x)\right|.$$

Estimate (25) is fundamental for what follows. Thus from here up to the end of the paper we assume that $\lambda \ge 1$.

Theorem 7 *Assume there exist a function $h : \mathbf{R}^+ \to \mathbf{R}^+$ such that for a suitable*

$$0 < \epsilon < \frac{\rho - \delta}{2m},$$

we have

$$V\left(\lambda + \mathcal{O}(\lambda^{1-\epsilon})\right) = V(\lambda)\left[1 + \mathcal{O}(h(\lambda))\right], \tag{26}$$

where

$$V(\lambda) = \int \chi(a(x,\xi),\lambda)\, dx\, đ\xi$$

is the Weyl term and χ is defined in (22). Then, for each $k \in \mathbf{R}$ we have

$$N(\lambda) = V(\lambda)\left[1 + \mathcal{O}(h(\lambda))\right] + \mathcal{O}\left(\lambda^{-k}\right). \tag{27}$$

Proof: We need three lemmas:

Lemma 1 *$E_\epsilon(\lambda)$ is a trace-class operator for which we have the* trace formula:

$$\operatorname{Tr} E_\epsilon(\lambda) = \int e_{\lambda,\epsilon}(x,\xi)\, dx\, đ\xi.$$

Lemma 2 *For each $k \in \mathbf{R}$ we have*

$$\left\| E_\epsilon(\lambda - E_\epsilon(\lambda)^2 \right\|_{\mathcal{B}_1(L^2)} = \mathcal{O}\left(\lambda^{-k}\left(V(\lambda + \lambda^{1-\epsilon}) - V(\lambda) + 1\right)\right).$$

Lemma 3 *We have*

$$\begin{aligned} C'_\epsilon &= \sup_{\lambda \geq 1} \lambda^{-(1-\epsilon)} \left\| E_\epsilon(\lambda)(\overline{A} - \lambda I) E_\epsilon(\lambda) \right\|_{\mathcal{B}(L^2)} < \infty \\ C''_\epsilon &= \inf_{\lambda \geq 1} \left\{ \lambda^{-(1-\epsilon)} \inf_{\|u\|_{L^2}=1} \left((I - E_\epsilon(\lambda))(\overline{A} - \lambda I)(I - E_\epsilon(\lambda))u, u \right)_{L^2} \right\} \\ &> -\infty. \end{aligned}$$

We shall prove Lemmas 1 to 3 in the next section. Now we prove the theorem. Let us denote by $\{\mu_j\}$ the sequence of the eigenvalues of $E_\epsilon(\lambda)$, repeated according to their multiplicity and let $\{\psi_j\}$ be the sequence of the corresponding eigenfuctions which belong to $\mathcal{S}(\mathbf{R}^n)$ because $E_\epsilon(\lambda)$ is regularizing. Moreover we may assume that $\{\psi_j\}$ is an orthonormal set.

Define

$$N_\epsilon(\lambda) = \sum_{\mu_j \in [1/2, 3/2]} 1.$$

Let us show that

$$|N_\epsilon(\lambda) - \operatorname{Tr} E_\epsilon(\lambda)| \leq 4 \left\| E_\epsilon(\lambda) - E_\epsilon(\lambda)^2 \right\|_{\mathcal{B}_1(L^2)}.$$

In fact we have

$$\begin{aligned} N_\epsilon(\lambda) - \operatorname{Tr} E_\epsilon(\lambda) &= \sum_{\mu_j \in [1/2,3/2]} 1 - \sum \mu_j \\ &= \sum_{\mu_j \in [1/2,3/2]} (1 - \mu_j) - \sum_{\mu_j \notin [1/2,3/2]} \mu_j. \end{aligned}$$

But

$$|1 - \mu_j| \geq \frac{1}{2}, \qquad \text{if } \mu_j \notin \left[\tfrac{1}{2}, \tfrac{3}{2}\right],$$

thus

$$\begin{aligned} \sum_{\mu_j \in [1/2,3/2]} |1 - \mu_j| + \sum_{\mu_j \notin [1/2,3/2]} |\mu_j| &\leq 4 \sum \left| \mu_j - \mu_j^2 \right| \\ &= 4 \left\| E_\epsilon(\lambda) - E_\epsilon(\lambda)^2 \right\|^2_{\mathcal{B}_1(L^2)}. \end{aligned}$$

From Lemma 2 we obtain for any $k \in \mathbf{R}$:

$$N_\epsilon(\lambda) - \operatorname{Tr} E_\epsilon(\lambda) = \mathcal{O}\left(\lambda^{-k}\left(V(\lambda+\lambda^{1-\epsilon}) - V(\lambda)+1\right)\right).$$

From (24) and Lemma 1 we have

$$V(\lambda) \leq \operatorname{Tr} E_\epsilon(\lambda) \leq V(\lambda+\lambda^{1-\epsilon}).$$

Therefore we have for any $k \in \mathbf{R}$:

$$N_\epsilon(\lambda) = V(\lambda) + \mathcal{O}\left(V(\lambda+\lambda^{1-\epsilon}) - V(\lambda)\right) + \mathcal{O}(\lambda^{-k}). \tag{28}$$

Now we prove that there exists a positive constant C such that

$$N(\lambda - C\lambda^{1-\epsilon}) \leq N_\epsilon(\lambda) \leq N(\lambda + C\lambda^{1-\epsilon}). \tag{29}$$

Let M be the space generated by the eigenfunctions ψ_j corresponding to $\mu_j \in [1/2, 3/2]$. Then

$$\dim M = N_\epsilon(\lambda).$$

Consider

$$u = \sum_{\mu_j \in [1/2,3/2]} u_j \psi_j \in M,$$

and let

$$v = \sum_{\mu_j \in [1/2,3/2]} \frac{u_j}{\mu_j} \psi_j \in M.$$

Then of course

$$E_\epsilon(\lambda) v = u,$$

and we have from Lemma 3

$$\begin{aligned}
\left(\overline{A}u, u\right)_{L^2} &= \left(\overline{A}E_\epsilon(\lambda)v, E_\epsilon(\lambda)v\right)_{L^2} \\
&= \left(E_\epsilon(\lambda)\overline{A}E_\epsilon(\lambda)v, v\right)_{L^2} \\
&= \lambda \|E_\epsilon(\lambda)v\|_{L^2}^2 + \left(E_\epsilon(\lambda)(\overline{A} - \lambda I)E_\epsilon(\lambda)v, v\right)_{L^2} \\
&\leq \lambda\|u\|_{L^2}^2 + C'_\epsilon \lambda^{1-\epsilon}\|v\|_{L^2}^2 \\
&\leq \|u\|_{L^2}^2 + 4C'_\epsilon \lambda^{1-\epsilon}\|u\|_{L^2}^2,
\end{aligned}$$

because

$$\|v\|_{L^2}^2 = \sum_{\mu_j \in [1/2,3/2]} \left|\frac{u_j}{\mu_j}\right|^2 \leq 4 \sum_{\mu_j \in [1/2,3/2]} |u_j|^2 = 4\|u\|_{L^2}^2.$$

Thus, if we choose

$$C > 4C'_\epsilon,$$

we have

$$\mathrm{Im}\left(I - E(\lambda + C\lambda^{1-\epsilon})\right) \cap M = 0,$$

because

$$\left(\overline{A}u, u\right)_{L^2} > \left(\lambda + C\lambda^{1-\epsilon}\right) \|u\|^2_{L^2},$$

for $u \in \mathrm{Im}\,(I - E\,(\lambda + C\lambda^{1-\epsilon}))$. This implies that

$$N_\epsilon(\lambda) = \dim M \leq \dim\{\mathrm{Im}\, E(\lambda + C\lambda^{1-\epsilon})\} = N(\lambda + C\lambda^{1-\epsilon}).$$

Let now consider the orthogonal complement $M^\perp$ to M. Given

$$u = \sum_{\mu_j \notin [1/2, 3/2]} u_j \psi_j \in M^\perp$$

we have

$$u = (I - E_\epsilon(\lambda))v,$$

with

$$v = \sum_{\mu_j \notin [1/2,3/2]} \frac{u_j}{1 - \mu_j} \psi_j.$$

Computing as before and using Lemma 3, we have

$$\begin{aligned}
\left(\overline{A}u, u\right)_{L^2} &= \left((I - E_\epsilon(\lambda))\overline{A}(I - E_\epsilon(\lambda))v, v\right)_{L^2} \\
&= \lambda \|(I - E_\epsilon(\lambda))v\|^2_{L^2} + \left((I - E_\epsilon(\lambda))(\overline{A} - \lambda I)(I - E_\epsilon(\lambda))v, v\right)_{L^2} \\
&\geq \lambda\|u\|^2_{L^2} - C''_\epsilon \lambda^{1-\epsilon}\|v\|^2_{L^2}.
\end{aligned}$$

But

$$\|v\|^2_{L^2} = \sum_{\mu_j \notin [1/2,3/2]} \left|\frac{u_j}{1 - \mu_j}\right|^2 \leq 4 \sum_{\mu_j \notin [1/2,3/2]} |u_j|^2 = 4\|u\|^2_{L^2}.$$

Therefore

$$\left(\overline{A}u, u\right)_{L^2} \geq \left(\lambda - 4C''_\epsilon \lambda^{1-\epsilon}\right) \|u\|^2_{L^2}.$$

This means that, if we choose

$$C > 4C''_\epsilon,$$

we have

$$\mathrm{Im}\, E(\lambda - C\lambda_{1-\epsilon}) \cap M^\perp = 0,$$

because

$$\left(\overline{A}u, u\right)_{L^2} \leq (\lambda - C\lambda^{1-\epsilon})\|u\|_{L^2}^2,$$

for $u \in \operatorname{Im} E(\lambda - C\lambda^{1-\epsilon})$. Therefore

$$N(\lambda - C\lambda^{1-\epsilon}) = \dim\{\operatorname{Im} E(\lambda - C\lambda^{1-\epsilon})\} \leq \dim M = N_\epsilon(\lambda).$$

From (29) we obtain that

$$N_\epsilon\left(\lambda + \mathcal{O}(\lambda^{1-\epsilon})\right) \leq N(\lambda) \leq N_\epsilon\left(\lambda + \mathcal{O}(\lambda^{1-\epsilon})\right). \tag{30}$$

But from (28) and (26) we have

$$\begin{aligned} N_\epsilon\left(\lambda + \mathcal{O}(\lambda^{1-\epsilon})\right) &= V\left(\lambda + \mathcal{O}(\lambda^{1-\epsilon})\right) \\ &+ \mathcal{O}\left\{V\left(\lambda + \mathcal{O}(\lambda^{1-\epsilon})\right) - V\left(\lambda + \mathcal{O}(\lambda^{1-\epsilon})\right)\right\} + \mathcal{O}\left(\lambda^{-k}\right) \\ &= V(\lambda)\left(1 + \mathcal{O}(h(\lambda))\right) + \mathcal{O}\left(\lambda^{-k}\right). \end{aligned}$$

Thus from (30) we can conclude that

$$N(\lambda) = V(\lambda)\left(1 + \mathcal{O}(h(\lambda))\right) + \mathcal{O}\left(\lambda^{-k}\right). \square$$

8. Trace-class operators and proof of Lemmas 1 to 3

In order to prove Lemmas 1 to 3 of the previous section, we need some results on trace-class operators.

For simplicity we limit ourselves to *self-adjoint compact operators on a Hilbert space H* (see [2] for general compact operators). Given such an operator T we can consider the sequence of its eigenvalues μ_j repeated according to their multiplicity and the corresponding eigenvectors ψ_j, we may assume to be ortonormal. For each $p \geq 1$ consider the norm

$$\|T\|_{\mathcal{B}_p(H)} = \left(\sum |\mu_j|^p\right)^{1/p}.$$

If $\|T\|_{\mathcal{B}_p(H)} < \infty$, we say that T is in the space $\mathcal{B}_p(H)$. Operators in $\mathcal{B}_1(H)$ are called *trace-class operators*, while operators in $\mathcal{B}_2(H)$ are called *Hilbert-Schmidt operators.*

As we have already done, for operators in $\mathcal{B}_1(H)$ we can define the *trace* as

$$\operatorname{Tr} T = \sum \mu_j.$$

We have

$$\|T\|_{\mathcal{B}_2(H)} \leq \|T\|_{\mathcal{B}_1(H)}. \tag{31}$$

Moreover $\mathcal{B}_2(H)$ is a Hilbert space with scalar product given by

$$(S,T)_{\mathcal{B}_2(H)} = \mathrm{Tr}(T^*S). \tag{32}$$

One can prove the following result

Proposition 9 *If T and S belong to $\mathcal{B}_2(H)$, then $ST \in \mathcal{B}_1(H)$ and*

$$\|ST\|_{\mathcal{B}_1(H)} \leq \|S\|_{\mathcal{B}_2(H)}\|T\|_{\mathcal{B}_2(H)}.$$

□

Moreover, in the special case $H = L^2(\mathbf{R}^n)$ we have

Proposition 10 *$T \in \mathcal{B}_2(L^2)$ if and only if it has a kernel $K_T \in L^2(\mathbf{R}^{2n})$. We have*

$$\|T\|_{\mathcal{B}_2(L^2)} = \|K_T\|_{L^2(\mathbf{R}^{2n})}.$$

In particular, if $T = \mathrm{Op}_W(t) \in L^m_{\mathcal{P},\rho,\delta}$, with $m < -n/\mu_0$, where

$$\mu_0 = \min_{(\alpha,\beta)\in V(\mathcal{P})\setminus(0,0)} |\alpha+\beta|,$$

then $T \in \mathcal{B}_2(L^2)$ and

$$\|T\|_{\mathcal{B}_2(L^2)} = \|K_T\|_{L^2(\mathbf{R}^{2n})} = (2\pi)^{-n}\|t\|_{L^2(\mathbf{R}^{2n})}.$$

□

Theorem 8 *Let $T = \mathrm{Op}_W(t) \in L^m_{\mathcal{P},\rho,\delta}$, with $m < -2n/\mu_0$, then $T \in \mathcal{B}_1(L^2)$ and we have the* trace formula

$$\mathrm{Tr}\, T = \int t(x,\xi)\, dx\, d\!\!{}^-\!\xi.$$

Moreover there exists M such that for each $s \leq -2$ we have

$$\|T\|_{\mathcal{B}_1(L^2)} \prec \sum_{|\alpha+\beta|\leq M} \int w_m(x,\xi)^s \left|\partial_\xi^\alpha \partial_x^\beta t(x,\xi)\right| dx\, d\!\!{}^-\!\xi, \tag{33}$$

for all $t \in L^m_{\mathcal{P},\rho,\delta}$, where

$$w_m = \sigma * \Lambda_{\mathcal{P}}^{-m/2}$$

and

$$\sigma(x,\xi) = \pi^{-n} e^{-\|x\|^2-\|\xi\|^2}.$$

Proof: Let

$$W_m = \mathrm{Op}_W(w_m) = \mathrm{Op}_{AW}(\Lambda_{\mathcal{P}}^{-m/2}).$$

From Theorem 5 we know that W_m is invertible. Thus we can write

$$T = W_m^{-1}(W_m T),$$

and W_m^{-1}, $W_m T \in L^{m/2}_{\mathcal{P},\rho,\delta}$. From Propositions 9 and 10 we obtain $T \in \mathcal{B}_1(L^2)$ and

$$\begin{aligned} \|T\|_{\mathcal{B}_1(L^2)} &\leq \|W_m^{-1}\|_{\mathcal{B}_2(L^2)} \|W_m T\|_{\mathcal{B}_2(L^2)} \\ &= (2\pi)^{-n} \|W_m^{-1}\|_{\mathcal{B}_2(L^2)} \|w_m \# t\|_{L^2}. \end{aligned} \tag{34}$$

Now it is easy to prove the trace formula. From (32) and Propositions 9 and 10 we have

$$\begin{aligned} \operatorname{Tr} T &= \left(W_m T, W_m^{-1}\right)_{\mathcal{B}_2(L^2)} \\ &= \int K_{W_m T}(x,y) K_{W_m^{-1}}(y,x)\, dx\, dy. \end{aligned}$$

But the kernel K_T is continuous because it is the Fourier transform of a L^1-function, so by Fubini Theorem we obtain

$$K_T(x,x) = \int K_{W_m T}(x,y) K_{W_m^{-1}}(y,x)\, dy$$

and from (18) we have

$$K_T(x,x) = \int t(x,\xi)\, đ\xi.$$

Eventually, we have

$$\operatorname{Tr} T = \int t(x,\xi)\, dx\, đ\xi.$$

Now we prove the estimate of $\|T\|_{\mathcal{B}_1(L^2)}$. This is done in several steps.

First we assume that t has support in the unit ball. Because $w_m \in S^{-m/2}_{\mathcal{P},\rho,\delta}$, from (19) we obtain that there exists M' such that

$$\|w_m \# t\|_{L^2} \prec \sum_{|\alpha+\beta| \leq M'} \left\| \partial_\xi^\alpha \partial_x^\beta t \right\|_{L^\infty},$$

for all $t \in S^{m/2}_{\mathcal{P},\rho,\delta}$ with support in the unit ball. Because t has compact support

$$t(x,\xi) = (2\pi)^n \int_{-\infty}^{x_1} \cdots \int_{-\infty}^{x_n} \int_{-\infty}^{\xi_1} \cdots \int_{-\infty}^{\xi_n} \partial_\eta^e \partial_y^e t(y,\eta)\, dy\, đ\eta,$$

where $e = (1, \dots, 1) \in \mathbf{R}^n$, so that, for example

$$\partial_\xi^e = \partial_{\xi_1} \cdots \partial_{\xi_n}.$$

It follows that

$$\left\| \partial_\xi^\alpha \partial_x^\beta t \right\|_{L^\infty} \prec \int \left| \partial_\xi^{e+\alpha} \partial_x^{e+\beta} t \right| \, dx \, d\!\!\!/\xi.$$

This shows that

$$\| w_m \# t \|_{L^2} \prec \sum_{|\alpha+\beta| \leq M} \int \left| \partial_\xi^\alpha \partial_x^\beta t \right| \, dx \, d\!\!\!/\xi,$$

with $M = M' + 2n$. Because t has support in the unit ball and $w_m > 0$, because $\Lambda_{\mathcal{P}}^{-m/2} \geq 1$ (see subsection 5.3), we have

$$\| w \# t \|_{L^2} \prec \sum_{|\alpha+\beta| \leq M} \int w^s \left| \partial_\xi^\alpha \partial_x^\beta t \right| \, dx \, d\!\!\!/\xi,$$

and (33) follows from (34).

Now we assume that t has support in a ball of radius 1 and center $(\tilde{x}, \tilde{\xi})$. Let

$$\tilde{t}(x, \xi) = t(x - \tilde{x}, \xi - \tilde{\xi}),$$

and

$$\tilde{T} = \mathrm{Op}_W(\tilde{t}).$$

Then

$$\tilde{T} = U^{-1} T U,$$

with

$$U u(x) = e^{-ix \cdot \tilde{\xi}} u(x + \tilde{x}).$$

U is unitary, hence

$$\left\| \tilde{T} \right\|_{\mathcal{B}_1(L^2)} = \| T \|_{\mathcal{B}_1(L^2)} \cdot$$

Let

$$\tilde{W}_m = \mathrm{Op}_W(\tilde{w}_m),$$

where

$$\tilde{w}_m(x, \xi) = w_m(x - \tilde{x}, \xi - \tilde{\xi}).$$

$\tilde{W}_m$ is the anti-Wick quantization of $\Lambda_{\mathcal{P}}(x - \tilde{x}, \xi - \tilde{\xi})^{-m/2}$ and, as before

$$\tilde{W}_m = U^{-1} W_m U.$$

Thus

$$\tilde{W}_m^{-1} = U^{-1} W_m^{-1} U$$

and

$$\left\|\tilde{W}_m^{-1}\right\|_{\mathcal{B}_2(L^2)} = \left\|W_m^{-1}\right\|_{\mathcal{B}_2(L^2)}.$$

It follows that

$$\begin{aligned}\|T\|_{\mathcal{B}_1(L^2)} &= \left\|\tilde{T}\right\|_{\mathcal{B}_1(L^2)} \\ &\prec \left\|\tilde{W}_m^{-1}\right\|_{\mathcal{B}_2(L^2)} \left\|\tilde{W}_m\tilde{T}\right\|_{\mathcal{B}_2(L^2)} \\ &= \left\|W_m^{-1}\right\|_{\mathcal{B}_2(L^2)} \left\|\tilde{W}_m\tilde{T}\right\|_{\mathcal{B}_2(L^2)}.\end{aligned}$$

But $\tilde{t}$ has support in the unit ball. Thus we have already proven that

$$\begin{aligned}\left\|\tilde{W}_n\tilde{T}\right\|_{\mathcal{B}_2(L^2)} &= (2\pi)^{-n}\left\|\tilde{w}_m\#\tilde{t}\right\|_{L^2(\mathbf{R}^{2n})} \\ &\prec \sum_{|\alpha+\beta|\le M}\int\left|\partial_\xi^\alpha\partial_x^\beta\tilde{t}\right|\,dx\,đ\xi \\ &= \sum_{|\alpha+\beta|\le M}\int\left|\partial_\xi^\alpha\partial_x^\beta t\right|\,dx\,đ\xi.\end{aligned}$$

Now we consider a general t with compact support. Let $\{\theta_j\}$ be a partition of unity of $\mathbf{R}^{2n}$ such that

1. each θ_j has support in a ball or radius 1,
2. there exists l such that each $\operatorname{supp}\theta_j$ intersects at most l $\operatorname{supp}\theta_k$,
3. for each $\alpha\,\beta \in \mathbf{N}^n$ there exists a positive constant $C_{\alpha,\beta}$ such that

$$\left\|\partial_\xi^\alpha\partial_x^\beta\theta_j\right\|_{L^\infty} \le C_{\alpha,\beta}, \qquad \text{for all } j.$$

For each j let $T_j = \operatorname{Op}_W(\theta_j t)$. Because $\operatorname{supp} t$ intersects only a finite number of $\operatorname{supp}\theta_j$, say for $j = j_1, \ldots, j_K$, we have that T is a finite sum of T_j's:

$$T = \sum_{k=1}^{K} T_{j_k}.$$

It follows that

$$\begin{aligned}\|T\|_{\mathcal{B}_1(L^2)} &\le \sum_{k=1}^{K}\|T_{j_k}\|_{\mathcal{B}_1(L^2)} \\ &\prec \sum_{k=1}^{K}\sum_{|\alpha+\beta|\le M}\int w_m^s\left|\partial_\xi^\alpha\partial_x^\beta\left(\theta_{j_k}t\right)\right|\,dx\,đ\xi\end{aligned}$$

$$\prec \sum_{k=1}^{K} \sum_{|\alpha+\beta|\leq M} \int_{\operatorname{supp}\theta_{j_k}} w_m^s \left|\partial_\xi^\alpha \partial_x^\beta t\right| \, dx \, đ\xi$$
$$\leq l \sum_{|\alpha+\beta|\leq M} \int w_m^s \left|\partial_\xi^\alpha \partial_x^\beta t\right| \, dx \, đ\xi.$$

Thus we obtained the estimate (33) for each t with compact support. Let now consider a general symbol t and let $\{t_j\}$ be a sequence of symbols with compact support tending to t with all the derivatives, uniformly on compact subsets. Choose $s \leq -2$. Then w^s is integrable and we obtain

$$\lim_{j\to\infty} \left\{ \sum_{|\alpha+\beta|\leq M} \int w^s \left|\partial_\xi^\alpha \partial_x^\beta t_j\right| \, dx \, đ\xi \right\} = \sum_{|\alpha+\beta|\leq M} \int w^s \left|\partial_\xi^\alpha \partial_x^\beta t\right| \, dx \, đ\xi.$$

Because each t_j has compact support

$$\|T_j\|_{\mathcal{B}_1(L^2)} \prec \sum_{|\alpha+\beta|\leq M} \int w^s \left|\partial_\xi^\alpha \partial_x^\beta t_j\right| \, dx \, đ\xi.$$

This implies in particular that $\{T_j\}$ is a Cauchy sequence in $\mathcal{B}_1\left(L^2(\mathbf{R}^n)\right)$. On the other side, $T_j \to T$ in $\mathcal{B}_2\left(L^2(\mathbf{R}^n)\right)$, because $t_j \to t$ in $L^2(\mathbf{R}^{2n})$. Thus from (31) we obtain that $T_j \to T$ in $\mathcal{B}_1\left(L^2(\mathbf{R}^n)\right)$, and the proof is complete. □

Now we prove Lemmas 1 to 3 of Section 7. Lemma 1 is obvious because $e_{\lambda,\epsilon}$ has compact support.

Let us prove Lemma 2. Let

$$w = w_{-\frac{n}{\mu_0}-\frac{1}{2}} = \sigma * \Lambda_{\mathcal{P}}^{\frac{n}{\mu_0}+\frac{1}{2}}.$$

From Theorem 8 we have that there exists M such that for any $s \leq -2$ we have

$$\left\|E_\epsilon(\lambda)^2 - E_\epsilon(\lambda)\right\|_{\mathcal{B}_1(L^2)} \prec \sum_{|\alpha+\beta|\leq M} \int w^s \left|\partial_\xi^\alpha \partial_x^\beta (e_{\lambda,\epsilon} \# e_{\lambda,\epsilon} - e_{\lambda,\epsilon})\right| \, dx \, đ\xi.$$

We have the asymptotic expansion:

$$\begin{aligned} e_{\lambda,\epsilon} \# e_{\lambda,\epsilon} - e_{\lambda,\epsilon} &= \sum_{0<|\phi+\psi|<N} \frac{(-1)^{|\phi+\psi|}}{\phi!\psi!2^{|\phi+\psi|}} \partial_\xi^\phi D_x^\psi e_{\lambda,\epsilon} \partial_\xi^\psi D_x^\phi e_{\lambda,\epsilon} + r_{\lambda,\epsilon,N} \\ &= p_{\lambda,\epsilon,N} + r_{\lambda,\epsilon,N}. \end{aligned}$$

From (24) and (25) we obtain

$$\left|\partial_\xi^\alpha \partial_x^\beta e_{\lambda,\epsilon}(x,\xi)\right| \prec \Lambda_{\mathcal{P}}(x,\xi)^{-\tilde{\rho}|\alpha|+\tilde{\delta}|\beta|}, \qquad \text{for all } (x,\xi) \text{ and } \lambda \geq 1, \tag{35}$$

where

$$\tilde{\rho} = \rho - m\epsilon, \qquad \text{and} \qquad \tilde{\delta} = \delta + m\epsilon. \tag{36}$$

Therefore we have that

$$p_{\lambda,\epsilon,N} \in S^0_{\mathcal{P},\tilde{\rho},\tilde{\delta}},$$

uniformly with respect to λ. Moreover

$$\partial_\xi^\alpha \partial_x^\beta e_{\lambda,\epsilon}(x,\xi) \prec \lambda^{-\epsilon'|\alpha+\beta|}\Lambda_{\mathcal{P}}(x,\xi)^{-\rho'|\alpha|+\delta'|\beta|},$$

with

$$\epsilon' = \frac{1}{2}\left(\frac{\rho-\delta}{2m} - \epsilon\right)$$

and

$$\rho' = \rho - (\epsilon+\epsilon')m, \qquad \delta' = \delta + (\epsilon+\epsilon')m.$$

Thus

$$\left|\partial_\xi^\phi \partial_x^\psi r_{\lambda,\epsilon,N}(x,\xi)\right| \prec \lambda^{-\epsilon'(N+|\phi+\psi|)}\Lambda_{\mathcal{P}}(x,\xi)^{-(\rho'-\delta')N-\rho'|\phi|+\delta'|\psi|}.$$

Finally we have

$$w(x,\xi)^{-s} \prec \Lambda_{\mathcal{P}}(x,\xi)^{-\left(\frac{n}{\mu_0}+1\right)s},$$

so (see (24))

$$w(x,\xi)^{-s} \prec \lambda^{-\left(\frac{n}{\mu_0}+\frac{1}{2}\right)ms}, \qquad \text{for } (x,\xi) \in \operatorname{supp}(e_{\lambda,\epsilon}).$$

Then, for any $s \leq -2$ and $N > 0$ we obtain the following estimate

$$\begin{aligned}
&\left\|E_\epsilon(\lambda)^2 - E_\epsilon(\lambda)\right\|_{\mathcal{B}_1(L^2)} \\
&= \mathcal{O}\left(\lambda^{-\left(\frac{n}{\mu_0}+\frac{1}{2}\right)ms}\int\left\{\chi(a(x,\xi),\lambda+\lambda^{1-\epsilon}) - \chi(a(x,\xi),\lambda)\right\}dx\,đ\xi\right) \\
&\quad + \mathcal{O}\left(\lambda^{-\epsilon'N}\right) \\
&= \mathcal{O}\left(\lambda^{-\left(\frac{n}{\mu_0}+\frac{1}{2}\right)ms}\left(V\left(\lambda+\lambda^{1-\epsilon}\right) - V(\lambda)\right)\right) + \mathcal{O}\left(\lambda^{-\epsilon'N}\right).
\end{aligned}$$

This proves Lemma 2. □

Let us prove Lemma 3. Consider $E_\epsilon(\lambda)(\overline{A}-\lambda I)E_\epsilon(\lambda)$. Its Weyl symbol is

$$e_{\lambda,\epsilon}\#(a-\lambda)\#e_{\lambda,\epsilon}.$$

On the support of $e_{\lambda,\epsilon}$ we have

$$a(x,\xi)-\lambda<\lambda^{1-\epsilon}$$

and, for $|\alpha+\beta|>0$

$$\begin{aligned}\left|\partial_\xi^\alpha\partial_x^\beta a(x,\xi)\right| &\prec a(x,\xi)\Lambda_{\mathcal{P}}(x,\xi)^{-\rho|\alpha|+\delta|\beta|}\\ &\prec \lambda\Lambda_{\mathcal{P}}(x,\xi)^{-\rho|\alpha|+\delta|\beta|}\\ &\prec \lambda^{1-\epsilon|\alpha+\beta|}\Lambda_{\mathcal{P}}(x,\xi)^{-\tilde\rho|\alpha|+\tilde\delta|\beta|},\end{aligned}$$

for $\|x\|+\|\xi\|\geq R\gg 0$. $\tilde\rho$ and $\tilde\delta$ are defined in (36). Thus we have

$$\left|\partial_\xi^\alpha\partial_x^\beta(a(x,\xi)-\lambda)\right|\prec\lambda^{1-\epsilon}\Lambda_{\mathcal{P}}(x,\xi)^{-\tilde\rho|\alpha|+\tilde\delta|\beta|},$$

for $(x,\xi)\in\mathrm{supp}(e_{\lambda,\epsilon})$ and $\|x\|+\|\xi\|\geq R\gg 0$. Hence, from (19) it follows that

$$\begin{aligned}e_{\lambda,\epsilon}\#(a-\lambda) &= \sum_{|\alpha+\beta|<n}\frac{(-1)^{|\beta|}}{\alpha!\beta!2^{|\alpha+\beta|}}\partial_\xi^\alpha D_x^\beta e_{\lambda,\epsilon}\partial_\xi^\beta D_x^\alpha(a-\lambda)+r_{\lambda,\epsilon,N}\\ &= p_{\lambda,\epsilon,N}+r_{\lambda,\epsilon,N},\end{aligned}$$

with

$$\lambda^{-(1-\epsilon)}p_{\lambda,\epsilon,N}\in S^0_{\mathcal{P},\tilde\rho,\tilde\delta},\qquad r_{\lambda,\epsilon,N}\in S^{m-(\tilde\rho-\tilde\delta)N}_{\mathcal{P},\tilde\rho,\tilde\delta},$$

uniformly with respect to λ. Thus we obtain that

$$\lambda^{-(1-\epsilon)}\left(e_{\lambda,\epsilon}\#(a-\lambda)\#e_{\lambda,\epsilon}\right)\in S^0_{\mathcal{P},\tilde\rho,\tilde\delta},$$

uniformly with respect to λ.

This proves that

$$\sup_{\lambda\geq 1}\lambda^{-(1-\epsilon)}\left\|E_\epsilon(\lambda)(\overline{A}-\lambda I)E_\epsilon(\lambda)\right\|_{\mathcal{B}(L^2)}<\infty.$$

Let now consider

$$(I-E_\epsilon(\lambda))\left(\overline{A}-\lambda I\right)(I-E_\epsilon(\lambda))$$

with Weyl symbol

$$(1-e_{\lambda,\epsilon})\#(a-\lambda)\#(1-e_{\lambda,\epsilon}).$$

As before we have

$$1 - e_{\lambda,\epsilon} \in S^0_{\mathcal{P},\tilde{\rho},\tilde{\delta}},$$

and

$$\lambda^{-(1-\epsilon)}\{(1-e_{\lambda,\epsilon})\#(a-\lambda)\#(1-e_{\lambda,\epsilon}) - (1-e_{\lambda,\epsilon})(a-\lambda)(1-e_{\lambda,\epsilon})\} \in S^0_{\mathcal{P},\tilde{\rho},\tilde{\delta}},$$

uniformly with respect to λ. Moreover

$$(1-e_{\lambda,\epsilon})(a-\lambda)(1-e_{\lambda,\epsilon}) \geq 0,$$

thus from Corollary 2 we have that

$$\mathrm{Op}_W\left(\lambda^{-(1-\epsilon)}(1-e_{\lambda,\epsilon})(a-\lambda)(1-e_{\lambda,\epsilon})\right)$$

is uniformly bounded from below with respect to λ and we obtain

$$\inf_{\lambda\geq 1}\left\{\lambda^{-(1-\epsilon)}\inf_{\|u\|_{L^2}=1}\left((I-E_\epsilon(\lambda))\left(\overline{A}-\lambda I\right)(I-E_\epsilon(\lambda))u,u\right)_{L^2}\right\} > -\infty.$$

□

9. Estimate of the Weyl term and proof of Theorem 6

Thanks to Theorem 7 in order to prove Theorem 6 we need only to show how to choose the function h. Of course it is of interest to choose h such that it goes to 0 as $\lambda \to \infty$. The following proposition shows how to do this.

Recall that

$$V(\lambda) = \int \chi(a(x,\xi),\lambda)\, dx\, đ\xi,$$

and

$$\chi(t,\lambda) = \begin{cases} 1, & \text{if } t \leq \lambda, \\ 0, & \text{if } t \geq \lambda. \end{cases}$$

Proposition 11 *Assume that for some constants V_0, p, q, p', q' such that $V_0 > 0$ and*

$$\textit{either} \quad p > p' \quad \textit{or} \quad p = p' \;\; \textit{and} \;\; q > q',$$

we have

$$V(\lambda) = V_0\lambda^p(\log\lambda)^q + \left(\mathcal{O}(\lambda^{p'}(\log\lambda)^{q'}\right).$$

Then (27) is valid, with

$$h(\lambda) = \max\left\{\lambda^{-\epsilon}, \lambda^{p'-p}(\log\lambda)^{q'-q}\right\}$$

and therefore also

$$\begin{aligned} N(\lambda) &= V(\lambda)\,(1 + \mathcal{O}(h(\lambda))) \\ &= V_0\lambda^p(\log\lambda)^q\,(1 + \mathcal{O}(h(\lambda)))\,. \end{aligned} \tag{37}$$

Proof: We have

$$\frac{V(\lambda + \mathcal{O}(\lambda^{1-\epsilon})) - V(\lambda)}{V(\lambda)} = \mathcal{O}\left[\lambda^{-\epsilon} + \lambda^{p-p'}(\log\lambda)^{q'-q}\right].$$

Then (37) follows from (27) with $k < p$. □

Now Theorem 6 is a consequence of this proposition and the following

Theorem 9 *Assume that $\mathcal{P}$ is non degenerate, so that there exist*

$$\nu = (\nu', \nu'') \in (\mathbf{R}^+)^n \times (\mathbf{R}^+)^n$$

and

$$0 < s \leq 1,$$

such that

$$\begin{aligned} &|x_1 \cdots x_n \cdot \xi_1 \cdots \xi_n|^{\frac{1-s}{|\nu|}} \cdot \\ &\cdot \left(|x_1|^{s/\nu_1'} + \cdots + |x_n|^{s/\nu_n'} + |\xi_1|^{s/\nu_1''} + \cdots + |\xi_n|^{s/\nu_n''}\right) \\ &\prec \Lambda_{\mathcal{P}}(x, \xi), \end{aligned}$$

for all (x, ξ).

Assume further that

$$a = a_0 + \tilde{a},$$

where

1. *a_0 is quasi-homogeneous:*

$$a_0\left(\lambda^{\nu_1'}x_1, \ldots, \lambda^{\nu_n'}x_n, \lambda^{\nu_1''}\xi_1, \ldots, \lambda^{\nu_n''}\xi_n\right) = \lambda^m a_0(x, \xi),$$

   for $\lambda > 0$ and $(x, \xi) \neq (0, 0)$,

2. *a_0 is non degenerate:*

$$\begin{aligned} &|x_1 \cdots x_n \cdot \xi_1 \cdots \xi_n|^{\frac{1-s}{|\nu|}} \cdot \\ &\cdot \left(|x_1|^{s/\nu_1'} + \cdots + |x_n|^{s/\nu_n'} + |\xi_1|^{s/\nu_1''} + \cdots + |\xi_n|^{s/\nu_n''}\right) \\ &\prec \Lambda_{\mathcal{P}}(x, \xi), \end{aligned}$$

3. there exists $0 \le \tilde{s} < 1$ *such that*

$$|\tilde{a}(x,\xi)|^{1/m} \prec 1 + |x_1|^{\tilde{s}/\nu_1'} + \cdots + |x_n|^{\tilde{s}/\nu_n'} + |\xi_1|^{\tilde{s}/\nu_1''} + \cdots + |\xi_n|^{\tilde{s}/\nu_n''}.$$

Then

$$V(\lambda) = \left(V_0 + \mathcal{O}(\tilde{V}(\lambda))\right) \lambda^{|\nu|/m},$$

where

$$V_0 = \int \chi(a_0(x,\xi),1)\, dx\, đ\xi$$

and

$$\tilde{V}(\lambda) = \begin{cases} \lambda^{-\epsilon_3} (\log \lambda)^{2n-2}, & \text{if } \epsilon_3 < \epsilon_2, \\ \lambda^{-\epsilon_2} (\log \lambda)^{2n-1}, & \text{if } \epsilon_2 = \epsilon_3, \\ \lambda^{-\epsilon_2}, & \text{if } \epsilon_2 > \epsilon_3, \end{cases}$$

where

$$\epsilon_2 = 1 - \tilde{s}, \qquad \epsilon_3 = \frac{(1-\tilde{s})s}{(1-s)\tilde{s}} \frac{|\nu|}{m}.$$

Outline of proof (see [2], Ch. 2, Thm. 6.1): We have to estimate

$$\int \chi(a(z),\lambda)\, dz - \int \chi(a_0(z),\lambda)\, dz,$$

where

$$z = (x,\xi).$$

Let

$$\nu_j' = \nu_j, \quad \text{and} \quad \nu_j'' = \nu_{j+n},$$

for $j = 1, \ldots, n$.

Of course we may limit ourselves to the first quadrant

$$z \ge 0, \quad \text{i.e. } z_j \ge 0, \text{ for } j = 1, \ldots, 2n.$$

Then we perform the following change of variables:

$$z_j = (\lambda u_j)^{\nu_j/m}, \quad \text{for } j = 1, \ldots, 2n,$$

and let

$$\begin{aligned} b_0(u) &= a_0\left(u_1^{\nu_1/m}, \ldots, u_{2n}^{\nu_{2n}/m}\right), \\ \tilde{b}_\lambda(u) &= \lambda^{-1} \tilde{a}\left((\lambda u_1)^{\nu_1/m}, \ldots, (\lambda u_{2n})^{\nu_{2n}/m}\right). \end{aligned}$$

$\lambda^{|\nu|/m}$ factors out and we reduce to prove that

$$\int\limits_{\substack{b_0+\tilde{b}_\lambda \le 1 \\ u \ge 0}} \prod_{j=1}^{2n} u_j^{\nu_j/m-1}\, du - \int\limits_{\substack{b_0 \le 1 \\ u \ge 0}} \prod_{j=1}^{2n} u_j^{\nu_j/m-1}\, du = \mathcal{O}(\tilde{V}(\lambda)), \qquad \text{as } \lambda \to +\infty.$$

We can limit ourselves to the sector

$$U = \left\{ u \in \mathbf{R}^{2n} : u_1 \geq u_2 \geq \cdots \geq u_{2n} \geq 0 \right\}.$$

Now we make a second change of co-ordinates:

$$u = \frac{t\omega(\theta)}{b_0(\omega(\theta))},$$

where

$$\begin{aligned} \theta &= (\theta_1, \ldots, \theta_{2n-1}), \\ \omega(\theta) &= (\omega_1(\theta), \ldots, \omega_{2n}(\theta)), \end{aligned}$$

and

$$\begin{aligned} \omega_1(\theta) &= \cos\theta_1, \\ \omega_k(\theta) &= \left(\prod_{j=1}^{k-1} \sin\theta_j \right) \cos\theta_k, \quad \text{for } 2 \leq k \leq 2n-1, \\ \omega_{2n}(\theta) &= \prod_{j=1}^{2n-1} \sin\theta_j. \end{aligned}$$

U is mapped over $\mathbf{R}^+ \times \Theta$, where

$$\Theta = \Big\{ \theta \in \mathbf{R}^{2n-1} : \\ 0 \leq \theta_j \leq \arctan(\sec\theta_{j+1}), \text{ for } 1 \leq j < 2n-1 \text{ and } 0 \leq \theta_{2n-1} \leq \pi/4 \Big\},$$

and the curve $b_0(u) = 1$ is mapped over $t = 1$.

So we reduce to estimate

$$\mathcal{R}(\lambda) = \int_{\substack{0<t\leq \max\{0,1-\tilde{r}_\lambda\} \\ \theta\in\Theta}} H(\theta) t^{|\nu|/m-1}\, dt\, d\theta - \int_{\substack{0<t\leq 1 \\ \theta\in\bar{\Theta}}} H(\theta) t^{|\nu|/m-1}\, dt\, d\theta,$$

where

$$\tilde{r}_\lambda(t,\theta) = \tilde{b}_\lambda \left(\frac{t\omega(\theta)}{b_0(\omega(\theta))} \right),$$

and

$$H(\theta) = (b_0(\omega))^{-|\nu|/m} \, \omega_1^{\nu_1/m} \prod_{j=2}^{2n} \frac{\omega_j^{\nu_j/m}}{\sin\theta_{j-1}\cos\theta_{j-1}}.$$

From hypoellipticity

$$b_0(u)^{l/m} \prec b_0(u) + \tilde{b}_\lambda(u), \qquad \text{for } \|u\| \geq R \gg 0 \text{ and } \lambda \geq 1.$$

Therefore

$$t^{l/m} \prec t + \tilde{r}_\lambda(t,\theta), \qquad \text{for } \frac{t}{b_0(\omega(\theta))} \geq R, \text{ and } \lambda \geq 1.$$

But on the domain of integration

$$t + \tilde{r}_\lambda(t,\theta) \leq 1,$$

thus, either

$$t \leq R \max_{\theta\in\Theta} b_0(\omega(\theta)) < \infty,$$

or

$$t \prec (t + \tilde{r}_\lambda(t,\theta))^{m/l} \leq 1.$$

Therefore, on the domain of integration t is bounded by a suitable constant T.

Form the hypotheses we have

$$(u^\nu)^{\frac{1-s}{|\nu|}} \left(u_1^{s/m} + \cdots + u_{2n}^{s/m}\right)^m \prec b_0(u),$$

and

$$|\tilde{b}_\lambda(u)| \prec \lambda^{-1} \left[1 + \lambda^{\tilde{s}/m} \left(u_1^{\tilde{s}/m} + \ldots + u_{2n}^{\tilde{s}/m}\right)\right]^m.$$

This yields

$$|\tilde{r}_\lambda(t,\theta)| \prec \lambda^{\tilde{s}-1} \left(\theta_2^{\nu_2+\cdots+\nu_{2n}} \cdots \theta_{2n-1}^{\nu_{2n}}\right)^{-\frac{(1-s)\tilde{s}}{|\nu|}},$$

for $\lambda \geq 1$, $-1 \leq t \leq T$, and $\theta \in \Theta$, and

$$H(\theta) \prec \left(\omega_1^{s/m} + \cdots + \omega_{2n}^{s/m}\right)^{-|\nu|} \omega_1^{s\nu_1/m} \prod_{j=2}^{2n} \frac{\omega_j^{s\nu_j/m}}{\sin\theta_{j-1}\cos\theta_{j-1}} \qquad \text{for } \theta \in \Theta.$$

From these estimates we obtain the result. □

References

1. Boggiatto P., Buzano E. (1995) Spectral asymptotics for multi-quasi-elliptic operators in $\mathbf{R}^n$, submitted to *Ann. Scuola Norm. Sup. Pisa.*
2. Boggiatto, P., Buzano, E. and Rodino, L. (1996) *Global Hypoellipticity and Spectral Theory,* Akademie Verlag, Berlin.

3. Boggiatto, P., Buzano, E. and Rodino, L. (1996) Spectral asymptotics for hypoelliptic operators, to be published in the *Proceedings of the conference "Partial Differential Equations"*, Potsdam, Germany, July 29–August 3.
4. Cattabriga, L. (1966–67) Su una classe di polinomi ipoellittici, *Rend. Sem. Mat. Univ. Padova,* **36**, 285–309; **37**, 60–74.
5. Friberg J. (1967) Multi-quasielliptic polynomials, *Ann. Scuola Norm. Sup. Pisa*, Cl. Sc., **21**, 239–260.
6. Helffer B., Robert D. (1981) *Comportement asymptotique précisé du spectre d'opérateurs globalement elliptiques dans* $\mathbf{R}^n$, C.R. Acad. Sc. Paris, **292**, 363–366.
7. Helffer B., Robert D. (1982) Propriétés asymptotiques du spectre d'opérateurs pseudodifférentiels sur $\mathbf{R}^n$, *Comm. in P.D.E.*, **7**, 795–882.
8. Hörmander, L. (1979) On the asymptotic distribution of the eigenvalues of pseudodifferential operators ir $\mathbf{R}^n$. *Arkiv för Mat.,* **17**, 297–313.
9. Kumano-go H. (1974) *Pseudo-Differential Operators,* The MIT Press, Cambridge, MA.
10. Mohamed A. (1989) Comportement asymptotique, avec estimation du reste, des valeurs propres d'une classe d'opérateurs pseudo-différentiels sur $\mathbf{R}^n$, *Math. Nachr.*, **140**, 127–186.
11. Shubin M.A. (1987) *Pseudodifferential operators and spectral theory*, Springer-Verlag, Berlin.
12. Tamura H. (1982) Asymptotic formulae with remainder estinates for eigenvalues of Schrödinger operators, *Comm. P.D.E.,* **7**, 1–53.
13. Taylor M. E. (1981) *Pseudodifferential operators,* Princeton Univ. Press, Princeton.
14. Tulovskiĭ V.N., Shubin M.A. (1973) On asymptotic distribution of eigenvalues of pseudodifferential operators in $\mathbf{R}^n$, Math. USSR Sbornik, **21**, 565–583.
15. Volevič L.R., Gindikin S.G. (1968) On a class of hypoelliptic polynomials, *Math. USSR Sbornik,* **4**, 369–383.

SPLITTING IN LARGE DIMENSION AND INFRARED ESTIMA[T

B. HELFFER
UA 760 du CNRS, Département de mathématiques,
Bât 425,
F-91405 Orsay Cédex, FRANCE

Abstract. These notes for the NATO ASI conference in Microlocal analysis and spectral theory consist in the analysis of the links between estimating the splitting between the two first eigenvalues for the Schrödinger operator and the proof of infrared estimates for quantities attached to Gaussian type measures. They are mainly based on the "old" contributions of Dyson, Fröhlich, Glimm, Jaffe, Lieb, Simon, Spencer (in the seventie's) in connection with more recent contributions of Pastur, Khoruzhenko, Barbulyak, Kondratev which treat in general more sophisticated models. We shall show how the recent semi-classical analysis permits sometimes to state more precise results.

1. Introduction

Our aim is to understand in the large dimension limit $m \to +\infty$ the splitting between the two lowest eigenvalues of the following Schrödinger operator

$$-h^2\Delta + \sum_{j=1}^{m} v(x_j) + \frac{\mathcal{J}}{2}\sum_{j=1}^{m} |x_j - x_{j+1}|^2 \tag{1.1}$$

in the case when the so-called one-particle potential v defines a double well (We take the convention that $x_{m+1} = x_1$). As well known, the first eigenvalue of the Schrödinger operator is simple, so the splitting is always strictly positive. Its behavior with respect to m as $m \to +\infty$ depends actually heavily on the nature of v and on the size on $\mathcal{J}$ which is always assumed positive.
More generally we are interested in the similar problem attached to a d-dimensional (periodic) lattice $\Lambda = (\mathbb{Z}/m\mathbb{Z})^d$, identified with $[1\ ,\ m]^d \cap$

L. Rodino (ed.), Microlocal Analysis and Spectral Theory, 307–347.

$\mathbb{Z}^d = \{1, \cdots, m\}^d$. In the odd case $m = 2L+1$, we shall sometimes identify Λ with $[-L, \cdots, +L]$).
So we consider

$$-h^2\Delta + \sum_{\ell \in \Lambda} v(x_\ell) + \frac{\mathcal{J}}{2} \sum_{i \sim j} |x_j - x_i|^2 , \tag{1.2}$$

where $i \sim j$ means that i and j are nearest neighbors in Λ considered as living on a torus.
The parameter $\mathcal{J}$ which measures the size of the interaction satisfies the condition

$$\mathcal{J} > 0 . \tag{1.3}$$

This model can also be written as

$$-h^2\Delta + \sum_{\ell \in \Lambda} \tilde{v}(x_\ell) - \mathcal{J} \sum_{i \sim j} x_i \cdot x_j , \tag{1.4}$$

with

$$\tilde{v}(y) = v(y) + \mathcal{J} d\, y^2 , \ \forall y \in \mathbb{R}^n . \tag{1.5}$$

Most of the time, we shall look at the case $n = 1$ but other results considered for example by [40] will deal with the case $n > 1$. In this case we shall usually assume that v is radial.

For a one dimensional lattice (d=1), we have analyzed with J. Sjöstrand [30] the case of the so called Kac model and it was shown in [21] how to adapt this proof to treat other cases containing (1.1) in order to get that

$$\lambda_2(m; h) - \lambda_1(m; h) \leq C \exp -(\frac{\epsilon m}{h}) , \tag{1.6}$$

for C, ϵ independent of m and for $h < h_0$, h_0 independent of m.
But we met an additional and unfortunate condition

$$m \leq C h^{-N_0} , \tag{1.7}$$

without to know if it is a technical or deep condition. We have proposed in [21] an example where the property was true without this condition,

$$-h^2\Delta + \sum_{j=-L}^{L} x_j^2 + |\sum_{j=-L}^{L} x_j| \tag{1.8}$$

but this example remains rather artificial, although it was motivated by the study of the small temperature limit of a Kac type model.
C. Albanese mentioned to us the following model, called "Ising model with

transverse field", where the double well problem is replaced by the 2×2 matrix and the space $\otimes^{2L+1} L^2(\mathbb{R}^n)$ by $\otimes^{2L+1}\mathbb{C}^2$.
This corresponds to the idea that we can replace, in the semi-classical context, the one-particle Schrödinger operator $-h^2\Delta + v$ by a two by two matrix M (the interaction matrix) representing the restriction to the spectral space attached to the two lowest eigenvalues.
This interaction can be expressed with help of the Pauli matrices

$$\sigma^{(3)} = \begin{pmatrix} 1 & 0 \\ 0 & -1 \end{pmatrix}, \sigma^{(1)} = \begin{pmatrix} 0 & 1 \\ 1 & 0 \end{pmatrix},$$

as

$$M = \nu Id + \lambda\sigma^{(1)} .$$

The parameter λ corresponds to the measure of the tunneling effect for the one particle problem and can, in the semiclassical limit and under suitable conditions, be computed as having the order

$$\lambda \approx \exp -\frac{S}{h}, \tag{1.9}$$

where $S > 0$ is the Agmon distance between the two minima (cf [27]).
Let us now describe the general hamiltonian. We denote by $\mathcal{H}$ our Hilbert space

$$\mathcal{H} = \bigotimes_{i=-L}^{+L} \mathbb{C}^2, \tag{1.10}$$

where our line lattice is

$$\Lambda = [-L, +L] \cap \mathbb{Z} .$$

$$\mathbf{H}_\lambda = \sum_{x\sim y} \frac{1}{2}(1 - \sigma_x^{(3)}\sigma_y^{(3)}) + \sum_x \lambda\sigma_x^{(1)} . \tag{1.11}$$

We shall also write it in the form

$$\mathbf{H}_\lambda = \mathbf{S} + \lambda\mathbf{K}, \tag{1.12}$$

with

$$\mathbf{S} = \sum_{x\sim y} \frac{1}{2}(1 - \sigma_x^{(3)}\sigma_y^{(3)}) := \sum_{x\sim y} S_{xy}, \tag{1.13}$$

and

$$\mathbf{K} = \sum_x \sigma_x^{(1)} . \tag{1.14}$$

Here $x \sim y$ means $x \neq y$ and x nearest neighbour of y (that is, in $\mathbb{Z}/(2L+1)\mathbb{Z}$, $x = y \pm 1$). Here the convention is that $\sigma_y^{(3)}$ acts only on the y component:

$$\sigma_y^{(3)}(e_{-L} \otimes \cdots \otimes e_y \otimes \cdots \otimes e_L) = e_{-L} \otimes \cdots \otimes \sigma^{(3)} e_y \otimes \cdots \otimes e_L \,.$$

The statement is

Theorem 1.1 *:*
If we denote by $E_\pm$ the two lowest levels, there exists C such that, for all L,

$$|E_+ - E_-| \leq (2L+1) C^{(2L+1)} \exp[(\ln \lambda)(2L+1)] \tag{1.15}$$

This theorem is due [37] (See [2] for another proof for an analogous more complicate model [3], TY83).
What is quite important is of course to have a good control with respect to the dimension. The standard perturbation theory gives of course a result for λ small because, according to symmetry arguments, we can consider the two lowest eigenvalues as simple if we restrict $\mathbf{H}_\lambda$ to suitable subspaces $\mathcal{H}_\pm$. We observe also that the norm of $\sum_x \sigma_x^{(1)}$ is a priori of order L and this makes the usual argument of perturbation true only for $|\lambda| \leq C/L$. This means that we have to do something more tricky.

If we have in mind that, in this model, the parameter λ satisfies (1.9), the condition gives

$$\exp -\frac{S}{h} \leq \frac{C}{L}\,, \tag{1.16}$$

and finally

$$L \leq C \exp \frac{S}{h}\,. \tag{1.17}$$

This is in any case better than the condition (1.7) which we met with J. Sjöstrand in [30].
On the other hand, for connected results concerning Laplace integrals, we have seen in [23] that this type of condition can be crucial and cannot be considered a priori as simply technical.

But the main object of these notes is to analyze another approach based on the so called infrared estimates which was developed by many authors in the late 70's Fröhlich-Simon-Spencer [12], Dyson-Lieb-Simon [9], and Glimm-Jaffe [14]. More recently Pastur-Khozurenko [40] and Barbulyak-Kondratev [4] look in the same spirit at other examples and it becomes clear that this infrared approach gives also information about the splitting for our initial questions. It will actually give a complete answer for all the lattices of dimension strictly greater than 1. In this case, the result does

not seem to be related to tunneling properties. In the case when the lattice is of dimension 1, we shall obtain only a partial result under this condition (1.17). All these results are obtained through relatively easy extensions of these contributions. Because, particularly in the paper by [4], semi-classical analysis is involved, we have also tried to be more precise that in the original paper, using our more precise knowledge of the tunneling [27].
But let us now present the main results obtained through the infrared estimates.
Barbulyak and Kondrat'ev look[1] in [4] at the d-dimensional extension of the quantum model above which is denoted by

$$H_\Lambda^{per} = -h^2 \sum_{k\in\Lambda} \frac{\partial^2}{\partial x_k^2} + \sum_{k\in\Lambda} \tilde{v}(x_k) - \mathcal{J} \sum_{k\sim j} x_k\, x_j\,. \tag{1.18}$$

Here Λ is a subset of $\mathbb{Z}^d$ as defined before (with the periodicity convention). The assumptions on $\tilde{v}$ are the following:

- (Ha) $\tilde{v} \in C^\infty(\mathbb{R})$.
- (Hb) $\tilde{v}(x) \geq ax^2 + b$, $a, b \in \mathbb{R}$, $a > \mathcal{J}d$, $x \in \mathbb{R}$.
- (Hc) $\tilde{v}(x) = \tilde{v}(-x)$.
- (Hd) For some $q_0 > 0$ the function $\tilde{v}$ attains its strong global non degenerate minimum at the points $\pm q_0$.

We could also consider potentials $\tilde{v}$ on $\mathbb{R}^n$. In this case, the potentials are assumed to be radial (invariance by $SO(n)$) and we take the natural extension of the conditions (Ha)-(Hd).
(Ha) and (Hb) are much stronger than necessary. (Hb) gives however the control of the interaction term $\mathcal{J}\sum_{k\sim j} x_k\, x_j$ at ∞ by the one particle term $\sum \tilde{v}(x_k)$. (Hd) will permit a detailed semi-classical analysis but a weaker assumption can still work.
(Hc) is finally an important assumption in the analysis of the splitting but seems to play only a technical role at the other steps.
Following [4], we now consider the operator $\exp -\beta H_\Lambda^{per}$ and the associated

$$\langle B\rangle_{\beta,\Lambda} := \frac{\mathrm{Tr}\,(B \exp(-\beta H_\Lambda^{per}))}{\mathrm{Tr}\,(\exp(-\beta H_\Lambda^{per}))}\,, \tag{1.19}$$

where $B \in \mathcal{A}_\Lambda$. $\mathcal{A}_\Lambda$ is a class of polynomials but we shall more specifically analyze the case when

$$B = B_\Lambda = (\frac{1}{|\Lambda|} \sum_{k\in\Lambda} x_k)^2\,.$$

[1] This is only a note without detailed proof.

In the case $n > 1$, this means that we consider with

$$B^{(j)} = (\frac{1}{|\Lambda|}\sum_{k\in\Lambda} x_k^{(j)})^2 ,$$

$$B = \sum_{j=1}^{n} B^{(j)} .$$

Physically the strictly positive parameter β corresponds to the inverse of the temperature.
We now introduce the so called parameter of long-range order

$$P(\beta) = \lim_{|\Lambda|\to\infty} P_\Lambda(\beta) , \tag{1.20}$$

with

$$P_\Lambda(\beta) = \left\langle (\frac{1}{|\Lambda|}\sum_{k\in\Lambda} x_k)^2 \right\rangle_{\beta,\Lambda} . \tag{1.21}$$

The presence of the long range order, i.e. the strict positivity of $P(\beta)$, will serve as a test for phase transition (cf [9] and [40]).
When the limit of $\langle\cdot\rangle_{\beta,\Lambda}$ as $|\Lambda| \to +\infty$ exists, we shall denote it by $\langle\cdots\rangle_\beta$.
Let

$$E(p) = \sum_{i=1}^{d}(1 - \cos p^{(i)}) , \tag{1.22}$$

where $\Lambda^\star$ is the dual lattice

$$\Lambda^\star = \{p = (p^{(1)}, \cdots, p^{(d)}) \mid p^{(i)} = 2\pi k^{(i)}/m \, , \; 0 \le k^{(i)} \le m-1 \, ; \; 1 \le i \le d\} . \tag{1.23}$$

The method of infrared estimates will permit to get [2] the following lower bound

$$P(\beta) \ge \langle x_k^2\rangle_\beta - \frac{n}{2(2\pi)^d}\int_{]-\pi,\pi[^d}(\frac{h^2}{\mathcal{J}E(p)})^{\frac{1}{2}} \coth\left[\left(h^2\beta^2\mathcal{J}E(p)\right)^{\frac{1}{2}}\right] dp . \tag{1.24}$$

This lower bound is deduced by a limiting argument (thermodynamic limit) from the actually more useful inequality (for our questions relative to the splitting), which is relative to the finite lattice case, and which is given by the following theorem.

[2] Cf [9], (52), p. 368 (cf also Theorem 3.2 and Theorem 5.1 in this article).

Theorem 1.2 :
Under the assumption (Ha)-(Hc) on $\tilde{v}$, we have, for any $k \in \Lambda$, the following universal estimates

$$P_\Lambda(\beta) \geq \langle x_k^2\rangle_{\beta,\Lambda} - \frac{1}{|\Lambda|}\frac{n}{2}\sum_{p\in\Lambda^*\setminus\{0\}}\left(\frac{h^2}{\mathcal{J}E(p)}\right)^{\frac{1}{2}}\coth\left[\left(\beta^2h^2\mathcal{J}E(p)\right)^{\frac{1}{2}}\right] . \quad (1.25)$$

Barbulyak and Kondrat'ev refer to [9] and [40]. The reference [40] treats actually (1.24) in the particular case when $v(x) = (x^2-1)^2$ and get in this case an universal lower bound for $\langle x_k^2\rangle_{\beta,\Lambda}$. The specific part of [4] is probably the semi-classical aspect which we shall develop further in these notes. We emphasize that [40] works also for $n = 1$ but only for the particular case. The nature at the semi-classical level of the splitting when $n > 1$ is completely changed for the model $v(x) = (|x|^2-1)^2$. The two first levels of the one-particle hamiltonian are separated from each other by $\mathcal{O}(h^2)$. This is indeed a Schrödinger operator with a uniformly degenerate well invariant by $SO(n)$. This theorem reduces the analysis of this long range order to the analysis of $\langle x_k^2\rangle_{\beta,\Lambda}$ and this will be one of our goals to explicit how it can be done in the semi-classical context. If we find indeed a lower bound for β large independent of the dimension and if the second term in the right hand side of (1.25) is small enough for h small enough, then we shall have a proof of the existence of the long range order.
One part of the analysis consists in using monotonicity argument based on Ginibre inequalities in order to reduce to an analysis of a one particle problems. This will be recalled in section 2.
The second part is the analysis of the one particle problem (symmetric double well problem) in the semi-classical limit. This will be presented in section 3.
Section 4 will be devoted to the opposite situation (single well case) when the potential is convex. The Brascamp-Lieb inequality then gives a rather explicit way to control the situation.
Section 5 recalls the links between the splitting and the study of the trace of $\langle B\rangle_{\beta,\Lambda}$. Taking the limit $\beta \to +\infty$ before taking the thermodynamic limit leads to some improvement of the results concerning the splitting. We get for example the following theorem

Theorem 1.3 :
Let us consider the family of Hamiltonians H_Λ^{per} defined in (1.18) where $\Lambda \subset \mathbb{Z}^d$ with $d \geq 2$. If the potential $\tilde{v}$ and $\mathcal{J}$ satisfy (Ha)-(Hd), then there exists h_0 independent of Λ such that we have, for $0 < h < h_0$,

$$\lim_{|\Lambda|\to+\infty}(\lambda_2^\Lambda - \lambda_1^\Lambda) = 0 , \quad (1.26)$$

where λ_1^Λ and λ_2^Λ are the two lowest eigenvalues of H_Λ^{per}.

The direct application of the result by Barbulyak-Kondrat'ev would have given, in the limit $\beta \to +\infty$, the condition $d \geq 3$.
Section 6 is devoted to a more precise analysis of the case β large and presents essentially the results obtained by Barbulyak-Kondratiev and Pastur-Khozurenko with some improvements.
The two last sections will be devoted to a short presentation of the infrared estimates obtained by Fröhlich-Spencer-Simon in the classical case ([12]) and by Dyson-Lieb-Simon in the quantum case ([9]).

From recent discussions with J. Fröhlich in Ascona (June 1996) where we presented some results contained in these notes, we learn that the infrared estimates are not the optimal approach when $n = 1$ and discrete symmetry is involved. J. Fröhlich indicates that, using the techniques developed by Glimm-Jaffe-Spencer [15], it is possible to prove that the splitting tends to 0 when $n = 1$ (Theorem 1.3) without any restriction on the dimension of the lattice. The alternative proof giving in principle better results when $d = 1$ and $n = 1$ is based on the so-called Peierls argument. But the proof in [15] is written in the framework of the Field theory and is difficult to understand for non-specialists in this field. A nice but non selfcontained presentation of the subject is also given in [10], particularly in Subsection 8 which presents many other models. Nethertheless the results of Dyson-Lieb-Simon and the results concerning the splitting are not explicitely analyzed in Fro#olich's lectures.

We heard also at this conference in Ascona of more recent contributions by Kondratev and Rebenko [41] using also the Peierls argument.

In any case, what seems still open is the complete analysis of the tunneling with control of the size of Λ.

2. Ginibre type inequalities

Ginibre type inequalities[3] make it possible to estimate $\langle x_k^2 \rangle_{\beta,\Lambda}$ from below by the average $\langle x_k^2 \rangle_{\beta,\hat{H}}$ taken over the measure corresponding to the "formal" Hamiltonian

$$\hat{H} = -h^2 \sum_{k \in \mathbb{Z}^d} \frac{\partial^2}{\partial x_k^2} + \sum_{k \in \mathbb{Z}^d} \tilde{v}(x_k) \equiv \sum_{k \in \mathbb{Z}^d} H_k \,, \tag{2.1}$$

[3]The authors refer to [43]. It is more explicitely written in [44]. This result of Ginibre generalizes previous inequalities of Griffiths extended by Kelly-Sherman.

with separate variables.
Less formally, we introduce

$$\hat{H}^\Lambda = \sum_{k \in \Lambda} H_k \tag{2.2}$$

and we shall get the

Proposition 2.1 :
Under the assumptions (Ha)-(Hc) on the potential $\tilde{v}$ and $\mathcal{J}$, the following inequality is true, for any $k \in \Lambda$,

$$\langle x_k^2 \rangle_{\beta,\Lambda} \geq \langle x_k^2 \rangle_{\beta,\Lambda,\hat{H}^\Lambda} \, . \tag{2.3}$$

Let us describe briefly the version of the Ginibre inequalities (we just treat for simplification the case $d = 1$) which is needed here. This is related to the control of the sign of the correlations attached to the more general measure:

$$Z^{-1} \exp \sum_{ij} J_{ij} x_i \, x_j \prod_{j=1}^{m} d\nu_j(x_j) \, . \tag{2.4}$$

We look at the partial derivative of $\langle (x_k)^2 \rangle$ with respect to J_{ij}. Here $\langle \cdot \rangle$ is taken with respect to the measure (2.4).
This gives

$$\partial_{J_{ij}} \langle (x_k)^2 \rangle = \langle x_i \, x_j \, (x_k)^2 \rangle - \langle (x_k)^2 \rangle \cdot \langle x_i \, x_j \rangle \, .$$

The right hand side appears as the pair correlation of the two functions $f = x_i \, x_j$ and $g = x_k^2$.
We shall deduce from the Ginibre's inequalities that this expression is positive.

$$\partial_{J_{ij}} \langle (x_k)^2 \rangle \geq 0 \, . \tag{2.5}$$

Let us briefly recall some elements of this theory (cf [13], [43], p. 271-279 and [44], p.119-124), for a nice exposition). We recall the

Theorem 2.2 :
Let $\mathcal{F}_1$ be the set of functions on $\mathbb{R}$ which are nonnegative and monotone increasing on $[0, +\infty)$ and either even or odd. Let $\mathcal{F}_m$ be the functions on $\mathbb{R}^m$ of the form $f_1(x_1) \cdots f_m(x_m)$ with $f_i \in \mathcal{F}_1$. Let $d\mu$ be a probability measure of the form (2.4) where $J_{ij} \geq 0$ and each $d\nu_j$ has the form $\exp(f_j(x)) d\lambda_j(x)$ with $f_j \in \mathcal{F}_1$ and $d\lambda_j$ even. Then

$$\begin{array}{ll} (GKS1) & \langle f \rangle \geq 0 \\ (GKS2) & \langle f \, g \rangle \geq \langle f \rangle \cdot \langle g \rangle \, , \end{array} \tag{2.6}$$

for all $f, g \in \mathcal{F}_m$.

The reader can find the proof of the theorem in [44].
For (GKS1), this is essentially obtained by expanding the exponentials $\exp f_j$ and $\exp \sum_{i \neq j} J_{ij} x_i x_j$ and controlling the sign of each term of the expansion.
The inequalities (GKS2) are obtained by a more sophisticated duplication method.

The measure we are actually considering, attached to the restriction to the diagonal of the distribution kernel of $\exp -\beta H_\Lambda^{per}$, where $\mathcal{J}$ becomes a parameter in $[0, \mathcal{J}_0]$, has not exactly the structure which is introduced above. It is consequently useful to use the Trotter product formula which describes the kernel of $\exp -\beta H_\Lambda^{per}$ as the limit in a weak sense of $(\exp -\frac{\beta V(x)}{N} \exp \frac{\beta \Delta}{N})^N$ and this kernel satisfies the assumptions of the theorem for $\mathcal{J} \geq 0$.
We have just to control the limit procedure in order to get the result.
This argument is sketched in [44] p. 120 and p.122. Let us recall the argument.

What we need to prove is some GKS-inequality for a measure which appears as a limit in a weak sense of a family of measures satisfying (GKS2). The measure $d\mu$ is the measure whose density with respect to the Lebesgue measure is the restriction to the diagonal of the kernel distribution of $\exp -\beta H_\Lambda^{per}$. The measures $d\mu_N$ are the measures whose density with respect to the Lebesgue measure is the restriction to the diagonal of the kernel distribution of $(\exp -\frac{\beta}{N} V \exp \frac{\beta . h^2}{N} \Delta)^N$.
The starting point is the Trotter-Kato product formula saying simply that

$$\lim_{N \to +\infty} \langle (\exp -\frac{\beta}{N} V \exp \frac{\beta . h^2}{N} \Delta)^N f \mid g \rangle_{L^2} = \langle \exp -\beta H_\Lambda^{per} f \mid g \rangle_{L^2} . \quad (2.7)$$

We have to verify two points.

- Verify that $d\mu_N$ satisfies the assumption of the Ginibre's Theorem.
- Go from a convergence property in the weak sense for the kernel to a convergence property for the trace.

The two points will be actually more intricate. We first observe that for the very specific kernels K (or K_N) which are involved we have

$$\begin{aligned} &\int_{\boldsymbol{R}^n} f(x) g(x) K(x,x) dx \\ &\quad = \lim_{\epsilon \to 0} \left(\epsilon^{-\frac{n}{2}} \int_{\boldsymbol{R}^n \times \boldsymbol{R}^n} f(x) g(y) K(x,y) \exp -\tfrac{1}{\epsilon} |x-y|^2 dx \cdot dy \right) , \end{aligned}$$

We have also for more general slowly increasing f or g

$$\int_{\boldsymbol{R}^n} f(x) g(x) K(x,x) dx = \lim_{\eta \to 0} \int_{\boldsymbol{R}^n} f(x) g(x) \exp -2\eta x^2 \, K(x,x) \, dx .$$

What is important here is that we can stay, in all the limiting procedures, inside the assumptions of the Ginibre's Theorem. This is indeed the case by our choice of regularization. (GKS2) has the following structure

$$\begin{aligned}&(\int_{\mathbb{R}^n} K(x,x)dx) \times (\int_{\mathbb{R}^n} f(x)g(x)K(x,x)dx) \\ &\quad \geq (\int_{\mathbb{R}^n} f(x)K(x,x)dx) \times (\int_{\mathbb{R}^n} g(x)K(x,x)dx)\end{aligned}$$

with $f(x) = x_i \cdot x_j$ and $g(x) = (x_k)^2$. We observe that f and g are in $\mathcal{F}_2$.

Application:
The Hamiltonian $\hat{H}^\Lambda$ describes a system of non interacting particles. Consequently, we get immediately, for $k \in \Lambda$,

$$\langle x_k^2 \rangle_{\beta,\Lambda;\hat{H}} = \langle x_k^2 \rangle_{\beta,H_k} = \langle x_0^2 \rangle_{\beta,H_0} , \tag{2.8}$$

for all $k \in \mathbb{Z}^d$.
Here we recall that H_k is the one-particle Hamiltonian at $k \in \mathbb{Z}^d$.
We consequently have obtained

Proposition 2.3 :
Under the assumption (Ha)-(Hc) on $\tilde{v}$ and $\mathcal{J}$, we have

$$\langle x_k^2 \rangle_{\beta,\Lambda} \geq \langle x_0^2 \rangle_{\beta,H_0} . \tag{2.9}$$

Remark 2.4 :
When $n > 1$, we can no more apply this technique directly. The case when we have a rotational symmetry can probably be treated by taking polar coordinates. The results in this case is mentioned in [14]. Pastur and Khoruzenko proceed differently in the case of the model $v(x) = (1 - |x|^2)^2$.

3. Semiclassical analysis of the one particle problem

We have seen in Section 2 how one can replace with help of the Ginibre inequalities the study of the quantity $\langle x_k^2 \rangle_{\beta,\Lambda}$ by the study of the simpler quantity

$$\langle x_0^2 \rangle_{\beta,H_0} = \frac{\text{Tr } (x_0^2 \exp -\beta H_0)}{\text{Tr } (\exp -\beta H_0)} \tag{3.1}$$

attached to the one particle Schrödinger operator $H_0 = -h^2 \frac{d^2}{dx^2} + \tilde{v}(x)$ where $\tilde{v}$ satisfies the condition (Ha)-(Hd). When considering the one particle problem, we sometimes write simply x instead of x_0.
The main topics of this section is the semi-classical analysis of $\langle x_0^2 \rangle_{\beta,H_0}$ as $h \to 0$ and $\beta \to +\infty$.
Two conditions on $\tilde{v}$ could be relaxed. The assumption $n = 1$ in (Ha) is not

important in this section although we shall write the results in this case. It is actually sufficient, instead of the strong (Hb), that $\exp -\beta H_0$ is trace class in order to perform the analysis of this section.
We shall be rather sketchy in this part and refer for example to [17] for the presentation of the semi-classical theory involved, which is mainly due to Simon [45], [46] and Helffer-Sjöstrand [27]. Actually, we need sometimes weaker results which are probably much older (particularly in the case when $n = 1$).
Let $\phi_1(x; h)$ be the ground state of H_0. Observing that the first eigenvalue is simple, the first remark is that

$$\lim_{\beta \to +\infty} \langle x_0^2 \rangle_{\beta, H_0} = \int_{\mathbf{R}} x^2 \phi_1(x; h)^2 dx \, . \tag{3.2}$$

The right hand side of (3.2) is clearly related to the localization of the first eigenfunction ϕ_1. In the case of the double well problem, one has indeed the

Proposition 3.1 :
Under the assumptions (Ha)-(Hd) on $\tilde{v}$,

$$\lim_{h \to 0} \int_{\mathbf{R}} x^2 \phi_1(x; h)^2 dx = q_0^2 \, . \tag{3.3}$$

We can actually prove by standard semi-classical analysis that there exists a constant C and h_0 such that

$$\int_{\mathbf{R}} x^2 \phi_1(x; h)^2 dx \geq q_0^2 - Ch \, , \; \forall h \leq h_0 \, . \tag{3.4}$$

More precisely, this analysis based on the harmonic approximation gives the existence of a complete expansion of the type

$$\int_{\mathbf{R}} x^2 \phi_1(x; h)^2 dx \sim q_0^2 + \sum_{j \geq 1} h^j \gamma_j \, . \tag{3.5}$$

An immediate consequence is

Proposition 3.2 :
For any $\epsilon > 0$, there exists h_0 such that, for $0 < h \leq h_0$, one can then find β_1 such that

$$\langle x_0^2 \rangle_{\beta, H_0} \geq (1 - \epsilon) q_0^2 \, , \tag{3.6}$$

for all β greater than β_1.

This statement was used by Barbulyak and Kondratev. It seems actually interesting to relate β_1 and h.

The formal expression for $\langle x^2\rangle_{\beta,H_0}$ is

$$\langle x^2\rangle_{\beta,H_0} = \frac{\sum_j \exp -\beta\lambda_j \ \left(\int x^2\phi_j(x;h)^2\, dx\right)}{\sum_j \exp -\beta\lambda_j} , \tag{3.7}$$

where λ_j is the sequence of the eigenvalues of H_0 arranged in increasing order and ϕ_j is the corresponding orthonormal basis of eigenvectors.
The semi-classical analysis says that near the first level of H_0 which is given modulo $\mathcal{O}(h^2)$ by the first level of the harmonic approximation at q_0

$$-h^2\frac{d^2}{dy^2} + \frac{1}{2}\tilde{v}''(q_0)y^2 , \tag{3.8}$$

they are two eigenvalues λ_1 and λ_2 which are exponentially close. There exists indeed S called the Agmon distance between q_0 and $-q_0$ and given (in the case $n = 1$) by

$$S = \int_{-q_0}^{q_0} \sqrt{\tilde{v}(x) - \tilde{v}(q_0)}\, dx , \tag{3.9}$$

such that

$$\lambda_2 - \lambda_1 \sim h^{\frac{1}{2}} \cdot \exp -\frac{S}{h} \cdot a(h) , \tag{3.10}$$

with $a(h) \neq 0$ and admitting a complete expansion in powers of h. The other point is that the third eigenvalues λ_3 is given modulo $\mathcal{O}(h^{\frac{3}{2}})$ by the second eigenvalue of the harmonic approximation leading for λ_3 to a splitting of order h

$$\lambda_3(h) - \lambda_2(h) \sim h\sqrt{2\tilde{v}''(q_0)} . \tag{3.11}$$

So we formulate the following natural question:
Do we have to assume

- $\beta_1 \sim \exp \frac{S}{h}$ corresponding to the inverse of the splitting between the two first eigenvalues,

- or the weaker $\beta_1 \sim \frac{1}{h}$ corresponding to the inverse of the splitting between the packet of the two first eigenvalues and the third one ?

We shall see that we are actually, under the assumption (Hd) (symmetric double well problem), in the second case of the alternative.

Let us indeed prove the following result.

Proposition 3.3 :
There exists constants C, h_0 and $\gamma_0 > 0$ such that, when $\beta h \geq \gamma_0$, $h \in]0, h_0]$, we have

$$\langle x_0^2\rangle_{\beta,H_0} \geq \left(\int x^2\phi_1^2\, dx\right)\cdot\left(1 - C\exp(-\frac{1}{Ch}) - C\exp(-\frac{1}{C}\beta h)\right) . \tag{3.12}$$

This leads immediately to

Corollary 3.4 :

For any $\epsilon > 0$, there exist constants $C_1(\epsilon)$ and $h_0(\epsilon)$ such that, for any $h \in]0, h_0(\epsilon)]$ and β, such that

$$\beta\, h \geq C_1(\epsilon)\ , \tag{3.13}$$

then we have

$$\langle x_0^2 \rangle_{\beta, H_0} \geq q_0^2(1-\epsilon)\ . \tag{3.14}$$

Proof of Proposition 3.3:

One observes indeed that

$$\begin{aligned} &\exp -\beta\lambda_1 \cdot (\textstyle\int x^2\phi_1^2\,dx) + \exp -\beta\lambda_2 \cdot (\textstyle\int x^2 \phi_2^2\,dx) \\ &= \exp -\beta\lambda_1 \cdot (\textstyle\int x^2\phi_1^2 dx)\, [1 + \exp -\beta(\lambda_2-\lambda_1)\, (\frac{\int x^2\phi_2^2 dx}{\int x^2 \phi_1^2\,dx})] \\ &= \exp -\beta\lambda_1 \cdot (\textstyle\int x^2\phi_1^2\,dx)[1 + \exp -\beta(\lambda_2 - \lambda_1)(1 + \mathcal{O}(\exp -\frac{\hat{S}}{h}))]\ , \end{aligned}$$

for any $\hat{S} < S$, where S is introduced in (3.9). Here we have used the more precise information coming from the semi-classical analysis that, in the double well problem, one can find normalized ϕ_1 and ϕ_2, such that $\phi_1 + \phi_2 := 2\phi^{(left)}$ is exponentially localized [4] in the left well and such that $\phi_1 - \phi_2 := 2\phi^{(right)}$ is exponentially localized in the right well. The second eigenvector ϕ_2 being odd, we have also the property

$$\phi^{(left)}(x) = \phi^{(right)}(-x)\ .$$

We consequently obtain

$$\int x^2\phi_1^2 dx = 2\int x^2(\phi^{(left)}(x))^2 dx + 2\int x^2 \phi^{(left)}(x)\phi^{(right)}(x) dx\ ,$$

and

$$\int x^2\phi_2^2 dx = 2\int x^2(\phi^{(left)}(x))^2 dx - 2\int x^2 \phi^{(left)}(x)\phi^{(right)}(x) dx\ .$$

But according to the exponential decay of $\phi^{(left)}$ and $\phi^{(right)}$, we get

$$\int x^2 \phi^{(left)}(x)\phi^{(right)}(x) dx = \int x^2\phi^{(left)}(x)\phi^{(left)}(-x) dx = \mathcal{O}(\exp -\frac{\hat{S}}{h})\ ,$$

[4] The decay is, for any $\epsilon > 0$, like $\mathcal{O}(exp - \frac{d(x)-\epsilon}{h})$, with

$$d(x) = \inf\{|\int_{-q_0}^{x} \sqrt{\tilde{v}(t) - \tilde{v}(-q_0)}\,dt|, S\}$$

for any $\hat{S}$ strictly less than the Agmon distance S between the two wells. We summarize what we have obtained in the following lemma

Lemma 3.5 *:*
For any $\hat{S} < S$, we have

$$\int x^2\phi_1^2 dx = \int x^2\phi_2^2 dx + \mathcal{O}(\exp -\frac{\hat{S}}{h}) . \tag{3.15}$$

We have now to control the other forgotten terms in the computation of $\langle x^2\rangle_{\beta,H_0}$. Because we are mainly interested by a lower bound of this quantity, we shall have to find an upper bound for Tr $(\exp -\beta H_0) = \sum_{j\geq 1}\exp -\beta\lambda_j$, and more precisely the quotient, as a function of (β, h),

$$(\beta,h)\mapsto (\exp -\beta\lambda_1(h) + \exp -\beta\lambda_2(h))^{-1}\left(\sum_{j\geq 1}\exp -\beta\lambda_j\right) , \tag{3.16}$$

which we want to be small in a suitable domain.

Here we recall from (3.11) that, for some strictly positive C_0, we have the estimate

$$\lambda_j(h) \geq \lambda_1(h) + C_0 h , \tag{3.17}$$

for $j \geq 3$.
It is sufficient to prove the existence of $C > 0$ such that

$$\exp\beta\lambda_1 \,(\sum_{j\geq 3}\exp -\beta\lambda_j) \leq C\exp -\frac{1}{C}\beta h , \tag{3.18}$$

for C large enough.
In order to control this expression, we can divide the sum in two parts

$$\begin{aligned}\exp\beta\lambda_1\,(\textstyle\sum_{j\geq 3}\exp -\beta\lambda_j) &= \exp\beta\lambda_1(\{\textstyle\sum_{j\geq 3\,,\,\lambda_j\leq\alpha\}}\exp -\beta\lambda_j)\\ &\quad + \exp\beta\lambda_1\,(\textstyle\sum_{\lambda_j\geq\alpha}\exp -\beta\lambda_j)\\ &= I_1 + I_2 ,\end{aligned} \tag{3.19}$$

for some $\alpha > 0$ (possibly h-dependent) to be determined.
We may assume without loss of generality that

$$\tilde{v}(q_0) = 0 . \tag{3.20}$$

The first part I_1 of the sum can be estimated by

$$\#\{j \mid \lambda_j \leq \alpha\}\exp -\beta(\lambda_3 - \lambda_1) \leq C\alpha h^{-1}\exp -\frac{\beta}{C}h . \tag{3.21}$$

Here we have used a very weak version of the Weyl Formula (compare for example with the harmonic oscillator) and the fact that, for α small, $\int_{p(x,\xi)\leq\alpha} dx d\xi \sim \alpha$, where $p(x,\xi) = \xi^2 + \tilde{v}(x)$. (See for example [26].)
Note here that, in order to look for an optimal result (by playing with α), we need an estimate which is uniform with respect to $\alpha \in [0, \alpha_0]$.
We have obtained

$$I_1 \leq C\alpha h^{-1} \exp -\frac{\beta}{C} h \,. \tag{3.22}$$

Let us now analyze the second part I_2 of the sum. In order to get this estimate, we first use the Golden Thompson inequality [38]) saying that, for any $t > 0$,

$$\mathrm{Tr}\ \exp -tH_0 \leq (\int \exp -th^2\xi^2 d\xi)(\int \exp -t\tilde{v}(x)dx) \,. \tag{3.23}$$

This leads to

$$\mathrm{Tr}\ \exp -tH_0 \leq C\frac{1}{t^{\frac{1}{2}}h}(\int \exp -t\tilde{v}(x)dx) \,. \tag{3.24}$$

If we use that $\tilde{v}$ has non degenerate minima (Hd) and (3.20), we obtain, for $t \geq 1$, the existence of a constant C such that, for all $h \in]0, h_0]$,

$$\mathrm{Tr}\ \exp -tH_0 \leq \frac{C}{th} \,. \tag{3.25}$$

Coming back to the definition of I_2 we write the inequality

$$I_2 \leq \exp -\frac{\beta\,\alpha}{2} \cdot (\exp \frac{\beta}{4}(4\lambda_1 - \alpha) \cdot \left(\sum_{\lambda_j \geq \alpha} \exp -\frac{\beta}{4}\,\lambda_j \right) \,.$$

When $\alpha > 4\lambda_1$, we obtain

$$I_2 \leq \exp -\frac{\beta\,\alpha}{2} \cdot \mathrm{Tr}\ [\exp -\frac{\beta}{4} H_0] \,. \tag{3.26}$$

Taking $\alpha = Dh$, with D large enough, we get, from (3.26) and (3.25) with $t = \frac{\beta}{4}$,

$$I_2 \leq \mathrm{const.}\ h^{-1}\ \beta^{-1}\ \exp -\frac{D}{4}\beta h \,. \tag{3.27}$$

Combining (3.27) and (3.22) with $\alpha = Dh$, we have obtained (3.12)under the condition $\beta h \geq \gamma_0 > 0$. This completes the proof of the proposition.

Remark 3.6 :
These arguments can be extended to any dimension. The semi-classical analysis involved in the argument is presented for example in [27].

Remark 3.7 :
In the case when the potential is $v(x) = (|x|^2 - 1)^2$ *then the Bogolyubov inequality permits to have a universal lower bound for* $\langle x_i^2 \rangle_{\beta,\Lambda}$. *This is mentioned in [40] which refers also to [6]. Again this argument is true for any* n. *We observe that when the dimension* n *is strictly larger than* 1, *we are no more in a double well situation. We have indeed a uniformly degenerate well given by* $|x|^2 = 1$ *(cf [28]) and no tunneling is involved. Note that the splitting* $\lambda_2 - \lambda_1$ *is of order* $\mathcal{O}(h^2)$ *as can be seen for example by taking polar coordinates.*

4. The strictly convex case

This section will be devoted to the case when the involved potential or phase is convex.

4.1. STRICTLY CONVEX CASE, CLASSICAL CASE

Let us first consider the "classical" case. The result is analyzed (for one dimensional lattice) for example in [31], where under suitable assumptions of strict convexity of the family of potential $\Phi^{(m)}$ on $I\!R^m$, it was shown that

$$\lim_{m \to +\infty} \frac{\int_{I\!R^m} (\frac{1}{m} \sum_i x_i)^2 \exp -\beta \Phi^{(m)}(x) dx}{\int \exp -\beta \Phi^{(m)}(x) dx} = 0 \,. \tag{4.1}$$

This is indeed a consequence of the following control of the correlations

$$\sum_i \left| \frac{\int_{I\!R^m} x_1 x_i \exp -\beta \Phi^{(m)}(x) dx}{\int \exp -\beta \Phi^{(m)}(x) dx} \right| \leq C_\beta \tag{4.2}$$

which leads to a convergence in $\mathcal{O}(\frac{1}{m})$.
Let us discuss these results more precisely. The starting point is the Brascamp-Lieb inequality.

Theorem 4.1 :
Let $F(x) = \exp(-\Phi(x)), x \in I\!R^m$, *with* Φ *in* C^2 *and strictly convex. We assume that* Φ *has a minimum and consequently* F *decays exponentially in all directions. Let* $f \in C^1(I\!R^m)$, *and let us assume that* $\text{var } f < \infty$. *Then*

$$\text{var } f \leq \langle \nabla f, (\Phi_{xx})^{-1} \nabla f \rangle \tag{4.3}$$

where ∇f *is the gradient of* f.

Here all the mean values $\langle \cdot \rangle$ are with respect to the measure $\exp -\Phi(x) dx$. An immediate consequence of (4.3) is

$$\operatorname{var} f \leq \frac{1}{\inf_{x \in \mathbb{R}^m} \lambda_{min}(\operatorname{Hess}\Phi(x))} \frac{\| \, |\nabla f|^2 \, \|_{L^2(\mathbb{R}^m \, ; \, \exp -\Phi(x)dx)}}{\int \exp -\Phi(x)dx} \,. \tag{4.4}$$

When Φ is even and when $f(x) = \frac{1}{m}\sum_{i=1}^m x_i$, we obtain

$$\frac{\int \left(\frac{1}{m}\sum_{i=1}^m x_i\right)^2 \exp -\Phi(x)dx}{\int \exp -\Phi(x)dx} \leq \frac{1}{m} \frac{1}{\inf \lambda_{min}(\operatorname{Hess}\Phi(x))} \,. \tag{4.5}$$

In particular, we obtain

Proposition 4.2 :

Let $\Phi = \Phi^{(m)}$ ($m \in \mathbb{N}$) a family of even strictly convex potentials on $\mathbb{R}^m$ such that there exists $\sigma > 0$ such that, for all $m \in \mathbb{N}$,

$$\operatorname{Hess}\Phi^{(m)} \geq \sigma \,, \tag{4.6}$$

then we have, for any $\beta > 0$,

$$\lim_{m \to +\infty} \frac{\int \left(\frac{1}{m}\sum_{i=1}^m x_i\right)^2 \exp -\beta\Phi^{(m)}(x)dx}{\int \exp -\beta\Phi^{(m)}(x)dx} = 0 \,. \tag{4.7}$$

As typical example, we can apply this result for

$$\Phi^{(m)}(x) = \sum_{j=1}^m v_j(x_j) + \frac{\mathcal{J}}{2}\sum_{j=1}^m |x_j - x_{j+1}|^2 \,, \tag{4.8}$$

where v_j is even and satisfies $v_j''(x) \geq \sigma > 0$.
We observe that, when $v = v_j$, we have also the property (4.2) in the non-convex case (using the approach of the transfer matrix [20]).

4.2. THE STRICTLY CONVEX CASE, QUANTUM CASE

Similarly to the classical case, it is interesting to observe the following proposition

Proposition 4.3 :

Let $\Phi = \Phi^{(m)}$ ($m \in \mathbb{N}$)be a family of strictly convex potentials on $\mathbb{R}^m$ such that (4.6) is satisfied for some $\sigma > 0$. Then, for $H^{(m)} = -h^2\Delta + \Phi^{(m)}$, we have

$$\lim_{m \to +\infty} \frac{\operatorname{Tr}\left[\left(\frac{1}{m}\sum_{i=1}^m x_i\right)^2 \exp -H^{(m)}\right]}{\operatorname{Tr}\, \exp -H^{(m)}} = 0 \,. \tag{4.9}$$

Corollary 4.4 :
Let $\tilde{v}$ satisfying (Ha)-(Hc) and, instead of (Hd),

$$v''(x) \geq 2\omega^2 > 0\,,\ \forall x \in I\!R\,,$$

for some $\omega > 0$, then,

$$P(\beta) := \lim_{|\Lambda| \to +\infty} P_\Lambda(\beta) = 0 \tag{4.10}$$

About the proof of the proposition:

We shall again apply the Brascamp-Lieb inequality [7] and techniques developed in their article.

We just give the proof for the case considered in the corollary. We know that the restriction to the diagonal of the distribution kernel of $\exp -\beta H_\Lambda^{(per)}$ is logconcave as a limit of logconcave densities associated to the distribution kernel of $\left(\exp -\frac{\beta V}{N} \exp \frac{\beta h^2 \Delta}{N}\right)^N$. What has to be verified is a quantitative control of the strict logconcavity in the procedure. The techniques developed by Brascamp-Lieb [7] as recalled in Simon [44] and also in [20] are actually relevant. Theorem 4.3 in Brascamp-Lieb (and the arguments presented in Section 6 of this paper for the proof of Theorem 6.1) say indeed that it is sufficient, for the study of the case when

$$V(x) = \sum_{\ell \in \Lambda} v(x_\ell) + \frac{J}{2} \sum_{i \sim j} |x_i - x_j|^2\,,$$

with $v(x) = \omega^2 x^2 + R(x)$ and R positive, convex, to analyze the quadratic case and that the distribution kernel $K_\Lambda(\beta)$ of $\exp -\beta H_\Lambda^{(per)}$ can be written as the product

- of the kernel obtained for $\exp -\beta \left(\sum_{\ell \in \Lambda}(-h^2 \Delta_{x_\ell} + \omega^2 |x_\ell|^2)\right)$.
- and of a logconcave kernel.

But the kernel of $\exp -\beta \left[\sum_{\ell \in \Lambda}(-h^2 \Delta_{x_\ell} + \omega^2 |x_\ell|^2)\right]$ is explicitely known using the Mehler's formula. We recall indeed that the Mehler's Formula, which is obtained by explicit computations (see the book of Simon [44] p. 36-38), expresses the kernel of $\exp -t(-\frac{1}{2}d^2/dx^2 + \frac{1}{2}x^2 - \frac{1}{2})$ by

$$\begin{aligned} H_t(x,y) = {} & \pi^{-1}(1 - \exp -2t)^{-\frac{1}{2}} \\ & \exp[-(1 - \exp -2t)^{-1}(\tfrac{1}{2}(1 + \exp -2t)(x^2 + y^2) - 2\exp -t\, xy)]\,. \end{aligned} \tag{4.11}$$

or

$$H_t(x,y) = \pi^{-1}(1-\exp -2t)^{-\frac{1}{2}} \exp[-(\tfrac{1}{2\tanh t})(x^2+y^2)+\tfrac{1}{\sinh t}\, xy]\,, \tag{4.12}$$

We consequently get the existence of $\sigma > 0$ and independent of $|\Lambda|$ such that

$$\mathrm{Hess}\,[-\ln k_\Lambda(\beta)](x) \geq \sigma > 0$$

where

$$k_\Lambda(\beta)(x) = K_\Lambda(\beta)(x,x)\,.$$

σ can be chosen as equal to

$$\sigma = \frac{\omega}{2h}\cdot\frac{\cosh t - 1}{\sinh t}\,,$$

or

$$\sigma = \frac{\omega}{2h}\cdot\tanh(\frac{t}{2})\,, \tag{4.13}$$

with

$$t = \beta h\omega\,. \tag{4.14}$$

This gives finally the stronger version of the corollary

Proposition 4.5 *:*
If V is strictly convex with $(\mathrm{Hess}\,V)_x \geq 2\omega^2$, $\forall x \in I\!R^m$, *then*

$$0 \leq P_\Lambda(\beta) \leq \frac{1}{\sigma|\Lambda|}\,, \tag{4.15}$$

with

$$\sigma = \frac{\omega}{2h}\cdot\tanh(\frac{\beta h\omega}{2})\,. \tag{4.16}$$

Remark 4.6 *:*
Let us consider two limiting cases.

When $\beta \to +\infty$, we find that

$$\lim_{\beta\to+\infty}\sigma = \frac{\omega}{2h}\,, \tag{4.17}$$

and consequently

$$\lim_{\beta\to+\infty} P_\Lambda(\beta) \leq \frac{1}{|\Lambda|}\,\frac{h}{2\omega}\,. \tag{4.18}$$

When $h \to 0$, we get

$$\lim_{h\to 0}\sigma = \beta\omega^2\,, \tag{4.19}$$

$$\lim_{h\to+0} P_\Lambda(\beta;h) \leq \frac{1}{\beta\omega^2|\Lambda|}\,. \tag{4.20}$$

This corresponds to the classical result for the measure $\exp -\beta V(x)dx$.

4.3. THE QUANTUM CASE, LIMIT AS $\beta \to +\infty$

Let us further analyze (4.17) and (4.18). If we perform the limit $\beta \to +\infty$, we get, with $B = (\frac{1}{|\Lambda|}\sum_i x_i)^2$,

$$\lim_{\beta\to+\infty}\left(\frac{\mathrm{Tr}\ B\exp -\beta H_\Lambda^{per}}{\mathrm{Tr}\ \exp -\beta H_{\Lambda^{per}}}\right) = \int (\frac{1}{|\Lambda|}\sum_i x_i)^2 \phi_1^\Lambda(x)^2\, dx\ . \tag{4.21}$$

We have used here that the lowest eigenvalue $\lambda_1 = \lambda_1^\Lambda$ of H_Λ^{per} is simple and ϕ_1^Λ is the associated normalized strictly positive eigenfunction.
When V is strictly convex, we can use Brascamp-Lieb and get

$$\int (\frac{1}{|\Lambda|}\sum_i x_i)^2 \phi_1^\Lambda(x)^2 dx \le \frac{1}{|\Lambda|}(\inf_x \lambda_{min}(\mathrm{Hess}\,[-\ln\phi_1^\Lambda(x)]))^{-1}\ . \tag{4.22}$$

We recall indeed a lower bound for $\lambda_{min}(\mathrm{Hess}\,[-\ln\phi_1^\Lambda](x)))$ through the maximum principle ([48]) by the square root of $\inf_{x\in\mathbb{R}^{|\Lambda|}}\lambda_{min}(\mathrm{Hess}\,V(x)))$. As expected, we recover (4.18).

5. Connection with the Splitting.

We analyze in this section different standard relations connecting the splitting between the two first eigenvalues of the Schrödinger operator and the different quantities introduced in the preceding sections. Our initial remark is that we can also write (without using Brascamp-Lieb or the condition of strict convexity but keeping V even) the general inequality

$$\int (\frac{1}{|\Lambda|}\sum_i x_i)^2 \phi_1^\Lambda(x^\Lambda)^2 dx^\Lambda \le \frac{1}{|\Lambda|}(\lambda_2^\Lambda - \lambda_1^\Lambda)^{-1}. \tag{5.1}$$

This is indeed just the minimax principle[5] saying that

$$(\lambda_2^\Lambda - \lambda_1^\Lambda) = \inf_{\substack{f\in C_0^\infty \\ \int f(x^\Lambda)\phi_1^\Lambda(x^\Lambda)^2 dx^\Lambda = 0}} \frac{\langle |\nabla f|^2\rangle}{\langle |f|^2\rangle}\ , \tag{5.2}$$

where $\langle\cdot\rangle$ is computed for the measure $\phi_1^\Lambda(x^\Lambda)^2 dx^\Lambda$:

$$\langle\cdot\rangle = \langle\cdot\rangle_{\infty,\Lambda}\ , \tag{5.3}$$

(remembering that it corresponds to the case $\beta = +\infty$).
This is then applied to the function

$$f(x^\Lambda) = \frac{1}{|\Lambda|}\sum_{i\in\Lambda} x_i\ . \tag{5.4}$$

[5]One can extend C_0^∞ to slowly increasing C^∞ functions in our case.

The condition of symmetry on V (which is a consequence of (Hc) for our specific example) gives the orthogonality of $f\Phi_1^\Lambda$ to Φ_1^Λ.
In the case when $x_i \in I\!R^n$ with $n > 1$, we have to introduce the functions $f^{(j)}(x^\Lambda) = \frac{1}{|\Lambda|}\sum_{i\in\Lambda} x_i^{(j)}$ (for $j = 1, \cdots, n$). This change the discussions by unimportant constants. We assume $n = 1$ in all this section.
Taking account of the lower bound for the splitting given by J. Sjöstrand [48] in the case of the Schrödinger operator with strictly convex potential, we recover (4.18).
Conversely, we can interpret this inequality as

$$\lambda_2^\Lambda - \lambda_1^\Lambda \leq \frac{1}{|\Lambda|}[\int (\frac{1}{|\Lambda|}\sum_i x_i)^2 \phi_1^\Lambda(x^\Lambda)^2 dx^\Lambda]^{-1} . \tag{5.5}$$

This means that if we can prove by other means the property that

$$[\int (\frac{1}{|\Lambda|}\sum_i x_i)^2 \phi_1^\Lambda(x^\Lambda)^2 dx^\Lambda] \geq \rho > 0 , \tag{5.6}$$

with ρ independent of Λ, then we get that

$$\lim_{|\Lambda|\to+\infty} |\lambda_2^\Lambda - \lambda_1^\Lambda| = 0 . \tag{5.7}$$

A lower bound for $P_\Lambda(+\infty)$ gives an upper bound for the splitting but Theorem 1.2 gives the starting point for finding this lower bound and this will then permit to prove Theorem 1.3. We shall mainly follow the proof of Barbulyak-Kondrat'ev but with a small change.
In [4], the authors take indeed first the limit $|\Lambda| \to +\infty$ and then the limit $\beta \to +\infty$. We shall proceed for the application to the splitting in the inverse order.

Proof of Theorem 1.3:

We start from (1.25)

$$\begin{aligned} &P_\Lambda(\beta) \\ &\geq \langle x_k^2\rangle_{\beta,\Lambda} - \tfrac{h}{2|\Lambda|}\textstyle\sum_{p\in\Lambda^*\setminus\{0\}}(\frac{1}{\mathcal{J}E(p)})^{\frac{1}{2}} \coth\left[(h^2\beta^2\mathcal{J}E(p))^{\frac{1}{2}}\right] . \end{aligned} \tag{5.8}$$

Taking the limit $\beta \to +\infty$, we first obtain

$$\begin{aligned} &\lim_{\beta\to+\infty} P_\Lambda(\beta) \\ &\geq \langle x_k^2\rangle_{+\infty,\Lambda} - \tfrac{h}{2|\Lambda|}\textstyle\sum_{p\in\Lambda^*\setminus\{0\}}(\frac{1}{\mathcal{J}E(p)})^{\frac{1}{2}} . \end{aligned} \tag{5.9}$$

Now, as $|\Lambda| \to +\infty$ and if $d \geq 2$, the right hand side is estimated from below by

$$P(\beta) \geq \lim \inf_{|\Lambda|\to+\infty} \langle x_k^2 \rangle_{+\infty,\Lambda} - \frac{h}{2} \mathcal{J}^{-\frac{1}{2}} I_d \,, \tag{5.10}$$

where we recall that

$$I_d = \frac{1}{(2\pi)^d} \int_{]-\pi,\pi[^d} E(p)^{-\frac{1}{2}} dp \,. \tag{5.11}$$

We now observe that I_d is finite for $d \geq 2$.
We have seen in Section 3 that

$$\langle x_k^2 \rangle_{\beta,\Lambda} \geq \langle x_0^2 \rangle_{\beta,H_0} \,. \tag{5.12}$$

We can then take the limit $\beta \to +\infty$ in this inequality and obtain

$$\langle x_k^2 \rangle_{\infty,\Lambda} \geq \langle x_0^2 \rangle_{\infty,H_0} \,. \tag{5.13}$$

The right hand side is consequently independent of Λ and we get, for any Λ,

$$\lim_{\beta\to+\infty} P_{\Lambda,\beta} \geq \left(\int x^2 \phi_1(x)^2 dx\right) - \frac{h}{2|\Lambda|} \sum_{p\in\Lambda^*\setminus\{0\}} \left(\frac{1}{\mathcal{J}E(p)}\right)^{\frac{1}{2}} \,, \tag{5.14}$$

where ϕ_1 is here the first normalized eigenfunction of the one particle hamiltonian $H_0 = -h^2 \frac{d^2}{dx^2} + \tilde{v}(x)$.

In the thermodynamical limit, we get first

$$\lim \inf_{|\Lambda|\to+\infty} \int \left(\frac{1}{|\Lambda|} \sum_i x_i\right)^2 \phi_1^\Lambda(x^\Lambda)^2 \, dx^\Lambda \geq \left(\int x_0^2 \phi_1(x)^2 dx\right) - \frac{h}{2}\left(\frac{1}{\mathcal{J}}\right)^{\frac{1}{2}} I_d \,. \tag{5.15}$$

Using the semi-classical analysis of Section 3, we obtain, observing also that the condition

$$q_0^2 - \frac{h}{2} \cdot \mathcal{J}^{-\frac{1}{2}} I_d > 0 \tag{5.16}$$

is satisfied for sufficiently small h, the following

Proposition 5.1 :
Let $\tilde{v}$ and $\mathcal{J}$ satisfying (Ha)-(Hd) and $d \geq 2$. Then there exists h_0 and $\rho_0 > 0$ such that, for $h \in]0, h_0]$, we have

$$\lim \inf_{|\Lambda|\to+\infty} \int \left(\frac{1}{|\Lambda|} \sum_i x_i\right)^2 \phi_1^\Lambda(x^\Lambda)^2 \, dx^\Lambda \geq \rho_0 \,. \tag{5.17}$$

In particular we have obtained the proof of Theorem 1.3 through (5.5).

We recall that q_0 corresponds to the minimum of the potential $\tilde{v}(x) = v(x) + \mathcal{J} dx^2$. This potential may become convex as J increases without the same property for v. In the case when v defines a symmetric double-well, the inequality $v''(0) < 0$ is satisfied in the most simple generic case and for $\mathcal{J}$ large enough, more precisely when $v''(0) + 2\mathcal{J}d > 0$, we get cases when $\tilde{v}$ has a unique minimum at 0. The mean value $\int x^2 \phi_1(x)^2 dx$ satisfies then semiclassically (using the harmonic approximation)

$$\int x^2 \phi_1(x)^2 dx \sim \frac{h}{\sqrt{\frac{1}{2}v''(0) + d\mathcal{J}}} \,. \tag{5.18}$$

The eigenfunction is indeed localized near the minimum of $\tilde{v}$, that is at 0. This changes of course the discussion but a part may remain true if the following inequality is satisfied

$$\frac{1}{\sqrt{\frac{1}{2}v''(0) + d\mathcal{J}}} - I_d \mathcal{J}^{-\frac{1}{2}} > 0 \,. \tag{5.19}$$

This is clearly satisfied when $v''(0) + 2d\mathcal{J}$ is a strictly positive sufficiently small number. This case is treated by the following proposition.

Proposition 5.2 :
Let $\tilde{v}$ and $\mathcal{J}$ satisfying (Ha)-(Hc) and $d \geq 2$. Let us assume that $\tilde{v}$ has a unique non degenerate minimum at 0. Let be satisfied the condition

$$(\frac{1}{2}\tilde{v}''(0))^{-\frac{1}{2}} - \frac{1}{2} \cdot \mathcal{J}^{-\frac{1}{2}} I_d > 0 \,. \tag{5.20}$$

Then there exists h_1 and $\rho_1 > 0$ such that, for $h \in]0, h_1]$, we have

$$\lim \inf_{|\Lambda| \to +\infty} \int (\frac{1}{|\Lambda|} \sum_i x_i)^2 \phi_1^\Lambda (x^\Lambda)^2 \, dx^\Lambda \geq \rho_1 \,. \tag{5.21}$$

When $v''(0) > 0$, we are happy to verify that the method does not work. We have indeed the inequality $I_d \geq \frac{1}{\sqrt{d}}$ (as a consequence of Cauchy-Schwarz) showing that (5.20) can not be true. (cf Theorem 7.1 in [9]).

Remark 5.3 :
Another case where $\int (\frac{1}{|\Lambda|} \sum_i x_i)^2 \phi_1^\Lambda (x^\Lambda)^2 \, dx^\Lambda$ is controlled is the case considered by Pastur and Kozurenko [40]. The potential is $(1 - |x|^2)^2$ and the proof is also valid for $n > 1$. We emphasize that no tunneling is involved as $n > 1$.

Remark 5.4 *:*
As will be clear in Section 6, we note here that, by taking the limit $\beta \to +\infty$ before to take the thermodynamic limit, we have eliminated a singularity at the origin which makes the argument valid for $d \geq 2$ instead of $d \geq 3$ in the proof of Barbulyak-Kondrat'ev. The direct study of the limit $\beta \to +\infty$ could also lead to weaker estimate on $\tilde{v}$ but we shall loose the control of the limit $\beta \to +\infty$.

The case: $d = 1$.

Another interesting point is to analyze the case $d = 1$. The sum

$$\frac{1}{|\Lambda|} \sum_{p \in \Lambda^* \setminus \{0\}} \left(\frac{1}{\mathcal{J} E(p)}\right)^{\frac{1}{2}}$$

is divergent as $|\Lambda| \to +\infty$, but this divergence is controlled in $\ln(|\Lambda|)$. This leads to the statement that the splitting remains in $\mathcal{O}(\frac{1}{|\Lambda|})$ for $h \leq h_0$ and $|\Lambda| \leq \exp \frac{T}{h}$, where T is explicitely computable.
This condition is much weaker than the condition (1.7) given in [30].
This may be connected to phenomena discussed in [23] and to the Ising model with transverse field discussed in the introduction. As $h \to 0$, the condition for the limit case is when

$$\ln |\Lambda| \sim \frac{q_0^2}{h} [2\mathcal{J}]^{\frac{1}{2}} . \tag{5.22}$$

We note also that the phenomenon is effectively related to $\mathcal{J} \neq 0$. To summarize, we prove in this case that

Theorem 5.5 *:*
Let $d = 1$ and $\tilde{v}$ satisfying (Ha)-(Hd), then there exists a constant C such that

$$\lambda_2^\Lambda - \lambda_1^\Lambda \leq \frac{C}{|\Lambda|} \frac{1}{(1 - Ch \log |\Lambda|)_+} . \tag{5.23}$$

As mentioned in the introduction and as was communicated to us by J. Fröhlich, this condition on $|\Lambda|$ is due to the method (when $n = 1$). An approach using the Peierls trick could be more effective in this case [10].

6. On the heat kernel for β large

This section describes mainly the results obtained by Barbulyak and Kondrat'ev but take account of the stronger results obtained by semi-classical

analysis in Section 3 concerning $\langle x^2\rangle_{\beta,H_0}$.
We assume that $d \geq 3$ and $n = 1$.
Let us start from (1.24) in the form

$$P(\beta) \geq \langle x_0^2\rangle_{\beta,H_0} - \frac{1}{2(2\pi)^d}\int_{]-\pi,\pi[^d}\left(\frac{h^2}{\mathcal{J}E(p)}\right)^{\frac{1}{2}}\coth\left[\left(h^2\beta^2\mathcal{J}E(p)\right)^{\frac{1}{2}}\right]\, dp\,. \tag{6.1}$$

We want just to precise under which conditions on β and h we can get the strict positivity of $P(\beta)$.
We observe as in (5.16) that

$$q_0^2 - \frac{1}{2}hI_d\,\mathcal{J}^{-\frac{1}{2}} > 0\,, \tag{6.2}$$

for sufficiently small h.
A natural critical value of β is the solution $\beta = \beta_1(h)$ of

$$q_0^2 = \frac{h}{2}\frac{1}{(2\pi)^d}\int_{]-\pi\,,\,\pi[^d}(\mathcal{J}E(p))^{-\frac{1}{2}}\coth\left[h\beta(\mathcal{J}E(p))^{\frac{1}{2}}\right]dp\,. \tag{6.3}$$

Clearly β_1 satisfies $\beta_1(h)h \leq C_0$.
Combining with (3.14) which analyzes the convergence of $\langle x^2\rangle_{\beta,H_0}$ to q_0^2 as $h \to 0$, we obtain the following theorem (due essentially to [4])

Theorem 6.1 :
If $d \geq 3$, then, for $\tilde{v}$ satisfying (Ha)-(Hd) and for any $\epsilon > 0$, there exists $h_0(\epsilon) > 0$ and $C_0(\epsilon)$ such that, for all $h < h_0$ and β such that $\beta h > C_0(\epsilon)$, we get

$$P(\beta) \geq q_0^2(1-\epsilon)\,. \tag{6.4}$$

Remark 6.2 :
It is interesting to look at the limit $\mathcal{J} \to 0$. We find

$$\lim_{\mathcal{J}\to 0}\int_{]-\pi,\pi[^d}(\mathcal{J}E(p))^{-\frac{1}{2}}\left(\coth\left[\beta h(\mathcal{J}E(p))^{\frac{1}{2}}\right]\right)dp = +\infty\,. \tag{6.5}$$

On the other hand, we know that the situation (say for the splitting) is quite different between the two cases:
For $\mathcal{J} = 0$, the splitting is in $\mathcal{O}(\exp -\frac{\hat{S}}{h})$ for any $\hat{S} < S$, while, for $\mathcal{J} > 0$, one hopes a splitting in $\mathcal{O}(\exp -\frac{|\Lambda|\hat{S}}{h})$.
In the same direction, let us observe that this strict positivity of $P(\beta)$ is no more true in the case $\mathcal{J} = 0$ as can be seen by direct computation. We have indeed, for $\Lambda = \{1, \cdots, m\}$,

$$P_\Lambda(\beta) = \frac{1}{m}\frac{\mathrm{Tr}\,(x^2\exp -\beta H_0)}{\mathrm{Tr}\,(\exp -\beta H_0)}\,, \tag{6.6}$$

which tends to 0 *as* $m \to +\infty$.
We have used here that, by symmetry of $\tilde{v}$,

$$\mathrm{Tr}\,(x \exp -\beta H_0) = 0\,.$$

About the Pastur and Khoruzenko results:

These authors discussed two different cases of operators in the case when $v(x) = -\frac{a}{2}|x|^2 + \frac{b}{4}|x|^4$ with $a > 0$, $b > 0$.

- The first case with $a > 2\mathcal{J}d$ is called ferroelectric model of the disorder type. The corresponding $\tilde{v}$ describes a double well (if $n = 1$)

$$\tilde{v}(x) = \frac{(2\mathcal{J}d - a)}{2}x^2 + \frac{b}{4}x^4\,. \tag{6.7}$$

- The second case corresponds to $0 < a < 2\mathcal{J}d$ and is called ferroelectric model of the displacement type. The potential $\tilde{v}$ is now convex.

They prove actually in this context and without the distinction between the two cases the following theorem for the above model,

Theorem 6.3 :
If $d \geq 3$ *and if*

$$\mathcal{J} > h^2(n+2)^2 b^2 I_d^2/4a^2\,, \tag{6.8}$$

then there exists a temperature β_0^{-1} *such that, for* $\beta > \beta_0$, *the corresponding* $P(\beta)$ *is strictly positive.*

The control of $\langle x_k^2 \rangle_{\beta,\Lambda}$ is obtained by using the Bogolyubov's inequality. We recall that this inequality (See [42], Lemma 5.5.1) gives in particular that

$$\langle [\,[C^\star, H], C] \rangle_{\beta,\Lambda} \geq 0\,. \tag{6.9}$$

In our context, this inequality is applied with

$$C = \frac{1}{i}\sum_{j\in\Lambda} \partial_{x_j}\,. \tag{6.10}$$

This leads, by considering the limit $\beta \to +\infty$ before to take the thermodynamic limit, to the following result for the splitting between the two first eigenvalues as $|\Lambda|$ tends to ∞.

Theorem 6.4 :
If $d \geq 2$, *then, for* $\tilde{v}$ *defined by (6.7) and* $h_0 = \frac{2a\mathcal{J}^{\frac{1}{2}}}{(n+2)bI_d}$, *we have, for* $h \in]0, h_0[$,

$$\lim_{|\Lambda| \to +\infty} |\lambda_2^\Lambda - \lambda_1^\Lambda| = 0\,. \tag{6.11}$$

Let us just detail a variant of the argument given by Pastur-Khozurenko for the case when $\beta = +\infty$.
In the limit $\beta \to +\infty$ and in the case when $H = -\Delta + V$ is a Schrödinger operator, the inequality (6.9) becomes simply

$$\int ([\,[C^*, H], C]\Phi)(x)\Phi(x)dx \geq 0\,, \tag{6.12}$$

where Φ is the first normalized positive eigenfunction of H. This is easily and directly obtained by the minimax principle. In particular, if C is given by (6.10), we get

$$\sum_{jk} \int (\partial^2 V/\partial x_j \partial x_k)(x)\Phi(x)^2 dx \geq 0\,. \tag{6.13}$$

When $V(x)$ has the form (we take for simplification $n = 1$)

$$V(x) = \sum_j v(x_j) + J \sum_{i \sim j} |x_i - x_j|^2\,,$$

we simply get

$$\sum_j \int v''(x_j)\Phi(x)^2 dx \geq 0\,. \tag{6.14}$$

By invariance of Φ in this case, we get that, for any j,

$$\int v''(x_j)\Phi(x)^2 dx \geq 0\,. \tag{6.15}$$

In the more specific case when $v(x) = \frac{b}{4}x^4 - \frac{a}{2}x^2$ we obtain

$$\int x_j^2 \Phi(x)^2 dx \geq \frac{a}{3b}\,. \tag{6.16}$$

The proof through the infrared estimates is then easy.

Remark 6.5 :
If we compare with the semi-classical lower bound obtained in Proposition 3.2, we note that when $n = 1$ *and* $a > 2\mathcal{J}d$, *we have* $q_0^2 = \frac{a - 2\mathcal{J}d}{b}$. *This suggests that the semi-classical result is far to be optimal and that we should be able to get the results with assumptions on* v *instead of* $\tilde{v}$.
An easy extension of the proof by Pastur- Khozurenko gives that Theorem 1.3 and some weak form of Theorem 6.1 is true if the pair $(\tilde{v}, v)$ *satisfies (Ha), (Hb), (Hc) and (H'd) with*

$$(H'd) \qquad v''(x) \leq -\gamma_0 + \gamma_1 x^2\,, \tag{6.17}$$

for real γ_0, γ_1 with $\gamma_0 > 0$. In many cases, we can take $\gamma_0 = -v''(0)$. The point 0 corresponds for example to the top between the two wells.
We note also that this approach does not make use of the Ginibre inequalities.
Extensions to $n > 1$ could in this spirit be also interesting, because avoiding some assumption occuring in the validity of these inequalities.

7. Infrared estimates: the classical case.

The basic reference is the paper by Fröhlich, Simon, and Spencer [12]. This is also presented in detail in the book by Glimm-Jaffe [14]. We treat actually here a rather simple example.
We change a little the notations in order to follow this last reference. The proof is more general in the reference. The interaction Hamiltonian $I(\Lambda)$ corresponds to the interaction term

$$I(\Lambda) = -\mathcal{J} \sum_{\substack{i \sim j \\ i,j \in \Lambda}} x_i \cdot x_j \, . \tag{7.1}$$

In this section, we take $\mathcal{J} = 1$ because $\mathcal{J}$ can be included in β. Here x_i belongs to $\mathbb{R}^n$ and $x_i \cdot x_j$ denotes the scalar product in $\mathbb{R}^n$. The other terms are put in the one-particle measure and we are considering the measure

$$d\mu_{\beta,\Lambda} = Z^{-1} \exp -\beta I(\Lambda) \prod_{i\in\Lambda} d\mu_i(x_i) \, , \tag{7.2}$$

with $d\mu_i(x_i) = \exp -\tilde{v}(x_i) dx_i$, where $\tilde{v}$ satisfies some natural condition (for example (Hb)) at ∞ permitting to control the interaction.
Moreover, $\tilde{v}(x) = \tilde{v}(-x)$ (condition (Hc)) when $n = 1$.
When $n > 1$, Glimm-Jaffe assume that $\tilde{v}$ is invariant by $SO(n)$.
This assumption of symmetry permits to have the property

$$\langle x_\ell \rangle_{\beta,\Lambda} := \int x_\ell \, d\mu_{\beta,\Lambda} = 0 \, . \tag{7.3}$$

We introduce

$$\tilde{g}_\Lambda(p,x) = \frac{1}{\sqrt{|\Lambda|}} \sum_{\ell\in\Lambda} \exp -i\, p \cdot \ell \; x_\ell \, . \tag{7.4}$$

This is a function defined on Λ^*, the dual lattice introduced in (1.23) of Λ, with value in $\mathbb{C}^n$. For $j \in \{1, \cdot, n\}$, we denote by $\tilde{g}_\Lambda^{(j)}(p,x)$ the j-th component. We now introduce

$$\tilde{S}_{\beta,\Lambda}(p) = \langle \tilde{g}(p,\cdot) \cdot \tilde{g}(-p,\cdot) \rangle_{\beta,\Lambda} = \sum_{j=1}^{n} \langle \tilde{g}^{(j)}(p,\cdot) \cdot \tilde{g}^{(j)}(-p,\cdot) \rangle_{\beta,\Lambda} = \langle |\tilde{g}_\Lambda(p,\cdot)|^2 \rangle \, . \tag{7.5}$$

We then have

$$\tilde{S}_{\beta,\Lambda}(p) = \sum_{\ell\in\Lambda} \exp -i\ell p \, \langle x_0 \cdot x_\ell \rangle_{\beta,\Lambda} \,. \tag{7.6}$$

This is simply the discrete Fourier transform of the correlation function $\Lambda \ni \ell \mapsto |\Lambda|^{-\frac{1}{2}} \langle x_0 \cdot x_\ell \rangle_{\beta,\Lambda}$.
The main result in the finite lattice version is

Theorem 7.1 :
For all $p \in \Lambda^* \setminus \{0\}$, we have

$$0 \le \tilde{S}_{\beta,\Lambda}(p) \le \frac{n}{4\beta \sum_{\alpha=1}^{d} \sin^2(\frac{p_\alpha}{2})} = \frac{n}{2\beta E(p)} \,. \tag{7.7}$$

Corollary 7.2 :
Let $d \ge 3$, and let β be sufficiently large, so that

$$\lim \inf_{|\Lambda|\to+\infty} \langle x_0^2 \rangle_{\beta,\Lambda} \cdot \beta \ge (2\pi)^{-d} \int_{]-\pi,+\pi[^d} \frac{n}{4\sum_{\alpha=1}^{d} \sin^2(\frac{p_\alpha}{2})} dp \,, \tag{7.8}$$

then,

$$P(\beta) := \lim_{|\Lambda|\to+\infty} \langle (\frac{1}{|\Lambda|} \sum_{\ell\in\Lambda} x_\ell)^2 \rangle_{\beta,\Lambda} > 0 \,. \tag{7.9}$$

Proof of the corollary:
We have:

$$\langle (\frac{1}{|\Lambda|} \sum_{\ell\in\Lambda} x_\ell)^2 \rangle_{\beta,\Lambda} = (\frac{1}{|\Lambda|})^2 \sum_{k,\ell\in\Lambda} \langle x_k \cdot x_\ell \rangle_{\beta,\Lambda} = \frac{1}{|\Lambda|} \tilde{S}_{\beta,\Lambda}(0) \,. \tag{7.10}$$

By the inverse discrete Fourier transform, we have also (Plancherel formula)

$$\sum_{p\in\Lambda^*} \frac{1}{|\Lambda|} \tilde{S}_{\beta,\Lambda}(p) = \langle x_0^2 \rangle_{\beta,\Lambda} \,. \tag{7.11}$$

We then write this identity in the form

$$\frac{1}{|\Lambda|} \tilde{S}_{\beta,\Lambda}(0) = \langle x_0^2 \rangle_{\beta,\Lambda} - \sum_{p\in\Lambda^*\setminus\{0\}} \frac{1}{|\Lambda|} \tilde{S}_{\beta,\Lambda}(p) \,. \tag{7.12}$$

The theorem gives

$$\frac{1}{|\Lambda|} \tilde{S}_{\beta,\Lambda}(0) \ge \langle x_0^2 \rangle_{\beta,\Lambda} - \frac{1}{|\Lambda|} \sum_{p\in\Lambda^*\setminus\{0\}} \frac{n}{4\beta \sum_{\alpha=1}^{d} \sin^2(\frac{p_\alpha}{2})} \,. \tag{7.13}$$

Taking the thermodynamic limit we then obtain

$$P(\beta) \geq \lim \inf_{|\Lambda| \to +\infty} \langle x_0^2 \rangle_{\beta,\Lambda} - (2\pi)^{-d} \int_{]-\pi,\pi[^d} \frac{n}{4\beta \sum_{\alpha=1}^d \sin^2(\frac{p_\alpha}{2})} dp \,. \quad (7.14)$$

In order to complete the proof, we need some control of $\langle x_0^2 \rangle_{\beta,\Lambda}$ in the limit $\beta \to +\infty$ and $|\Lambda| \to +\infty$. This may lead to conditions permitting to apply the Ginibre inequalities as analyzed in Section 2.

The proof of Theorem 7.1 is based on the following lemma

Lemma 7.3 :
Let ∂ be the forward finite difference quotient. Let $f_\alpha \in \ell^2(\Lambda; I\!R^n)$ ($\alpha = 1, \cdots, d$), $f = (f_\alpha)_{\alpha=1,\cdots,d} \in \ell^2(\Lambda \times \{1, \cdots, d\}); I\!R^n)$. Then

$$\langle \exp\left(x \cdot (\sum_{\alpha=1}^d \partial_\alpha f_\alpha)\right) \rangle \leq \exp\left((2\beta)^{-1} ||f||_{\ell^2}^2\right) , \quad (7.15)$$

where

$$||f||_{\ell^2}^2 = \sum_{\ell,\alpha} f_\alpha(\ell)^2 = \sum_{\ell,\alpha,j} |f_\alpha^{(j)}(\ell)|^2 \,. \quad (7.16)$$

This inequality is sometimes called the "Gaussian domination estimate". We recall that, for $f \in \ell^2(Z\!\!Z^d; I\!R^n)$,

$$(\partial_\alpha f)(\ell) = f(\ell + e_\alpha) - f(\ell) , \quad (7.17)$$

where e_α is the unit vector in the α-th coordinate direction. We recall also that

$$x \cdot g = \sum_{\ell \in Z\!\!Z^d} x_\ell \cdot g(\ell) , \quad (7.18)$$

for $g \in \ell^2(\Lambda)$.

Proof of Theorem 7.1, assuming Lemma 7.3:

We substract 1 from both sides of (7.15), substitute ϵf for f, multiply by ϵ^{-2} and let $\epsilon \to 0$.
Thus we obtain[6]

$$\langle \left(x \cdot \sum_{\alpha=1}^d \partial_\alpha f_\alpha \right)^2 \rangle \leq \beta^{-1} ||f||_{\ell^2}^2 \,. \quad (7.19)$$

[6]We shall need the natural extension of this inequality for functions with values in $C\!\!\!\!I^n$.

Note here that we have used the invariance by translation of $\langle\cdot\rangle$ in order to obtain $\langle x\cdot\partial_\alpha f_\alpha\rangle = 0$. We note also that this inequality can be extended to complex valued functions.
With $\partial^\star$ the negative of the backward lattice difference quotient, we define, for a given $p\in\Lambda^\star\setminus\{0\}$, for $j=1,\cdots,n$, $f_{\cdot,j}\in\ell^2(\Lambda\times\{1,\cdots,d\};\mathbb{R}^n)$ by

$$f_{\alpha,j}=\partial_\alpha^\star(-\Delta^{per})^{-\frac{1}{2}}\chi_p\delta_j\ , \tag{7.20}$$

where δ_j is the vector in $\mathbb{R}^n$

$$\delta_j=(\delta_{j1},\cdots,\delta_{jn})\ .$$

We recall that the periodic Lattice Laplace operator Δ^{per} on Λ has as eigenvalues

$$-4\sum_{\alpha=1}^{d}(\sin(\frac{p_\alpha}{2}))^2$$

with

$$p_\alpha\in\Lambda^\star:=\left\{\frac{\pm 2\pi n_\alpha}{m}\ :\ 0\leq n_\alpha\leq[\frac{m}{2}]\right\}\ .$$

The corresponding eigenvectors being defined as

$$\Lambda\ni\ell\mapsto\chi_p(\ell)=|\Lambda|^{-\frac{1}{2}}\exp ip\cdot\ell\ .$$

This gives the bound (7.7), by easy computations (cf also [14]).

Remark 7.4 :
If we consider, say in the case $n=1$, for g constant and orthogonal to $Ker\Delta^{per}$, the vector $f_\alpha=\partial_\alpha^\star(-\Delta^{per})^{-1}g$, the inequality (7.19) gives

$$\langle(x\cdot g)^2\rangle\leq\beta^{-1}\langle\left((-\Delta^{per})^{-1}g\right)\cdot g\rangle\ . \tag{7.21}$$

This is interesting to compare with some forms of the Brascamp-Lieb inequality (See [7], [20], [51] and [39]).

About the proof of Lemma 7.3:

The argument for this lemma is called a "multiple reflection bound". We first observe that

$$x\cdot\partial_\alpha g=\sum_{\ell\in\Lambda}g(\ell)(-x_\ell+x_{\ell-e_\alpha})\ .$$

Taking $g=f_\alpha$ and summing over α, we get

$$\begin{aligned} I &\equiv\langle\exp(\textstyle\sum_\alpha x\cdot\partial_\alpha f_\alpha)\exp(-\frac{1}{2\beta}||f||^2_{\ell^2})\rangle\\ &=\frac{\int\exp\left(-\sum_{\ell,\alpha}\frac{\beta}{2}((-x_\ell+x_{\ell-e_\alpha}+\beta^{-1}f_\alpha(\ell))^2\right)\prod d\mu_\ell}{\int\exp\left(-\sum_{\ell,\alpha}\frac{\beta}{2}(-x_\ell+x_{\ell-e_\alpha})^2\right)\prod d\mu_\ell}\ . \end{aligned}$$

The desired inequality is then

$$I \leq 1 \, .$$

This is obtained by reflection arguments. The first step is an easy but basic inequality:

Lemma 7.5 :
Under weak assumptions on the measures μ and ν defined on $I\!R^n$, we have for any a in $I\!R^n$

$$\begin{aligned}&\left(\int \exp[-\tfrac{1}{2}(x-y-a)^2]d\mu(x)\cdot d\nu(y)\right)^2 \\ &\quad \leq \left(\int \exp[-\tfrac{1}{2}(x-y)^2]d\mu(x)\cdot d\mu(y)\right)\left(\int \exp[-\tfrac{1}{2}(x-y)^2]d\nu(x)\cdot d\nu(y)\right) \, .\end{aligned} \tag{7.22}$$

The proof is simple. We just use the Plancherel formula

$$\begin{aligned}&\left(\int \exp[-\tfrac{1}{2}(x-y-a)^2]d\mu(x)\cdot d\nu(y)\right) \\ &\quad = \text{const.}\left(\int \exp -\tfrac{1}{2}\xi^2 \, \exp i\xi a \, \hat{\nu}(\xi)\cdot \overline{\hat{\mu}(\xi)}d\xi\right) \, .\end{aligned} \tag{7.23}$$

We then Cauchy-Schwarz the right-hand side and get

$$\begin{aligned}&\left(\int \exp -\tfrac{1}{2}\xi^2 \, \exp i\xi a \, \hat{\nu}(\xi)\cdot \overline{\hat{\mu}(\xi)}d\xi\right)^2 \\ &\quad \leq \left(\int \exp -\tfrac{1}{2}\xi^2 \, |\hat{\nu}(\xi)|^2 d\xi\right)\cdot\left(\int \exp -\tfrac{1}{2}\xi^2 \, |\hat{\mu}(\xi)|^2 \, d\xi\right) \, .\end{aligned} \tag{7.24}$$

The inequality (7.22) is then clear, using again the Plancherel formula.
A quantum variant of this argument is proposed in Lemma 4.1 in [9]. It involves the Trotter product formula. We shall come back to this point in Section 8.

The reflection argument :

For any $h \in \ell^2(\Lambda \times \{1,\cdots,d\}; I\!R^n)$, we introduce

$$Z(\{h_\alpha(\ell)\}) = \int \exp -\frac{\beta}{2}\sum_{\ell\in\Lambda}\sum_{\alpha=1}^{d}(x_\ell - x_{\ell+e_\alpha} + \beta^{-1}h_\alpha(\ell))^2 \prod_{\ell\in\Lambda} d\mu_\ell \, . \tag{7.25}$$

The proof is then reduced to the proof of the inequality:

$$Z(\{h_\alpha(\ell)\}) \leq Z_0 \, , \tag{7.26}$$

where Z_0 is by definition $Z(\{h_\alpha(\ell)\})$ for $h_\alpha(\ell) = 0$:

$$Z_0 = Z(\{0\}) \, .$$

We assume that $\Lambda = [0, m-1]^d$ with m even. We first observe that $\sup_h Z(\{h_\alpha(\ell)\})$ is attained ($Z(h)$ tends to 0 as $||h|| \to +\infty$). Inside the class of the maximizers, we can choose h^0 such that $Z(h^0) = \sup_h Z(\{h_\alpha(\ell)\})$ and which has a maximal number of components equal to 0 inside this class. Let us show that this leads to a contradiction if h^0 is not identically 0. After easy manipulations we can assume that $h_1^0(m-1, 0, \cdots, 0) \neq 0$. We now rewrite the measure

$$\left(\exp -\frac{\beta}{2} \sum_{\ell \in \Lambda} \sum_{\alpha=1}^{d} (x_\ell - x_{\ell+e_\alpha} + \beta^{-1} h_\alpha(\ell))^2\right) \prod_{\ell \in \Lambda} d\mu_\ell$$

as the tensor product of two measures multiplied by an interaction density. The first measure is

$$\begin{aligned} \mu' &= \left(\exp -\tfrac{\beta}{2} \left[\textstyle\sum_{\mathcal{I}'} (x_\ell - x_{\ell+e_\alpha} + \beta^{-1} h_\alpha(\ell))^2\right]\right) \\ &\quad \times \textstyle\prod_{\ell \in \Lambda, \ell_1 \in [0, \frac{m}{2}-1]} d\mu_\ell \,, \end{aligned}$$

with

$$\mathcal{I}' := \left\{ \begin{array}{c} (\ell, \alpha) \in \Lambda \times \{1, \cdots, d\} \\ \ell_1 \in [0, \frac{m}{2} - 1], \ell_1 + e_\alpha \in [0, \frac{m}{2} - 1] \end{array} \right\}$$

and the second one is

$$\begin{aligned} \mu'' &= \left(\exp -\tfrac{\beta}{2} \left[\textstyle\sum_{\mathcal{I}''} (x_\ell - x_{\ell+e_\alpha} + \beta^{-1} h_\alpha(\ell))^2\right]\right) \\ &\quad \times \textstyle\prod_{\ell \in \Lambda, \ell_1 \in [\frac{m}{2}, m-1]} d\mu_\ell \,, \end{aligned}$$

with

$$\mathcal{I}'' := \left\{ \begin{array}{c} (\ell, \alpha) \in \Lambda \times [1, \cdots, d] \\ \ell_1 \in [\frac{m}{2}, m-1], \ell_1 + e_\alpha \in [\frac{m}{2}, m-1] \,. \end{array} \right\}$$

The interaction density is then given by

$$\begin{aligned} I(x) &= \exp -\tfrac{\beta}{2} \left(\textstyle\sum_{\ell'} \left[(x_{\frac{m}{2}-1,\ell'} - x_{\frac{m}{2},\ell'} + \beta^{-1} h_1(\tfrac{m}{2} - 1, \ell'))^2\right]\right) \\ &\quad \times \exp -\tfrac{\beta}{2} \left(\textstyle\sum_{\ell'} \left[(x_{m-1,\ell'} - x_{0,\ell'} + \beta^{-1} h_1(m-1, \ell'))^2\right]\right) \end{aligned}$$

If we change the name of the variables by posing $y_\ell = x_{\sigma(\ell)}$ where σ denotes the symmetry around the hyperplane: $\ell_1 = \frac{m-1}{2}$. The interaction takes the form of some product of $\exp -\frac{\beta}{2}(x_\ell - y_\ell - h'_\ell)^2$, corresponding to some $\ell \in \Lambda$ such that $\ell_1 = 0$ or $\ell_1 = \frac{m}{2} - 1$. We now apply the inequality given by the lemma and obtain

$$Z_0 = Z(\{h_\alpha^0(\ell)\})^2 \leq Z(\{h_\alpha^1(\ell)\}) Z(\{h_\alpha^2(\ell)\}) \,.$$

Here $h_\alpha^1(\ell)$ and $h_\alpha^2(\ell)$ satisfy

$$h_\alpha^j(\ell) = h_\alpha^j(\ell)(\sigma(\ell)) \,,$$

and are equal to 0 when $\alpha_1 = 1$, $\ell_1 = \frac{m}{2} - 1$ and when $\alpha_1 = 1$, $\ell_1 = m - 1$.
We observe also that necessarily

$$Z_0 = Z(\{h^j_\alpha(\ell)\}) \ , \ j = 1, 2 \ .$$

But at least h^1 or h^2 has a larger number of vanishing components than h^0 and has the same property as h^0.
This gives the contradiction.

8. Infrared estimates: the quantum case.

8.1. INTRODUCTION

We mainly describe the results given by Dyson-Lieb-Simon [9] with some additional remarks given by Pastur, Khozurenko, Barbulyak, Kondrat'ev.
We shall consider unbounded operators on $\otimes_{i\in\Lambda} L^2(\mathbb{R}^n; \mathbb{R})$ of the form $H = H^{per}_\Lambda$ with

$$\begin{aligned} H^{per}_\Lambda &= \sum_{\ell\in\Lambda} \hat{H}_\ell + \tfrac{1}{2} J \sum_{i\sim j} |x_i - x_j|^2 \\ &= \sum_{\ell\in\Lambda} H_\ell - J \sum_{i\sim j} x_i \cdot x_j \ . \end{aligned} \tag{8.1}$$

The operator $\hat{H}_\ell$ works only in the variable x_ℓ:

$$\hat{H}_\ell = -h^2 \Delta_{x_\ell} + v(x_\ell) \ . \tag{8.2}$$

and H_ℓ has the same form with v replaced by $\tilde{v}$.
In all this section, $\tilde{v}$ satisfies (Ha)- (Hc), but n is not necessarily equal to 1. When $n > 1$, we assume that $\tilde{v}$ is invariant by $SO(n)$ which implies

$$\mathrm{Tr}\,[x_\ell \exp - \beta H^{per}_\Lambda] = 0 \ . \tag{8.3}$$

8.2. THE DUHAMEL TWO-POINT FUNCTION

The first point is the introduction of the Duhamel two-point function. For quantum systems in finite volume (attached with some selfadjoint hamiltonian H on $L^2(\mathbb{R}^{n|\Lambda|})$) with partition function $Z = \mathrm{Tr}\,(\exp - \beta H)$, we define the Duhamel two-point function (DTF) by

$$(A, B) = Z^{-1} \int_0^1 \mathrm{Tr}\,[\exp - x\beta H \ A \ \exp - (1 - x)\beta H \ B] \, dx \ . \tag{8.4}$$

A and B are non-bounded operators and we assume that (8.4) has a sense.
This will be clear in our applications.
This (DTF) has the following properties.

$$(A, B) = (B, A) \ , \tag{8.5}$$

by cyclicity of the trace and change of variable in the integral $x \mapsto (1-x)$.

$$(A^*, A) \geq 0 \,. \tag{8.6}$$

We can indeed, with $C_x = \exp - \left((\frac{1-x}{2})\beta H\right) \; A \; \exp -(\frac{x}{2}\beta H)$, write (A^*, A) as

$$\mathrm{Tr}\,[\exp -x\beta H \; A^* \; \exp -(1-x)\beta H \; A)] = \mathrm{Tr}\,[C_x^* C_x] \,.$$

Let us also observe that one recovers the mean value of A through

$$\langle A \rangle = (A, 1) \,. \tag{8.7}$$

We shall use also the following property which actually explains the introduction of the (DTF) function. If

$$\langle B \rangle_\mu \equiv \{\mathrm{Tr}\,[\exp -\beta H + \mu A)]\}^{-1}\, \mathrm{Tr}\,[B \exp(-\beta H + \mu A)] \tag{8.8}$$

then

$$\frac{\partial \langle B \rangle_\mu}{\partial \mu}_{/\mu=0} = (A, B) - \langle A \rangle \langle B \rangle \,. \tag{8.9}$$

In particular when $A = B$, the term $\frac{1}{2}\mu^2 (A, A) Z$ is the second order term in a perturbation expansion of $\mu \mapsto \mathrm{Tr}\,[\exp(-\beta H + \mu A)]$.

8.3. A TRICKY FUNCTION

The second point is a tricky lemma

Lemma 8.1 :
There exists a function f from $[0, +\infty]$ to $[0, 1]$ defined implicitely by

$$f(x \tanh x) = x^{-1} \tanh x \,.$$

This function is convex, monotone, decreasing and satisfies

$$\lim_{x \to 0} f(x) = 1 \,, \quad \lim_{x \to +\infty} f(x) = 0 \,.$$

We now introduce three thermodynamical quantities.

–

$$g(A) = \frac{1}{2} \langle A^* A + A A^* \rangle = \frac{1}{2} Z^{-1}\, \mathrm{Tr}\,[(A^* A + A A^*) \exp -\beta H] \,. \tag{8.10}$$

–

$$b(A) = (A^*, A) \,. \tag{8.11}$$

–

$$c(A) = \langle [A^*, [H, A]] \rangle \,. \tag{8.12}$$

From essentially the Jensen's inequality, one gets under suitable assumptions the following theorem [9]

Theorem 8.2 *:*

$$b(A) \geq g(A) f\left(\frac{c(A)}{4g(A)}\right) . \tag{8.13}$$

This is extended by convexity to a finite sum A_i of operators in the form

$$\sum_i b(A_i) \geq \left(\sum_i g(A_i)\right) \cdot f\left(\frac{\sum_i c(A_i)}{4\sum_i g(A_i)}\right) . \tag{8.14}$$

We can now deduce from an estimate on b and c an estimate on g through the following Theorem

Theorem 8.3 *:*
Suppose $b \geq gf(\frac{c}{g})$, $b, g, c \geq 0$ *and* $b \leq b_0$, $c \leq c_0$. *Then we have*

$$g \leq g_0 \tag{8.15}$$

where

$$\left\{ \begin{array}{ll} g_0 & = \frac{1}{2}(c_0 b_0)^{\frac{1}{2}} \coth x_0 \\ x_0^2 & = \frac{c_0}{4b_0} \end{array} \right. . \tag{8.16}$$

8.4. GAUSSIAN DOMINATION IN THE QUANTUM CASE

We recall a small extension of Lemma 7.5 which was used in the classical case.

Lemma 8.4 *:*
Let $\mathcal{H}_1$ *be a finite-dimensional vector space and let* $\mathcal{H} = \mathcal{H}_1 \otimes \mathcal{H}_1$. *If* $A, B, \cdots$ *are operators on* $\mathcal{H}_1$, *we use the same symbol for* $A \otimes Id$, $B \otimes Id, \cdots$, *and the symbols* $\tilde{A}, \tilde{B}, \cdots$ *for* $Id \otimes A$, $Id \otimes B$, $\cdots$. *Then for any selfadjoint operator* $A, B, C_i \cdots$ *with real matrix representations and real numbers* $h_1, \cdots, h_k$

$$\begin{array}{l} \left(\mathrm{Tr}\left\{\exp\left[A + \tilde{B} - \sum_{i=1}^k (C_i - \tilde{C}_i - h_i)^2\right]\right\}\right)^2 \\ \quad \leq \mathrm{Tr}\left\{\exp\left[A + \tilde{A} - \sum_{i=1}^k (C_i - \tilde{C}_i - h_i)^2\right]\right\} \\ \qquad \times \mathrm{Tr}\left\{\exp\left[B + \tilde{B} - \sum_{i=1}^k (C_i - \tilde{C}_i - h_i)^2\right]\right\} . \end{array} \tag{8.17}$$

We now apply the Gaussian domination argument. Lemma 7.3 is now replaced by

Lemma 8.5 :
Let H be a Hamiltonian of the form (8.1). Let $\{h_\alpha(\ell)\}$, $\ell \in \Lambda, \alpha = 1, \cdots, d$ be $d|\Lambda|$ vectors in $\mathbb{R}^n$. Let Λ be $[1, \cdots, m]^d$ with m even. Let $\phi(h) = \sum_\ell h(\ell) \cdot x_\ell$.

$$\frac{\mathrm{Tr}\ \exp[-\beta H + \phi(\sum_\alpha \partial_\alpha h_\alpha)]}{\mathrm{Tr}\ [\exp -\beta H]} \leq \exp \frac{||h||^2}{2\beta} , \tag{8.18}$$

where $||h||^2 = \sum_{\ell,\alpha} |h_\alpha(\ell)|^2$.

Similarly to the way we get (7.19) from (7.15) and taking account of (8.9), we deduce from (8.18) the inequality

$$b(x \cdot \sum_{\alpha=1}^{d} \partial_\alpha f_\alpha) := \left(x \cdot \sum_{\alpha=1}^{d} \partial_\alpha f_\alpha, x \cdot \sum_{\alpha=1}^{d} \partial_\alpha f_\alpha\right) \leq \beta^{-1} ||f||_{\ell^2}^2 . \tag{8.19}$$

We now introduce

$$b_p^{(j)} = b(A_j) = (g_p^{(j)}, g_{-p}^{(j)}) , \tag{8.20}$$

with

$$A_j = g_p^{(j)} := \partial_\alpha^\star \chi_p \delta_j . \tag{8.21}$$

The analog of Theorem 7.1 is then true when one uses the (DTF).

Theorem 8.6 :
For hamiltonians of the form (8.1) in square boxes Λ of size m (m even integer) and under the same conditions as in Lemma 8.5 holds,

$$b_p^{(j)} \leq (2\beta \mathcal{J} E(p))^{-1} , \ j = 1, \cdots, n . \tag{8.22}$$

8.5. END OF THE PROOF OF THE INFRARED ESTIMATE

We can now finish our sketch of the proof of Theorem 1.2.

In order to use Theorem 8.2, we have also to control the bracket: $[g_p^{(j)}, [\beta H, g_{-p}^{(j)}]]$. We immediately get

$$[g_p^{(j)} , [\beta H , g_{-p}^{(j)}]] = 2\beta h^2 . \tag{8.23}$$

We can then apply Theorem 8.3 with

$$b_0 = \frac{n}{2\beta \mathcal{J} E(p)} , \tag{8.24}$$

and

$$c_0 = 2n\beta h^2 . \tag{8.25}$$

The corresponding x_0 is given by

$$x_0 = \sqrt{\beta^2 h^2 \mathcal{J} E(p)}\,. \tag{8.26}$$

Coming back to the definition of $g(A) = \sum_j g(A_j)$, we apply Theorem 8.3 and get the equivalent of Theorem 7.1.

Theorem 8.7 :
For all $p \in \Lambda^* \setminus \{0\}$, *we have*

$$g(A) = \tilde{S}_{\beta,\Lambda}(p) \le \frac{n}{2}\left(\frac{h^2}{\mathcal{J}E(p)}\right)^{\frac{1}{2}} \coth\left[\left(h^2\beta^2\mathcal{J}E(p)\right)^{\frac{1}{2}}\right]\,. \tag{8.27}$$

The end of the proof is then similar to the proof of (7.13).

Acknowledgements :

I would like to thank F. Klopp for communicating the paper of V.S. Barbulyak and Y. Kondrat'ev, J. Fröhlich for his informations and J.P. Solovej for motivating discussions. The redaction of these notes was also stimulated by the preparation of the proposal for the european contract "Postdoctoral training program in Partial Differential Equations and applications in Quantum Mechanics" and by L. Rodino who initiated this NATO ASI conference.

References

1. Agmon, S. (1982) *Lecture on exponential decay of solutions of second order elliptic equations.* Math. Notes, t.29, Princeton University Press .
2. Albanese, C. (1989) On the spectrum of the Heisenberg Hamiltonian, *Journal of statistical Physics* **55** (1-2), 297-309.
3. Albanese, C. (1990) Unitary dressing transformations and exponential decay below Threshold for Quantum spin systems, *Comm. in Math. Physics* **134**, 1-7.
4. Barbulyak, V.S. and Kondrat'ev, Y. (1992) The quasiclassical limit for the Schrödinger operator and phase transitions in Quantum statistical physics, *Functional Analysis and applications*, **26**(2), 61-64.
5. Barbulyak, V.S. and Kondrat'ev, Y. (1990) Methods of functional analysis in Problems of mathematical Physics (in Russian), Institut Mat. AN USSR, Kiev, 30-41.
6. Bouziane, M. and Martin, Ph. A. (1976) Bogolyubov inequality for unbounded operators and the Bose gas, *J. Math. Phys.* **17**(10), 1848-1851.
7. Brascamp, H.J. and Lieb, E. (1976) On extensions of the Brunn-Minkovski and Prékopa-Leindler Theorems, including inequalities for Logconcave functions, and with an application to diffusion equation, *Journal of Functional Analysis* **22** 366-389.
8. Brunaud, M. and Helffer, B. (1991) Un problème de double puits provenant de la théorie statistico-mécanique des changements de phase (ou relecture d'un cours de M.Kac), *Preprint LMENS.*
9. Dyson, F.J., Lieb,E.H. and Simon, B. (1978) Phase transitions in quantum spins systems with isotropic and nonisotropic interactions, *J. Statistic. Phys.* **18**(4), 335-383.

10. Fröhlich, J. (1976) Phase transitions, Goldstone bosons and topological superselection rules, *Acta Phys. Austriaca* Suppl. XV, p. 133-269.
11. Fröhlich, J., Israel, R., Lieb, E.H. and Simon, B. (1978) Phase transition and reflection positivity, I General theory and long range lattice models, *Comm. in Math. Physics* **62** (1), 1-34.
12. Fröhlich, J., Simon, B. and Spencer, T. (1976) Infrared bounds, phase transitions, and continuous symmetry breaking, *Comm. Math. Phys.* **50**, 79-85.
13. Ginibre, J. (1970) General formulation of Griffiths inequalities, *Comm. in Math. Phys.* **16**, 310-328.
14. Glimm, J. and Jaffe, A. (1987) *Quantum physics (a functional integral point of view),* Springer Verlag, Second edition.
15. Glimm, J., Jaffe, A. and Spencer, T. (1975) Phase transitions for ϕ_2^4 quantum fields, *Comm. in Math. Phys.* **45**, 203-216.
16. Globa, S.A. and Kondrat'ev, Yu. G. (1987) Application of functional analysis in Problems of mathematical Physics (in Russian), Institut Mat. AN USSR, Kiev, 4-16.
17. Helffer, B. (1988) *Introduction to the semiclassical analysis for the Schrödinger operator and applications,* Springer lecture Notes in Math., n^0 1336.
18. Helffer, B. (1992) Problèmes de double puits provenant de la théorie statistico-mécanique des changements de phase, II Modèles de Kac avec champ magnétique, étude de modèles près de la température critique. *Preprint LMENS.*
19. Helffer, B. (1992) Décroissance exponentielle des fonctions propres pour l'opérateur de Kac: le cas de la dimension > 1, in *Operator calculus and spectral theory*, Symposium on operator calculus and spectral theory, Lambrecht (Germany), December 1991, edited by Demuth, M., Gramsch, B. and Schulze, B.-W., Operator Theory: Advances and Applications **57**; Birkhäuser Verlag Basel, 99-116.
20. Helffer, B. (1993) Spectral properties of the Kac operator in large dimension. Proceedings on Mathematical Quantum Theory II: Schrödinger Operators, edited by J. Feldman, R. Froese, and L.M. Rosen. *CRM Proceedings and Lecture Notes.*
21. Helffer, B. (1996) Recent results and open problems on Schrödinger operators, Laplace integrals and transfer operators in large dimension, to appear in Advances in PDE, Akademie Verlag.
22. Helffer, B. (1995) *Semiclassical analysis for Schrödinger operators, Laplace integrals and transfer operators in large dimension: an introduction,* Paris 11- Edition.
23. Helffer, B. (1996) On Laplace integrals and transfer operators in large dimension: examples in the non convex case, to appear in Letters in Math. Physics (1996).
24. Helffer, B. (1996) Semi-classical analysis for the transfer operator: WKB constructions in dimension 1, in Partial differential equations and mathematical physics, The danish-swedish analysis seminar (1995), Birkhäuser (1996).
25. Helffer, B. (1996) Semi-classical analysis for the transfer operator: formal WKB constructions in large dimension, Preprint Université Paris 11.
26. Helffer, B. and Robert, D. (1984) Puits de potentiel généralisés et asymptotique semi-classique, *Annales de l'IHP (section Physique théorique)* **41** (3), 291-331.
27. Helffer, B. and Sjöstrand, J. (1984) Multiple wells in the semiclassical limit, *Comm. in PDE* **9** (4), 337-408.
28. Helffer, B. and Sjöstrand, J. (1987) Puits multiples en limite semi-classique VI-le cas des puits variétés-, *Annales de l'IHP (Section Physique théorique)* **46** (4), 353-373.
29. Helffer, B. and Sjöstrand, J. (1992) Semiclassical expansions of the thermodynamic limit for a Schrödinger equation, *Astérisque* **210**, 135-181.
30. Helffer, B. and Sjöstrand, J. (1992) Semiclassical expansions of the thermodynamic limit for a Schrödinger equation II, *Helvetica Physica Acta* **65**, 748-765 and Erratum **67** (1994), 1-3.
31. Helffer, B. and Sjöstrand, J. (1994) On the correlation for Kac like models in the convex case, *J. Statist. Physics*, **74** (1-2), 349-369.

32. Kac, M. (1962) Statistical mechanics of some one-dimensional systems, Studies in mathematical analysis and related topics in *Essays in honor of Georges Polya*, Stanford University Press, Stanford, California, 165-169.
33. Kac, M. (1966) Mathematical mechanisms of phase transitions, Brandeis lectures, Gordon and Breach.
34. Kac, M. and Helfand, E. (1963) Study of several lattice systems with long range forces, *J. Math. Phys.* **4**, 1078-1088.
35. Kac, M. and Thompson, C. (1966) On the mathematical mechanism of phase transition, *Proc. N. A. S.* **55**, 676-683 and Erratum **56**, 1625.
36. Kac, M. and Thompson, C. (1969) Phase transition and eigenvalue degeneracy of a one dimensional anharmonic oscillator, *Studies in Applied Mathematics* **48**, 257-264.
37. Kirkwood, J. and Thomas, L. (1983) Expansions and phase transitions for the ground state in quantum spin systems, *Comm. in Math. Physics* **88**, 569-580.
38. Lenard, A. (1971) Generalization of the Golden Thompson inequality, *Indiana Univ. J. Math.* **21**, 457-468.
39. Naddaf, A. and Spencer, T. (1995) On homogeneization and scaling limit of some gradient perturbations of a massless free field, Preprint Princeton.
40. Pastur, L.A. and Khoruzhenko, B. A. (1987) Phase transitions in quantum models of rotators and ferroelectrics, *Teor. Mat. Fiz.*, **73** (1), 111-124 .
41. Rebenko, A.L. (1996) Peierls argument and long range behavior of quantum lattice systems with unbounded spins. Lecture in Ascona (June 1996).
42. Ruelle, D. (1969) *Statistical mechanics,* Math. Physics monograph series, W.A.Benjamin, Inc..
43. Simon, B. (1974) *The $P(\phi)_2$ Euclidean quantum field theory,* Princeton Series in Physics.
44. Simon, B. (1979) *Functional integration and quantum physics.* Pure and Applied mathematics, $n^0$86, Academic press, New york.
45. Simon, B. (1983) Instantons, double wells and large deviations. *Bull. AMS.* **8**, 323-326.
46. Simon, B. (1983) Semi-classical Analysis of low lying eigenvalues.
I *Ann. Inst. H. Poincaré* **38** (1983), 295-307.
II *Annals of Mathematics* **120** (1984), 89-118.
47. Sjöstrand, J. (1993) Potential wells in high dimensions I, *Ann. Inst. Poincaré,* Section Physique théorique **58** (1), 1-41.
48. Sjöstrand, J. (1993) Potential wells in high dimensions II, more about the one well case. *Ann. Inst. H. Poincaré,* Section Physique théorique **58** (1), 42-53.
49. Sjöstrand, J. (1994) Evolution equations in a large number of variables, Mathematische Nachrichten **166**, 17-53.
50. Sjöstrand, J. (1994) Ferromagnetic integrals, correlations and maximum principles, *Ann. Inst. Fourier,* **44** (2), 601-638.
51. Sjöstrand, J. (1996) Correlation asymptotics and Witten laplacians, to appear in *St Petersburg Mathematical Journal.*
52. Sokal, A.D. (1982) Mean-field bounds and correlation inequalities. *Journal of statistical Physics,* **28** (3), 431-439.
53. Thomas, L. and Yin, Z. (1983) Quantum Ising lattice systems, *Comm. in Math. Phys.* **91**, 405-417.
54. Thompson, C. (1972) *Mathematical statistical mechanics,* The Macmillan company New York.

MICROLOCAL EXPONENTIAL ESTIMATES AND APPLICATIONS TO TUNNELING

A. MARTINEZ
Université Paris-Nord
Institut Galilée - Département de Mathématiques
Av. Jean-Baptiste Clément
F-93430 Villetaneuse (FRANCE)

1. Introduction

The purpose of this lecture is to present a technique related to the study of the behavior as $h \to 0_+$ of the solutions $u \in L^2(\mathbf{R}^n)$ of partial differential equations of the type:

$$P(x, hD_x; h)u = 0 \tag{1.1}$$

where $D_x = \frac{1}{i}\frac{\partial}{\partial x}$ and the operator $P(x, hD_x; h) = \sum_{k=0}^{N} h^k p_k(x, hD_x)$ is assumed to have analytic coefficients. In particular, the eigenfunctions of semiclassical operators (such as the Schrödinger operator: $-h^2\Delta + V(x)$ with V analytic) can be investigated in this way.

It is a well known fact in microlocal analysis that the behavior of u is strongly related to the geometric properties of the principal symbol $p_0(x, \xi)$ of $P(x, hD_x; h)$, where $(x, \xi) \in \mathbf{R}^{2n}$. Actually, if $P(x, hD_x; h)$ has analytic coefficients, then some phenomena occurring for (x, ξ) complex can also give rise to particular properties of the solution u. As a simple example, if P is elliptic at some point x_0 (in the sense that $p_0(x, \xi)$ never vanishes for x close to x_0 and ξ real with some uniformity as $|\xi| \to \infty$), and u is normalized by $\|u\|_{L^2} = 1$, then there exists a positive constant δ such that $u = \mathcal{O}(e^{-\delta/h})$ uniformly near x_0. Not much is known about this δ, but one expects that it is a reflect of the distance between $\{x_0\} \times \mathbf{R}^n$ and the complex characteristic set of P:

$$\text{Char}(P) = \{(x, \xi) \in \mathbf{C}^{2n} \; ; \; p_0(x, \xi) = 0\}.$$

L. Rodino (ed.), Microlocal Analysis and Spectral Theory, 349–376.

This is the case e.g. when P is the Schrödinger operator (see [2]): outside the classically allowed region $U = \{V(x) \leq 0\}$, the solution is known to decay like $\exp(-d(U,x)/h)$ where d is the so-called "Agmon distance" (or "Lithner-Agmon distance"), that is the pseudodistance associated to the degenerate metric $\text{Max}(V(x),0)dx^2$. Here this distance lives on the "position-space" $\{\xi = 0\}$, but as we shall see, different situations can be considered in which the decay is described by an (x,ξ)-dependent function.

This is precisely for studying such situations that the tool we present here is made, but we shall start by showing how it can also be applied to recover in a particularly easy way (but in a somehow simplified context) several well-known results of microlocal analytic singularities, in the spirit of the book of J.Sjöstrand [16].

Actually, we shall remain very close to the considerations of [16] [17], in particular by working with a so-called "Fourier-Bros-Iagolnitzer" (in short: FBI) transform, which has been intensively studied in [16]. From this point of view, the way in which we recover some results of [16] does not really contain new ideas, but constitute to our opinion a quite simpler presentation. Our main originality lies on the fact that we derive all the main features of analytic microlocal analysis from a single a priori estimate, the proof of which, moreover, turns out to be elementary.

Anyway, this a priori estimate also permits to work with essentially arbitrarily large exponential weights, and therefore gives access to phenomena occurring far in the complex domain. As applications, we use it to investigate several spectral problems involving (microlocal) tunneling.

At first, we apply our technique to adiabatic theory, that is to evolution equations of the type:

$$i\varepsilon\frac{\partial}{\partial t}\varphi = H(t)\varphi$$

where $H(t)$ is a selfadjoint operator depending analytically on the time t, and where we investigate the behavior of the solutions as $\varepsilon \to 0_+$. Assuming that the spectrum of $H(t)$ admits a gap which depends continuously on t, we show that the transition probability from $-\infty$ to $+\infty$ between the two separated parts of the spectrum, can be upper-bounded by $\mathcal{O}(e^{-\Sigma/\varepsilon})$ where $\Sigma > 0$ is a geometrical constant explicitly related to $H(t)$. Therefore this result (the details of which can be found in [8]) permits to specify the previous upper bounds given in [5], [15].

Next, we consider a problem where two semiclassical Schrödinger operators interact to create resonances. The two potentials are assumed to have no crossing on the real, so that at a classical level the interaction can take place only in the complex. Applying our a priori estimate, we get that the width of the resonances can be upper bounded by $\mathcal{O}(e^{-S/h})$ where, here again, $S > 0$ is related to geometrical quantities associated with the symbols of the two operators. Such a problem has been studied in [9], and then the one dimensional case has been specified in [13], [1].

Finally, we show how our a priori estimate can be used to justify WKB constructions, in the case of a semiclassical operator P whose (real-analytic) symbol admits a non degenerate minimum at some point (x_0, ξ_0) of $\mathbf{R}^{2n}$. After a convenient linear change of symplectic variables, we prove that the FBI-transform Tu of the first eigenfunction of P admits a WKB expansion in a neighborhood Ω of (x_0, ξ_0). Moreover, Ω can be described in terms of deformations of Lagrangian manifolds, and one can show that it contains at least a ball centered at (x_0, ξ_0) with radius explicitly given in terms of some constants attached to the symbol of P.

2. Microlocalization

For $h > 0$ small, $(x, \xi) \in \mathbf{R}^{2n}$, and u (possibly h-dependent) in $L^2(\mathbf{R}^n)$, we define:

$$Tu(x, \xi; h) = c(n, h) \int e^{i(x-y)\xi/h-(x-y)^2/2h} u(y)dy \tag{2.1}$$

where $c(n, h) = 2^{-n/2}(\pi h)^{-3n/4}$ is chosen in such a way that:

$$\|Tu\|_{L^2(\mathbf{R}^{2n})} = \|u\|_{L^2(\mathbf{R}^n)}. \tag{2.2}$$

T is called the Fourier-Bros-Iagolnitzer transform, and has been studied by J.Sjöstrand in [16] and [17]. In some sense, $Tu(x, \xi; h)$ describes the behavior of u both in the space variable x and in the momentum (or Fourier) variable ξ. The behavior of Tu as h tends to zero is called the *microlocal* behavior of u. In particular, one can consider the *microsupport* of u which is the closed subset $MS(u)$ of $\mathbf{R}^{2n}$ defined by:

Definition 2.1

$$(x_0, \xi_0) \notin MS(u) \quad \Leftrightarrow \left\{ \begin{array}{l} \textit{There exists } \delta > 0 \textit{ such that} \\ Tu = \mathcal{O}(e^{-\delta/h}) \textit{ uniformly for} \\ (x, \xi) \in \mathbf{R}^{2n} \textit{ close enough to} \\ (x_0, \xi_0) \textit{ and } h > 0 \textit{ small enough.} \end{array} \right.$$

Noticing that Tu satisfies the equation:

$$(hD_x - \xi - ihD_\xi)Tu = 0 \tag{2.3}$$

we see that $e^{\xi^2/2h}Tu(x, \xi; h)$ is an holomorphic function of $z = x - i\xi$. As a consequence, we get for $(x, \xi), (t, \tau) \in \mathbf{R}^{2n}$:

$$Tu(x + it, \xi + i\tau; h) = e^{(t^2+\tau^2-2\xi(t+i\tau))/2h} Tu(x + \tau, \xi - t; h). \tag{2.4}$$

Therefore we see that in the definition of $MS(u)$, one can equivalently take (x,ξ) in a *complex* neighborhood of (x_0,ξ_0), and also, using Cauchy formulas, replace the uniformity with respect to (x,ξ) by a local L^2 (or even L^p, $p\geq 1$) -norm.

In the case when u does not depend itself on h, this microlocal behavior is closely related to the analytic singularities of u: in fact, denoting $WF_a(u)\subset T^*\mathbf{R}^n\backslash 0$ the analytic wave front set of u (see [16]), one can prove:

$$MS(u)=WF_a(u)\ \cup\ \text{Supp}\,u\times\{0\}.$$

Moreover, in many instances one can recover from $MS(u)$ the points x where u is (or is not) exponentially small. Actually, if one has an a priori estimate of the type

$$\|Tu\|_{L^2(K\times\{|\xi|\geq C\})}=\mathcal{O}(e^{-\delta/h}) \tag{2.5}$$

for any compact $K\subset\mathbf{R}^n$ and for some positive constants $C=C_K$, $\delta=\delta_K$, then one can show that the x-projection of $MS(u)$ is precisely the complementary of the points near which u is uniformly exponentially small as h tends to 0 (see [7]). Note that an estimate such as (2.5) is automatically satisfied when u is solution of a partial differential equation which is elliptic in the classical sense (that is with symbol $p(x,\xi)$ polynomial of degree m with respect to ξ such that, locally with respect to x, the quantity $|\xi|^m p(x,\xi)^{-1}$ is uniformly bounded as $|\xi|$ tends to infinity).

One of the interests of working with T is that one can write easily and explicitly how it transforms the pseudodifferential operators acting on $L^2(\mathbf{R}^n)$. More precisely, denote

$$\mathcal{S}_n(1)=\{p\in C^\infty(\mathbf{R}^{2n})\ ;\ \forall\,\alpha\in\mathbf{N}^{2n},\ \partial^\alpha p=\mathcal{O}(1)\text{ uniformly}\} \tag{2.6}$$

and for $p\in\mathcal{S}_n(1)$ and $t\in[0,1]$, consider the t-semiclassical quantization of p defined by:

$$\text{Op}_{h,t}(p)u(x,h)=\frac{1}{(2\pi h)^n}\int e^{i(x-y)\xi/h}p(tx+(1-t)y,\xi)u(y)dyd\xi. \tag{2.7}$$

Then, by the Calderon-Vaillancourt theorem, $\text{Op}_{h,t}(p)$ defines a bounded operator on $L^2(\mathbf{R}^n)$, and we have:

Proposition 2.1 *For all $p\in\mathcal{S}_n(1)$ and $t\in[0,1]$, one has*

$$T\circ\text{Op}_{h,t}(p)=\text{Op}_{h,t}(\tilde{p})\circ T$$

where $\tilde{p}\in\mathcal{S}_{2n}(1)$ is defined by (denoting (x^*,ξ^*) the dual variables of (x,ξ)):

$$\tilde{p}(x,\xi,x^*,\xi^*)=p(x-\xi^*,x^*).$$

Proof- For $u \in C_0^\infty(\mathbf{R}^n)$, we have:

$$\begin{aligned}&\mathrm{Op}_{h,t}(p(x-\xi^*,x^*))Tu(x,\xi) \qquad (2.8)\\ &= \frac{c(n,h)}{(2\pi h)^{2n}}\int_{\mathbf{R}^{5n}} e^{i\Phi/h}p(tx+(1-t)x'-\xi^*,x^*)u(y)dydx'd\xi'dx^*d\xi^*\end{aligned}$$

with

$$\Phi = (x-x')x^* + (\xi-\xi')\xi^* + (x'-y)\xi' + i(x'-y)^2/2.$$

Then, integrating first with respect to ξ' and using the fact that

$$\int e^{i(x'-y-\xi^*)\xi'/h}d\xi' = (2\pi h)^n\delta_{\xi^*=(x'-y)}$$

we get from (2.8):

$$\begin{aligned}&\mathrm{Op}_{h,t}(p(x-\xi^*,x^*))Tu(x,\xi)\\ &= \frac{c(n,h)}{(2\pi h)^n}\int e^{i\Phi_1/h}p(tx-tx'+y,x^*)u(y)dydx'dx^* \qquad (2.9)\end{aligned}$$

with

$$\Phi_1 = (x'-y)\xi + (x-x')x^* + i(x'-y)^2/2.$$

Finally, making the change of variables $x' \mapsto z = x - x' + y$ in (2.9), the result follows easily. ◇

3. Exponential weighted estimates

In this section we state and proof the basic a priori estimate from which we shall derive all the results of this lecture.

Let $a, b > 0$ and $p \in \mathcal{S}_n(1)$ such that p extends holomorphically to the complex strip $S(a,b) = \{(x,\xi) \in \mathbf{C}^{2n}\ ;\ |\mathrm{Im}x| < a\ ,\ |\mathrm{Im}\xi| < b\}$, and satisfies:

$$\forall\alpha \in \mathbf{N}^{2n}\ ,\ \partial^\alpha p = \mathcal{O}(1) \text{ uniformly in } S(a,b). \qquad (3.1)$$

Let also $\psi = \psi(x,\xi) \in \mathcal{S}_n(1)$ be a real-valued function on $\mathbf{R}^{2n}$ satisfying:

$$\operatorname*{Sup}_{\mathbf{R}^{2n}}|\nabla_x\psi| < b \quad \text{and} \quad \operatorname*{Sup}_{\mathbf{R}^{2n}}|\nabla_\xi\psi| < a. \qquad (3.2)$$

We denote

$$\partial_z = \frac{1}{2}(\nabla_x + i\nabla_\xi) \qquad (3.3)$$

which corresponds to the usual holomorphic derivation with respect to $z = x - i\xi$.

Theorem 3.1 *Assume (3.1) and (3.2) and denote* $P = \mathrm{Op}_{h,t}(p)$ *where* $t \in [0,1]$ *is fixed. Let also* $a \in \mathcal{S}_n(1)$. *Then for all* $u, v \in L^2(\mathbf{R}^n)$, *one has:*

$$\langle a e^{\psi/h} T P u, e^{\psi/h} T v\rangle_{L^2} = \langle (q(x,\xi;h) + R(h)) e^{\psi/h} T u, e^{\psi/h} T v\rangle_{L^2}$$

where $q(x,\xi;h)$ *(which depends also on* t*) has an asymptotic expansion of the form:*

$$q(x,\xi;h) \sim \sum_{j\geq 0} h^j q_j(x,\xi)$$

with

$$q_0(x,\xi) = a(x,\xi) p(x - 2\partial_z \psi(x,\xi), \xi + 2i\partial_z \psi(x,\xi))$$

and all the q_j *are smooth bounded functions on* $\mathbf{R}^{2n}$. *Moreover,* $R(h)$ *is a bounded operator on* $L^2(\mathbf{R}^{2n})$ *satisfying:*

$$\|R(h)\|_{\mathcal{L}(L^2)} = \mathcal{O}(h^\infty)$$

as h *tends to zero.*

Sketch of the proof- Using proposition 2.1, we have $a e^{\psi/h} T P = Q e^{\psi/h} T$ with $Q = a e^{\psi/h} \mathrm{Op}_{h,t}(p(x - \xi^*, x^*)) e^{-\psi/h}$. Then by standard arguments, we see that Q is a classical h-pseudodifferential operator with principal symbol $q(x,\xi,x^*,\xi^*) = a(x,\xi) p(x - \xi^* - i\partial_\xi \psi, x^* + i\partial_x \psi)$. Now, denoting

$$\tilde{q}(x,\xi,x^*,\xi^*) = q(x,\xi,x^*,\xi^*) - q(x,\xi,\xi - \partial_\xi \psi, \partial_x \psi)$$

and using standard h-pseudodifferential calculus, we get in particular the existence of two pseudodifferential operators Q_1 and Q_2 satisfying:

$$\mathrm{Op}_{h,t}(\tilde{q}) = \frac{1}{2}(Q_1 A + A Q_1) + \frac{1}{2}(Q_2 B + B Q_2) + R_1$$

with

$$A = hD_x - \xi + \partial_\xi \psi \quad ; \quad B = hD_\xi - \partial_x \psi \quad ; \quad \|R_1\|_{\mathcal{L}(L^2)} = \mathcal{O}(h).$$

Moreover as a consequence of (2.3), we have

$$A \circ e^{\psi/h} T = iB \circ e^{\psi/h} T.$$

Using the fact that both A and B are symmetric on $L^2(\mathbf{R}^{2n})$, we then obtain

$$\langle (Q_1 A + A Q_1) e^{\psi/h} T u, e^{\psi/h} T v\rangle = i\langle [Q_1, B] e^{\psi/h} T u, e^{\psi/h} T v\rangle$$

and

$$\langle (Q_2 B + B Q_2) e^{\psi/h} T u, e^{\psi/h} T v\rangle = i\langle [A, Q_2] e^{\psi/h} T u, e^{\psi/h} T v\rangle.$$

Since $[Q_1, B]$ and $[A, Q_2]$ are both pseudodifferential operator with symbol uniformly $\mathcal{O}(h)$, and $q(x,\xi,\xi-\partial_\xi\psi,\partial_x\psi) = a(x,\xi)p(x-2\partial_z\psi(x,\xi),\xi+2i\partial_z\psi(x,\xi))$, we have proved the result up to a remaining operator of order $\mathcal{O}(h)$ instead of $\mathcal{O}(h^\infty)$. However, an iteration argument plus a resummation procedure permits to get a $\mathcal{O}(h^\infty)$ remainder term, and therefore to finish the proof. We refer to [7] or [14] for more details. ◇

Let us immediately state two corollaries of theorem 3.1, that will be useful in the sequels.

Corollary 3.1 *Under the assumptions of theorem 3.1 one has*

$$\|ae^{\psi/h}TPu\|^2 = \|p(x-2\partial_z\psi,\xi+2i\partial_z\psi)ae^{\psi/h}Tu\|^2 + \mathcal{O}(h)\|e^{\psi/h}Tu\|^2$$

uniformly for $h>0$ small enough and $u \in L^2(\mathbf{R}^n)$.

Corollary 3.2 *If moreover p is real on the real, then there exists a constant $C>0$ such that*

$$\|e^{\varepsilon\psi/h}TPu\|^2 \geq \varepsilon^2\|(H_p\psi)e^{\psi/h}Tu\|^2 - C(h+\varepsilon^3)\|e^{\varepsilon\psi/h}Tu\|^2$$

uniformly for $\varepsilon, h>0$ small enough and $u \in L^2(\mathbf{R}^n)$.
Here, $H_p = \partial_\xi p\partial_x - \partial_x p\partial_\xi$ is the Hamiltonian field associated to p.

Corollary 3.1 is an easy consequence of the proof of theorem 3.1, and corollary 3.2 is obtained from corollary 3.1 by taking the imaginary part of the first order Taylor expansion of $p(x-2\varepsilon\partial_z\psi,\xi+2i\varepsilon\partial_z\psi)$ with respect to ε. In view of the applications to tunneling, let us also mention that corollary 3.1 has the following immediate consequence:

Corollary 3.3 *Assume in addition that $Pu=0$ and, for $\delta>0$, denote*

$$A_\delta = \{(x,\xi)\in\mathbf{R}^{2n}\ ;\ |p(x-2\partial_z\psi,\xi+2i\partial_z\psi)| \geq \delta\}.$$

Then, for all $\delta>0$ one has:

$$\|e^{\psi/h}Tu\|_{L^2(A_\delta)} = \mathcal{O}(\sqrt{h}\|e^{\psi/h}Tu\|_{L^2(\mathbf{R}^{2n}\backslash A_\delta)})$$

uniformly for $h>0$ small enough. In particular, if $A_\delta \supset \operatorname{Supp}\psi$ for some positive δ, then $\|e^{\psi/h}Tu\| = \mathcal{O}(\sqrt{h}\|Tu\|)$.

As we shall see, it is sometimes useful to introduce an extra parameter $\mu > 0$ into the definition of T, by setting:

$$T_\mu u(x,\xi;h) = \mu^{-n/2} Tu(x, \frac{\xi}{\mu}; \frac{h}{\mu}). \tag{3.4}$$

Of course, all the previous estimates on T have analog for T_μ, obtained by a change of variable in ξ and a modification of the parameter h.

We conclude this section with a few remarks.

First, by restricting the space where u and v are taken, we can allow some polynomial growth at infinity for $p(x,\xi)$. This permits to consider directly partial differential operators also. However, one can always reduce to the case where p is bounded by composing P with a suitable regularizing elliptic operator. Also, adding to ψ a function of the type $h\ln(\langle x\rangle^{-s_1}\langle\xi\rangle^{-s_2})$, one can replace the space $L^2(\mathbf{R}^n)$ by any polynomial weighted Sobolev space.

Next, concerning the assumption of analyticity, one can show that it can be locally relaxed in the regions which are sufficiently far (depending on the smallness of the quantities one wants to study) from $\mathrm{Supp}\nabla\psi$ (see e.g. [9]). It is also clear from our proof that if $\psi = \psi(x)$ depends only on x, the analyticity of p is required in the ξ-variables only, and vice versa. In particular, the result applies without any analyticity assumption when $\psi \equiv 0$, and permits e.g. to recover the semiclassical version of the Sharp Garding inequality, or even the unicity part of a theorem of Levy-Mizohata (see [7] for more details).

Finally, let us notice the existence of a Gevrey version of this estimate, which permits to work with weights of the type $e^{\psi(x,\xi)/h^{1/s}}$ (with $s > 1$): see [6]. Actually, the method (which relies on the almost analytic extensions introduced by Melin and Sjöstrand in [12]) can also probably be adapted in the C^∞ case, with weights of the type $h^{-\psi(x,\xi)}$.

4. Microsupport of solutions of P.D.E.

As a first application of theorem 3.1, we are going to recover many results concerning the microlocal behavior of the solutions of partial differential equations of the type

$$P(x,hD_x;h)u = 0. \tag{4.1}$$

As we have already mentioned, one can always compose $P(x,hD_x;h)$ to the left by an elliptic regularizing pseudodifferential operator, so that we can actually reduce us to the case where

$$P(x, hD_x; h) = P = \mathrm{Op}_h(p) \tag{4.2}$$

with $p = \sum_{k=0}^{N} h^k p_k(x,\xi)$, $p_k \in \mathcal{S}_n(1)$ (here we do not specify which quantization we use, since this will have no relevance in the sequels). Assuming that the initial partial differential operator has its coefficients holomorphic in a complex strip, we also get that p satisfies (3.1) for convenient $a, b > 0$.

We first have:

Proposition 4.1 *Assume (4.2) and let $u \in L^2(\mathbf{R}^n)$ be a solution of (4.1) normalized by $\|u\|_{L^2} = 1$. Then:*
(i) $MS(u) \subset \mathrm{Char}(P)$;
(ii)(Hanges theorem) For any real integral curve γ of H_{p_0}, either $\gamma \subset MS(u)$, or $\gamma \cap MS(u) = \emptyset$.

Proof - For part (i), we fix (x_0, ξ_0) outside $\mathrm{Char}(P)$, and we apply corollary 3.3 with ψ non negative, ψ supported near (x_0, ξ_0), $\psi(x_0,\xi_0) > 0$, and ψ flat enough so that $p_0(x - 2\partial_z\psi, \xi + 2i\partial_z\psi) \neq 0$ on Supp ψ. We then get

$$\|e^{\psi/h} Tu\|_{L^2(\mathrm{Supp}\psi)} = \mathcal{O}(1)$$

from which the result follows.

Concerning (ii), let us first prove it when p_0 is real on the real. In this case, we are going to apply corollary 3.2 with a function ψ adapted to γ, which, in some sense, will permit to 'slide' along γ. More precisely, assume there is a point (x_0, ξ_0) of γ which is not in $MS(u)$. Without loss of generality, we can also assume that $p_0 |_\gamma = 0$ and $H_{p_0} \neq 0$ on γ. Then, denote (y_1, y') a system of coordinates centered at (x_0, ξ_0) such that $H_{p_0} = \partial/\partial y_1$ near γ (and therefore γ is given by $y' = 0$), and define $\psi(x,\xi) = f(y_1)\chi(|y'|)$ with $\chi \in C_0^\infty(\mathbf{R}_+)$ supported near 0, $\chi(0) = 1$, $\chi' \leq 0$, and where $f \in C_0^\infty$ satisfies for some $y_1^0 > 0$ arbitrary:

$$\begin{aligned}
&f(0) = \delta_1 > 0 \text{ small enough;}\\
&f' \geq 0 \text{ on } (-\infty, 0] \text{ and } f' \leq 0 \text{ on } [0, +\infty);\\
&|f'(y_1)| \geq \delta_2 > 0 \text{ for } |y_1| \in [\delta_3, y_1^0]\,,\ \ \delta_3 > 0 \text{ small enough;}\\
&f(\pm y_1^0) = \delta_1/2.
\end{aligned}$$

Then, we choose $\varepsilon > 0$ small enough so that $\varepsilon^2\delta_2^2 \geq 4C\varepsilon^3$ where C is the constant appearing in corollary 3.2, and we apply corollary 3.2. Since by construction $|H_{p_0}\psi| \geq \delta_2$ on $\gamma \cap \{|y_1| \in [\delta_3, y_1^0]\}$, and (having chosen δ_1 and δ_3 in a convenient way) $e^{\varepsilon\psi/h}Tu = \mathcal{O}(1)$ on $\gamma \cap \{|y_1| \leq \delta_3\}$, we get for some $\delta > 0$ small enough, denoting $V_\delta = \{|y_1| \leq y_1^0 \, ; \, |y'| \leq \delta\}$:

$$\|e^{\varepsilon\psi/h}Tu\|_{V_\delta} = \mathcal{O}\left(1 + \|e^{\varepsilon\psi/h}Tu\|_{\mathbf{R}^{2n}\setminus V_\delta}\right) \tag{4.3}$$

Looking more carefully, we see that we can also arrange in such a way that $\chi(|y'|) \leq 1/4$ on $|y'| \geq \delta$, and then we can deduce from (4.3) that

$$e^{\varepsilon\delta_1/8h}Tu = \mathcal{O}(1) \ \text{ near } \ \gamma \cap \{y_1 = \pm y_1^0\}.$$

Since y_1^0 has been taken arbitrarily, the result follows in this case.

In the general case where p_0 is not necessarily real on the real, the proof is more subtle. One has to consider the solution $\psi_t = \psi_t(x, \xi; \varepsilon)$ of the system:

$$\begin{cases} \varepsilon\partial_t\psi_t & = & -\chi(x,\xi)\mathrm{Im}p_0(x - 2\varepsilon\partial_z\psi_t, \xi + 2i\varepsilon\partial_z\psi_t) \\ \psi_t\,|_{t=0} & = & \psi_0 \end{cases}$$

where $\chi \in C_0^\infty$ is a cut-off function supported around a fixed segment of γ containing (x_0, ξ_0), and ψ_0 is supported near (x_0, ξ_0) and is chosen in such a way that $\psi_0(x_0, \xi_0) > 0$ and $\|e^{\varepsilon\psi_0/h}Tu\| = \mathcal{O}(1)$. Then one can see that for small enough values of ε and t (but independently of the choice of (x_0, ξ_0)), and for (x, ξ) on γ, ψ_t behaves like $\psi_0(\exp(-tH_{p_0}(x, \xi)))$. In particular $\psi_t(\exp(tH_{p_0}(x_0, \xi_0))) > 0$, and if we denote

$$f(t) = \|e^{\varepsilon\psi_t/h}Tu\|^2$$

we obtain by applying corollary 3.1:

$$\begin{aligned} hf'(t) &= -2\mathrm{Im}\langle \chi e^{\varepsilon\psi_t/h}TPu, e^{\varepsilon\psi_t/h}Tu\rangle + \mathcal{O}(h)\|e^{\varepsilon\psi_t/h}Tu\|^2 \\ &= \mathcal{O}(h)f(t). \end{aligned}$$

As a consequence, we get for some positive constant C

$$f(t) = \mathcal{O}(e^{C|t|})f(0) = \mathcal{O}(1)$$

from which the result follows. ◇

As another application of theorem 3.1 to the study of the microsupport, we have the following celebrated theorem of Kawai and Kashiwara. To the previous assumptions, we add that there exists a real C^∞ function $\phi(x, \xi)$

defined near some point $(x_0, \xi_0) \in \mathrm{Char}(P) \cap \mathbf{R}^{2n}$ such that $\phi(x_0, \xi_0) = 0$ and for all (x, ξ) in a neighborhood of (x_0, ξ_0) and all $\varepsilon > 0$ small enough:

$$p_0(x + i\varepsilon\partial_\xi\phi(x_0, \xi_0), \xi - i\varepsilon\partial_x\phi(x_0, \xi_0)) \neq 0. \tag{4.4}$$

When (4.4) is satisfied, P is said to be microhyperbolic at (x_0, ξ_0) in the direction H_ϕ. Then we have

Theorem 4.1 *(Kawai-Kashiwara) Assume (4.2) and (4.4), and let $u \in L^2(\mathbf{R}^n)$ be a solution of (4.1) normalized by $\|u\|_{L^2} = 1$. Assume also that there exists a neighborhood $\mathcal{V}$ of (x_0, ξ_0) such that*

$$MS(u) \cap \{(x, \xi) \in \mathcal{V}\,;\ \phi(x, \xi) < 0\} = \emptyset.$$

Then, $(x_0, \xi_0) \notin MS(u)$.

Proof - Apart from the fact that we use theorem 3.1 (which makes the proof simpler), the idea is taken from [16]: we fix $\delta > 0$ small enough and $\chi \in C_0^\infty(\mathbf{R}^{2n})$ $(0 \leq \chi \leq 1)$, such that $\chi = 1$ on the ball B_δ centered at (x_0, ξ_0) of radius δ, and χ is supported in a small enough neighborhood of (x_0, ξ_0). Then we set

$$\phi_\delta(x, \xi) = \chi(x, \xi)\left(\phi(x, \xi) - \frac{\delta^3}{2} + \delta(x - x_0)^2 + \delta(\xi - \xi_0)^2\right)$$

so that $\nabla\phi_\delta(x_0, \xi_0) = \nabla\phi(x_0, \xi_0)$, $\phi_\delta(x_0, \xi_0) < 0$, and $\phi_\delta \geq (\phi + \delta^3/2)\chi$ outside B_δ. Applying corollary 3.3 with the weight $e^{-\varepsilon\phi_\delta/h}$ ($\varepsilon > 0$ small enough), and using the microhyperbolicity (as well as a standard "Bochner' tubes" theorem), one finds:

$$\|e^{-\varepsilon\phi_\delta/h}Tu\|_{L^2(B_\delta)} = \mathcal{O}\left(1 + \|e^{-\varepsilon\phi_\delta/h}Tu\|_{L^2(\mathbf{R}^{2n}\backslash B_\delta)}\right). \tag{4.5}$$

Now, since $(\mathbf{R}^{2n}\backslash B_\delta) \cap \{\phi_\delta < 0\} \subset (\mathbf{R}^{2n}\backslash B_\delta) \cap \{\phi < -\delta^3/2\}$, we see that the assumption on $MS(u)$ implies that

$$\|e^{-\varepsilon\phi_\delta/h}Tu\|_{L^2(\mathbf{R}^{2n}\backslash B_\delta)} = \mathcal{O}(1)$$

so that the results follows from (4.5) and the fact that $\phi_\delta(x_0, \xi_0) < 0$. ◇

As the last application of this section, let us show how theorem 3.1 can also be applied to boundary problems, and permits for instance to recover in an easy way the so-called 'microlocal Holmgren theorem' (see [16]).

Let $P = P(x, hD_x)$ be a partial differential operator on $\mathbf{R}^n$, with bounded analytic coefficients (in a complex strip as before), such that the hyperplane $\{x_n = 0\}$ is non characteristic for P, which means that P can be put under the form:

$$P = (hD_{x_n})^m + \sum_{k=0}^{m-1} A_k(x, hD_{x'})(hD_{x_n})^k \tag{4.6}$$

where for any k, A_k is a x_n-dependent partial differential operator in the variables $x' = (x_1, ..., x_{n-1})$ of order at most $m - k$. Here we shall work locally near 0 with respect to x_n, but for simplicity we remain global with respect to x'. We also extend the definition of $MS(u)$ in an obvious way for those u which are defined only for small x_n (e.g. by inserting a cut-off function with respect to y_n in the definition of Tu). Then the theorem is:

Theorem 4.2 *Let P given as in (4.6), and let u be a solution in $L^2(\mathbf{R}^{n-1}\times I_0)$ (where I_0 is a small interval around 0) of the equation $Pu = 0$, normalized by $\|u\|_{L^2(\mathbf{R}^{n-1}\times I_0)} = 1$. Let also $(x_0', \xi_0') \in \mathbf{R}^{2(n-1)}$ such that:*

$$(x_0', \xi_0') \notin MS(u\,|_{x_n=0}) \cup MS(\partial_{x_n} u\,|_{x_n=0}) \cup ... \cup MS(\partial_{x_n}^{m-1} u\,|_{x_n=0}).$$

Then, there exists $\delta > 0$ such that

$$(x_0', x_n, \xi_0', \xi_n) \notin MS(u)$$

for all $x_n \in [-\delta, \delta]$ and all $\xi_n \in \mathbf{R}$.

Proof - By setting $v = (u, hD_{x_n}u, ..., (hD_{x_n})^{m-1}u)$, the equation becomes

$$hD_{x_n} v = A(x, hD_{x'})v \tag{4.7}$$

where A is a $m \times m$ matrix of partial differential operators of order at most m. Let T' denotes the partial FBI-transform defined as in (2.1) but acting in the variables x' only. In particular, $T'v$ is a (vectorial) function of $(x', x_n, \xi'; h)$, and it is enough to prove that $T'v$ is exponentially small near $(x_0', 0, \xi_0')$ uniformly as h tends to 0.

Let $\psi = \psi(x', \xi') \in C_0^\infty(\mathbf{R}^{2(n-1)})$ real such that $\psi(x_0', \xi_0') > 0$ and

$$\|e^{\psi/h} T'v\,|_{x_n=0}\,\|_{L^2(\mathbf{R}^{2(n-1)})} = \mathcal{O}(1). \tag{4.8}$$

For any $\lambda > 0$, denote $\psi_\lambda(x, \xi') = \psi(x', \xi') - \lambda|x_n|$ and set

$$f_\lambda(x_n) = \|\langle\xi'\rangle^{-m/2} e^{\psi_\lambda(x,\xi')/h} T'v\|^2_{L^2(\mathbf{R}^{2(n-1)})}$$

Then, using theorem 3.1, we get on $I_0 \cap \{x_n \geq 0\}$:

$$\begin{aligned} hf'_\lambda(x_n) &= 2\mathrm{Re}\langle \frac{\tilde{A}(x,\xi')-\lambda}{\langle\xi'\rangle^m} e^{\psi_\lambda(x,\xi')/h}T'v\ ,\ e^{\psi_\lambda(x,\xi')/h}T'v\rangle \\ &\quad +\mathcal{O}(h)\|e^{\psi_\lambda(x,\xi')/h}T'v\|^2 \end{aligned} \tag{4.9}$$

with $\tilde{A}(x,\xi') = iA(x'-2\partial_{z'}\psi, x_n, \xi'+2i\partial_{z'}\psi) = \mathcal{O}(\langle\xi'\rangle^m)$ (here $z' = x'-i\xi'$). In particular, taking λ large enough so that (in the sense of $m\times m$ selfadjoint matrices) $\mathrm{Re}\tilde{A}(x,\xi')-\lambda \leq -\delta_0 < 0$ on $\mathrm{Supp}\psi\times I_0$, we get for $h>0$ small enough:

$$hf'_\lambda(x_n) \leq C\|T'v\|^2_{L^2(\mathbf{R}^{2(n-1)})}$$

where C is a positive constant. Integrating from 0 to $x_n \geq 0$ and using the fact that $f_\lambda(0) = \mathcal{O}(1)$ uniformly as h tends to 0, this gives

$$f_\lambda(x_n) = \mathcal{O}(h^{-1}).$$

Using a similar argument for $x_n \leq 0$, we therefore get for any $x_n \in I_0$:

$$\|\langle\xi'\rangle^{-m/2}e^{\psi(x',\xi')/h}T'v\|^2_{L^2(\mathbf{R}^{2(n-1)})} = \mathcal{O}(h^{-1}e^{2\lambda|x_n|/h})$$

and since $\psi(x'_0,\xi'_0) > 0$, the result follows for $|x_n| \leq \delta$ by taking $\delta > 0$ sufficiently small. ◇

5. Adiabatic transition probabilities

As our first application to tunneling, let us look at some evolution equation of the type:

$$i\varepsilon\frac{\partial}{\partial t}\varphi = H(t)\varphi \tag{5.1}$$

where for all $t\in\mathbf{R}$, $H(t)$ is a selfadjoint operator (uniformly semibounded from below) on a Hilbert space $\mathcal{H}$, and where we investigate the behavior of the solutions as $\varepsilon \to 0_+$ (the so-called adiabatic limit).

We see that equation (5.1) involves the operator $P = \varepsilon D_t + H(t)$ which can be interpreted as a semiclassical operator (ε playing the role of h) with (operator-valued) symbol $\tau + H(t)$.

In this context, the characteristic set becomes

$$\begin{aligned} \mathrm{Char}(P) &= \{(t,\tau)\in\mathbf{C}^2\ ;\ \tau + H(t) \text{ is not invertible }\} \\ &= \{(t,\tau)\in\mathbf{C}^2\ ;\ -\tau\in\sigma(H(t))\} \end{aligned}$$

where $\sigma(H(t))$ denotes the spectrum of $H(t)$.

Now, assume $\sigma(H(t))$ admits a gap, that is there exists two bounded continuous functions $e_j(t)$ $(j = 1,2)$ such that

$$\Gamma_0 := \operatorname*{Inf}_{t\in\mathbf{R}} (e_2(t) - e_1(t)) > 0 \tag{5.2}$$

and

$$\sigma(H(t)) \cap [e_1(t), e_2(t)] = \emptyset. \tag{5.3}$$

Denote $\Pi_1(t)$ the spectral projection of $H(t)$ associated to $\sigma(H(t)) \cap (-\infty, e_1(t)]$, and $\Pi_2(t) = 1 - \Pi_1(t)$. Then for $s, t \in \mathbf{R}$ one can define the so-called transition probability between $\mathrm{Ran}\Pi_1(s)$ and $\mathrm{Ran}\Pi_2(t)$ by:

$$\mathcal{P}_{1,2}(s,t) := \|\Pi_2(t)U(t,s)\Pi_1(s)\|^2$$

where $U(t,s)$ is the unitary evolution operator defined by:

$$i\varepsilon\frac{\partial}{\partial t}U(t,s) = H(t)U(t,s) \quad ; \quad U(s,s) = \mathbf{1}_{\mathcal{H}}$$

and the norms are those of the bounded operators on $\mathcal{H}$. Physically, the quantity $\mathcal{P}_{1,2}(s,t)$ represents the probability for a particle with energy in $\sigma(H(s)) \cap (-\infty, e_1(s)]$ at time s, to have its energy in $\sigma(H(t)) \cap [e_2(t), +\infty)$ at time t.

Now, denoting $E(t) = \frac{1}{2}(e_1 + e_2)(t)$, assume in addition that $H(t) - E(t)$ depends analytically on t in a complex strip $S_a = \{|\mathrm{Im}\ t| < a\}$, and that on each side of S_a it tends sufficiently rapidly towards a limit, in the sense that there exist two operators $H_\pm$ such that for some $\rho > 1$:

$$\operatorname*{Sup}_{\substack{t\in S_a \\ \pm\mathrm{Re}\ t\geq 0}} (1 + |t|)^\rho \|H(t) - E(t) - H_\pm\| < +\infty. \tag{5.4}$$

Here the norm of the operators are those of the bounded operators from the domain $\mathcal{H}_1$ of $H(t)$ (which is assumed not to depend on t) to $\mathcal{H}$. Possibly by taking a smaller, we also assume that for $t \in S_a$ the spectrum of $H(t) - E(t)$ remains separated into two disjoint parts, which deform continuously into $\sigma(H(t) - E(t)) \cap (-\infty, \frac{1}{2}(e_1 - e_2)(t)]$ and $\sigma(H(t) - E(t)) \cap [\frac{1}{2}(e_2 - e_1)(t), +\infty)$ as t becomes real (see [8] for a more precise statement of this property).

Under theses assumptions, one can show that that $\mathcal{P}_{1,2}(s,t)$ has a limit $\mathcal{P}_{1,2}(-\infty, +\infty)$ as $s \to -\infty$ and $t \to +\infty$ (see e.g. [15]), and the problem is to know its behavior as ε becomes small. This problem has been studied by many authors, and we send the reader to the bibliography of [8] for references about it.

Now, for $\tau \in (-\frac{\Gamma_0}{2}, \frac{\Gamma_0}{2})$, we denote

$$\kappa(\tau) = \mathrm{Sup}\ \{\kappa \in (0,a)\ ;\ \tau + H(t) - E(t) \text{ is invertible for all } t \in S_\kappa\} \tag{5.5}$$

and we set

$$\Sigma_0 = \int_{-\frac{\Gamma_0}{2}}^{\frac{\Gamma_0}{2}} \kappa(\tau)d\tau. \tag{5.6}$$

Then $\Sigma_0 > 0$ and we have:

Theorem 5.1 *Under the preceding assumptions, for any $\delta > 0$ arbitrarily small there exists a constant $C_\delta > 0$ such that:*

$$\mathcal{P}_{1,2}(-\infty, +\infty) \leq C_\delta e^{-2(\Sigma_0-\delta)/\varepsilon}$$

uniformly as $\varepsilon \to 0$.

Idea of the proof - First of all, one can reduce to $E(t) = 0$ and show that

$$\mathcal{P}_{1,2}(-\infty, +\infty) = \operatorname*{Sup}_{\|\varphi^\pm(0)\|_{\mathcal{H}}=1} |\langle \varphi^-(t), \varphi^+(t)\rangle_{\mathcal{H}}|^2 \tag{5.7}$$

where $\varphi^-(t)$ and $\varphi^+(t)$ are $\mathcal{H}_1$-valued functions solutions of:

$$\begin{cases} (\frac{\varepsilon}{i}\partial_t + H(t))\varphi^\pm = 0\,; \\ \Pi_2(t)\varphi^-(t) \to 0 \text{ as } t \to -\infty\,; \\ \Pi_1(t)\varphi^+(t) \to 0 \text{ as } t \to +\infty. \end{cases} \tag{5.8}$$

Setting $\varphi_j^\pm(t) = \Pi_j(t)\varphi^\pm$ $(j = 1, 2)$ and denoting

$$\tilde{D}_t = \frac{\varepsilon}{i}\frac{\partial}{\partial t}$$

the system (5.8) becomes:

$$\begin{cases} (\tilde{D}_t + H(t) + i\varepsilon[\dot{\Pi}_1, \Pi_1])\varphi_1^\pm = i\varepsilon\dot{\Pi}_2(t)\varphi_2^\pm\,; \\ (\tilde{D}_t + H(t) + i\varepsilon[\dot{\Pi}_1, \Pi_1])\varphi_2^\pm = i\varepsilon\dot{\Pi}_1(t)\varphi_1^\pm\,; \\ \varphi_1^+(t) \to 0 \text{ as } t \to +\infty\ ;\ \varphi_2^-(t) \to 0 \text{ as } t \to -\infty. \end{cases} \tag{5.9}$$

(where $\dot{\Pi}_j$ denotes the derivative of Π_j with respect to t). When ε tends to 0, one can consider (5.9) as two semiclassical systems with (operator-valued) principal symbol given by:

$$p(t, \tau) = (\tau + H(t))\mathbf{I}_2.$$

Instead of working with the FBI transform of the previous section, here we prefer to work with T_μ given in (3.4), which in this case becomes:

$$T_\mu\varphi(t, \tau; \varepsilon) = 2^{-1/2}(\pi\varepsilon)^{-3/4}\mu^{1/4} \int e^{i(t-t')\tau/\varepsilon-\mu(t-t')^2/2\varepsilon}\varphi(t')dt'. \tag{5.10}$$

The extra parameter $\mu > 0$ will be fixed small enough later on. Setting

$$\partial_\mu = \frac{1}{\mu}\frac{\partial}{\partial t} + i\frac{\partial}{\partial \tau}$$

we have the following analog of corollary 3.1:

Proposition 5.1 *Let $g = g(t, \tau) \in C_b^\infty(\mathbf{R}^2; \mathbf{R})$ such that Sup $|\partial_\tau g| < a$. Then for all $\varphi \in L^2(\mathbf{R}; \mathcal{H}_1)$ one has:*

$$\begin{aligned} &\|e^{g/\varepsilon} T_\mu (\tilde{D}_t + H(t))\varphi\|_{L^2(\mathbf{R}^2;\mathcal{H})} \\ &= \|e^{g/\varepsilon}(\tau + i\mu\partial_\mu g + H(t - \partial_\mu g))T_\mu \varphi\|_{L^2(\mathbf{R}^2;\mathcal{H})} \\ &\qquad\qquad + \mathcal{O}(\sqrt{\varepsilon})\|e^{g/\varepsilon} T_\mu \varphi\|_{L^2(\mathbf{R}^2;\mathcal{H}_1)} \end{aligned}$$

uniformly with respect to $\varepsilon > 0$ small enough and $\varphi \in L^2(\mathbf{R}; \mathcal{H}_1)$.

Then, in the same spirit as for corollary 3.3, this proposition permits to show that the $T_\mu \varphi_j^\pm$'s are exponentially small in the elliptic region $\{|\tau| < \Gamma_0/2\}$ with a rate of decay given by

$$g_1(\tau) = \mathbf{1}_{(-\frac{\Gamma_0}{2}, \frac{\Gamma_0}{2})}(\tau) \mathrm{Min}\Big\{ \int_{-\Gamma_0/2}^{\tau} \kappa(s)ds \; ; \; \int_{\tau}^{\Gamma_0/2} \kappa(s)ds \Big\}$$

where the precise meaning is the following: for all $\delta > 0$, $k, l \in \mathbf{N}$, and $\mu > 0$ sufficiently small, one has

$$\|\langle\tau\rangle^{-k-l-1}\langle t\rangle^{-\rho/2} e^{g_1/\varepsilon} \tilde{D}_t^k \tilde{D}_\tau^l T_\mu \varphi_j^\pm\|_{L^2(\mathbf{R}^2;\mathcal{H}_1)} = \mathcal{O}(e^{\delta/\varepsilon}). \tag{5.11}$$

Denoting τ_1 the point where g_1 reaches its maximum (the value of which is $\Sigma_0/2$), the estimate (5.9) permits to study $T_\mu \varphi_j^\pm$ separately in the two regions $\{\tau \le \tau_1\}$ and $\{\tau \ge \tau_1\}$, in the following way : Let χ be a C^∞ function on $\mathbf{R}$ such that $\chi = 1$ on $\{\tau \le \tau_1\}$, and $\chi = 0$ on $\{\tau \ge \tau_1 + \nu\}$ where $\nu > 0$ is arbitrarily small. Then one can deduce from (5.9) the following result:

Lemma 5.1 *For all $\delta > 0$ and μ, ν both positive and small enough, one has*

$$\|\langle t\rangle^{\rho/2}[H(t - \tilde{D}_\tau), \chi(\tau)]T_\mu \varphi_j^\pm\| = \mathcal{O}\left(e^{-(\Sigma_0 - \delta)/2\varepsilon}\right)$$

uniformly with respect to $\varepsilon > 0$ small enough.

This lemma permits to insert the function χ into the system (5.9) (after having transformed this one with T_μ), and to control the error terms that it makes appear. Forgetting these error terms, we are reduced to a new system that acts on functions supported in $\{\tau \le \tau_1 + \nu\}$. Since in this region the operator $\tilde{D}_t + H(t)\Pi_1(t)$ is elliptic, the first equation of this system permits to get the estimate:

$$\|\langle t\rangle^{-\rho/2}\chi(\tau)T_\mu\varphi_1^-\| = \mathcal{O}\left(\varepsilon\|\langle t\rangle^{-\rho/2}\chi(\tau)T_\mu\varphi_2^-\| + e^{-(\Sigma_0-\delta)/2\varepsilon}\right). \qquad (5.12)$$

Inserting (5.12) into the second equation of the system, one finds:

$$\begin{aligned}&\|\langle t\rangle^{\rho/2}(\tilde{D}_t + H_A(t - \tilde{D}_\tau))\chi(\tau)T_\mu\varphi_2^-\| \\ &\quad = \mathcal{O}\left(\varepsilon^2\|\langle t\rangle^{-\rho/2}\chi(\tau)T_\mu\varphi_2^-\| + e^{-(\Sigma_0-\delta)/2\varepsilon}\right)\end{aligned} \qquad (5.13)$$

where $H_A = H + i\varepsilon[\dot{\Pi}_1, \Pi_1]$. Then, evaluating $\partial_t\|\chi(\tau)T_\mu\varphi_2^-\|^2_{L^2(\mathbf{R}_\tau;\mathcal{H})}$ and using the fact that φ_2 vanishes at $-\infty$, one can deduce from this an estimate on $\chi(\tau)T_\mu\varphi_2^-$ which finally gives by (5.12):

$$\|\langle t\rangle^{-\rho/2}\chi(\tau)T_\mu\varphi^-\| = \mathcal{O}(e^{-(\Sigma_0-\delta)/2\varepsilon}). \qquad (5.14)$$

A similar estimate holds for $\tilde{\chi}T_\mu\varphi^+$, where this time $\tilde{\chi}$ cuts off in $\{\tau \geq \tau_1\}$.

Now, coming back to the elliptic region $\{|\tau| < \Gamma_0/2\}$, we see that (5.14) permits to apply again proposition 5.1 to φ^- with a new weight g_2 which is constant on $\{\tau \leq \Gamma_0/2\}$ with value $\Sigma_0/2$. As a consequence, one obtains an improvement of (5.9) in the region $\{\tau \leq \tau_1\}$, that one can again propagate along the negative values of τ as before. Iterating this procedure and working in a similar way for φ^+, one finds that for all $\delta > 0$, there exists $\mu = \mu(\delta) > 0$ arbitrarily small such that:

$$\begin{aligned}\|\langle t\rangle^{-\rho/2}e^{g_-/\varepsilon}T_\mu\varphi^-\| &= \mathcal{O}(e^{\delta/\varepsilon}) \\ \|\langle t\rangle^{-\rho/2}e^{g_+/\varepsilon}T_\mu\varphi^+\| &= \mathcal{O}(e^{\delta/\varepsilon})\end{aligned} \qquad (5.15)$$

where (extending the function κ by zero outside $(-\Gamma_0/2, \Gamma_0/2)$):

$$\begin{aligned}g_-(\tau) &= \mathrm{Max}\Big(0, \int_\tau^{\Gamma_0/2}\kappa(\tau)d\tau\Big) \\ g_+(\tau) &= \mathrm{Max}\Big(0, \int_{-\Gamma_0/2}^{\tau}\kappa(\tau)d\tau\Big).\end{aligned}$$

Finally, noticing that $\langle\varphi^-(t), \varphi^+(t)\rangle_\mathcal{H} = \langle T_\mu\varphi^-(t,.), T_\mu\varphi^+(t,.)\rangle_{L^2(\mathbf{R}_\tau,\mathcal{H})}$, theorem 5.1 is a consequence of (5.15) and of the fact that $g_- + g_+ = \Sigma_0$. ◇

6. Widths of resonances

Now, we investigate another problem of tunneling, related to the resonances of molecules. Let $P(x, hD_x)$ be the matricial system:

$$P = \begin{pmatrix} -h^2\Delta + V_1(x) & hR(x, hD_x) \\ hR^*(x, hD_x) & -h^2\Delta + V_2(x)\end{pmatrix} \qquad (6.1)$$

acting on $L^2(\mathbf{R}^n) \oplus L^2(\mathbf{R}^n)$, where $R(x, hD_x)$ is a (pseudo)differential operator of order less than 2, and V_1, V_2 are real-analytic functions on $\mathbf{R}^n$.

Such kind of systems occurs for the study of molecules when one considers the so-called Born-Oppenheimer approximation, in which the nuclei are supposed to be much heavier than the electrons. In this situation, the parameter h represents the inverse of the square root of the mean mass of the nuclei.

Here, we assume that the symbol of P extends holomorphically in a complex domain of the type

$$D = \{(x,\xi) \in \mathbf{C}^{2n} \ ; \ |\text{Im } x| \leq a + \delta_0 |\text{Re } x| \ , \ |\text{Im } \xi| \leq b + \delta_0 |\text{Re } \xi|\}$$

(where $a, b, \delta_0 > 0$) and that V_2 admits a compact well at some fixed energy $E_0 \in \mathbf{R}$, that is:

$$U_0 := \{x \in \mathbf{R}^n \ ; \ V_2(x) \leq E_0\} \ \text{ is compact.} \tag{6.2}$$

We also assume

$$U_0 \subset \{x \in \mathbf{R}^n \ ; \ V_1(x) < E_0\} \tag{6.3}$$

(so that the two characteristic sets $\{\xi^2 + V_1(x) = E_0\}$ and $\{\xi^2 + V_2(x) = E_0\}$ do not intersect on the real) and that there is no trapped trajectories of energy close to E_0 for $p_1(x,\xi) = \xi^2 + V_1(x)$, that is:

$$\forall \ E \text{ close to } E_0, \ \forall \ (x,\xi) \in p_1^{-1}(E), \ |\exp tH_{p_1}(x,\xi)| \to \infty \text{ as } |t| \to \infty. \tag{6.4}$$

In this situation, there exists near E_0 a discrete subset of $\mathbf{C}$ (whose elements are called *resonances* of P), which can be characterized by the following property: ρ is a resonance of P if there exists $u \in [C^\infty(\mathbf{R}^n)]^2$ such that:

$$\begin{aligned} &Pu = \rho u \\ &Tu \in [L^2((1+it)\mathbf{R}^n \times (1-it)\mathbf{R}^n)]^2 \text{ for some } t \in (0, \delta_0). \end{aligned} \tag{6.5}$$

Then, the quantity $|\text{Im}\rho|$ is called the width of the resonance ρ, and its inverse can be physically interpreted as the life-time of an unstable state associated to u. As a consequence, any upperbound on $|\text{Im}\rho|$ permits to predict how long such a "molecule" will exist at least. This is precisely the purpose of this section.

Denote also $p_2(x,\xi) = \xi^2 + V_2(x)$, $\Sigma_j = p_j^{-1}(E_0)$ $(j = 1, 2)$, and for any $\delta > 0$, $\Sigma_j(\delta) = p_j^{-1}([E_0 - \delta, E_0 + \delta])$. For $\mu > 0$, we consider the set $\mathcal{E}_\mu$ of functions $\psi \in C^\infty(\mathbf{R}^{2n})$ such that $\nabla\psi \in C_0^\infty(\{\xi^2 + V_1(x) < E_0\}$, $\psi|_{U_0} = 0$, $\text{Sup}|\partial_x \psi| < b$, $\text{Sup}|\partial_\xi \psi| < a$, and:

$$\forall \ \delta > 0, \ \exists C_\delta > 0 \text{ such that } |p_2(x - \partial_\mu \psi, \xi + i\mu \partial_\mu \psi)| \geq \frac{1}{C_\delta} \text{ on } \mathbf{R}^n \backslash \Sigma_2(\delta);$$

$$\exists C > 0 \text{ such that } |p_1(x - \partial_\mu \psi, \xi + i\mu\partial_\mu \psi)| \geq \frac{1}{C} \text{ on } \operatorname{Supp} \nabla\psi.$$

Then the number:

$$S_0 = \operatorname*{Sup}_{\mu>0} \operatorname*{Sup}_{\psi\in\mathcal{E}_\mu} \operatorname*{Inf}_{(x,\xi)\in\Sigma_1} \psi(x,\xi) \tag{6.6}$$

is positive, and the result is:

Theorem 6.1 *Assume (6.2)-(6.4), and let $\rho = \rho(h)$ be a resonance of P tending to E_0 as $h \to 0$. Then for any $\delta > 0$ there exists a constant $C(\delta)$ such that:*

$$|\text{Im } \rho| \leq C(\delta)e^{-2(S_0-\delta)/h}$$

uniformly for $h > 0$ small enough.

In this result, the constant S_0 can seem rather abstract since it rests on a good choice (not necessarily easy to make) of a function ψ in $\mathcal{E}_\mu$. However, there are situations where it can be specified, for instance if U_0 is a non degenerate point-well (that is $U_0 = \{x_0\}$ with Hess $V_2(0) > 0$): see [9]. In the particular case $V_2(x) = x^2$ and $V_1(x) = -x_n - 1$, the best choice for ψ is of the form (see [10]):

$$\psi(x,\xi) = \chi(x,\xi)\frac{\mu x^2 + \xi^2}{2(\mu+1)} + (1 - \chi(x,\xi))S_0$$

with $\mu = \frac{1+\sqrt{5}}{4}$, $S_0 = \frac{3-\sqrt{5}}{4}$ and where χ is a cut-off supported in $\{p_1(x,\xi) < E_0\}$.

Notice that when $n = 1$, this estimate can be improved by making a convenient symplectic change of variables (see [13]), and, using exact WKB expansions, it is even possible to understand the precise nature of the tunneling (see [1]).

Here we do not sketch the proof of theorem 6.1, but it is based on the same idea as e.g. corollary 3.3: if u is a resonant state (suitably normalized) associated to ρ, the definition of $\mathcal{E}_\mu$ permits to estimate $||e^{\psi/h}\mathbf{T}u||_t$ for every $\psi \in \mathcal{E}_\mu$, where $\mathbf{T}$ is a *FBI* transform introduced by Helffer and Sjöstrand in [3] which (for technical reasons attached to the fact that we deal with resonances) is more complicated to write down than our previous T_μ, but whose properties are essentially similar (see [9] for more details), and where $|| \, . \, ||_t$ is a norm similar to the L^2-norm outside a neighborhood of Σ_1, but slightly modified near Σ_1. Then, writing

$$\text{Im } \rho = \frac{\text{Im } \langle \mathbf{T}Pu, \mathbf{T}u\rangle_{L^2(\Omega)}}{||\mathbf{T}u||^2_{L^2(\Omega)}}$$

for any $\Omega \subset\subset \{p_1(x,\xi) < E_0\}$, and using the fact that P is formally selfadjoint on L^2, the result can be deduced by choosing Ω in a proper way.

7. Microlocal WKB expansions

As our last application of theorem 3.1, we present a joint work in preparation with V. Sordoni, where we investigate the existence of microlocal WKB expansions for the eigenfunctions of pseudodifferential operators whose symbol admits a non degenerate minimum at some point (x_0, ξ_0) of $\mathbf{R}^{2n}$. This is e.g. the case for the electromagnetic Schrödinger operator

$$P_A(x, hD_x) = \sum_{j=1}^{n} \left(hD_{x_j} - A_j(x)\right)^2 + V(x)$$

when V admits a non degenerate minimum at some point x_0. In the case where the A_j's can be taken small enough (that is when the magnetic field is small enough) and everything is analytic, it has been shown by Helffer and Sjöstrand [4] that the first eigenfunction u of P_A admits near x_0 a WKB expansion of the form

$$u(x,h) \sim e^{-\phi_A(x)/h} \sum_{k\geq 0} h^k a_k(x)$$

where ϕ_A and the a_k's are smooth functions. Moreover, the set of x's where such an expansion is valid can be estimated geometrically by means of the minimal geodesics starting from x_0, relatively to the so-called Agmon distance (i.e. the distance associated to the degenerate metric $(V(x) - V(x_0))dx^2$). But the problem remains entirely open for greater magnetic fields.

Here, we are going to show that in any case (but still under assumptions of analyticity), a similar WKB expansion exists near (x_0, ξ_0) for the FBI-transform Tu of u, at least if one choose convenient symplectic coordinates in $\mathbf{R}^{2n}$. Moreover, the set of (x,ξ)'s where the expansion is valid can be estimated by means of simple constants attached to the symbol of P, and a more general notion of "admissible open set" will be given, in terms of deformation properties.

Now, let us specify our assumptions.

For the sake of simplicity we take p in $S_n(1)$ (although everything could be generalized to symbols with polynomial growth at infinity), and we assume that p satisfies (3.1) for some positive a, b. We also assume that $p(x,\xi)$ is real non negative for real (x,ξ), $p^{-1}(0) = (0,0)$, Hess $p(0,0)$ is positive definite, and there exists $\delta_1 > 0$ such that $p(x,\xi) \geq \delta_1$ outside some neighborhood of 0. It is standard to show that there exists a linear symplectic change of variables such that, in the new coordinates, p satisfies:

$$p(x,\xi) = \sum_{j=1}^{n} \mu_j (x_j^2 + \xi_j^2) + \mathcal{O}(|x,\xi|^3) \tag{7.1}$$

where $0 < \mu_1 \leq ... \leq \mu_n$. Moreover, since p can be written under the form $p = \hat{p} + \kappa$ with Inf $\hat{p} > 0$ and $\kappa \in C_0^\infty(\mathbf{R}^{2n})$, the Weyl' theorem of perturbation implies that the spectrum of $P :=\mathrm{Op}_{h,\frac{1}{2}}(p)$ is discrete near 0. Also, because of (7.1) one can use the same arguments as Helffer and Sjöstrand in [2], which show that the first eigenvalue E of P is simple and has an asymptotic expansion of the form

$$E \sim h \sum_{k \geq 0} e_k h^k$$

as h tends to 0. Denote u the first eigenfunction of P, normalized by $\|u\|_{L^2} = 1$. We are going to show that, in suitable neighborhoods of $(0,0)$, Tu admits a WKB asymptotics of the form:

$$Tu(x,\xi;h) \sim h^{-m_0} e^{-\xi^2/2h - \varphi(x-i\xi)/h} \sum_{j \geq 0} h^j a_j(x - i\xi) \tag{7.2}$$

where φ and the a_j's are holomorphic near $0 \in \mathbf{C}^n$. Notice that in the particular case where $p(x,\xi)$ in (7.1) is even with respect to ξ, then the method of [2] can be directly generalized and gives a WKB expansion for $u(x;h)$. However, if it is not the case (as for the general electromagnetic Schrödinger operator), then the usual Agmon estimates fail to give the expected expansion.

Since $TPu = \tilde{P}Tu$ with $\tilde{P} = \mathrm{Op}_{h,\frac{1}{2}}(p(x - \xi^*, x^*))$, if we look for a solution Tu of the form (7.2) we are lead to solve the eiconal equation:

$$p(z - \partial_z\varphi, i\partial_z\varphi) = 0 \tag{7.3}$$

where $z = x - i\xi$. To solve it, we proceed in an analogous way as in [2]: Define

$$q(z,\zeta) = -p(z - \zeta, i\zeta)$$

so that (7.3) becomes $q(z,\partial_z\varphi) = 0$. Near $(0,0) \in \mathbf{C}^{2n}$ we have:

$$q(z,\zeta) = \sum_{j=1}^{n} \mu_j (2z_j\zeta_j - z_j^2) + \mathcal{O}(|z,\zeta|^3)$$

and therefore the fundamental matrix of q at $(0,0)$ is

$$F_q = \begin{pmatrix} 2\mu & 0 \\ 2\mu & -2\mu \end{pmatrix} \tag{7.4}$$

where $\mu = \operatorname{diag}(\mu_1, ..., \mu_n)$. The spectrum of F_q is $\{\pm 2\mu_j \ ; \ j = 1, ..., n\}$, and the direct sum of the eigenspaces associated to $\{+2\mu_j \ ; \ j = 1, ..., n\}$ (resp. to $\{-2\mu_j \ ; \ j = 1, ..., n\}$) is the Lagrangian space $\mathcal{E}_+ = \{\zeta = z/2\}$ (resp. $\mathcal{E}_- = \{z = 0\}$). Then by adapting the analytic version of the 'stable-unstable manifold theorem' which is in the appendix of [18], one can show that there exist two holomorphic complex Lagrangian manifolds $\Lambda_\pm$ containing $(0,0)$, stable under the action of H_q, and such that $T_{(0,0)}\Lambda_\pm = \mathcal{E}_\pm$. In particular, Λ_+ projects bijectively on the base $\{\zeta = 0\}$, and therefore there exists a holomorphic function φ such that in a complex neighborhood of 0, Λ_+ is given by:

$$\Lambda_+ = \{\zeta = \varphi'(z)\}.$$

Since $q(0,0) = 0$ and q is constant on Λ_+, we see in particular that φ solves (7.3). Notice that if we normalize φ by setting $\varphi(0) = 0$, we also have:

$$\varphi(z) = \frac{1}{4}z^2 + \mathcal{O}(|z|^3)$$

and therefore

$$\frac{\xi^2}{2} + \operatorname{Re}\varphi(x - i\xi) = \frac{x^2 + \xi^2}{4} + \mathcal{O}(|x, \xi|^3).$$

Now, first working with z real, one can construct as in [2] an analytic symbol $a(z,h) \sim \sum_{k \geq 0} h^k a_k(z)$ defined near 0, such that formally

$$\operatorname{Op}_{h,\frac{1}{2}}(p(z + i\zeta, \zeta) - E)(a(z,h)e^{-\varphi(z)/h}) \sim 0$$

and after resummation, this means that there exists $\varepsilon > 0$ such that, for (x, ξ) small enough:

$$(\tilde{P} - E)\left(a(x - i\xi, h)e^{-\xi^2/2h - \varphi(x - i\xi)/h}\right) = \mathcal{O}\left(e^{-\xi^2/2h - \operatorname{Re}\varphi(x - i\xi)/h - \varepsilon/h}\right). \tag{7.5}$$

where the action of the pseudodifferential operator $\tilde{P}$ on the function $a(x - i\xi, h)e^{-\xi^2/2h - \varphi(x - i\xi)/h}$ (which is defined only near $(0,0)$) is defined via a formal stationary phase expansion. Moreover, the estimate (7.5) is valid locally uniformly in the maximal connected open set Ω_0 where both φ and the a_k's extend holomorphically and $|\operatorname{Im}(z - \partial_z\varphi)| < a$, $|\operatorname{Re}\, \partial_z\varphi| < b$.

By an abstract spectral argument (still as in [2]), one can deduce from (7.5) that for convenient constants m_0 and α_0, one has near 0:

$$Tu - \alpha_0 h^{-m_0} a(x-i\xi,h) e^{-\xi^2/2h - \varphi(x-i\xi)/h} = \mathcal{O}(e^{-\varepsilon'/h})$$

with some $\varepsilon' > 0$. In particular, denoting

$$v(x,\xi,h) = \alpha_0 h^{-m_0} a(x-i\xi,h) e^{-\xi^2/2h-\varphi(x-i\xi)/h}$$

we have

$$e^{\xi^2/2h+\mathrm{Re}\varphi(x-i\xi)/h}(Tu - v) = \mathcal{O}(e^{-\varepsilon'/2h}) \tag{7.6}$$

for $|x,\xi|$ sufficiently small compared with ε'. Since not much is known about this ε', the problem is now to extend (7.6) in a neighborhood of $(0,0)$ that one can control in a more geometrical way.

Coming back to the variables (z,ζ), we set:

$$\Lambda_0 = \{\zeta = i\mathrm{Im}z\}$$

(so that $(z,\zeta) \in \Lambda_0$ if and only if $(z-\zeta, i\zeta)$ is real), and for $t \geq 0$:

$$\Lambda_t = \exp tH_q\ (\Lambda_0).$$

Since Λ_0 is $\mathbf{R}$-Lagrangian (that is Lagrangian for the real symplectic form $\mathrm{Re}(d\zeta \wedge dz)$ on $\mathbf{C}^{2n} \approx \mathbf{R}^{4n}$), and the map $\exp tH_q$ is a complex canonical transformation, we have that Λ_t is $\mathbf{R}$-Lagrangian for all t. Moreover, approximating H_q by its linearization at $(0,0)$, one can see that Λ_t is transversal to $\{z = 0\}$ at $(0,0)$. As a consequence, Λ_t admits near $(0,0)$ an equation of the form:

$$\Lambda_t\ :\qquad \zeta = \frac{\partial \phi_t(z,\overline{z})}{\partial z}$$

where ϕ_t is a real C^∞ function defined in a neighborhood of 0 and vanishing at 0. Looking carefully at the proof of [18]Appendix, one can also see that there exists a fix neighborhood Ω of 0 in $\mathbf{C}^n$ such that for all $t \geq 0$, ϕ_t is smooth in Ω, and

$$\phi_t(z,\overline{z}) \to 2\mathrm{Re}\varphi(z) \ \text{ as } \ t \to +\infty \tag{7.7}$$

in $C^\infty(\Omega)$.

Now, for $(z,\zeta) = (z, i\mathrm{Im}z) \in \Lambda_0$, we have $-q(z,\zeta) = p(\mathrm{Re}z, -\mathrm{Im}z)$ and is therefore real non negative, with a non degenerate minimum at $z = 0$. Since q is constant along $\{\exp tH_q(z)\ ;\ t \geq 0\}$, we deduce from this that for any $t \geq 0$, $-q\,|_{\Lambda_t}$ is real non negative and there exists a constant $C_t > 0$ such that:

$$\forall z \in \Omega\,,\; -q\left(z, \frac{\partial \phi_t(z,\overline{z})}{\partial z}\right) \geq \frac{1}{C_t}|z|^2. \tag{7.8}$$

One can also prove that on Ω, ϕ_t satisfies the evolution equation:

$$\frac{\partial \phi_t}{\partial t} = -2q(z, \partial_z \phi_t)$$

and therefore, $\phi_t(z)$ is an increasing function of t for $z \neq 0$. As a consequence, we have for all $t > 0$ and $z \neq 0$:

$$\phi_0(z,\overline{z}) < \phi_t(z,\overline{z}) < 2\mathrm{Re}\varphi(z) \tag{7.9}$$

where $\phi_0 = -(\mathrm{Im} z)^2$.

Then we introduce the following notion of "admissible open set":

Definition 7.1 *Let Ω_1 be an open subset of Ω containing 0. Then Ω_1 is said to be "admissible" if for any compact subset K of Ω_1 there exist $\varepsilon_K > 0$ and a neighborhood $\mathcal{V}_K$ of $\partial\Omega$ with the following property:*
For every $t > 0$ large enough, there exists $\psi_t \in C^\infty(\mathbf{C}^n)$ real such that

$\psi_t = \phi_t$ on K ;

$\psi_t - \phi_0$ is constant outside Ω ;

$\psi_t \leq 2\mathrm{Re}\varphi$ everywhere;

$\psi_t \leq 2\mathrm{Re}\varphi - \varepsilon_K$ on $\mathcal{V}_K$;

$\mathrm{Sup}\,|\mathrm{Im}(z - \partial_z \psi_t)| < a$ and $\mathrm{Sup}\,|\mathrm{Re}\,\partial_z \psi_t| < b$;

$\exists\, C_t' > 0$ such that $|q(z, \partial_z \psi_t)| \geq \frac{1}{C_t'}|z|^2$ on Ω.

In terms of deformation of **R**-Lagrangian manifolds, this means that one can deform (in a somehow non increasing sense) Λ_t into Λ_0 within an arbitrarily small neighborhood of $\Omega\backslash\Omega_1$, in such a way that q remains elliptic along the deformation. Moreover, the deformed weight is required to be smaller than $2\mathrm{Re}\varphi$ near $\partial\Omega$, since the WKB constructions may cease to exist there.

Before proving that the estimate (7.6) remains valid locally uniformly in any admissible open set, let us state a result of existence of such an admissible open set, which actually permits to exhibit such a set in terms of some constants attached to q and easy to compute.

First of all, working with the quadratic approximation of q it is easy to see that for all $t \geq 0$, one has:

$$\phi_t(z,\overline{z}) = \frac{1}{2}\sum_{j=1}^{n}\left((1 - e^{-4t\mu_j})(\mathrm{Re} z_j)^2 - (1 + e^{-4t\mu_j})(\mathrm{Im} z_j)^2\right) + \mathcal{O}(|z|^3)$$

where the $\mathcal{O}(|z|^3)$ is uniform with respect to t.

Then, noticing that $2\text{Re}\varphi(z) - \phi_0(z, \bar{z}) = |z|^2/2 + \mathcal{O}(|z|^3)$, we set

$$|z|_\varphi = (2\text{Re}\varphi(z) - \phi_0(z, \bar{z}))^{\frac{1}{2}} \tag{7.10}$$

and we define the five constants γ_0', $\gamma_j > 0$ $(j = 0, 1, 2, 3)$ in the following way:

$$\gamma_0' = \operatorname*{Sup}_{z\in\Omega} \frac{|z|}{|z|_\varphi}$$

$$\gamma_0 = \operatorname*{Sup}_{z\in\Omega} \frac{|z|_\varphi}{|z|}$$

$$\gamma_1 = \operatorname*{Sup}_{\substack{t\geq 0\\ z\in\Omega}} \frac{|\partial_z \phi_t(z, \bar{z})|}{|z|_\varphi}$$

$$\gamma_2 = \operatorname*{Sup}_{\substack{t\geq 0\\ z\in\Omega}} \frac{|\partial_z \phi_t(z, \bar{z}) - (z - e^{-4t\mu}\bar{z})/2|}{|z|_\varphi^2}$$

$$\gamma_3 = \operatorname*{Sup}_{\substack{z\in\Omega\\ |\zeta|\leq\gamma_1|z|_\varphi}} \frac{|\partial_\zeta q(z, \zeta) - 2\mu z|}{|z|_\varphi^2}.$$

Then we have:

Proposition 7.1 *If $r > 0$ satisfies:*

$$\gamma_0(2\gamma_2\mu_n + \frac{1}{2}\gamma_3)\sqrt{r} + \gamma_0'^2\gamma_2\gamma_3 r \leq \mu_1 \tag{7.11}$$

and

$$\gamma_1\sqrt{r} \leq b \quad ; \quad (\gamma_2 + \frac{\gamma_0'}{2})\sqrt{r} \leq a \tag{7.12}$$

then the set

$$\mathcal{B}_r = \{z \in \mathbf{C}^n \; ; \; (\text{Im}z)^2 + 2\text{Re}\varphi(z) < r\} = \{|z|_\varphi^2 < r\}$$

is an admissible open set in the sense of definition 7.1.

We refer to [11] for the proof of this proposition, and now we state the main result of this section:

Theorem 7.1 *Assume (7.1) and let Ω_1 be any admissible open set in the sense of definition 7.1. Then for any compact set $K \subset \Omega_1$, there exists $\varepsilon > 0$ such that*

$$e^{\xi^2/2h + \text{Re}\varphi(x - i\xi)/h}(Tu(x, \xi; h) - v(x, \xi; h)) = \mathcal{O}(e^{-\varepsilon/h})$$

uniformly for $x - i\xi \in K$ *and* $h > 0$ *small enough. Here* u *is the first normalized eigenfunction of* $P = \mathrm{Op}_{h,\frac{1}{2}}(p)$, *and*

$$v(x,\xi;h) = \alpha_0 h^{-m_0} a(x - i\xi, h) e^{-\xi^2/2h - \varphi(x-i\xi)/h}$$

is the WKB solution constructed at the beginning of this section.

Proof - Using (7.6), let $\varepsilon_0 > 0$ and V_0 be a neighborhood of $0 \in \mathbf{R}^{2n}$ such that

$$\|e^{\xi^2/2h + \mathrm{Re}\varphi(x-i\xi)/h}(Tu - v)\|^2_{L^2(V_0)} = \mathcal{O}(e^{-\varepsilon_0/h}). \tag{7.13}$$

Then, fix $K \subset\subset \Omega_1$, and let $\varepsilon_K > 0$ and ψ_t ($t > 0$ large enough) be given by definition 7.1. By (7.7), we can fix t_0 sufficiently large so that:

$$|\psi_{t_0} - 2\mathrm{Re}\varphi| \leq \frac{1}{2}\mathrm{Min}(\varepsilon_0, \varepsilon_K) \quad \text{on } K. \tag{7.14}$$

Let also $\chi \in C_0^\infty(\Omega)$ be such that $\chi = 1$ on K and $\mathrm{Supp}\nabla\chi$ is included in the interior of the neighborhood $\mathcal{V}_K$ of $\partial\Omega$ where $\psi_{t_0} \leq 2\mathrm{Re}\varphi - \varepsilon_K$, and define

$$w = \chi v - Tu.$$

Then by construction, we have

$$\begin{aligned}(hD_x - \xi - ihD_\xi)w &= h(D_x\chi - iD_\xi\chi)\, v + \mathcal{O}(e^{-(\xi^2/2 + \mathrm{Re}\varphi + \varepsilon)/h}) \\ &= \mathcal{O}(e^{-(\xi^2 + \psi_{t_0} + \varepsilon_K)/2h})\end{aligned} \tag{7.15}$$

since $D_x\chi - iD_\xi\chi$ is supported in $\mathcal{V}_K$. Moreover

$$\begin{aligned}(\tilde{P} - E)w &= (\tilde{P} - E)\chi v \\ &= [\tilde{P}, \chi]v + \mathcal{O}(e^{-\xi^2/2h - \mathrm{Re}\varphi(x-i\xi)/h - \varepsilon_1/h})\end{aligned} \tag{7.16}$$

with $\varepsilon_1 > 0$. We set

$$\psi = \frac{1}{2}\left(\xi^2 + \psi_{t_0} + \frac{1}{2}\mathrm{Min}(\varepsilon_0, \varepsilon_K)\right)$$

(which is constant outside Ω), and we plan to apply theorem 3.1 with this ψ, but with Tu replaced by w. Actually, since w does not satisfy (2.3) but only (7.15), following the proof of theorem 3.1 we see than an extra error term appears, namely:

$$\begin{aligned}\|e^{\psi/h}(\tilde{P} - E)w\|^2 &= \|p(x - 2\partial_z\psi, \xi + i\partial_z\psi)e^{\psi/h}w\|^2 + \mathcal{O}(h)\|e^{\psi/h}w\|^2 \\ &\quad + \mathcal{O}\left(h\|e^{\psi/h}v\|_{L^2(\mathcal{V}_K)} + e^{-\varepsilon_K/4h}\right)\|e^{\psi/h}w\|.\end{aligned} \tag{7.17}$$

However, since $\psi \leq \frac{\xi^2}{2} + \mathrm{Re}\varphi - \frac{\varepsilon_K}{4}$ on $\mathcal{V}_K$, we have

$$\|e^{\psi/h}v\|_{L^2(\mathcal{V}_K)} = \mathcal{O}(e^{-\varepsilon_K/8h})$$

and therefore, using also the fact that $p(x-2\partial_z\psi, \xi+2i\partial_z\psi) = -q(z, \partial_z\psi_{t_0})$ is elliptic outside 0, we get from (7.17) and (7.16):

$$\|e^{\psi/h}w\|^2 = \mathcal{O}\left(1 + \|e^{\psi/h}w\|^2_{L^2(V_0)} + \|e^{\psi/h}[\tilde{P}, \chi]v\|^2\right). \tag{7.18}$$

Using (7.14), we also have

$$|\psi - \frac{\xi^2}{2} - \mathrm{Re}\varphi| \leq \frac{1}{2}\mathrm{Min}(\varepsilon_0, \varepsilon_K) \tag{7.19}$$

and thus, by (7.13):

$$\|e^{\psi/h}w\|^2_{L^2(V_0)} = \mathcal{O}(e^{-\varepsilon_0/2h}). \tag{7.20}$$

In view of (7.18), it remains to study the term $\|e^{\psi/h}[\tilde{P}, \chi]v\|^2$. We write:

$$[\tilde{P}, \chi]v(X) = \frac{1}{(2\pi h)^{2n}} \int e^{i(X-Y)X^*/h}\tilde{p}(\frac{X+Y}{2}, X^*)(\chi(Y) - \chi(X))v(Y)dYdX^*$$

and, denoting $X = (x, \xi)$, $Y = (y, \eta)$ and $X^* = (x^*, \xi^*)$, we make in (7.21) the change of contour of integration:

$$\mathbf{R}^{2n} \ni (x^*, \xi^*) \mapsto (x^* + ib'\frac{x-y}{|x-y|}, \xi^* + ia'\frac{\xi - \eta}{|\xi - \eta|}) \tag{7.21}$$

with $\mathrm{Sup}\,|\mathrm{Im}(z - \partial_z\varphi)| < a' < a$ and $\mathrm{Sup}\,|\mathrm{Re}\,\partial_z\varphi| < b' < b$. Then, denoting $\varepsilon'_K = \mathrm{Min}(\varepsilon_0, \varepsilon_K)$, we get:

$$\|e^{\psi/h}[\tilde{P}, \chi]v\| = \mathcal{O}\left(1 + \sup_{\substack{Y \notin \mathcal{V}_K \\ \chi(X) \neq \chi(Y)}} e^{-\delta|X-Y|/h + \varepsilon'_K/h}\right) \tag{7.22}$$

with some $\delta > 0$ depending only on K. Now we have

$$\inf_{\substack{Y \notin \mathcal{V}_K \\ \chi(X) \neq \chi(Y)}} |X - Y| > 0$$

and we see that ε_K can possibly be taken smaller without modifying the set $\{Y \notin \mathcal{V}_K\,,\ \chi(X) \neq \chi(Y)\}$. Then we deduce from (7.22) that $\|e^{\psi/h}[\tilde{P}, \chi]v\|$ is exponentially small, and in view of (7.18) and (7.20), we finally get

$$\|e^{\psi/h}w\| = \mathcal{O}(1).$$

Since $\psi(x,\xi) \geq \frac{1}{2}\xi^2 + \mathrm{Re}\varphi(x - i\xi) + \frac{1}{8}\mathrm{Min}(\varepsilon_0, \varepsilon_K)$ on K, this completes the proof of theorem 7.1. ◇

References

1. [Ba] H. Baklouti, *Asymptotique de largeurs de résonances pour un modèle d'effet tunnel microlocal,* Thèse Université Paris-Nord (1995)
2. [HeSj1] B. Helffer, J. Sjöstrand, *Multiple Wells in the Semiclassical Limit I,* Comm. P.D.E., vol. 9, (4), 1984, p. 337-408
3. [HeSj2] B. Helffer, J. Sjöstrand, *Résonances en limite semi-classique,* Bull. Soc. Math. France, Mémoire n. 24/25, tome 114, (1986)
4. [HeSj3] B. Helffer, J. Sjöstrand, *Effet tunnel pour l'équation de Schrödinger avec champs magnétique,* Ann. Sc. Norm. Sup. di Pisa, Ser. IV, 14(4), 625-657 (1987)
5. [JoPf] A. Joye, C.-E. Pfister, *Exponentially small adiabatic invariant for the Schrödinger equation,* Commun. Math. Phys. 140, p. 15-41 (1991)
6. [Ju] K. Jung, *Phase space tunneling in Gevrey class regularity,* preprint (1995)
7. [Ma1] A. Martinez, *An introduction to semiclassical analysis,* book in preparation
8. [Ma2] A. Martinez, *Precise exponential estimates in adiabatic theory,* J. Math. Phys. 35 (8), (1994)
9. [Ma3] A. Martinez, *Estimates on complex interactions in phase space,* Math. Nachr. 167 (1994)
10. [Ma4] A. Martinez, *Estimations sur l'effet tunnel microlocal,* Séminaire E.D.P. de l'Ecole Polytechnique 1991-92
11. [MaSo] A. Martinez, V.Sordoni, paper in preparation
12. [MeSj] A. Melin, J. Sjöstrand, *Fourier integral operators with complex valued phase functions,* Springer Lecture Notes in Math., No.459, 120-223 (1976)
13. [Na1] S. Nakamura, *On an example of phase-space tunneling,* Annales Inst. H. Poincaré, Vol. 63, 211-229 (1995)
14. [Na2] S. Nakamura, *On Martinez' method on phase space tunneling,* Rev. math. Phys. vol 7, p.431-441 (1995)
15. [Ne] G. Nenciu, *Linear Adiabatic Theory. Exponential Estimates,* Commun. Math. Phys. 152, 479-496, (1993)
16. [Sj1] J. Sjöstrand, *Singularités analytiques microlocales,* Astérisque 95 (1982)
17. [Sj2] J. Sjöstrand, *Function spaces associated to global I-Lagrangian manifolds,* Preprint Ecole Polytechnique de Palaiseau No.1111, (1995)
18. [Sj3] J. Sjöstrand, *Analytic wavefront sets and operators with multiple characteristics,* Hokkaido Mathematical Journal, Vol. XII No.3, 392-433 (1983)

A TRACE FORMULA AND REVIEW OF SOME ESTIMATES FOR RESONANCES

J. SJÖSTRAND
Centre de Mathématiques
Ecole Polytechnique
F-91128 Palaiseau, France
and UA 169 CNRS

Abstract. The main part of theses notes from the NATO ASI on microlocal analysis and spectral theory at Il Ciocco, Sept.-Oct. 1996, is devoted to a new trace formula for resonances, which is valid for long range perturbations of the Laplacian in all dimensions. We work in the frame work of complex scaling and have a natural opportunity to review that method. We also review some lower bounds and some upper bounds on the density of resonances near the real axis, mainly following joint works with M.Zworski. The lower bounds however, are new in the case of even dimensions and form a first application of the new trace formula.

1. Introduction

The original plan for these notes was to explain some estimates for the density of resonances for compactly supported perturbations of the Laplacian, largely obtained in collaboration with M.Zworski, and closely related to work of G.Vodev as far as the upper bounds are concerned. For the lower bounds a well-known Poisson type trace formula, valid in odd dimensions, plays an important role. This formula has been elaborated in the frame work of the Lax-Phillips theory successively by Lax-Phillips, Bardos-Guillot-Ralston, Melrose, Sjöstrand-Zworski. During the preparation of some lectures at Ecole Polytechnique in Spring 1996, I tried to obtain "my own" approach to this formula. Instead however, I obtained a new trace formula with a remainder, valid in many situations where the resonances can only be defined in some neighborhood of the real axis. This formula seems to have new applications for instance to even-dimensional

L. Rodino (ed.), Microlocal Analysis and Spectral Theory, 377–437.

cases and to long range scattering for the Schrödinger equation, and despite the fact that there has not yet been enough time to develop most of these applications, I found it natural to give a detailed account here. The sections 2–8 are devoted to the statement and the proof of the new result. In those sections we also have the occasion to review the method of complex scaling, (though our proof is of such a general nature that it should be easily adaptable to some other frameworks for the study of resonances, such as the one developed with B.Helffer in [14]). It would no doubt be useful to understand better the links with a recent approach by Guillopé and Zworski to a Poisson type formula on hyperbolic surfaces, which is based on general scattering theory and especially on the Birman-Krein formula for the scattering phase. Our approach uses no scattering theory.

In section 9 we compare our trace formula with the Poisson type formula of Lax-Phillips theory in the more restrictive situations, where the latter one applies.

In section 10, we review some lower bounds on the density of resonances (which are new in the case of even dimensions), and in section 11, we review some upper bounds.

2. The main result

Our trace formula will concern a pair of self-adjoint operators P_0, P_1, but much of the work will concern each of these two operators individually, so in order to ease the notation, we will often suppress the subscript $j = 0, 1$ and simply write $P_.$ or P, and similarly for the various quantities attached to $P_.$. We shall use essentially the abstract setting introduced with M.Zworski in [34], but our assumptions will be weaker in the sense that we do not assume that $P_.$ be equal to $-\Delta$ near infinity, and we do not assume the dimension to be odd. It will also be convenient to adopt a semiclassical framework from the very beginning, so that $P_.$ and the corresponding quantities, depend on a Planck's constant $h \in]0, h_0]$, where $h_0 > 0$.

Let $\mathcal{H}_.$ be a complex separable Hilbert space with an orthogonal decomposition:

$$\mathcal{H}_. = \mathcal{H}_{.,R_0} \oplus L^2(\mathbf{R}^n \setminus B(0, R_0)), \tag{2.1}$$

where $R_0 > 0$ is some fixed constant and $B(x, R) = \{y \in \mathbf{R}^n; |x - y| < R\}$. The corresponding orthogonal projections will be denoted by $u \mapsto u_{|B(0,R_0)}$, and $u \mapsto u_{|\mathbf{R}^n \setminus B(0,R_0)}$ or simply by the characteristic function (1_L) of the corresponding set (L). We consider an unbounded self-adjoint operator

$$P_. : \mathcal{H}_. \to \mathcal{H}_. \text{ with domain } \mathcal{D}_. = \mathcal{D}(P_.). \tag{2.2}$$

Assume that

$$1_{\mathbf{R}^n \setminus B(0,R_0)} \mathcal{D}_. = H^2(\mathbf{R}^n \setminus B(0, R_0)), \tag{2.3}$$

uniformly with respect to h in the following topological sense: Equip $H^2(\mathbf{R}^n \setminus B(0,R_0))$, with the norm $\|\langle hD\rangle^2 u\|_{L^2}$, where $\langle hD\rangle = (1+(hD)^2)^{\frac{1}{2}}$, $(hD)^2 = \sum_1^n (hD_{x_j})^2$, and equip $\mathcal{D}_\cdot$ with the norm $\|(i+P)u\|_{\mathcal{H}_\cdot}$. Then in (2.3), we require that the restriction map from $\mathcal{D}$ to $H^2(\mathbf{R}^2 \setminus B(0,R_0))$ is bounded uniformly with respect to h and has a uniformly bounded right inverse (that we may call an extension map).

Assume

$$1_{B(0,R_0)}(P_\cdot + i)^{-1} \text{ is compact.} \tag{2.4}$$

We also assume that,

$$\begin{gathered} 1_{\mathbf{R}^n\setminus B(0,R_0)} P_\cdot u = Q_\cdot u,\ Q_\cdot u = \textstyle\sum_{|\alpha|\le 2} a_{\cdot,\alpha}(x;h)(hD_x)^\alpha u, \\ a_{\cdot,\alpha}(x;h) = a_{\cdot,\alpha}(x) \text{ is independent of } h \text{ for } |\alpha| = 2, \\ \text{and } a_{\cdot,\alpha} \in C_b^\infty(\mathbf{R}^n) \text{ are uniformly bounded w.r.t. } h. \end{gathered} \tag{2.5}$$

Here $C_b^\infty(\mathbf{R}^n)$ denotes the space of C^∞ functions on $\mathbf{R}^n$ which are bounded together with all derivatives. Observe that if $\psi \in C_b^\infty(\mathbf{R}^n)$ is constant near $\overline{B(0,R_0)}$, then there is a natural way of defining the multiplication: $\mathcal{H}_\cdot \ni u \mapsto \psi u \in \mathcal{H}_\cdot$, and we have $\psi u \in \mathcal{D}_\cdot$ if $u \in \mathcal{D}_\cdot$.

It is further assumed that Q is formally self-adjoint on $\mathbf{R}^n$ with:

$$\begin{gathered} \textstyle\sum_{|\alpha|=2} a_{\cdot,\alpha}(x)\xi^\alpha \ge \frac{1}{C}|\xi|^2, \\ \sum a_{\cdot,\alpha}(x;h)\xi^\alpha \to \xi^2,\ |x| \to \infty, \\ \text{uniformly with respect to } h. \end{gathered} \tag{2.6}$$

It is quite likely that the second part of this assumption can be weakened, so that we could allow operators of the form $-h^2\Delta + V(x)$, where $V(x)$ may be unbounded. In some cases like that, the final trace formula is much easier to obtain and results very quickly from Lidskii's theorem. This fact has been used by L.Nedelec [25] for Schrödinger operators with linear matrix-valued potentials.

Assume

$$|a_{1,\alpha}(x;h) - a_{0,\alpha}(x;h)| \le C\langle x\rangle^{-\tilde{n}},\ \tilde{n} > n. \tag{2.7}$$

This assumption will guarantee that $f(P_1) - f(P_0)$ is "of trace class near infinity" , when $f \in C_0^\infty(\mathbf{R})$.

Let $R > R_0$ and $M = (\mathbf{R}/\tilde{R}\mathbf{Z})^n$, where $\tilde{R} > 2R$. Then we can view $B(0,R)$ as a subset of M, and as in [34], we can define an unbounded self-adjoint operator $P_\cdot^\#$ in $\mathcal{H}_\cdot^\# = H_{R_0} \oplus L^2(M \setminus B(0,R_0))$, which coincides with $P_\cdot$ (in the natural sense) near $\overline{B(0,R)}$ and which outside $B(0,R_0)$ is of the form $Q_\cdot^\#$ and has the same properties (except for the behaviour at infinity) as in (2.5),(2.6). As in [34], we see that $P_\cdot^\#$ has discrete spectrum.

Let $N(P_\cdot^\#; I)$ denote the number of eigenvalues of $P_\cdot^\#$ in the interval I. We assume:

$$N(P_\cdot^\#;[-\lambda,\lambda]) = \mathcal{O}((\lambda/h^2)^{n_\cdot/2}),\ \lambda \geq 1, \tag{2.8}$$

for some number $n_\cdot \geq n$. As in [34], this assumption does not depend on the choice of R, $\widetilde{R}$, or $Q_\cdot^\#$. We briefly explain why. Let $\lambda_1 \leq \lambda_2 \leq \lambda_3 \leq ..$ be the eigenvalues of $P^\# = P_\cdot^\#$, so that, if $\mu_j(K)$ denotes the j:th characteristic (or singular) value of the compact operator K, $\mu_j((P^\# - i)^{-1}) = |i-\lambda_j|^{-1} = \langle\lambda_j\rangle^{-1}$. (See [9].) Then it is easy to see that the property (2.8) is equivalent to the property:

$$\mu_j((i-P_\cdot^\#)^{-1}) \leq \frac{\mathcal{O}(1)}{(1+h^2 j^{2/n_\cdot})}.$$

Recall that if K is a compact operator on some separable Hilbert space, then the characteristic values, $\mu_1(K) \geq \mu_2(K) \geq ..$ are defined to be the eigenvalues of $\sqrt{K^*K}$.

If $\widetilde{M}$ is a second torus and $\widetilde{P}_\cdot^\#$ a corresponding operator analogous to $P_\cdot^\#$, we can identify M and $\widetilde{M}$ by means of a diffeomorphism which is the identity map near $\overline{B(0,R_0)}$, and achieve that the two operators act in the same Hilbert space and coincide near $\overline{B(0,R_0)}$. The invariance of the assumption (2.8) then follows from the resolvent identity

$$(i-P_\cdot^\#)^{-1} = (i-\widetilde{P}_\cdot^\#)^{-1} + (i-P_\cdot^\#)^{-1}(P_\cdot^\# - \widetilde{P}_\cdot^\#)(i-\widetilde{P}_\cdot^\#)^{-1}$$

and the general identities for characteristic values:

$$\mu_{j+k-1}(A+B) \leq \mu_j(A) + \mu_k(B),$$
$$\mu_{j+k-1}(AB) \leq \mu_j(A)\mu_k(B).$$

(See [9], where both identities are easily derived from the Ky Fan identity

$$\mu_j(A) = \inf_{\mathrm{rank}\,(R)\leq j-1} \|A-R\|.$$

Here $\|\cdot\|$ denotes standard operator norm.)

Let S^m denote the standard symbol space of functions $a \in C^\infty(\mathbf{R})$, satisfying $a^{(k)}(t) = \mathcal{O}(\langle t\rangle^{m-k})$ for every $k \in \mathbf{N}$. In the next section, we shall prove:

Proposition 2.1 *Let $f \in S^{-m(f)}$ be independent of h, where $m(f)\frac{2}{n_{\max}} > 1$, $n_{\max} =_{\mathrm{def}} \max(n_0, n_1)$. Let $\chi \in C_0^\infty(\mathbf{R}^n)$ be equal to 1 near $\overline{B(0,R_0)}$. Then $\chi f(P_\cdot)$, $f(P_\cdot)\chi$, $(1-\chi)f(P_1)(1-\chi) - (1-\chi)f(P_0)(1-\chi)$ are of trace class and*

$$\begin{aligned}&\text{"tr}\,(f(P_1) - f(P_0))\text{"} = \\ &= [\mathrm{tr}\,\chi f(P_j)\chi + \ \mathrm{tr}\,(1-\chi)f(P_j)\chi + \mathrm{tr}\,\chi f(P_j)(1-\chi)]_{j=0}^1 \\ &\qquad + \mathrm{tr}\,[(1-\chi)f(P_j)(1-\chi)]_{j=0}^1\end{aligned} \tag{2.9}$$

is independent of the choice of χ *and is* $\mathcal{O}(h^{-n_{\max}})$. *Here we write* $[a_j]_{j=0}^1 = a_1 - a_0$.

This proposition will be proved in section 3. Notice that $f(P_1) - f(P_0)$ is not a well defined operator in general and this is the reason for the use of quotation marks. Also notice that $[(1-\chi)f(P_j)(1-\chi)]_{j=0}^1$ is a well defined operator in $L^2(\mathbf{R}^n)$, $\mathcal{H}_0$, and $\mathcal{H}_1$ and has the same trace in all these spaces (as soon as it is of trace class as an operator in one of the spaces).

Our trace formula will involve resonances of P_0, P_1 and for simplicity, we will use the frame work of complex scaling, or complex distorsion ([1], [18]), and we will follow the presentation in [34]. For that, we shall use the following assumption:

$$\begin{gathered}\text{There exist } \theta_0 \in [0,\pi[,\ \epsilon > 0, \text{ and } R \geq R_0, \text{ such that}\\ \text{the coefficients } a_{\cdot,\alpha}(x;h) \text{ of } Q_\cdot \text{ extend holomorphically in } x\\ \text{to } \{r\omega;\ \omega \in \mathbf{C}^n,\ \mathrm{dist}\,(\omega, \mathbf{S}^{n-1}) < \epsilon,\ r \in \mathbf{C},\ |r| > R,\ \arg r \in [-\epsilon, \theta_0 + \epsilon[\}\\ \text{and (2.7) and the second half of (2.6) remain valid}\\ \text{in this larger set.}\end{gathered} \tag{2.10}$$

It is quite likely the main result below remains valid in other frame works, such as the one in [14]. We can now define the resonances $\lambda_{\cdot,j}$ of $P_\cdot$ in the sector $S_{\theta_0} = \{z \in \mathbf{C} \setminus \{0\}; 0 \leq -\arg z < 2\theta_0\}$ as the eigenvalues of $P_\cdot$ on a suitable contour in $\mathbf{C}^n$ (see [34], and section 5) and it follows from section 8 and also from the methods of [30], [34]) that if $\Omega \subset\subset S_{\theta_0}$ is independent of h, then the number of resonances of $P_\cdot$ in Ω is $\mathcal{O}(h^{-n_\cdot})$. The same estimate holds for the number of eigenvalues in any fixed compact subinterval of $]-\infty, 0[$.

Let $W \subset\subset \Omega$ be open relatively compact subsets of $e^{i]-2\theta_0,\epsilon_0]}]0,+\infty[$, where $\epsilon_0 > 0$ may be arbitrarily small, such that the intersections I, J with the open positive half axis are intervals, and denote by Ω_-, W_-, the intersections with $e^{]-2\theta_0,0]}]0,+\infty[$. Also assume for simplicity, that Ω is simply connected.

Theorem 2.2 *We make the assumptions above and assume that* W, Ω *are independent of* h. *Let* $f = f(z;h)$ *be holomorphic in* z *on* Ω, *and satisfy* $|f(z;h)| \leq 1$ *for* $z \in \Omega \setminus W$. *Let* $\chi \in C_0^\infty(J)$ *be h-independent and satisfy* $\chi = 1$ *near* $\bar{I}$. *Then*

$$\text{"tr}\,((\chi f)(P_1;h) - (\chi f)(P_0;h))\text{"} =$$

$$[\sum_{\lambda_{\cdot,j} \in W_-} f(\lambda_{\cdot,j};h) - \sum_{\mu \in \sigma(P_\cdot) \cap]-\infty,0[\cap W_-} f(\mu;h)]_0^1 + \mathcal{O}(h^{-n_{\max}}) \tag{2.11}$$

The proof of the above theorem will occupy the sections 3–8. We end this section by showing that the theorem is still valid under slightly different assumptions about f. This will be useful in section 9.

Proposition 2.3 *Let $f = f(z;h)$ satisfy the assumptions of Threorem 2.2 with the following modification: Instead of assuming that $|f(z;h)| \leq 1$ in $\Omega \setminus W$, we assume that for all α:*

$$|\partial^\alpha f(z;h)| \leq C_\alpha e^{C(\arg z)_+/h}, \quad z \in \Omega \setminus W,$$

for some constant $C > 0$. Let $\widetilde{W} \subset \Omega$ be (relatively) open with $W \subset\subset \widetilde{W} \subset\subset \Omega$. Then there exist $0 < \widetilde{\epsilon}_0 < \epsilon_0$, and a holomorphic function $\widetilde{f}$ in $\Omega \cap \{\arg z \leq \widetilde{\epsilon}_0\}$ such that $\widetilde{f}(z;h) = \mathcal{O}(1)$, when $z \in \Omega \setminus \widetilde{W}$, $\arg z \leq \widetilde{\epsilon}_0$, and such that $f - \widetilde{f} = \mathcal{O}(1)$, $z \in \Omega_-$, $\partial^\alpha(\widetilde{f} - f) = \mathcal{O}_\alpha(1)$, $z \in J$.

Proof. Let $\chi \in C_0^\infty(\widetilde{W})$ be equal to 1 near $\overline{W}$. Treating separately the equations

$$\overline{\partial} g_1 = \overline{\partial}(\chi f) 1_{[0,\widetilde{\epsilon}_0]}(\arg z),$$

$$\overline{\partial} g_2 = \overline{\partial}(\chi f) 1_{]-2\theta_0,0]}(\arg z),$$

we get a function $g = g_1 + g_2$ in $\Omega \cap \{\arg z \leq \widetilde{\epsilon}_0\}$, such that $\overline{\partial} g = \overline{\partial}(\chi f)$, with $g = \mathcal{O}(1) \exp \widetilde{C}(\arg z)_+/h$, such that $\partial^k_{\mathrm{Re}\, z} g = \mathcal{O}_k(1)$ in J, and such that $g(z) = \mathcal{O}(1)$ in $(\Omega \setminus \widetilde{W}) \cap \arg^{-1}([0, \widetilde{\epsilon}_0])$, if $\widetilde{\epsilon}_0$ is small enough.

Indeed, we can treat the $\overline{\partial}$-equations in "polar" coordinates ζ given by $z = e^\zeta$ and solve the equation for g_2 using the convolution with the standard kernel $1/\pi\zeta$. For g_1, we use the convolutor $(\pi\zeta)^{-1} e^{q(\zeta)/h}$, where $q(\zeta)$ is a suitable quadratic polynomial with $q(0) = 1$. To understand the choice of q, consider for instance the case $\widetilde{\epsilon}_0 = 1$, so we have an equation $\frac{\partial}{\partial \overline{\zeta}} g = k$, where $|k| \leq e^{C \mathrm{Im}\, \zeta}$, $\mathrm{supp}\, k \subset \{0 \leq \mathrm{Im}\, \zeta \leq 1\}$. Take $q(\zeta) = -\alpha\zeta^2 + i^{-1}(C+1)\zeta$, where $\alpha > 0$ will be small. The convolution at a point ζ is then bounded by $\mathcal{O}(1) e^{r(\zeta)/h}$, where

$$r(\zeta) = \sup_{\widetilde{\zeta} \in \mathrm{supp}\, k} \mathrm{Re}\, (-\alpha(\zeta - \widetilde{\zeta})^2 - i(C+1)(\zeta - \widetilde{\zeta})) + C \mathrm{Im}\, \widetilde{\zeta}.$$

If K is the projection to the real axis of $\mathrm{supp}\, k$, we get

$$r(\zeta) \leq (C+1)\mathrm{Im}\, \zeta - \alpha \mathrm{dist}\, (\mathrm{Re}\, \zeta, K)^2 + \alpha(\mathrm{Im}\, \zeta)^2 + \max(0, -2\alpha \mathrm{Im}\, \zeta + \alpha - 1).$$

Taking into account that $\mathrm{Im}\, \zeta$ is bounded for $z \in \Omega$, we choose $\alpha > 0$ small enough and get,

$$r(\zeta) \leq (C+1)\mathrm{Im}\, \zeta + \alpha(\mathrm{Im}\, \zeta)^2 - \alpha \mathrm{dist}\, (\mathrm{Re}\, \zeta, \mathrm{K})^2.$$

We then get g_1 with the required properties.

We finally take $\widetilde{f} = \chi f - g$ and check that $\widetilde{f}$ has the required properties.
□

If f is as in the last proposition, then we can still apply Theorem 2.2. In fact, we first check that "tr $[(\chi f)(P_\cdot)]_0^1$" only changes by $\mathcal{O}(h^{-n_{\max}})$, if we replace f by $\widetilde{f}$. In section 8 we shall see that the number of eigenvalues and resonances that appear in the sums to the right in (2.11) is $\mathcal{O}(h^{-n_{\max}})$, so the sums also only change by this amount, if we replace f by $\widetilde{f}$.

3. Trace class estimates before complex scaling

We start by estimating the characteristic values of certain resolvents and truncated resolvents. In doing so, we let the spectral parameter belong to the open set in $\mathbf{C}$, defined by:

$$|\mathrm{Im}\, z| > \frac{1}{C}(|\mathrm{Re}\, z| - C)_+. \tag{3.1}$$

From (2.8), we get:

$$\mu_j((P_\cdot^\# - i)^{-1}) \leq C(1 + h^2 j^{2/n_\cdot})^{-1}. \tag{3.2}$$

Using the resolvent identity, we get for a more general z in the set (3.1):

$$\mu_j((P_\cdot^\# - z)^{-1}) \leq \begin{cases} C(|z| + h^2 j^{2/n_\cdot})^{-1}, \text{ when } |\mathrm{Im}\, z| \geq \text{Const.} > 0, \\ \frac{C}{|\mathrm{Im}\, z|}(1 + h^2 j^{2/n_\cdot})^{-1}, \text{ when } |\mathrm{Im}\, z| \leq \text{Const.}. \end{cases} \tag{3.3}$$

Let $\Gamma \subset \mathbf{R}^n$ be a sufficiently widely spaced lattice and let $0 \leq \psi_\nu \in C_0^\infty(\mathbf{R}^n)$, $\nu \in \Gamma$, be a translation invariant partition of unity ($\psi_\nu(x) = \psi_0(x - \nu)$) with $\psi_0 = 1$ near $\overline{B(0, R_0)}$.

Lemma 3.1 *Let χ vary in some bounded subset of $C_b^\infty(\mathbf{R}^n)$ with $\chi =$ Const. near $\overline{B(0, R_0)}$ and let ν vary in Γ. Then for z satisfying (3.1):*

$$\mu_j(\chi(P_\cdot - z)^{-1}\psi_\nu),\ \mu_j(\psi_\nu(P_\cdot - z)^{-1}\chi) \leq \frac{C\langle z\rangle}{|\mathrm{Im}\, z|(\langle z\rangle + h^2 j^{2/n_\cdot})} e^{-\frac{|\mathrm{Im}\, z|}{Ch}(\mathrm{dist}\,(\mathrm{supp}\,\chi, \mathrm{supp}\,\psi_\nu) - 1)_+}, \tag{3.4}$$

where $\langle z\rangle = \sqrt{1 + |z|^2}$.

Proof. Assume first that $\nu = 0$. If f, g are real functions on $\mathbf{R}^n$, we write $f \prec g$, if supp f is contained in the interior of the region where $g = 1$. Let $\widetilde{\psi}_0 \in C_0^\infty(\mathbf{R}^n)$ with $\psi_0 \prec \widetilde{\psi}_0$. If we choose $P_\cdot^\#$ and the corresponding torus suitably, we have:

$$(P_\cdot - z)^{-1}\psi_0 = \widetilde{\psi}_0(P_\cdot^\# - z)^{-1}\psi_0 - (P_\cdot - z)^{-1}[P_\cdot, \widetilde{\psi}_0](P_\cdot^\# - z)^{-1}\psi_0. \tag{3.5}$$

$$\psi_0(P_\cdot - z)^{-1} = \psi_0(P_\cdot^\# - z)^{-1}\widetilde{\psi}_0 + \psi_0(P_\cdot^\# - z)^{-1}[P_\cdot, \widetilde{\psi}_0](P_\cdot - z)^{-1}. \quad (3.6)$$

Here $(P_\cdot - z)^{-1}[P_\cdot, \widetilde{\psi}_0]$, $[P_\cdot, \widetilde{\psi}_0](P_\cdot - z)^{-1}$ are bounded operators of norm $\mathcal{O}(\frac{h\langle z\rangle}{|\mathrm{Im}\, z|})$. (We use the ellipticity of $P_\cdot$ near $\mathrm{supp}\,[P_\cdot, \widetilde{\psi}_0]$.) Assume first that $|\mathrm{Im}\, z| \geq \frac{1}{\mathrm{Const.}}$. Then (3.3) implies that

$$\mu_j((P_\cdot - z)^{-1}\psi_0),\ \mu_j(\psi_0(P_\cdot - z)^{-1}) \leq \frac{C}{(\langle z\rangle + h^2 j^{2/n_\cdot})}, \quad (3.7)$$

and we get (3.4) without the exponential factor. If $\mathrm{dist}\,(\mathrm{supp}\,\chi, \mathrm{supp}\,\psi_0) \geq 1$, choose $\widetilde{\psi}_0$ with support sufficiently close to that of ψ_0 and notice that by a Combes-Thomas argument,

$$\chi(P_\cdot - z)^{-1}[P_\cdot, \widetilde{\psi}_0],\ [P_\cdot, \widetilde{\psi}_0](P_\cdot - z)^{-1}\chi = \mathcal{O}(h)e^{-\frac{1}{Ch}(\mathrm{dist}\,(\mathrm{supp}\,\chi, \mathrm{supp}\,\psi_0) - 1)_+},$$

in operator norm. (A typical Combes-Thomas argument is employed in the proof of (7.8) below.) When multiplying (3.5) to the left or (3.6) to the right by χ, only the last terms survive in the right hand sides, and we get (3.4).

Still in the case $\nu = 0$ it remains to treat the case when $|\mathrm{Im}\, z|$ is bounded (so that $|z|$ is also bounded). Using the resolvent identities,

$$(P_\cdot - z)^{-1}\psi_0 = (P_\cdot - i)^{-1}\psi_0 + (P_\cdot - z)^{-1}(z - i)(P_\cdot - i)^{-1}\psi_0, \quad (3.8)$$

$$\psi_0(P_\cdot - z)^{-1} = \psi_0(P_\cdot - i)^{-1} + \psi_0(P_\cdot - i)^{-1}(z - i)(P_\cdot - z)^{-1}, \quad (3.9)$$

and (3.7) with $z = i$, we get,

$$\mu_j((P_\cdot - z)^{-1}\psi_0),\ \mu_j(\psi_0(P_\cdot - z)^{-1}) \leq \frac{C}{|\mathrm{Im}\, z|(1 + h^2 j^{2/n_\cdot})}, \quad (3.10)$$

which gives (3.4) without the exponential factor.

Assume that $d = \mathrm{dist}\,(\mathrm{supp}\,\chi, \mathrm{supp}\,\psi_0) \geq 1$. Let $\widetilde{\chi} \in C_b^\infty$ be equal to 1 on

$$\{x \in \mathbf{R}^n; \mathrm{dist}\,(x, \mathrm{supp}\,\chi) \leq d/3\},$$

and equal to 0 on $\{x \in \mathbf{R}^n; \mathrm{dist}\,(x, \mathrm{supp}\,\psi_0) \leq d/3\}$. From (3.8) we get

$$\begin{aligned}\chi(P_\cdot - z)^{-1}\psi_0 = \ & \chi(P_\cdot - i)^{-1}\psi_0 + (z - i)\chi(P_\cdot - z)^{-1}\widetilde{\chi}(P_\cdot - i)^{-1}\psi_0 \\ & +(z - i)\chi(P_\cdot - z)^{-1}(1 - \widetilde{\chi})(P_\cdot - i)^{-1}\psi_0, \quad (3.11)\end{aligned}$$

and from (3.9) we get a similar relation:

$$\begin{aligned}\psi_0(P_\cdot - z)^{-1}\chi = \ & \psi_0(P_\cdot - i)^{-1}\chi + (z - i)\psi_0(P_\cdot - i)^{-1}\widetilde{\chi}(P_\cdot - z)^{-1}\chi \\ & +(z - i)\psi_0(P_\cdot - i)^{-1}(1 - \widetilde{\chi})(P_\cdot - z)^{-1}\chi. \quad (3.12)\end{aligned}$$

We can estimate the characteristic values of the first two terms of the RHS of (3.11) and (3.12) by using (3.4) with $z = i$ (already established). A Combes-Thomas argument shows that the norm of $\chi(P_\cdot - z)^{-1}(1 - \tilde{\chi})$ is

$$\mathcal{O}(1)|\mathrm{Im}\, z|^{-1} \exp -d/(Ch|\mathrm{Im}\, z|).$$

Combining this with (3.7) for $z = i$, we can estimate the characteristic values of the last terms of the RHS of (3.11) and (3.12) and we get (3.4) for $\nu = 0$.

Consider finally the case $\nu \neq 0$. The proof is the same except that in the formulas (3.5), (3.6), we replace "0" by "ν" and now let $P_\cdot^{\#}$ be an operator on a torus containing the support of $\tilde{\psi}_\nu$ and be equal to $P_\cdot$ in a neighborhood of the latter support. □

As a special case of Lemma 3.1, we have

$$\mu_j(\psi_\nu(P_\cdot - z)^{-1}\psi_\mu) \leq \frac{C\langle z\rangle e^{-\frac{|\mathrm{Im}\, z|}{Ch}(|\nu-\mu|-C)_+}}{|\mathrm{Im}\, z|(\langle z\rangle + h^2 j^{2/n_\cdot})}. \tag{3.13}$$

When $\nu, \mu \in \Gamma \setminus \{0\}$, then $[\psi_\nu(P_\cdot - z)^{-1}\psi_\mu]_0^1$ is a well defined operator in any of the spaces $\mathcal{H}_0$, $\mathcal{H}_1$, $L^2(\mathbf{R}^n)$ and we have the following result, where the assumption (2.7) is used for the first time:

Lemma 3.2 *For $\nu, \mu \in \Gamma \setminus \{0\}$, we have*

$$\mu_j([\psi_\nu(P_\cdot - z)^{-1}\psi_\mu]_0^1) \leq$$
$$\frac{C\langle z\rangle^2(\max(|\mu|,|\nu|)^{-\tilde{n}} + e^{-\frac{|\mathrm{Im}\, z|}{Ch}(\max(|\nu|,|\mu|)-C)_+})}{|\mathrm{Im}\, z|^2(\langle z\rangle + h^2 j^{2/n_{\max}})} e^{-\frac{|\mathrm{Im}\, z|}{Ch}(|\nu-\mu|-C)_+}. \tag{3.14}$$

Proof. It suffices to treat the case when $|\nu|, |\mu| >> 1$, since we otherwise get (3.14) from (3.4). For the same reason, we may restrict the attention to the case when $|\nu - \mu| << |\mu|$.

We need a resolvent identity with cut-offs. Let $\chi \in C_0^\infty(\mathbf{R}^n)$ be equal to 1 near $\overline{B(0, R_0)}$ and let $v \in L^2(\mathbf{R}^n \setminus B(0, R_0))$ so that v can be considered as an element of $\mathcal{H}_j$, $j = 0, 1$. Let $u_j \in \mathcal{H}_j$ be the solution of $(P_j - z)u_j = v$. Then

$$(P_j - z)(1 - \chi)u_j = (1 - \chi)v - [P_j, \chi]u_j.$$

Rewrite the equation for $j = 1$ as

$$(P_0 - z)(1 - \chi)u_1 = (1 - \chi)v - [P_1, \chi]u_1 - (P_1 - P_0)(1 - \chi)u_1.$$

Then,

$$(P_0 - z)((1-\chi)u_1 - (1-\chi)u_0) = -([P_1, \chi]u_1 - [P_0, \chi]u_0 + (P_1 - P_0)(1-\chi)u_1),$$

which we can write

$$
\begin{aligned}
(1-\chi)(P_1-z)^{-1}v-(1-\chi)(P_0-z)^{-1}v= \\
-(P_0-z)^{-1}([P_1,\chi](P_1-z)^{-1}v- \\
[P_0,\chi](P_0-z)^{-1}v+(P_1-P_0)(1-\chi)(P_1-z)^{-1}v).
\end{aligned}
$$

Let χ_1, χ_2 have the same properties as χ. Then, if $(1-\chi_1)(1-\chi)=1-\chi_1$,

$$
\begin{aligned}
[(1-\chi_1)(P_\cdot-z)^{-1}(1-\chi_2)]_0^1= \\
-[(1-\chi_1)(P_0-z)^{-1}[P_\cdot,\chi](P_\cdot-z)^{-1}(1-\chi_2)]_0^1- \\
(1-\chi_1)(P_0-z)^{-1}(P_1-P_0)(1-\chi)(P_1-z)^{-1}(1-\chi_2).
\end{aligned}
\tag{3.15}
$$

Multiplying this identity from the left by ψ_ν and from the right by ψ_μ, we get the same relation with $1-\chi_1$ replaced by ψ_ν and $1-\chi_2$ by ψ_μ. Choose $\chi=\sum_{\widetilde{\mu}\in\Gamma;|\widetilde{\mu}|\leq|\nu|/2}\psi_{\widetilde{\mu}}$. In estimating the characteristic values of the first term of the RHS of (3.15) (with the substitutions: $1-\chi_1\mapsto\psi_\nu$, $1-\chi_2\mapsto\psi_\mu$) we use (3.4) to estimate the characteristic values of $\psi_\nu(P_0-z)^{-1}$ and estimate the norm of $[P_\cdot,\chi](P_\cdot-z)^{-1}\psi_\mu$ by

$$
\frac{C\langle z\rangle}{|\mathrm{Im}\, z|}e^{-\frac{|\mathrm{Im}\, z|}{Ch}(|\mu|-C)_+}.
$$

It follows that the characteristic values of the first term of the RHS of (3.15) satisfy (3.14). In estimating the characteristic values of the second term of the RHS of (3.15), we observe that the coefficients of P_1-P_0 are $\mathcal{O}(\langle\mu\rangle^{-\widetilde{n}})$ on the support of $1-\chi$, and it follows that the characteristic values of this term satisfy (3.14) without the exponential factor. In order to get also the exponential factor, when $d_{\nu,\mu}=_{\mathrm{def}}\mathrm{dist}\,(\mathrm{supp}\,\psi_\nu,\mathrm{supp}\,\psi_\mu)\geq 1$, we split the term into

$$
\sum_0^1\psi_\nu(P_0-z)^{-1}f_k(P_1-P_0)(1-\chi)(P_1-z)^{-1}\psi_\mu,
$$

where $f_1+f_2=1$, $f_k\in C_b^\infty$, $f_1(x)=1$ when $\mathrm{dist}\,(x,\mathrm{supp}\,\psi_\nu)\leq d_{\nu,\mu}/3$, $f_2(x)=1$ when $\mathrm{dist}\,(x,\mathrm{supp}\,\psi_\mu)\leq d_{\nu,\mu}/3$. Let $\widetilde{f}_k\in C_b^\infty$ be equal to 1 on $\mathrm{supp}\, f_k$ and have its support in $\mathrm{supp}\,(f_k)+B(0,1)$ and write the term for $k=1$ as

$$
\psi_\nu(P_0-z)^{-1}f_1(P_1-P_0)(1-\chi)\widetilde{f}_1(P_1-z)^{-1}\psi_\mu. \tag{3.16}
$$

Here the norm of $\psi_\nu(P_0-z)^{-1}f_1(P_1-P_0)(1-\chi)$ is $\mathcal{O}(\frac{\langle z\rangle}{|\mathrm{Im}\, z|}|\mu|^{-\widetilde{n}})$ while the characteristic values of $\widetilde{f}_1(P_1-z)^{-1}\psi_\mu$ can be estimated by (3.4). The characteristic values of (3.16) then obey (3.14). It remains to treat $\psi_\nu(P_0-$

$z)^{-1}f_2(P_1-P_0)(1-\chi)(P_1-z)^{-1}\psi_\mu$, and we do essentially the same thing, estimating now the norm of $(P_1-P_0)(1-\chi)(P_1-z)^{-1}\psi_\mu$ as before and the characteristic values of $\psi_\nu(P_0-z)^{-1}f_2$ by means of (3.4). Putting all the estimates together, we get (3.14) for the second term of the RHS of (3.15). □

Lemma 3.3 *Let $m\in\mathbf{N}$ with $(1+m)\frac{2}{n_{\max}}>1$ and let $z_0\in\mathbf{C}\setminus\mathbf{R}$ be some fixed point. Then for z as in (3.1), and $\nu,\mu\in\Gamma$, $\psi_\nu(P_\cdot-z_0)^{-m}(P_\cdot-z)^{-1}\psi_\mu$ is of trace class and*

$$\|\psi_\nu(P_\cdot-z_0)^{-m}(P_\cdot-z)^{-1}\psi_\mu\|_{\mathrm{tr}}\leq$$
$$Ch^{-n_\cdot}\langle z\rangle^{-\min(1,m+1-n_\cdot/2)}(1+\delta_{m,n_\cdot/2}\log\langle z\rangle)$$
$$\frac{\langle z\rangle}{|\mathrm{Im}\, z|}e^{-\min(1,|\mathrm{Im}\, z|)(|\nu-\mu|-C)_+/(Ch)}. \tag{3.17}$$

Here we use the standard notation for the Kronecker delta. For $\nu,\mu\in\Gamma\setminus\{0\}$, we have

$$\|[\psi_\nu(P_\cdot-z_0)^{-m}(P_\cdot-z)^{-1}\psi_\mu]_0^1\|_{\mathrm{tr}}\leq$$
$$Ch^{-n_{\max}}\frac{\langle z\rangle^2}{|\mathrm{Im}\, z|^2}\langle z\rangle^{-\min(1,m+1-n_{\max}/2)}(1+\delta_{m,n_{\max}/2}\log\langle z\rangle)\times$$
$$(|\nu|^{-\tilde{n}}+e^{-\frac{\min(1,|\mathrm{Im}\, z|)}{Ch}(|\nu|-C)_+})e^{-\frac{\min(1,|\mathrm{Im}\, z|)}{Ch}(|\nu-\mu|-C)_+} \tag{3.18}$$

Proof. Write

$$\psi_\nu(P_\cdot-z_0)^{-m}(P\cdot-z)^{-1}\psi_\mu=$$
$$\sum_{\alpha\in\Gamma^m}\psi_\nu(P_\cdot-z_0)^{-1}\psi_{\alpha_1}(P_\cdot-z_0)^{-1}\psi_{\alpha_2}..(P_\cdot-z_0)^{-1}\psi_{\alpha_m}(P_\cdot-z)^{-1}\psi_\mu, \tag{3.19}$$

with $\alpha=(\alpha_1,..,\alpha_m)$. Using (3.4) we can estimate the j:th characteristic value of the general term in the sum by

$$\frac{C\langle z\rangle}{|\mathrm{Im}\, z|}(1+h^2j^{2/n_\cdot})^{-m}(\langle z\rangle+h^2j^{2/n_\cdot})^{-1}\times$$
$$e^{-\frac{1}{Ch}((|\nu-\alpha_1|-C)_++(|\alpha_1-\alpha_2|-C)_++..+(|\alpha_{m-1}-\alpha_m|-C)_++|\mathrm{Im}\, z|(|\alpha_m-\mu|-C)_+)}, \tag{3.20}$$

and hence the trace class norm of the term is bounded by the sum over j of the values (3.20). We estimate

$$S=\sum_1^\infty(1+h^2j^{2/n_\cdot})^{-m}(\langle z\rangle+h^2j^{2/n_\cdot}))^{-1}$$
$$\leq\int_0^\infty(1+h^2x^{2/n_\cdot})^{-m}(\langle z\rangle+h^2x^{2/n_\cdot})^{-1}dx. \tag{3.21}$$

After a change of variables, we get

$$S \leq \int_0^\infty (1+t)^{-m}(\langle z\rangle + t)^{-1} t^{n./2-1} dt \times \frac{n.}{2} h^{-n.}. \tag{3.22}$$

The integral is convergent, since $m+1 > n./2$ and if we treat separately the integrals over the intervals $[0, \langle z\rangle]$ and $[\langle z\rangle, \infty[$, we get

$$S \leq \mathcal{O}(1) h^{-n.} \langle z\rangle^{-\min(1, m+1-n./2)} (1 + \delta_{m,n./2} \log\langle z\rangle). \tag{3.23}$$

The trace class norm of the general term in (3.19) is bounded by

$$\mathcal{O}(1) \frac{\langle z\rangle}{|\mathrm{Im}\, z|} \langle z\rangle^{-\min(1,m+1-n./2)} (1 + \delta_{m,n./2} \log\langle z\rangle) h^{-n.} \times$$
$$e^{-\frac{1}{Ch}((|\nu-\alpha_1|-C)_+ + (|\alpha_1-\alpha_2|-C)_+ + .. + (|\alpha_{m-1}-\alpha_m|-C)_+ + |\mathrm{Im}\, z|(|\alpha_m-\mu|-C)_+)}. \tag{3.24}$$

Summing over α, we get (3.17).

Turning to the proof of (3.18), we write first for $\nu, \mu \in \Gamma \setminus \{0\}$:

$$[\psi_\nu (P_. - z_0)^{-m} (P_. - z)^{-1} \psi_\mu]_0^1 =$$
$$[\sum_{\alpha \in \Gamma^m \setminus (\Gamma\setminus\{0\})^m} \psi_\nu (P_. - z_0)^{-1} \psi_{\alpha_1} (P_. - z_0)^{-1} ..$$
$$\psi_{\alpha_{m-1}} (P_. - z_0)^{-1} \psi_{\alpha_m} (P_. - z)^{-1} \psi_\mu]_0^1$$
$$+ \sum_{\alpha \in (\Gamma\setminus\{0\})^m} [\psi_\nu (P_. - z_0)^{-1} \psi_{\alpha_1} .. \psi_{\alpha_{m-1}} (P_. - z_0)^{-1} \psi_{\alpha_m} (P_. - z)^{-1} \psi_\mu]_0^1. \tag{3.25}$$

In each term in one of the two sums constituting the first term of the RHS of (3.25) at least one of the components α_j is equal to 0, and the proof of (3.17) shows that the trace class norm of the first term of the RHS of (3.25) is bounded by

$$\mathcal{O}(1) \frac{\langle z\rangle}{|\mathrm{Im}\, z|} \langle z\rangle^{-\min(1,m+1-n_{\max}/2)} (1 + \delta_{m,n_{\max}/2} \log\langle z\rangle) \times$$
$$e^{-\frac{\min(|\mathrm{Im}\, z|,1)}{Ch} (\max(|\mu|,|\nu|)-C)_+} h^{-n_{\max}}. \tag{3.26}$$

The general term of the last sum in (3.25) is a sum of m expressions of the form

$$\psi_\nu (P_1 - z_0)^{-1} \psi_{\alpha_1} .. (P_1 - z_0)^{-1} [\psi_{\alpha_k} (P_. - z_0)^{-1} \psi_{\alpha_{k+1}}]_0^1$$
$$(P_0 - z_0)^{-1} \psi_{\alpha_{k+2}} .. \psi_{\alpha_m} (P_0 - z)^{-1} \psi_\mu, \tag{3.27}$$

for $k = 0, .., m-1$ with the convention that $\alpha_0 = \nu$, and a term

$$\psi_\nu (P_1 - z_0)^{-1} \psi_{\alpha_1} .. (P_1 - z_0)^{-1} [\psi_{\alpha_m} (P_. - z)^{-1} \psi_\mu]_0^1. \tag{3.28}$$

Combining (3.4) and (3.14), we can estimate the j:th characteristic value of the operator (3.27) by

$$\frac{\mathcal{O}(1)\langle z\rangle(\max(|\alpha_k|,|\alpha_{k+1}|)^{-\widetilde{n}}}{|\mathrm{Im}\, z|(1+h^2j^{2/n_{\max}})^m(\langle z\rangle+h^2j^{2/n_{\max}})}\times$$
$$e^{-\frac{1}{Ch}((|\nu-\alpha_1|-C)_+ +(|\alpha_1-\alpha_2|-C)_+ +..+(|\alpha_{m-1}-\alpha_m|-C)_+ +|\mathrm{Im}\, z|(|\alpha_m-\mu|-C)_+)}, \tag{3.29}$$

and of the operator (3.28) by

$$\frac{\mathcal{O}(1)\langle z\rangle^2(\max(|\alpha_m|,|\mu|)^{-\widetilde{n}}+e^{-\frac{|\mathrm{Im}\, z|}{Ch}(\max(|\alpha_m|,|\mu|)-C)_+})}{|\mathrm{Im}\, z|^2(1+h^2j^{2/n_{\max}})^m(\langle z\rangle+h^2j^{2/n_{\max}})}\times A, \tag{3.30}$$

where A denotes the same exponential factor as in (3.29).

As in the proof of (3.17) it follows that the trace class norm of (3.27) is

$$\mathcal{O}(1)\frac{\langle z\rangle\langle z\rangle}{|\mathrm{Im}\, z|}^{-\min(1,m+1-n_{\max}/2)}(1+\delta_{m,n_{\max}/2})\log\langle z\rangle)$$
$$\max(|\alpha_k|,|\alpha_{k+1}|)^{-\widetilde{n}}h^{-n_{\max}}\times A, \tag{3.31}$$

and that the trace class norm of (3.28) is

$$\mathcal{O}(1)\frac{\langle z\rangle^2}{|\mathrm{Im}\, z|^2}\langle z\rangle^{-\min(1,m+1-n_{\max}/2)}(1+\delta_{m,n_{\max}/2}\log\langle z\rangle)$$
$$(\max(|\alpha_m|,|\mu|)^{-\widetilde{n}}+e^{-\frac{|\mathrm{Im}\, z|}{Ch}(\max(|\alpha_m|,|\mu|)-C)_+})h^{-n_{\max}}\times A. \tag{3.32}$$

Possibly after modifying the constant C in the exponential factor A, we may replace $\max(|\alpha_k|,|\alpha_{k+1}|)$ in (3.31) by $|\nu|$. In (3.32), we may similarly replace $|\alpha_m|$ (at two places) by $|\nu|$ (noticing that the term $e^{-\frac{|\mathrm{Im}\, z|}{Ch}(\max(|\alpha_m|,|\mu|)-C)_+}$ can be ignored all together when $|\mathrm{Im}\, z|\geq 1$). It then follows that the trace class norm of the last term in (3.25) is

$$\mathcal{O}(1)\frac{\langle z\rangle^2}{|\mathrm{Im}\, z|^2}\langle z\rangle^{-\min(1,m+1-n_{\max}/2)}(1+\delta_{m,n_{\max}/2}\log\langle z\rangle)(|\nu|^{-\widetilde{n}}+e^{-\frac{|\mathrm{Im}\, z|}{Ch}(|\nu|-C)_+})\times$$
$$h^{-n_{\max}}e^{-\frac{1}{Ch}\min(1,|\mathrm{Im}\, z|)(|\nu-\mu|-C)_+}. \tag{3.33}$$

Combining this with (3.26), we get (3.18). □

Proof of Proposition 2.1. The first part of the proof will be to treat the case when $f\in C_0^\infty(\mathbf{R}^n)$. We shall use the operator version of the Cauchy-Green-Riemann-Stokes formula ([15]):

$$f(P.)=\frac{1}{\pi}\int\frac{\partial\widetilde{f}}{\partial\bar{z}}(P.-z)^{-1}L(dz), \tag{3.34}$$

where $L(dz)$ is the standard Lebesgue measure on $\mathbf{C} \cong \mathbf{R}^2$ and $\tilde{f} \in C_0^\infty(\mathbf{C})$ is an almost analytic extension of f, so that $\tilde{f}_{|\mathbf{R}} = f$, and $\frac{\partial \tilde{f}}{\partial \bar{z}}$ vanishes to infinite order on $\mathbf{R}$. Almost analytic extensions were introduced by Hörmander [16]. To get trace class operators under the sign of integration, we fix some $z_0 \in \mathbf{C} \setminus \mathbf{R}$ and write

$$f(z) = (z - z_0)^{-m} g(z), \; g \in C_0^\infty, \tag{3.35}$$

$$f(P.) = (P. - z_0)^{-m} g(P.) = \frac{1}{\pi} \int \frac{\partial \tilde{g}(z)}{\partial \bar{z}} (P. - z_0)^{-m} (P. - z)^{-1} L(dz). \tag{3.36}$$

Here we choose m as in Lemma 3.3.

Let $\chi \in C_0^\infty(\mathbf{R}^n)$ be equal to 1 near $\overline{B(0, R_0)}$. It follows from (3.17) that for $\operatorname{Im} z \neq 0$:

$$\begin{aligned} &\|\chi (P. - z_0)^{-m} (P. - z)^{-1} \chi\|_{\mathrm{tr}} \leq \\ &C h^{-n.} \langle z \rangle^{-\min(1, m+1-n./2)} (1 + \delta_{m, n./2} \log \langle z \rangle) \frac{\langle z \rangle}{|\operatorname{Im} z|}, \end{aligned} \tag{3.37}$$

$$\begin{aligned} &\|(1 - \chi)(P. - z_0)^{-m} (P. - z)^{-1} \chi\|_{\mathrm{tr}}, \; \|\chi (P. - z_0)^{-m} (P. - z)^{-1} (1 - \chi)\|_{\mathrm{tr}} \\ &\leq \; C h^{-n.} \langle z \rangle^{-\min(1, m+1-n./2)} (1 + \delta_{m, n./2} \log \langle z \rangle) \frac{\langle z \rangle}{|\operatorname{Im} z|} \\ &(1 + \frac{h}{\min(1, |\operatorname{Im} z|)})^n. \end{aligned} \tag{3.38}$$

From (3.18) we get (since $\tilde{n} > n$):

$$\begin{aligned} &\|[(1 - \chi)(P. - z_0)^{-m} (P. - z)^{-1} (1 - \chi)]_0^1\|_{\mathrm{tr}} \leq \\ &C h^{-n_{\max}} \frac{\langle z \rangle^2}{|\operatorname{Im} z|^2} \langle z \rangle^{-\min(1, m+1-n_{\max}/2)} (1 + \delta_{m, n_{\max}/2} \log \langle z \rangle) \\ &\qquad (1 + \frac{h}{\min(1, |\operatorname{Im} z|)})^{2n}. \end{aligned} \tag{3.39}$$

From these three estimates and (3.36), we get,

$$\begin{aligned} &\|\chi f(P.) \chi\|_{\mathrm{tr}}, \; \|(1 - \chi) f(P.) \chi\|_{\mathrm{tr}}, \; \|\chi f(P.)(1 - \chi)\|_{\mathrm{tr}}, \\ &\|[(1 - \chi) f(P.)(1 - \chi)]_0^1\|_{\mathrm{tr}} \leq \mathcal{O}(h^{-n_{\max}}), \end{aligned} \tag{3.40}$$

and hence also the corresponding estimates for the traces. The RHS of (2.9) is therefore $\mathcal{O}(h^{-n_{\max}})$ and it is straight forward to see that it is independent of the choice of χ.

We now turn to the general case and let $f \in S^{-m(f)}$ as in the proposition. Let $\chi \in C_0^\infty(\mathbf{R} \setminus \{0\})$ and consider for $\lambda \geq 1$:

$$f(P_.)\chi(\lambda^{-2}P_.) = f(\lambda^2(\lambda^{-2}P_.))\chi(\lambda^{-2}P_.).$$

The function $t \mapsto f(\lambda^2 t)\chi(t)$ has its support in a bounded λ-independent interval and

$$|\partial_t^k(f(\lambda^2 t)\chi(t))| \leq C_k \lambda^{-2m(f)}. \tag{3.41}$$

Consider the operator $\widetilde{P}_. = \lambda^{-2}P_.$. It is then straight forward to check that $\widetilde{P}_.$ satisfies the assumptions for $P_.$ in the proposition, provided that we replace h by $\widetilde{h} = h/\lambda$.

In view of (3.41), the uniform control of the support of the function $t \mapsto f(\lambda^2 t)\chi(t)$ and the fact that we already have established the proposition in the case of C_0^∞ functions, we get

$$\text{"tr}\,[f(P_.; h)\chi(\lambda^{-2}P_.)]_0^1\text{"} = \mathcal{O}(\lambda^{-2m(f)}\widetilde{h}^{-n_{\max}}) = \mathcal{O}(\lambda^{-2m(f)+n_{\max}})h^{-n_{\max}}.$$

Here $-2m(f) + n_{\max} < 0$ so we can decompose $f(P_.)$ into terms of the type $f(P_.)\chi(\lambda^{-2}P_.)$, with a sequence of λ's which grow exponentially, and the corresponding estimates above can then be summed and we obtain the proposition in the general case. □

4. Review of functional calculus for $P_.^\#$

In this section we essentially only use material from [34] in a straight forward way. Let M be a torus containing $B(0, R)$ for some $R > R_0$, and define $P_.^\# : \mathcal{H}_.^\# \to \mathcal{H}_.^\#$ as an unbounded operator with domain $\mathcal{D}_.^\#$ as in section 2 and in [34]. Then if $\chi \in C_0^\infty(M \setminus \overline{B(0, R_0)})$, we get as in Proposition 4.1 in [34], that χ is a uniformly bounded operator with respect to h: $\mathcal{D}_.^{\#,k} \to H^{2k}(M)$, $H^{2k}(M) \to \mathcal{D}_.^{\#,k}$, for every $k \in \mathbf{R}$, if we equip $H^{2k}(M)$ with the norm $\|\langle hD\rangle^{2k}u\|_{L^2}$ and let $\mathcal{D}_.^{\#,k}$ denote the domain of $\langle P_.\rangle^k$, for $k \geq 0$, and the dual of $\mathcal{D}_.^{\#,-k}$ for $k < 0$. An operator $A = A(z; h) : \mathcal{H}_.^\# \to \mathcal{H}_.^\#$, defined for $0 < h \leq h_0$ and for z in some subset of $\mathbf{C} \setminus \mathbf{R}$ will be called negligible if for every $N \in \mathbf{N}$, there exists $M > 0$, such that

$$A(z; h) = \mathcal{O}(h^N |\mathrm{Im}\, z|^{-M}) : \mathcal{D}_.^{\#,-N} \to \mathcal{D}_.^{\#,N}. \tag{4.1}$$

Lemma 4.1 *Let $\psi_1, \psi_2 \in C^\infty(M)$, be constant near $\overline{B(0, R_0)}$ and have disjoint supports. If we restrict z to a bounded subset of $\mathbf{C}\setminus\mathbf{R}$, then $\psi_1(P_.^\# - z)^{-1}\psi_2$ is negligible.*

Proof. We follow the proof of the corresponding statements (Propositions 5.1 and 6.1) in [34]: Let $\psi_2 \prec \psi_3 \prec \psi_4 \prec .. \prec \psi_N \prec 1 - \psi_1$ and consider the identity:

$$\psi_1(P^\#_\cdot - z)^{-1}\psi_2 = (-1)^N\psi_1(P^\#_\cdot - z)^{-1}[P^\#_\cdot, \psi_N](P^\#_\cdot - z)^{-1}..$$
$$(P^\#_\cdot - z)^{-1}[P^\#_\cdot, \psi_3](P^\#_\cdot - z)^{-1}\psi_2.$$

Notice that $(P^\#_\cdot - z)^{-1} = \mathcal{O}(|\mathrm{Im}\, z|^{-1}) : \mathcal{D}^{\#,k}_\cdot \to \mathcal{D}^{\#,k+1}_\cdot$ (since z varies in a bounded set), and that $[P^\#_\cdot, \psi_j] = \mathcal{O}(h) : \mathcal{D}^{\#,k} \to \mathcal{D}^{\#,k-\frac{1}{2}}$. Then the lemma follows. □

Lemma 4.2 *Let $Q_\cdot = \sum_{|\alpha|\le 2} b_{\cdot,\alpha}(x;h)(hD_x)^\alpha$, $b_{\cdot,\alpha} \in C^\infty(M)$ satisfy the same assumptions on M as $P^\#_\cdot$ outside $B(0,R_0)$. Also assume that $Q_\cdot$ is self-adjoint (necessarily with domain H^2). Let $\Omega \subset M\setminus\overline{B(0,R_0)}$ be an open set where $Q_\cdot$ and $P^\#_\cdot$ coincide. Then for every $\chi \in C_0^\infty(\Omega)$, the operator $\chi(Q_\cdot - z)^{-1}\chi - \chi(P^\#_\cdot - z)^{-1}\chi$ is negligible when z is restricted to some bounded set.*

Proof. Let $\chi \prec \psi \in C_0^\infty(\Omega)$. Write

$$(P^\#_\cdot - z)^{-1}\chi = \psi(Q_\cdot - z)^{-1}\chi - (P^\#_\cdot - z)^{-1}[P^\#_\cdot, \psi](P^\#_\cdot - z)^{-1}\chi,$$

so that

$$\chi(P^\#_\cdot - z)^{-1}\chi - \chi(Q_\cdot - z)^{-1}\chi = -\chi(P^\#_\cdot - z)^{-1}[P^\#_\cdot, \psi](P^\#_\cdot - z)^{-1}\chi.$$

Here the RHS is negligible since $[P^\#_\cdot, \psi](P^\#_\cdot - z)^{-1}\chi$ is, as we see from Lemma 4.1, if we notice that $[P^\#_\cdot, \psi] = \mathcal{O}(h) : \mathcal{D}^{\#,k} \to \mathcal{D}^{\#,k-\frac{1}{2}}$. □

Notice that if $f \in C_0^\infty(\mathbf{R})$ is independent of h or varies with h in a bounded subset of C_0^∞, then under the assumptions of Lemma 4.2, $\chi f(Q_\cdot)\chi - \chi f(P^\#_\cdot)\chi$ is negligible in the sense of operators independent of z: We say that $A = A(h)$ is negligible if $A : \mathcal{O}(h^N) : \mathcal{D}^{\#,-N} \to \mathcal{D}^{\#,N}$ for every $N \in \mathbf{N}$.

Further it follows from the results of Helffer-Robert, see [28] (and also [30], [5] for a presentation based on the operator Cauchy-Green-Riemann-Stokes formula,) that $f(Q_\cdot)$ is a an h-pseudodifferential operator (,from now on h-pseudor for short,) with leading symbol $f(q_0)$.

5. Review of complex scaling in the semi-classical case

Complex scaling or analytic distorsion is a standard technique in resonance theory since the work of Aguilar-Combes [1]. Among the numerous later works, we can mention the work of Hunziker [18]. Here we shall follow [34] since we also need large angle distorsion. More precisely, we give a quick review of section 3 of [34] with some minor modifications, due to the fact that our operators are slightly more general. We refer to [34] for more details.

A smooth submanifold $\Gamma \subset \mathbf{C}^n$ is said to be totally real if $T_x\Gamma \cap iT_x\Gamma = \{0\}$ for every $x \in \Gamma$, where $T_x\Gamma$ is viewed as a real linear subspace of $T_x\mathbf{C}^n \simeq \mathbf{C}^n$, and i denotes (multiplication by) the imaginary unit. We say that Γ is maximally totally real (m.t.r.) if Γ is totally real and of maximal (real) dimension n. The standard example of such a manifold is $\Gamma = \mathbf{R}^n$. Let $\Gamma \subset \mathbf{C}^n$ be a locally closed m.t.r. manifold and indentify $T^*\Gamma$ with a submanifold of $\mathbf{C}^n \times \mathbf{C}^n$, via the map $T^*\Gamma \ni (x, du(x)) \mapsto (x, \partial_x\widetilde{u}(x))$, where $\widetilde{u}$ is an almost analytic extension of the real valued smooth function u on Γ, and $\partial\widetilde{u}$ is the holomorphic part of the differential of $\widetilde{u}$, here identified with the corresponding n-vector of holomorphic partial derivatives. By almost analytic extension we mean a smooth extension such that $\overline{\partial}\widetilde{u}$ vanishes to infinte order on Γ. (Here $\overline{\partial}\widetilde{u}$ denotes the antiholomorphic part of the differential of $\widetilde{u}$, so that $d\widetilde{u} = \partial\widetilde{u} + \overline{\partial}\widetilde{u}$. In [34], we reviewed the existence and quasi uniqueness of almost analytic extensions of functions on m.t.r. manifolds, due to Hörmander [16] and Hörmander-Wermer [17].)

Let $\Omega \subset \mathbf{C}^n$ be an open neighborhood of Γ such that Γ is closed in Ω, and let

$$P(x, D_x) = \sum_{|\alpha| \leq m} a_\alpha(x) D^\alpha \tag{5.1}$$

be a differential operator on Ω with holomorphic coefficients. Define $P_\Gamma : C^\infty(\Gamma) \to C^\infty(\Gamma)$ by

$$P_\Gamma u = (P\widetilde{u})_{|\Gamma}, \tag{5.2}$$

where $\widetilde{u}$ is an almost analytic extension of u as above. P_Γ is a then a differential operator on Γ with smooth coefficients and for the principal symbols, we have the relation:

$$p_\Gamma = p_{|T^*\Gamma}. \tag{5.3}$$

It is well known, that if P_Γ is elliptic and $P_\Gamma u = v$, $u \in \mathcal{D}'(\Gamma)$, where v has a holomorphic extension to a neighborhood of Γ, then the same holds for u. Lemma 3.1 in [34] gives a deformation version of this and says roughly that if Γ_t, $t \in [0,1]$ is a smooth family of m.t.r. manifolds which are independent of t outside a compact in Ω, and with the property that P_{Γ_t} is elliptic, then if $P_{\Gamma_0} u = v$ and v has a holomorphic extension to a neighborhood of the union of all the Γ_t, then the same holds for u.

Let

$$P(x, hD_x; h) = \sum_{|\alpha| \leq m} a_\alpha(x; h)(hD_x)^\alpha, \tag{5.4}$$

where a_α are holomorphic on Ω and uniformly bounded with respect to h for x in any fixed compact subset of Ω. Then P_Γ has an analogous form with C^∞ coefficients, for every choice of local coordinates on Γ, and the coefficients are locally uniformly bounded. The semi-classical principal symbol is then defined modulo $\mathcal{O}(h\langle\xi\rangle^m)$ by

$$p(x,\xi;h) = \sum_{|\alpha|\leq m} a_\alpha(x;h)\xi^\alpha, \tag{5.5}$$

and similarly for P_Γ. The relation (5.3), also holds (modulo $\mathcal{O}(h\langle\xi\rangle^m)$) for the semiclassical symbols. If $a_\alpha(x;h) = a^0_\alpha(x) + \mathcal{O}(h)$, then we can make p and p_Γ h-independent by choosing $p(x,\xi) = \sum_{|\alpha|\leq m} a^0_\alpha(x)\xi^\alpha$ and similarly for P_Γ, and (5.3) holds without any remainder term.

For given $\epsilon_0 > 0$ and $R_1 > R_0$, we can construct a smooth function $[0,\pi]\times[0,\infty[\ni(\theta,t)\mapsto f_\theta(t)\in\mathbf{C}$, injective for every θ, with the following properties,

(i) $f_\theta(t) = t$ for $0\leq t\leq R_1$,
(ii) $0\leq \arg f_\theta(t)\leq\theta$, $\partial_t f_\theta\neq 0$,
(iii) $\arg f_\theta(t)\leq \arg\partial_t f_\theta(t)\leq \arg f_\theta(t)+\epsilon_0$,
(iv) $f_\theta(t) = e^{i\theta}t$ for $t\geq T_0$, where T_0 only depends on ϵ_0 and R_1.

For later use we shall give an explicit construction and derive a fifth property, which will be convenient though probably not essential. We look for $f(t) = f_{\theta,\epsilon_0}(t)$ of the form $f(t) = te^{i\Theta(s)}$, where $s=\log t$ and $\Theta = \Theta_{\theta,\epsilon_0}$ depends smoothly on θ, ϵ_0. Then $f'(t) = (1+i\Theta'(s))e^{i\Theta(s)}$, so it is enough to take Θ smooth in all variables with $0\leq\Theta'(s)\leq\epsilon_0$, $\Theta(s)=0$ for $s\leq\log R_1$, $\Theta(s)=\theta$ for s large enough. This is easy, but we make an explicit choice of Θ: Let $0\leq\phi\in C_0^\infty(]0,1[)$ with $\int\phi(s)ds = 1$, $\phi_\epsilon(s) = \epsilon^{-1}\phi(s/\epsilon)$. Let Θ be the solution which vanishes far to the left, of the equation,

$$\Theta'(s) = \epsilon_0(\phi_{\epsilon_0} * 1_{[0,\theta/\epsilon_0]})(s-\log R_1).$$

Then (i)-(iv) hold and moreover:

(v) $\arg f_\theta(t)$ is an increasing function of θ and of t and if $\theta_1\leq\theta_2$ and $f_{\theta_1}(t)\neq f_{\theta_2}(t)$, then $(\log t - \log R_1)\geq \theta_1/\epsilon_0$ and we have $\theta_1-\epsilon_0^2\leq\arg f_{\theta_1}(t)$.

Consider the map

$$\kappa_\theta : \mathbf{R}^n\ni x = t\omega\mapsto f_\theta(t)\omega\in\mathbf{C}^n,\ t=|x|.$$

The image is a m.t.r. manifold which coincides with $\mathbf{R}^n$ along $B(0,R_1)$. Let $\mathcal{H}_{\cdot,\theta} = \mathcal{H}_{\cdot,R_0}\oplus L^2(\Gamma_\theta\setminus B(0,R_0))$. If $\chi\in C_0^\infty(B(0,R_1))$ is equal to 1 near $\overline{B(0,R_0)}$, let $\mathcal{D}_{\cdot,\theta} = \{u\in\mathcal{H}_\theta;\chi u\in\mathcal{D}(P_\cdot),\ (1-\chi)u\in H^2(\Gamma_\theta)\}$, where $H^2(\Gamma_\theta)$ is equipped with the natural semi-classical norm. Let $P_{\cdot,\theta}$ be the unbounded operator $\mathcal{H}_{\cdot,\theta}\to\mathcal{H}_{\cdot,\theta}$ with domain $\mathcal{D}_{\cdot,\theta}$, defined by

$$P_{\cdot,\theta}u = P_\cdot(\chi u) + P_{\cdot,\Gamma}(1-\chi)u.$$

These definitions do not depend on the choice of (the h-independent) χ. Parametrizing Γ_θ by means of κ_θ, we get outside the origin:

$$-\Delta_{\Gamma_\theta} = (f'(t)^{-1}D_t)^2 - (f(t)f'(t))^{-1}(n-1)iD_t + (f(t))^{-2}D_\omega^2, \qquad (5.6)$$

where $-D_\omega^2$ is the Laplacian on $\mathbf{S}^{n-1}$. If ω^{*2} denotes the principal symbol of D_ω^2 and we let τ be the dual variable of t, then the principal symbol of $-\Delta_{\Gamma_\theta}$ is

$$p_{0,\theta} = (\tau/f'(t))^2 + (\omega^*/f(t))^2, \; f = f_\theta, \qquad (5.7)$$

so pointwise on Γ_θ, $-\Delta_{\Gamma_\theta}$ is elliptic and the principal symbol takes its values in an angle of size $\leq 2\epsilon_0$, while globally, $p_{0,\theta}$ takes its values in the sector,

$$-2(\theta + \epsilon_0) \leq \arg z \leq 0. \qquad (5.8)$$

In the following we shall always take $\theta \leq \theta_0 \, (< \pi)$ so when ϵ_0 is small enough, the angle $2(\theta + \epsilon_0)$ of the sector (5.8) is $< \pi$.

Choosing R_1 large enough, we get the following facts in view of the assumptions (2.6), (2.10):

In $\mathbf{R}^n \setminus B(0, R_0)$, $h^{-2}P_{\cdot,\theta}$ is an elliptic differential operator whose principal symbol (in the classical sense) over each fixed point in Γ_θ takes its values in an angle of size $\leq 3\epsilon_0$, and globally in a sector $-2\theta - 3\epsilon_0 \leq \arg z \leq \epsilon_0$. (5.9)

In $\mathbf{R}^n \setminus B(0, R_1)$, the difference between the semiclassical principal symbol of $P_{\cdot,\theta}$ and the principal symbol of $h^{-2}P_{\cdot,\theta}$ is $o(1)\langle\xi\rangle^2$, when $R_1 \to \infty$. (5.10)

The coefficients of $P_{\cdot,\theta} - e^{-2i\theta}(-h^2\Delta)$ and all their derivatives tend to zero uniformly with respect to h when $\Gamma_\theta \ni x \to \infty$, and we identify Γ_θ and $\mathbf{R}^n$, by means of κ_θ. (5.11)

Here we write the operators semiclassically as in (5.4)

Lemma 5.1 *If $z \in \mathbf{C} \setminus \{0\}$, $\arg z \neq -2\theta$, then $P_{\cdot,\theta} - z : \mathcal{D}_{\cdot,\theta} \to \mathcal{H}_{\cdot,\theta}$ is a Fredholm operator of index 0.*

This is essentially a consequence of a certain ellipticity near infinity and the proof is the same as the one of Lemma 3.2 in [34]. The only difference is that to the operators $K(z)$, $L(z)$ there, we have to add operators with arbitrarily small norm (depending in the choice of the partition of unity).

It follows from Lemma 5.1 that if $\arg z \neq -2\theta$, $z \neq 0$, then z belongs to the spectrum of $P_{\cdot,\theta}$ iff $\operatorname{Ker}(P_{\cdot,\theta} - z) \neq 0$.

Lemma 5.2 *Assume that* $0 \leq \theta_1 < \theta_2 \leq \theta_0$ *and let* $z_0 \in \mathbf{C} \backslash e^{-2i[\theta_1,\theta_2]}[0,\infty[$. *Then* $\dim \operatorname{Ker}(P_{\cdot,\theta_1} - z_0) = \dim \operatorname{Ker}(P_{\cdot,\theta_2} - z_0)$.

This is practically identical to Lemma 3.4 of [34] and the proof is the same as there, using (the extension to our present situation of) Lemma 3.1 of [34] evoked after (5.3).

The lemma above and analytic Fredholm theory (as developed for instance in the appendix of [14]) show that the spectrum of $P_{\cdot,\theta}$ in $\mathbf{C} \setminus e^{-2i\theta}[0,\infty[$ is discrete and in particular (when $\theta = 0$) that the spectrum of $P_{\cdot}$ in $]-\infty, 0[$ is discrete. If $\theta < \frac{\pi}{2}$, this discrete set consists of the negative eigenvalues of $P_{\cdot}$ plus a discrete set in the sector $e^{-2i[0,\theta[}]0,\infty[$. If $\theta \geq \frac{\pi}{2}$, then the spectrum of $P_{\cdot,\theta}$ in $\mathbf{C} \setminus e^{-2i\theta}[0,\infty[$ is contained in $e^{-2i[0,\theta[}]0,\infty[$. Lemma 5.2 tels us that the spectrum in $e^{-2i[0,\theta_0[}]0,\infty[$ is independent of θ in the following sense: We say that $z \in e^{-2i[0,\theta_0[}]0,\infty[$ is a resonance for $P_{\cdot}$ if and only if $z \in \sigma(P_{\cdot,\theta})$ for some (and hence for all) $\theta \in]0,\theta_0]$ with $\zeta \in e^{-2i[0,\theta[}]0,\infty[$.

By analytic Fredholm theory (see for instance the appendix of [14]), we know that if $z_0 \in e^{-2i[0,\theta[}]0,\infty[$ is a resonance, then the spectral projection

$$\pi_{\cdot,\theta,z_0} = \frac{1}{2\pi i}\int_\gamma (z - P_{\cdot,\theta})^{-1} dz, \tag{5.12}$$

with $\gamma : [0,2\pi] \ni s \mapsto z_0 + \epsilon e^{is}$, and $\epsilon > 0$ small enough, is of finite rank. The image $F_{\cdot,\theta,z_0}$ is contained in the domain of any power of $P_{\cdot,\theta}$ and is invariant under $P_{\cdot,\theta}$. Moreover the restriction of $P_{\cdot,\theta} - z_0$ to $F_{\cdot,\theta,z_0}$ is nilpotent, so $F_{\cdot,\theta,z_0} = \operatorname{Ker}(P_{\cdot,\theta} - z_0)^{k_0}$ for some $k_0 \in \mathbf{N}$. If $\widetilde{\theta} \in [0,\theta_0]$ is a second number with $z_0 \in e^{-2i[0,\widetilde{\theta}[}]0,\infty[$, then since Lemma 5.2 can be extended to "$\dim \operatorname{Ker}(P_{\cdot,\theta_1} - z_0)^k = \dim \operatorname{Ker}(P_{\cdot,\theta_2} - z_0)^k$ for all k", $\pi_{\cdot,\theta,z_0}$ and $\pi_{\cdot,\widetilde{\theta},z_0}$ have the same rank, which by definition is the multiplicity of the resonance z_0. Further, in Lemma 5.2 and in the above mentioned extension, we have invariance not only under changes of θ, but also under changes of the family f_θ.

One can also define the resonances as the poles in $e^{-2i[0,\theta_0[}]0,\infty[$ of the meromorphic continuation from the upper half plane across $]0,\infty[$ of $(P_{\cdot} - z)^{-1} : \mathcal{H}_{\cdot,\mathrm{comp}} \to \mathcal{D}_{\cdot,\mathrm{loc}}$. (See [34] for the definition of $\mathcal{H}_{\cdot,\mathrm{comp/loc}}$, $\mathcal{D}_{\cdot,\mathrm{comp/loc}}$.) We shall not use that point of view in the following, so we do not give details.

In the following result we use the special family f_θ, given after (i)-(iv) with $\epsilon_0 > 0$ small enough.

Proposition 5.3 *Let* $K \subset e^{i]0,\min(\pi,2\pi-2\theta_0)[}]0,\infty[$ *be compact, and let* $m \in \mathbf{N}$ *with* $(1+m)\frac{2}{n_{\max}} > 1$. *Then for* $0 \leq \theta \leq \theta_0$ *and* $z_0, z \in K$, *we can define* "$\operatorname{tr}[(P_{\cdot,\theta} - z_0)^{-m}(P_{\cdot,\theta} - z)^{-1}]_0^1$" *as in the case* $\theta = 0$ *and this quantity is independent of* θ.

Proof. Identifying Γ_θ with $\mathbf{R}^n$ by means of κ_θ, we can use the same cutoffs and partitions of unity on Γ_θ as on $\mathbf{R}^n$. We also choose $\epsilon_0 > 0$ in the construction of Γ_θ sufficiently small depending on K. Then $(P_{\cdot,\theta} - z)^{-1} : \mathcal{H}_{\cdot,\theta} \to \mathcal{D}_{\cdot,\theta}$ is uniformly bounded for $z \in K$, $h \in]0, h_0]$, $0 \le \theta \le \theta_0$ and we can go through the first part of the proof of Lemma 3.1 and replace $P_\cdot$ everywhere by $P_{\cdot,\theta}$. We get the same estimates and obtain with χ and ψ_ν as in Lemma 3.1:

$$\mu_j(\chi(P_{\cdot,\theta} - z)^{-1}\psi_\nu),\ \mu_j(\psi_\nu(P_{\cdot,\theta} - z)^{-1}\chi) \le \frac{C}{(1+h^2 j^{2/n_\cdot})} e^{-\frac{1}{Ch}(\mathrm{dist}\,(\mathrm{supp}\,\chi, \mathrm{supp}\,\psi_\nu)-1)_+}, \tag{5.13}$$

$$\mu_j(\psi_\nu(P_{\cdot,\theta} - z)^{-1}\psi_\mu) \le \frac{C}{(1+h^2 j^{2/n_\cdot})} e^{-\frac{1}{Ch}(|\nu-\mu|-C)_+}, \tag{5.14}$$

uniformly for $0 < h \le h_0$, $0 \le \theta \le \theta_0$, $\nu, \mu \in \Gamma$.

Similarly, we can replace P_0, P_1, by $P_{0,\theta}$, $P_{1,\theta}$ in the proof of Lemma 3.2 (avoiding the more delicate discussion there of the case when $\mathrm{Im}\, z \to 0$,) and obtain:

$$\mu_j([\psi_\nu(P_{\cdot,\theta} - z)^{-1}\psi_\mu]_0^1) \le \frac{C}{(1+h^2 j^{2/n_{\max}})} \max(|\nu|, |\mu|)^{-\widetilde{n}} e^{-\frac{1}{Ch}(|\nu-\mu|-C)_+}, \tag{5.15}$$

for $\nu, \mu \in \Gamma \setminus \{0\}$, $0 < h \le h_0$, $z \in K$, $0 \le \theta \le \theta_0$.

The proof of Lemma 3.3 gives:

$$\|\psi_\nu(P_{\cdot,\theta} - z_0)^{-m}(P_{\cdot,\theta} - z)^{-1}\psi_\mu\|_{\mathrm{tr}} \le Ch^{-n_\cdot} e^{-\frac{1}{Ch}(|\nu-\mu|-C)_+},\ \nu, \mu \in \Gamma, \tag{5.16}$$

$$\|[\psi_\nu(P_{\cdot,\theta} - z_0)^{-m}(P_{\cdot,\theta} - z)^{-1}\psi_\mu]_0^1\|_{\mathrm{tr}} \le Ch^{-n_{\max}} \max(|\nu|, |\mu|)^{-\widetilde{n}} e^{-\frac{1}{Ch}(|\nu-\mu|-C)_+},\ \nu, \mu \in \Gamma \setminus \{0\}, \tag{5.17}$$

with h, z, θ as in (5.15).

By applying some of the arguments (see in particular (3.36), (3.37), (3.38)) of the proof of Proposition 2.1, we see that "$\mathrm{tr}\,[(P_{\cdot,\theta} - z_0)^{-m}(P_{\cdot,\theta} - z)^{-1}]_0^1$" can be defined and is $\mathcal{O}(h^{-n_{\max}})$.

To prove the independence of θ of "$\mathrm{tr}\,[(P_{\cdot,\theta} - z_0)^{-m}(P_{\cdot,\theta} - z)^{-1}]_0^1$", it suffices to show that we get the same value for $\theta = \theta_1$ and for $\theta = \theta_2$ provided that $\theta_1, \theta_2 \in [0, \theta_0]$ and $|\theta_1 - \theta_2|$ is small enough. Let $\chi \in C_0^\infty([0, 1[; [0, 1])$ be equal to 1 on $[0, \frac{1}{2}]$, and consider for $R \ge 1$, the intermediate contours $\Gamma_{\theta_1,\theta_2,R} = \kappa_{\theta_1,\theta_2,R}(\mathbf{R}^n)$, where

$$\kappa_{\theta_1,\theta_2,R} : \mathbf{R}^n \ni x = t\omega \mapsto f_{\theta_1,\theta_2,R}(t)\omega \in \mathbf{C}^n,\ t = |x|, \tag{5.18}$$

$$f_{\theta_1,\theta_2,R}(t) = f_{\theta_1}(t) + \chi(\frac{t}{R})(f_{\theta_2}(t) - f_{\theta_1}(t)). \tag{5.19}$$

For $R \leq R_1$ (given in (i) after (5.5),) we have $f_{\theta_1,\theta_2,R} = f_{\theta_1}$, and $\Gamma_{\theta_1,\theta_2,R}$ converges to Γ_{θ_2} pointwise when R tends to ∞. We can define $P_{\cdot,\theta_1,\theta_2,R} = P_{\cdot|\Gamma_{\theta_1,\theta_2,R}}$ in the obvious way. We claim that

$$\text{"tr}\,[(P_{\cdot,\theta_1,\theta_2,R} - z_0)^{-m}(P_{\cdot,\theta_1,\theta_2,R} - z)^{-1}]_0^{1}\text{"} \to \text{"tr}\,[(P_{\cdot,\theta_2} - z_0)^{-m}(P_{\cdot,\theta_2} - z)^{-1}]_0^{1}\text{"} \tag{5.20}$$

when $R \to \infty$, provided that $|\theta_2 - \theta_1|$ is small enough. In fact, this follows from the following two statements, which are easily verified:

(A) (5.15)-(5.17) are uniformly valid if we replace $P_{\cdot,\theta}$ by $P_{\cdot,\theta_1,\theta_2,R}$ with $|\theta_1 - \theta_2|$ and $R \geq 1$.

(B) $\psi_\nu(P_{\cdot,\theta_1,\theta_2,R} - z_0)^{-m}(P_{\cdot,\theta_1,\theta_2,R} - z)^{-1}\psi_\mu \to \psi_\nu(P_{\cdot,\theta_2} - z_0)^{-m}(P_{\cdot,\theta_2} - z)^{-1}\psi_\mu$ in trace norm when $R \to \infty$.

It now only remains to prove:

$$\text{"tr}\,[(P_{\cdot,\theta_1,\theta_2,R} - z_0)^{-m}(P_{\cdot,\theta_1,\theta_2,R} - z)^{-1}]_0^{1}\text{" is independent of } R. \tag{5.21}$$

Let R vary in some compact interval I contained in $[R_1, +\infty[$. Then $f_{\theta_1,\theta_2,R}(t)$ is independent of R except for t in some compact interval $J \subset]R_0, +\infty[$ and on this interval we have

$$\arg f_{\theta_1,\theta_2,R}(t),\ \arg f'_{\theta_1,\theta_2,R}(t) = \theta_1 + \mathcal{O}(\epsilon_0) + \mathcal{O}(|\theta_1 - \theta_2|).$$

By abuse of notation, we write $P_{\cdot,R}$ for $P_{\cdot,\theta_1,\theta_2,R}$ and similarly for the contours. Let $\chi = \chi(t) \in C_0^\infty(\Gamma_R)$ have support disjoint from $\overline{B(0, R_0)}$ and be equal to 1 near $t \in J$. (Strictly speaking, χ will depend on R, since we let χ live on Γ_R, but we arranged so that χ is identically equal to one on the part of Γ_R which varies when R varies in I.)

The operators $\chi_1(z - P_{\cdot,R})^{-1}\chi_2$ are independent of R, if $\chi_1, \chi_2 \in C_0^\infty(\Gamma_R)$ are constant on $\overline{B(0, R_0)}$ and supp (χ_j) are disjoint from the set where $\chi = 1$. In fact, this follows from the same principle of non-characteristic deformation as the one which is behind Lemma 5.2. It is now clear that (5.21) will follow from

$$\text{tr}\,(\chi \circ (P_{\cdot,R} - z_0)^{-m}(P_{\cdot,R} - z)^{-1} \circ \chi) \text{ is independent of } R \in I. \tag{5.22}$$

Let $\widetilde{\chi} \in C_0^\infty(\Gamma_R)$ be $= 1$ near supp χ. Put $G_\epsilon(x) = C_n \epsilon^{-n/2} e^{-\frac{1}{\epsilon}(e^{-i\theta}x)^2}$, $\theta = \theta_1$ with C_n chosen so that

$$\int_{\mathbf{R}^n} e^{i\theta} G_\epsilon(x)\,dx = 1.$$

Denote by $G_\epsilon *$ the operator with kernel $G_\epsilon(x-y)dy$ on any one of Γ_R. $\widetilde{\chi}(G_\epsilon *)\chi$, $\chi(G_\epsilon *)\widetilde{\chi}$ tend strongly to (multiplication by) χ in the space of L^2 bounded operators over each of these contours, so

$$\begin{aligned}\chi(P_{\cdot,R}-z_0)^{-m}(P_{\cdot,R}-z)^{-1}\chi = \\ \lim_{\epsilon\to 0}\widetilde{\chi}(G_\epsilon *)\chi(P_{\cdot,R}-z_0)^{-m}(P_{\cdot,R}-z)^{-1}\chi(G_\epsilon *)\widetilde{\chi}\end{aligned} \tag{5.23}$$

in the space of trace class operators.

For $u \in C_0^\infty(\Gamma_R)$:

$$\begin{aligned}&\widetilde{\chi}(G_\epsilon *)\chi(P_{\cdot,R}-z_0)^{-m}(P_{\cdot,R}-z)^{-1}\chi(G_\epsilon *)\widetilde{\chi}u(x) \\ &= \textstyle\int_{\Gamma_R}\widetilde{\chi}(x)K_\epsilon(x,y)\widetilde{\chi}(y)u(y)dy,\ x\in\Gamma_R,\end{aligned} \tag{5.24}$$

where $K_\epsilon(x,y)$ is an entire function which is independent of R. The trace of the operator (5.24) is therefore

$$\int_{\Gamma_R}\widetilde{\chi}(x)^2 K_\epsilon(x,x)dx, \tag{5.25}$$

which is independent of R. From this and (5.23), we get (5.22) and the proof is complete. □

Simpler proofs are likely to exist. For instance, one might try to prove that the θ-derivative of the "trace" in the proposition is the trace of a commutator, hence 0.

6. Grushin problem for the scaled operator

In the following, we write θ instead of θ_0. Let F be a smooth mapping from a neighborhood of $e^{-i[-\epsilon_0,2(\theta+\epsilon_0)]}[0,\infty[$ into itself, such that

$$F(z)=z \text{ for } |z| \text{ sufficiently large}, \tag{6.1}$$

$$F(z)=z \text{ for } z \text{ in a neighborhood of } e^{-2i\theta}[0,\infty[, \tag{6.2}$$

$$\overline{\Omega} \text{ is disjoint from the range of } F. \tag{6.3}$$

Here Ω is the same set as in Theorem 2.2

Let $f = F_{|\mathbf{R}}$. Choosing Γ_θ conveniently as in section 5, let $p_{\cdot,\theta}$ be the semi-classical principal symbol of $P_{\cdot,\theta}$, defined on $T^*(\Gamma_\theta\setminus B(0,R_0))$. Then $F\circ p_{\cdot,\theta}$ is a well-defined smooth function with values away from $\overline{\Omega}$ such that $(F\circ p_{\cdot,\theta})(x,\xi)=p_{\cdot,\theta}(x,\xi)$, when $|x|\geq R_2$, if $R_2>R_0$ is large enough. (Here we identify Γ_θ with $\mathbf{R}^n$ by means of κ_θ.)

Using functional calculus, we shall first construct a finite rank perturbation of P_θ for which $\overline{\Omega}$ belongs to the resolvent set, when $h>0$ is small enough. Then using this perturbation, we construct a Grushin problem for $P_{\cdot,\theta} - z$ which is well posed for z near $\overline{\Omega}$.

Let $R_1 \in]R_0, R_2[$ have the property that Γ_θ coincides with $\mathbf{R}^n$ near $B(0, R_1)$. Let $P^\#_\cdot$ be the same operator as in the sections 3, 4, realized on a torus which contains $B(0, R_1)$. Then

$$f(P^\#_\cdot) = P^\#_\cdot + K^\#_\cdot, \tag{6.4}$$

where $K^\#_\cdot$ is a uniformly bounded operator (w.r.t. h) with

$$\operatorname{rank} K^\#_\cdot = \mathcal{O}(h^{-n_\cdot}). \tag{6.5}$$

Moreover, $\overline{\Omega}$ belongs to the resolvent set of $f(P^\#_\cdot)$ and

$$(z - f(P^\#_\cdot))^{-1} = \mathcal{O}(1) \text{ for } z \text{ in an } h\text{-independent neighborhood of } \overline{\Omega}. \tag{6.6}$$

Let $Q_\cdot$ be an h–differential operator on the same torus M where $P^\#_\cdot$ is defined, which is elliptic self-adjoint and which coincides with $P^\#_\cdot$ outside $\overline{B(0, R_0)}$. If $\chi_1, \chi_2 \in C^\infty(M)$ are constant on $\overline{B(0, R_0)}$ and have disjoint supports, then in [34] we showed that $\chi_1(z - P^\#_\cdot)^{-1}\chi_2$ is negligible for z in a fixed compact set, in the sense that for every $N \in \mathbf{N}$, there exists $M = M(N)$ such that

$$\|\chi_1(z - P^\#_\cdot)^{-1}\chi_2\|_{\mathcal{L}(\mathcal{H}^\#_\cdot, \mathcal{D}(P^\#_\cdot))} = \mathcal{O}_N(h^N|\operatorname{Im} z|^{-M}). \tag{6.7}$$

The same holds for $Q_\cdot$ with $\mathcal{D}(P^\#_\cdot)$ replaced by $H^2(M)$. As in [34] we also obtain that $\chi(z - P^\#_\cdot)^{-1}\chi - \chi(z - Q_\cdot)^{-1}\chi$ is negligible, if $\chi \in C_0^\infty(M \setminus \overline{B(0, R_0)})$. Using some standard integral formula, we obtain that $\chi f(P^\#_\cdot)\chi - \chi f(Q^\#_\cdot)\chi$ is negligible, i.e. of norm $\mathcal{O}_N(h^N)$ in $\mathcal{L}(L^2, H^2)$ for every N.

Now according to the functional calculus for h-pseudors, due to Helffer-Robert (see [28] and also [31], [5]), we know that $\chi f(Q^\#_\cdot)\chi$ is an h-pseudor on M of the natural class with leading symbol $\chi(x)^2 f(p^\#_\cdot(x, \xi))$, where $p^\#_\cdot$ is the (semi-classical) leading symbol of $P^\#_\cdot$.

Let $1 = \chi_0 + \chi_1 + \chi_2$, where $\chi_j \geq 0$ are smooth, χ_0 is equal to 1 on $\overline{B(0, R_0)}$ and has its support close to that set, $\chi_1 \in C_0^\#$ has support disjoint from $\overline{B(0, R_0)}$ and $\chi_0 + \chi_1 = 1$ near $\overline{B(0, R_2)}$, so supp χ_2 is disjoint from $\overline{B(0, R_2)}$ and $\chi_2 = 1$ near infinity. Let $\chi_j \prec \widetilde{\chi}_j$, where supp $\widetilde{\chi}_j$ is close to supp χ_j and consider

$$\widetilde{P}_{\cdot,\theta} = \widetilde{\chi}_0 f(P^\#_\cdot)\chi_0 + \widetilde{\chi}_1 R_{\cdot,F}\chi_1 + \widetilde{\chi}_2 P_{\cdot,\theta}\chi_2, \tag{6.8}$$

where $R_{\cdot,F}$ is an h-pseudor with leading symbol $F(p_{\cdot,\theta})$ and such that the total symbol of $R_{\cdot,F} - P_{\cdot,\theta} = S_{\cdot,F}$ has compact support in ξ.

It is easy to see that

$$(z - \widetilde{P}_{\cdot,\theta})^{-1}\text{is well defined and}$$
$$\mathcal{O}(1) : \mathcal{H}_{\cdot,\theta} \to \mathcal{D}_{\cdot,\theta} \text{ for } z \in \Omega, \tag{6.9}$$

Write,

$$\widetilde{P}_{\cdot,\theta} = P_{\cdot,\theta} + \widetilde{\chi}_0(f(P_{\cdot}^{\#}) - P_{\cdot}^{\#})\chi_0 + \widetilde{\chi}_1 S_{\cdot,F}\chi_1.$$

In view of the properties of $S_{\cdot,F}$ we can find $T_{\cdot,F}$ of finite rank $\mathcal{O}(h^{-n})$ such that $\widetilde{\chi}_1(S_{\cdot,F} - T_{\cdot,F})\chi_1 = \mathcal{O}(h^{\infty})$. Put

$$\widehat{P}_{\cdot,\theta} = P_{\cdot,\theta} + \widetilde{\chi}_0(f(P_{\cdot}^{\#}) - P_{\cdot}^{\#})\chi_0 + \widetilde{\chi}_1 T_{\cdot,F}\chi_1. \tag{6.10}$$

Then

$$(z - \widehat{P}_{\cdot,\theta})^{-1}\text{is well defined and}$$
$$\mathcal{O}(1) : \mathcal{H}_{\cdot,\theta} \to \mathcal{D}_{\cdot,\theta} \text{ for } z \in \Omega, \tag{6.11}$$

and

$$\widehat{P}_{\cdot,\theta} = P_{\cdot,\theta} + K_{\cdot}, \text{ where}$$
$$K_{\cdot} = \mathcal{O}(1) \text{ in } \mathcal{L}(\mathcal{H}_{\cdot,\theta}, \mathcal{H}_{\cdot,\theta}) \text{ and } \operatorname{rank} K_{\cdot} = \mathcal{O}(h^{-n_{\cdot}}). \tag{6.12}$$

In the following we identify Γ_θ with $\mathbf{R}^n$ by means of κ_θ, and consequently we shall not always write the subscript θ for the spaces $\mathcal{H}_{\cdot}$ and $\mathcal{D}_{\cdot}$. On $\mathcal{D}_{\cdot}$, we use the scalar product,

$$(u|v)_{\mathcal{D}_{\cdot}} = ((P_{\cdot}^2 + 1)u|v)_{\mathcal{H}_{\cdot}}, \tag{6.13}$$

where $P_{\cdot}$ is the original (unscaled) operator. Let $e_{\cdot,1}, e_{\cdot,2}, ..$ be an O.N. basis in $\mathcal{D}_{\cdot}$ such that

$$e_{\cdot,1}, .., e_{\cdot,N_{\cdot}} \text{ span Im } K_{\cdot}^{\otimes},\ N_{\cdot} = \mathcal{O}(h^{-n_{\cdot}}), \tag{6.14}$$

where $K_{\cdot}^{\otimes} : \mathcal{H}_{\cdot} \to \mathcal{D}_{\cdot}$ is the adjoint of $K_{\cdot} : \mathcal{D}_{\cdot} \to \mathcal{H}_{\cdot}$. Notice that

$$K_{\cdot}^{\otimes} = \langle P_{\cdot}\rangle^{-2} K_{\cdot}^{*}, \tag{6.15}$$

where $K_{\cdot}^{*}$ is the adjoint of $K_{\cdot} : \mathcal{H}_{\cdot} \to \mathcal{H}_{\cdot}$. It follows from (6.14) that $e_{\cdot,N_{\cdot}+1}, ... \in (\text{Im } K_{\cdot}^{\otimes})^{\perp} = \operatorname{Ker} K_{\cdot}$.

Put

$$R_{\cdot,+}u(j) = (u|e_{\cdot,j})_{\mathcal{D}_{\cdot}},\ 1 \leq j \leq N_{\cdot}, \tag{6.16}$$

$$R_{\cdot,-}(z)u_- = \sum_1^{N.} u_-(j) f_{\cdot,j},\ f_{\cdot,j} = (P_{\cdot,\theta} + K_\cdot - z)e_{\cdot,j},\ z \in \Omega, \tag{6.17}$$

so that $R_{\cdot,+} : \mathcal{D}_\cdot \to \mathbf{C}^{N_\cdot}$, $R_{\cdot,-} : \mathbf{C}^{N_\cdot} \to \mathcal{H}_\cdot$. Consider the Grushin problem:

$$\begin{cases} (P_{\cdot,\theta} - z)u + R_{\cdot,-}u_- = v,\ u \in \mathcal{D}_\cdot,\ u_- \in \mathbf{C}^{N_\cdot} \\ R_{\cdot,+}u = v_+,\ v \in \mathcal{H}_\cdot,\ v_+ \in \mathbf{C}^{N_\cdot}. \end{cases} \tag{6.18}$$

This problem is of index 0, so in order to show that it is well posed, it suffices to show the injectivity. Suppose that

$$\begin{cases} (P_{\cdot,\theta} - z)u + R_{\cdot,-}u_- = 0, \\ R_{\cdot,+}u = 0. \end{cases} \tag{6.19}$$

If $u = \sum_1^\infty u_j e_{\cdot,j}$, we first get $u_1 = .. = u_N = 0$, and the first equation in (6.19) becomes:

$$\sum_{N_\cdot+1}^{\infty} u_j (P_{\cdot,\theta} - z)e_{\cdot,j} + \sum_1^{N_\cdot} u_-(j)(P_{\cdot,\theta} + K_\cdot - z)e_{\cdot,j} = 0. \tag{6.20}$$

Since $K_{\cdot,j} e_{\cdot,j} = 0$ for $j = 1, .., N_\cdot$, we can write this (dropping temporarily the subscript "$\cdot$"):

$$(P_\theta + K - z)(\sum_1^N u_-(j)e_j + \sum_{N+1}^\infty u_j e_j) = 0 \tag{6.21}$$

and the bijectivity of $P_\theta + K - z$ implies $u_-(j) = 0$, $u_k = 0$, $k \geq N+1$, and we have shown the injectivity of (6.19) and the wellposedness of (6.18).

We also need a priori estimates for (6.18): Writing $u = u' + \sum_1^N v_+(j)e_j$, $u' = \sum_{N+1}^\infty e_j$, the first equation in (6.18) becomes:

$$(P_\theta - z)u' + R_- u_- = v - \sum_1^N v_+(j)(P_\theta - z)e_j,$$

and as in the proof of the injectivity, we get:

$$(P_\theta + K - z)(\sum_1^N u_-(j)e_j + \sum_{N+1}^\infty u_j e_j) = v - (P_\theta - z)(\sum_1^N v_+(j)e_j),$$

which gives

$$\|u\|_{\mathcal{D}} + \|u_-\|_{\mathbf{C}^n} \leq C(\|v\|_{\mathcal{H}} + \|v_+\|_{\mathbf{C}^N}). \tag{6.22}$$

$e_1, .., e_N$ do not necessarily have compact support, but (6.15) shows that $\langle P_\cdot\rangle^2 e_j$ have support in some fixed compact subset of $B(0, R_2)$ and by a Combes-Thomas argument, we infer that

$$\|e_j\|_{H^2(\Gamma_\theta \setminus B(0,R_2))} = \mathcal{O}(e^{-\frac{1}{Ch}}),\ 1 \le j \le N.$$

Let $\tilde{\chi} \in C_0^\infty(\Gamma_\theta)$ be equal to 1 near $\overline{B(0, R_2)}$. We still have a well-posed Grushin problem, if we replace e_j by $\tilde{\chi} e_j$ and f_j by $(P_\theta + K - z)(\tilde{\chi} e_j)$, and the preceding estimates remain valid. In the following, we shall refer to this modified problem. After increasing R_2 slightly, we may then assume that (for the new modified quantities):

$$\operatorname{supp} e_j,\ \operatorname{supp} f_j \subset B(0, R_2),\ 1 \le j \le N. \tag{6.23}$$

7. Trace class estimates for the inverse of the Grushin problem

For $z \in \Omega$, let

$$\mathcal{E}_\cdot(z) = \begin{pmatrix} E_\cdot(z) & E_{\cdot,+}(z) \\ E_{\cdot,-}(z) & E_{\cdot,-+}(z) \end{pmatrix} \tag{7.1}$$

denote the inverse of

$$\mathcal{P}_\cdot(z) = \begin{pmatrix} P_{\cdot,\theta} - z & R_{\cdot,-} \\ R_{\cdot,+} & 0 \end{pmatrix}. \tag{7.2}$$

Here $\theta = \theta_0$ and we identify $\mathbf{R}^n$ with Γ_θ by means of κ_θ. Our estimates will be essentially the same as those of section 3 for $(P_\cdot - w)^{-1}$, when w is in a compact set disjoint from $\mathbf{R}$.

Let M be a large torus containing $B(0, R_2)$ and define $P^\# = P_\cdot^\#$ as in the preceding section, so that $P^\# = P$ in $B(0, R_2)$. Since $e_1, .., e_N, f_1, .., f_N$ have their support in $B(0, R_2)$, according to the last modifications in section 6, we can define $R_+^\# : \mathcal{D}^\# \to \mathbf{C}^N$, $R_-^\# : \mathbf{C}^N \to \mathcal{H}^\#$ as before and get

$$\mathcal{P}^\#(z) = \begin{pmatrix} P^\# - z & R_-^\# \\ R_+^\# & 0 \end{pmatrix} : \mathcal{D}^\# \times \mathbf{C}^N \to \mathcal{H}^\# \times \mathbf{C}^N,$$

with a uniformly bounded inverse

$$\mathcal{E}^\# = \begin{pmatrix} E^\# & E_+^\# \\ E_-^\# & E_{-+}^\# \end{pmatrix}.$$

(We sometime drop the subscript $\cdot$.)

Let $w \in \Omega$ with $\operatorname{Im} w \ge \frac{1}{\text{Const.}}$. Then

$$E^\#(w) = (P^\# - w)^{-1} - (P^\# - w)^{-1} R_-^\#(w) E_-^\#(w),$$

where the last term is uniformly bounded and of rank $\mathcal{O}(h^{-n.})$. Then

$$\mathcal{E}^{\#}(w) = \begin{pmatrix} (P^{\#} - w)^{-1} & 0 \\ 0 & 0 \end{pmatrix} + R$$

where R is uniformly bounded of rank $\mathcal{O}(h^{-n.})$. It follows that,

$$\mu_j(\mathcal{E}^{\#}(w)) \leq C(1 + h^2 j^{2/n.})^{-1}.$$

If $z \in \Omega$, we use the relation

$$\mathcal{E}^{\#}(z) - \mathcal{E}^{\#}(w) = -\mathcal{E}^{\#}(z)(\mathcal{P}^{\#}(z) - \mathcal{P}^{\#}(w))\mathcal{E}^{\#}(w),$$

which together with the previous estimate gives:

$$\mu_j(\mathcal{E}^{\#}(z)) \leq C(1 + h^2 j^{2/n.})^{-1}, \quad z \in \Omega. \tag{7.3}$$

Let $\chi \in C_0^\infty(\mathbf{R}^n)$ be $= 1$ near $\overline{B(0, R_2)}$ and choose $\tilde{\chi} \in C_0^\infty(\mathbf{R}^n)$ with $\chi \prec \tilde{\chi}$, and assume that we have taken M large enough, so that $\tilde{\chi}$ also lives on M and $P^{\#} = P_\theta$ on $\operatorname{supp} \tilde{\chi}$. Then for $z \in \Omega$:

$$\begin{aligned}\mathcal{E}(z)\begin{pmatrix} \chi & 0 \\ 0 & I \end{pmatrix} &= \begin{pmatrix} \tilde{\chi} & 0 \\ 0 & I \end{pmatrix}\mathcal{E}^{\#}(z)\begin{pmatrix} \chi & 0 \\ 0 & I \end{pmatrix} \\ &-\mathcal{E}(z)\begin{pmatrix} [P_\theta, \tilde{\chi}] & 0 \\ 0 & 0 \end{pmatrix}\mathcal{E}^{\#}(z)\begin{pmatrix} \chi & 0 \\ 0 & I \end{pmatrix},\end{aligned} \tag{7.4}$$

since $R_+\tilde{\chi} = R_+$, $\chi R_- = R_-$, so that,

$$\left[\mathcal{P}(z), \begin{pmatrix} \tilde{\chi} & 0 \\ 0 & I \end{pmatrix}\right] = \begin{pmatrix} [P_\theta, \tilde{\chi}] & 0 \\ 0 & 0 \end{pmatrix}. \tag{7.5}$$

We also used that

$$\mathcal{P}^{\#}(z)\begin{pmatrix} \tilde{\chi} & 0 \\ 0 & I \end{pmatrix} = \mathcal{P}(z)\begin{pmatrix} \tilde{\chi} & 0 \\ 0 & I \end{pmatrix},$$

$$\begin{pmatrix} \tilde{\chi} & 0 \\ 0 & I \end{pmatrix}\mathcal{P}(z) = \begin{pmatrix} \tilde{\chi} & 0 \\ 0 & I \end{pmatrix}\mathcal{P}^{\#}(z)$$

in the obvious sense.

Similarly,

$$\begin{aligned}\begin{pmatrix} \chi & 0 \\ 0 & I \end{pmatrix}\mathcal{E}(z) &= \begin{pmatrix} \chi & 0 \\ 0 & I \end{pmatrix}\mathcal{E}^{\#}(z)\begin{pmatrix} \tilde{\chi} & 0 \\ 0 & I \end{pmatrix} + \\ &\begin{pmatrix} \chi & 0 \\ 0 & I \end{pmatrix}\mathcal{E}^{\#}(z)\begin{pmatrix} [P_\theta, \tilde{\chi}] & 0 \\ 0 & 0 \end{pmatrix}\mathcal{E}(z).\end{aligned} \tag{7.6}$$

From this and (7.3), we obtain

$$\mu_j(\mathcal{E}(z)\begin{pmatrix}\chi & 0\\ 0 & I\end{pmatrix}), \mu_j(\begin{pmatrix}\chi & 0\\ 0 & I\end{pmatrix}\mathcal{E}(z)) \leq C(1+h^2 j^{2/n_\cdot})^{-1}, \; z\in\Omega. \quad (7.7)$$

Actually, when using (7.4) to estimate one half of the characteristic values above, we need to estimate $\mathcal{E}(z)h\mathcal{A}$, where $\mathcal{A} = \begin{pmatrix}A & 0\\ 0 & 0\end{pmatrix}$, $A = h^{-1}[P_\theta, \widetilde{\chi}]$. We have $A = \sum_{|\alpha|\leq 1} a_\alpha(x;h)(hD)^\alpha$, with $a_\alpha(\cdot;h)$ bounded in C_b^∞ and $\operatorname{supp} a_\alpha(\cdot;h)$ bounded away from $\overline{B(0,R_2)}$. Then $[\mathcal{P},\mathcal{A}] = \mathcal{O}(h) : \mathcal{D}\to\mathcal{H}$ and hence, $[\mathcal{E},\mathcal{A}] = -\mathcal{E}[\mathcal{P},\mathcal{A}]\mathcal{E} = \mathcal{O}(h) : \mathcal{H}\to\mathcal{D}$. (For the special $\mathcal{A}$ appearing in (7.4) we will also be able to gain exponential decay at large distances.) It follows that $\mathcal{E}\mathcal{A} = \mathcal{A}\mathcal{E} + [\mathcal{E},\mathcal{A}] = \mathcal{O}(1) : \mathcal{H}\to\mathcal{D}^{1/2}$.

Next we derive exponentially weighted estimates for $E(z)$ in the usual way. Let $\phi\in C^\infty(\mathbf{R}^n)$ with $\phi =$ Const. near $\overline{B(0,R_2)}$ and $\nabla\phi, \nabla^2\phi = \mathcal{O}(\epsilon)$ in sup norm, and with $\nabla\phi\in C_b^\infty$. Then in $\mathcal{L}(\mathcal{D}\times\mathbf{C}^N; \mathcal{H}\times\mathbf{C}^N)$:

$$\begin{pmatrix}e^{-\phi/h} & 0\\ 0 & e^{-\phi(0)/h}\end{pmatrix}\begin{pmatrix}P_\theta - z & R_-\\ R_+ & 0\end{pmatrix}\begin{pmatrix}e^{\phi/h} & 0\\ 0 & e^{\phi(0)/h}\end{pmatrix} = \begin{pmatrix}P_\theta - z + \mathcal{O}(\epsilon) & R_-\\ R_+ & 0\end{pmatrix}$$

has a uniformly bounded inverse if $\epsilon > 0$ is sufficiently small. Approximating ϕ with functions that are constant near infinity and passing to the limit, we see that:

$$\begin{pmatrix}e^{-\phi/h} & 0\\ 0 & e^{-\phi(0)/h}\end{pmatrix}\mathcal{E}(z)\begin{pmatrix}e^{\phi/h} & 0\\ 0 & e^{\phi(0)/h}\end{pmatrix} = \mathcal{O}(1) : \mathcal{H}\times\mathbf{C}^N \to \mathcal{D}\times\mathbf{C}^N, \quad (7.8)$$

or more explicitly:

$$\begin{aligned} e^{-\phi/h}E(z)e^{\phi/h} &= \mathcal{O}(1) \text{ in } \mathcal{L}(\mathcal{H};\mathcal{D}),\\ e^{-\phi/h}E_+(z)e^{\phi(0)/h} &= \mathcal{O}(1) \text{ in } \mathcal{L}(\mathbf{C}^N;\mathcal{D}),\\ e^{-\phi(0)/h}E_-(z)e^{\phi/h} &= \mathcal{O}(1) \text{ in } \mathcal{L}(\mathcal{H};\mathbf{C}^N). \end{aligned} \quad (7.9)$$

Let χ_1 belong to a bounded set in $C_b^\infty(\mathbf{R}^n)$, such that dist $(\operatorname{supp}\chi_1, \operatorname{supp}\chi) \geq 1/\text{Const.}$. Combining (7.3), (7.4), (7.6), (7.8) with ϕ properly chosen, we get

$$\mu_j(\begin{pmatrix}\chi_1 & 0\\ 0 & 0\end{pmatrix}\mathcal{E}(z)\begin{pmatrix}\chi & 0\\ 0 & I\end{pmatrix}), \mu_j(\begin{pmatrix}\chi & 0\\ 0 & I\end{pmatrix}\mathcal{E}(z)\begin{pmatrix}\chi_1 & 0\\ 0 & 0\end{pmatrix}) \leq C(1+h^2 j^{2/n_\cdot})^{-1} e^{-\frac{1}{Ch}\operatorname{dist}(\operatorname{supp}\chi_1, \operatorname{supp}\chi)}. \quad (7.10)$$

Let now χ belong to some subset of C_0^∞ with uniform bounds on $\partial^\alpha\chi$ for every α, on the diameter of the support of χ and with dist $(\operatorname{supp}\chi, \overline{B(0,R_2)}) \geq 1/\text{Const.}$. Then near $\operatorname{supp}\chi$ we approximate P_θ by a (new) operator $P^\#$ defined on a torus M et c., so that $(P^\# - z)^{-1}$ is $\mathcal{O}(1)$ for $z \in \Omega$. If $\chi \prec \widetilde{\chi} \in C_0^\infty$, where $\widetilde{\chi}$ has only slightly larger support, then

$$\mathcal{E}(z)\begin{pmatrix} \chi & 0 \\ 0 & 0 \end{pmatrix} = \begin{pmatrix} \widetilde{\chi} & 0 \\ 0 & 0 \end{pmatrix}\begin{pmatrix} (P^\# - z)^{-1} & 0 \\ 0 & 0 \end{pmatrix}\begin{pmatrix} \chi & 0 \\ 0 & 0 \end{pmatrix} - \mathcal{E}(z)\begin{pmatrix} [P_\theta, \widetilde{\chi}] & 0 \\ 0 & 0 \end{pmatrix}\begin{pmatrix} (P^\# - z)^{-1} & 0 \\ 0 & 0 \end{pmatrix}\begin{pmatrix} \chi & 0 \\ 0 & 0 \end{pmatrix}, \tag{7.11}$$

$$\begin{pmatrix} \chi & 0 \\ 0 & 0 \end{pmatrix}\mathcal{E}(z) = \begin{pmatrix} \chi & 0 \\ 0 & 0 \end{pmatrix}\begin{pmatrix} (P^\# - z)^{-1} & 0 \\ 0 & 0 \end{pmatrix}\begin{pmatrix} \widetilde{\chi} & 0 \\ 0 & 0 \end{pmatrix} + \begin{pmatrix} \chi & 0 \\ 0 & 0 \end{pmatrix}\begin{pmatrix} (P^\# - z)^{-1} & 0 \\ 0 & 0 \end{pmatrix}\begin{pmatrix} [P_\theta, \widetilde{\chi}] & 0 \\ 0 & 0 \end{pmatrix}\mathcal{E}(z). \tag{7.12}$$

As before, we get

$$\mu_j(\mathcal{E}(z)\begin{pmatrix} \chi & 0 \\ 0 & 0 \end{pmatrix}),\ \mu_j(\begin{pmatrix} \chi & 0 \\ 0 & 0 \end{pmatrix}\mathcal{E}(z)) \leq C(1 + h^2 j^{2/n.})^{-1},\ z \in \Omega, \tag{7.13}$$

and if χ_1 is chosen as before (with respect to the new χ), and constant near $\overline{B(0,R_2)}$, we get

$$\mu_j(\begin{pmatrix} \chi_1 & 0 \\ 0 & * \end{pmatrix}\mathcal{E}(z)\begin{pmatrix} \chi & 0 \\ 0 & 0 \end{pmatrix}),\ \mu_j(\begin{pmatrix} \chi & 0 \\ 0 & 0 \end{pmatrix}\mathcal{E}(z)\begin{pmatrix} \chi_1 & 0 \\ 0 & * \end{pmatrix}) \leq C(1 + h^2 j^{2/n.})^{-1} e^{-\frac{1}{Ch}\operatorname{dist}(\operatorname{supp}\chi_1, \operatorname{supp}\chi)}, \tag{7.14}$$

where $* = 1$ when $\operatorname{supp}\chi_1 \cap \overline{B(0,R_2)} \neq \emptyset$ and 0 otherwise.

Let $0 \leq \psi_\nu \in C_0^\infty$, $\nu \in \Gamma$ be a partition of unity as in section 2, with $\psi_0 = 1$ near $\overline{B(0,R_2)}$. Put

$$\Psi_\nu = \begin{pmatrix} \psi_\nu & 0 \\ 0 & \delta_{\nu,0} I \end{pmatrix}. \tag{7.15}$$

Then (7.10), (7.14) imply that

$$\mu_j(\Psi_\nu \mathcal{E}.(z)\Psi_\mu) \leq C(1 + h^2 j^{2/n.})^{-1} e^{-\frac{1}{Ch}(|\nu-\mu|-C)_+},\ \nu, \mu \in \Gamma. \tag{7.16}$$

We also wish to estimate the characteristic values of $\Psi_\nu(\mathcal{E}_1(z) - \mathcal{E}_0(z))\Psi_\mu$ when $\nu, \mu \neq 0$. Let $\widetilde{\psi}_0 \in C_0^\infty(\mathbf{R}^n)$ with $1_{\overline{B(0,R_2)}} \prec \widetilde{\psi}_0$ and put

$$\widetilde{\Psi}_0 = \begin{pmatrix} \widetilde{\psi}_0 & 0 \\ 0 & 1 \end{pmatrix}.$$

Consider,

$$V = (1 - \widetilde{\Psi}_0)\mathcal{E}_1\Psi_\mu - (1 - \widetilde{\Psi}_0)\mathcal{E}_0\Psi_\mu.$$

Here

$$(1 - \widetilde{\Psi}_0)\mathcal{E}.\Psi_\mu = \begin{pmatrix} 1-\widetilde{\psi}_0 & 0 \\ 0 & 0 \end{pmatrix}\begin{pmatrix} E. & E._{,+} \\ E._{,-} & E._{,-+} \end{pmatrix}\begin{pmatrix} \psi_\mu & 0 \\ 0 & 0 \end{pmatrix}$$
$$= \begin{pmatrix} (1-\widetilde{\psi}_0)E.\psi_\mu & 0 \\ 0 & 0 \end{pmatrix},$$

and since $R._{,+}(1 - \widetilde{\psi}_0) = 0$:

$$\mathcal{P}_j(1 - \widetilde{\Psi}_0)\mathcal{E}_k\Psi_\mu = \begin{pmatrix} (P_{j,\theta} - z)(1-\widetilde{\psi}_0)E_k\psi_\mu & 0 \\ 0 & 0 \end{pmatrix}.$$

Consequently

$$\mathcal{P}_0 V = \begin{pmatrix} A & 0 \\ 0 & 0 \end{pmatrix} \tag{7.17}$$

$$\begin{aligned} A &= (P_{0,\theta} - z)(1 - \widetilde{\psi}_0)E_1\psi_\mu - (P_{0,\theta} - z)(1 - \widetilde{\psi}_0)E_0\psi_\mu \\ &= (P_{1,\theta} - z)(1 - \widetilde{\psi}_0)E_1\psi_\mu - (P_{0,\theta} - z)(1 - \widetilde{\psi}_0)E_0\psi_\mu \\ &\quad -(P_{1,\theta} - P_{0,\theta})(1 - \widetilde{\psi}_0)E_1\psi_\mu. \end{aligned} \tag{7.18}$$

Here

$$\begin{aligned} (P._{,\theta} - z)(1 - \widetilde{\psi}_0)E. &= -[P._{,\theta}, \widetilde{\psi}_0]E. + (1 - \widetilde{\psi}_0)(P._{,\theta} - z)E. \\ &= -[P._{,\theta}, \widetilde{\psi}_0]E. + (1 - \widetilde{\psi}_0), \end{aligned}$$

since $(P._{,\theta} - z)E. + R._{,-}E._{,-} = I$ and $(1 - \widetilde{\psi}_0)R._{,-} = 0$. It follows that

$$A = [-[P._{,\theta}, \widetilde{\psi}_0]E.]_0^1\psi_\mu - (P_{1,\theta} - P_{0,\theta})(1 - \widetilde{\psi}_0)E_1\psi_\mu, \tag{7.19}$$

and applying E_0 to the left in (7.17), we get,

$$V = \mathcal{E}_0\begin{pmatrix} A & 0 \\ 0 & 0 \end{pmatrix} = \begin{pmatrix} E_0A & 0 \\ E_{0,-}A & 0 \end{pmatrix},$$

$$\Psi_\nu V = \begin{pmatrix} \psi_\nu E_0 A & 0 \\ 0 & 0 \end{pmatrix}.$$

We have shown that

$$[\Psi_\nu\mathcal{E}.\Psi_\mu]_0^1 =$$
$$\begin{pmatrix} -[\psi_\nu E_0[P._{,\theta}, \widetilde{\psi}_0]E.\psi_\mu]_0^1 - \psi_\nu E_0(P_{1,\theta} - P_{0,\theta})(1 - \widetilde{\psi}_0)E_1\psi_\mu & 0 \\ 0 & 0 \end{pmatrix} \tag{7.20}$$

This is analogous to (3.15) with $(1-\chi_1)$, $(1-\chi_2)$, χ replaced by ψ_ν, ψ_μ, $\widetilde{\psi}_0$ and the same discussion as after (3.15) leads to

$$\mu_j([\Psi_\nu \mathcal{E}.\Psi_\mu]_0^1) \leq C\max(|\mu|,|\nu|)^{-\widetilde{n}} e^{-\frac{1}{Ch}(|\nu-\mu|-C)_+}(1+h^2 j^{2/n_{\max}})^{-1}. \tag{7.21}$$

Actually, to obtain this is, there is a minor technical difficulty, similar to the one we encountered in order to obtain (7.7). We do that in the following way: For $j=1,2$, let $A_j = \sum_{|\alpha|\leq 1} a_{j,\alpha}(x;h)(hD)^\alpha$, $a_{j,\alpha}(\cdot;h)$ be uniformly bounded in C_b^∞ with support in x uniformly at a distance >0 from $\overline{B(0,R_2)}$. Put $\mathcal{A}_j = \begin{pmatrix} A_j & 0 \\ 0 & 0 \end{pmatrix}$. Using the identity $[\mathcal{E},\mathcal{A}_j] = -\mathcal{E}[\mathcal{P},\mathcal{A}_j]\mathcal{E}$, we get:

$$\begin{aligned} &\mathcal{E}\mathcal{A}_1\mathcal{A}_2 = \\ &\mathcal{A}_1\mathcal{A}_2\mathcal{E} - \mathcal{A}_1\mathcal{E}[\mathcal{P},\mathcal{A}_2]\mathcal{E} - \mathcal{A}_2\mathcal{E}[\mathcal{P},\mathcal{A}_1]\mathcal{E} + \mathcal{E}[\mathcal{P},\mathcal{A}_1]\mathcal{E}[\mathcal{P},\mathcal{A}_2]\mathcal{E} \\ &+\mathcal{E}[\mathcal{P},\mathcal{A}_2]\mathcal{E}[\mathcal{P},\mathcal{A}_1]\mathcal{E} - \mathcal{E}[[\mathcal{P},\mathcal{A}_1],\mathcal{A}_2]\mathcal{E}, \end{aligned}$$

and we get exponentially weighted estimates under suitable assumptions on ϕ:

$$e^{-\phi/h}\mathcal{E}\mathcal{A}_1\mathcal{A}_2 e^{\phi/h} = \mathcal{O}(1) : \mathcal{H}\times\mathbf{C}^{N.} \to \mathcal{H}\times\mathbf{C}^{N.}.$$

We can then replace $\mathcal{A}_1\mathcal{A}_2$ by a more general operator $\mathcal{B} = \begin{pmatrix} B & 0 \\ 0 & 0 \end{pmatrix}$, $B = \sum_{|\alpha|\leq 2} b_\alpha(x;h)(hD)^\alpha$, where b_α have the same properties as $a_{j,\alpha}$ above, and we then have the additional control over weighted norms that is needed to obtain (7.21).

The proof of Lemma 3.3 gives

Lemma 7.1 *Let m, z_0 be as in Lemma 3.3, $z\in\Omega$ and $\nu,\mu\in\Gamma$. Then $\psi_\nu(P_{\cdot,\theta}-z_0)^{-m}E.(z)\psi_\mu$ is of trace class and*

$$\|\psi_\nu(P_{\cdot,\theta}-z_0)^{-m}E.(z)\psi_\mu\|_{\mathrm{tr}} \leq Ch^{-n.}e^{-\frac{1}{Ch}(|\nu-\mu|-C)_+}. \tag{7.22}$$

For $\nu,\mu\neq 0$, we have

$$\|[\psi_\nu(P_{\cdot,\theta}-z_0)^{-m}E.(z)\psi_\mu]_0^1\|_{\mathrm{tr}} \leq Ch^{-n_{\max}}\max(|\nu|,|\mu|)^{-\widetilde{n}}e^{-\frac{1}{Ch}(|\nu-\mu|-C)_+}. \tag{7.23}$$

In particular we get,

Proposition 7.2 *"$\mathrm{tr}\,[(P_{\cdot,\theta}-z_0)^{-m}E.(z)]_0^1$" is well defined and $=\mathcal{O}(h^{-n.})$.*

$$\frac{1}{2\pi i}\int_\gamma g(z)(P_{\cdot,\theta}-z_0)^{-k}(z-z_0)^{-\ell}(P_{\cdot,\theta}-z)E_{\cdot,+}(z)E_{\cdot,-+}(z)^{-1}E_{\cdot,-}(z)dz$$
$$=-\frac{1}{2\pi i}\int_\gamma g(z)(P_{\cdot,\theta}-z_0)^{-k}(z-z_0)^{-\ell}R_{\cdot,-}(z)E_{\cdot,-}(z)dz. \tag{8.17}$$

In the last integral we can replace γ by $\tilde{\gamma}$ and we get an operator which is $\mathcal{O}(h^{-n_{\max}})$ in trace norm. Consequently,

$$M_\cdot = N_\cdot + \mathcal{O}(h^{-n_{\max}}) \text{ in trace norm,} \tag{8.18}$$

where

$$N_\cdot = \frac{1}{2\pi i}\int_\gamma f(z)E_{\cdot,+}(z)E_{\cdot,-+}(z)^{-1}E_{\cdot,-}(z)dz. \tag{8.19}$$

When taking the trace of $N_\cdot$ we can pass the trace inside the integral and use the cyclicity of the trace,

$$\operatorname{tr} E_{\cdot,+}(z)E_{\cdot,-+}(z)^{-1}E_{\cdot,-}(z) = \operatorname{tr} E_{\cdot,-+}(z)^{-1}E_{\cdot,-}(z)E_{\cdot,+}(z). \tag{8.20}$$

From

$$\frac{\partial}{\partial z}\mathcal{E}(z) = -\mathcal{E}(z)\frac{\partial \mathcal{P}}{\partial z}(z)\mathcal{E}(z), \tag{8.21}$$

we get

$$E_-(z)E_+(z) = E'_{-+}(z) + E_-(z)R'_-(z)E_{-+}(z), \tag{8.22}$$

so N has the same trace as $Q+R$, where

$$Q = \frac{1}{2\pi i}\int_\gamma f(z)E_{-+}(z)^{-1}E'_{-+}(z)dz, \tag{8.23}$$

$$R = \frac{1}{2\pi i}\int_\gamma f(z)E_-(z)R'_-(z)dz. \tag{8.24}$$

In (8.24) we can replace γ by $\tilde{\gamma}$ and see that

$$R = \mathcal{O}(h^{-n_{\max}}) \text{ in trace norm.} \tag{8.25}$$

We sum up the discussion so far in:

$$\text{"}\operatorname{tr}[\chi f(P_\cdot)]_0^1\text{"} = [\operatorname{tr} Q_\cdot]_0^1 - [\sum_{\sigma(P_\cdot)\cap\mathbf{R}_-\cap W} f(\mu)]_0^1 + \mathcal{O}(h^{-n_{\max}}). \tag{8.26}$$

It remains to study $\operatorname{tr} Q_\cdot$. In Ω, E_{-+} is a matrix of size $\mathcal{O}(h^{-n_\cdot})$ and of norm $\mathcal{O}(1)$, so

$$D(z;h) =_{\text{def}} \det E_{-+}(z) = \mathcal{O}(e^{C/h^{n_\cdot}}). \tag{8.27}$$

For $\delta_0 > 0$ small but independent of h, let $\Omega_{+,\delta_0} = \{z \in \Omega_+; \operatorname{Im} z \geq \delta_0\}$. Choosing Γ_θ conveniently, we have

$$(z - P_\theta)^{-1} = \mathcal{O}(1), \ z \in \Omega_{+,\delta_0}, \tag{8.28}$$

and using that $E_{-+}(z)^{-1} = -R_+(P_\theta - z)^{-1}R_-$, we get

$$E_{-+}(z)^{-1} = \mathcal{O}(1), \ z \in \Omega_{+,\delta_0}. \tag{8.29}$$

Consequently,

$$|D(z;h)| \geq e^{-C/h^{n.}}, \ z \in \Omega_{+,\delta_0}. \tag{8.30}$$

It is easy to see as in [14] (or as in [33], where the Dirichlet to Neumann operator plays the same role as E_{-+} here,) that the resonances in Ω, i.e. the eigenvalues of P_θ in Ω, coincide with the zeros of D in Ω and that the multiplicities agree. Let $N(P, \Omega; h)$ be the number of resonances in Ω.

(8.27) remains valid for z in a slightly larger domain and combining this with (8.30) and Jensen's inequality, we get

$$N(P, \Omega; h) \leq Ch^{-n.}. \tag{8.31}$$

(Cf. [30], [34], [23], [44], [41].)

Let z_j be the resonances in Ω repeated according to their multiplicity and put

$$D_w(z;h) = \Pi_j(z - z_j). \tag{8.32}$$

Then thanks to (8.31), we have

$$|D_w(z;h)| \leq e^{C/h^{n.}}, \ z \in \Omega. \tag{8.33}$$

Since the z_j are confined to $\overline{\Omega_-}$, we also have,

$$|D_w(z;h)| \geq e^{-C/h^{n.}}, \ z \in \Omega_{+,\delta_0}. \tag{8.34}$$

In $\Omega \setminus \Omega_{+,\delta_0}$, we can get the same lower bound, if we avoid to go too close to the resonances. To see that, we first establish the simple

Lemma 8.1 *Let $x_1, .., x_N \in \mathbf{R}$ and let $I \subset \mathbf{R}$ be an interval of length $|I| \in]0, \infty[$. Then there exists $x \in I$, such that $\Pi_1^N |x - x_j| \geq \exp[-N(1 + \log \frac{2}{|I|})]$.*

Proof. Consider $F(x) = \sum_1^N \log \frac{1}{|x - x_j|}$. Notice that

$$\int_I \log \frac{1}{|x - x_j|} dx \leq 2 \int_0^{|I|/2} \log \frac{1}{t} dt = |I|(1 + \log \frac{2}{|I|}),$$

so that

$$\int_I F(x)dx \leq N|I|(1+\log\frac{2}{|I|}).$$

We can therefore find $x \in I$ such that $F(x) \leq N(1+\log\frac{2}{|I|})$, i.e.

$$\Pi_1^N |x - x_j| = e^{-F(x)} \geq e^{-N(1+\log(2/|I|))}.$$

□

Now make $\widetilde{\gamma} = \widetilde{\gamma}_\alpha$ depend smoothly on a parameter α in some bounded interval J but with fixed end points b, a, so that $\widetilde{\gamma}$ moves transversally in $\Omega \setminus \Omega_{+,\delta_0/2}$, in such a way that if $z_j \in \widetilde{\gamma}_{\alpha_j}$, then dist $(z_j, \widetilde{\gamma}_\alpha) \geq |\alpha - \alpha_j|$ for all $\alpha \in J$, and if z_j belongs to no $\widetilde{\gamma}_\alpha$, then dist $(z_j, \widetilde{\gamma}_\alpha) \geq$ dist $(\alpha, \partial J)$. It follows from Lemma 8.1 , that if I is a non-trivial subinterval of J, then we can find $\alpha \in I$ such that

$$|D_w(z;h)| \geq e^{-Ch^{-n.}}, \ z \in \widetilde{\gamma}_\alpha, \tag{8.35}$$

where $C = C_{|I|}$. We factorize D:

$$D(z;h) = G(z;h)D_w(z;h), \ z \in \Omega, \tag{8.36}$$

where G and $1/G$ are holomorphic in Ω. Combining (8.27), (8.34), (8.35), we get (with a new constant)

$$|G(z;h)| \leq e^{C/h^{n.}}, \ z \in \Omega_{+,\delta_0} \cup \widetilde{\gamma}_\alpha. \tag{8.37}$$

The maximum principle gives

$$|G(z;h)| \leq e^{Ch^{-n.}}, \ z \in \widetilde{\Omega}, \tag{8.38}$$

where $\widetilde{\Omega} \subset\subset \Omega$ is any simply connected relatively open h-independent set with $\overline{W}$ in its interior, provided that we choose the family $\widetilde{\gamma}_\alpha$ such that $\widetilde{\gamma}_\alpha \cap \widetilde{\Omega} \cap \Omega_- = \emptyset$.

(8.30), (8.33) imply that

$$|G(z;h)| \geq e^{-C/h^{n.}}, \ z \in \Omega_{+,\delta_0}. \tag{8.39}$$

Consider the harmonic function on $\widetilde{\Omega}$:

$$0 \leq \ell(z;h) = Ch^{-n.} - \log|G(z;h)|. \tag{8.40}$$

Harnack's inequality for non-negative harmonic functions tells us that for every $K \subset\subset \widetilde{\Omega}$, we have

$$\sup_K \ell \leq C_K \inf_K \ell, \tag{8.41}$$

i.e. that the function $\ell(z;h)$ is uniformly of constant order of magnitude on K. After an arbitrarily small decrease of $\widetilde{\Omega}$, we have (8.41) with $K=\widetilde{\Omega}$ and if we use (8.39), we get $\ell(z;h)\leq Ch^{-n_\cdot}$ on Ω_{+,δ_0} and hence by (8.41) with $K=\widetilde{\Omega}$:

$$\ell(z;h)\leq Ch^{-n_\cdot},\text{ on }\widetilde{\Omega}. \tag{8.42}$$

We conclude that $\log|G(z;h)|\geq -Ch^{-n_\cdot}$ on $\widetilde{\Omega}$ and with (8.38), we get

$$|\log|G(z;h)||\leq Ch^{-n_\cdot},\ z\in\widetilde{\Omega}. \tag{8.43}$$

Since $\log|G(z;h)|=\mathrm{Re}\,\log G(z;h)$ is harmonic we get after an arbitrarily small decrease of $\widetilde{\Omega}$:

$$\nabla\mathrm{Re}\,\log G=\mathcal{O}(1)h^{-n_\cdot}. \tag{8.44}$$

The Cauchy-Riemann equations then give the same estimate for $\nabla\mathrm{Im}\,\log G$ and consequently

$$\frac{d}{dz}\log G=\mathcal{O}(1)h^{-n_\cdot}. \tag{8.45}$$

Choose a family $\widetilde{\gamma}_\alpha$, $\alpha\in J$ in $\widetilde{\Omega}\setminus W$ with the same properties as after Lemma 8.1. If $z_j\in\widetilde{\gamma}_{\alpha_j}$, we see that

$$\int_{\widetilde{\gamma}_\alpha}\frac{1}{|z-z_j|}|dz|=\mathcal{O}(1)\log\frac{1}{|\alpha-\alpha_j|}. \tag{8.46}$$

Consequently we can apply Lemma 8.1 or rather its proof, to see that there is a $\widetilde{\gamma}=\widetilde{\gamma}_\alpha$, such that:

$$\int_{\widetilde{\gamma}}|\frac{d}{dz}\log D_w(z)||dz|=\mathcal{O}(1)h^{-n_\cdot}. \tag{8.47}$$

Since $\frac{d}{dz}\log D=\frac{d}{dz}\log G+\frac{d}{dz}\log D_w$, we get from (8.47), (8.45) that

$$\int_{\widetilde{\gamma}}|\frac{d}{dz}\log D||dz|=\mathcal{O}(1)h^{-n_\cdot}. \tag{8.48}$$

From this and (8.23), we get since $\frac{d}{dz}\log D_w=\sum\frac{1}{z-z_j}$:

$$\begin{aligned}\mathrm{tr}\,Q=\frac{1}{2\pi i}\int_\gamma f(z)\frac{d}{dz}\log D(z)dz=\\ \sum_{z_j\text{between }\widetilde{\gamma}\text{ and }\gamma}f(z_j)+\frac{1}{2\pi i}\int_{\widetilde{\gamma}}f(z)\frac{d}{dz}\log D(z)dz,\end{aligned} \tag{8.49}$$

where the last integral is $\mathcal{O}(1)h^{-n_\cdot}$. Combining this and (8.26), we get the theorem. □

9. Comparison with the Poisson formula for resonances in Lax-Phillips theory.

Let P_0, P_1 be independent of h and satisfy the assumptions of Theorem 2.2 with $h = 1$. Assume in addition that P_0, P_1 are semi-bounded in the sense of self-adjoint operators: $P_. \geq -C$. We can then define

$$u(t) = 2\text{"tr}\,[\cos t\sqrt{P_.}]_0^1\text{"} \in \mathcal{D}'(\mathbf{R}) \tag{9.1}$$

by the following formal computation: For $\phi \in C_0^\infty(\mathbf{R})$, we should have

$$\langle u, \phi\rangle = 2\int \phi(t)\text{"tr}\,[\cos t\sqrt{P_.}]_0^1\text{"}dt = \text{"tr}\,[2\int \phi(t)\cos t\sqrt{P_.}dt]_0^1\text{"}. \tag{9.2}$$

Here $2\int \phi(t)\cos t\sqrt{P_.}dt = \int \phi(t)(e^{it\sqrt{P_.}} + e^{-it\sqrt{P_.}})dt = \widetilde{\phi}(P_.)$, where $\widetilde{\phi}(z) = \widehat{\phi}(\sqrt{z}) + \widehat{\phi}(-\sqrt{z})$ (with $\widehat{\phi}(\tau) = \int e^{-it\tau}\phi(t)dt$ denoting the Fourier transform) is an entire function, independent of the choice of sign for $\sqrt{z}$, and of Schwartz class $\mathcal{S}$ on $[-C, +\infty[$ for every $C \geq 0$. According to Proposition 2.1, and the semiboundedness of $P_.$, we can then define $u \in \mathcal{D}'(R)$, by

$$\langle u, \phi\rangle = \text{"tr}\,[\widetilde{\phi}(P_.)]_0^1\text{"}, \ \phi \in C_0^\infty(\mathbf{R}). \tag{9.3}$$

In the case when n is odd ≥ 3, $P_0 = -\Delta$ and P_1 coincides with $-\Delta$ outside a compact set, it is known that resonances are naturally defined as a discrete set of numbers μ_j (repeated at most finitely many times according to multiplicity) in the closed upper halfplane or alternatively as the the conjugate numbers $\overline{\mu}_j$ in the lower half-plane. The relation with the resonances defined by complex scaling as in section 5 is given by: $\overline{\mu}_j = \sqrt{\lambda_j}$. (See [34] and [13] and further references, given there.) It is then known that we have the following Poisson type formula,

$$u(t) = \sum e^{i\mu_j t} \text{ in } \mathcal{D}'(]0, +\infty[), \tag{9.4}$$

where the sum converges in the sense of distributions, in view of known polynomial bounds on the number of resonances in large discs. For more concrete classes of operators P_1 and on an interval of the type $]R, \infty[$, this relation was established by Lax-Phillips [20], Bardos-Guillot-Ralston [4]. Melrose [21] extended the validity of (9.4) to the full interval $]0, +\infty[$ and Sjöstrand-Zworski [36] further extended it to the case of operators P_1 like the ones in Theorem 2.2, but still with the restriction that $P_1 = -\Delta$ outside a compact set and that the dimension is odd ≥ 3. The proofs in these works are based on the Lax-Phillips theory (cf [20]).

In the case of certain hyperbolic surfaces a formula of the type (9.4) was recently established by Guillopé-Zworski [11] and their proof uses more general scattering theory including the Birman-Krein formula for the scattering phase. According to private communications from the authors it is quite possible that their proof can be adapted to give (9.4) under the assumptions described above.

When the Poisson formula is valid, we have for every $\chi \in C_0^\infty(]0,+\infty[)$, $\lambda \in \mathbf{R}$:

$$\widehat{\chi u}(\lambda) = \sum \widehat{\chi}(\lambda - \mu_j), \tag{9.5}$$

where we notice that the Paley-Wiener theorem assures the convergence of the sum. This version of the Poisson type formula was used in [35] together with asymptotic informations about the LHS when λ tends to infinity, to get lower bounds on the density of resonances in certain neighborhoods of the real axis.

Under the more general assumptions of the beginning of this section, we shall now see how to get an asymptotic version of (9.5) which is strong enough to recover the lower bounds in [35] and extend them to the case of even dimensions. (Lack of time in preparing these notes has prevented us from developing other applications of Theorem 2.2, and we have the intention to return to these questions in some future work(s).)

For the remainder of this section, we let the self-adjoint operators P_j, $j = 0, 1$ be semi-bounded from below, independent of h and satisfy the assumptions of Theorem 2.2 with $h = 1$. Define $u(t) \in \mathcal{D}'(\mathbf{R})$ as in the beginning of this section. Let $\chi \in C_0^\infty(]a,b[)$, where $0 < a < b$. For $\lambda \in \mathbf{R}$ with $|\lambda|$ large, we consider

$$\widehat{\chi u}(\lambda) = \text{"tr}\,[\widehat{\chi}(\lambda - \sqrt{P_.}) + \widehat{\chi}(\lambda + \sqrt{P_.})]_0^1\text{"}.$$

Write $\lambda = \frac{\mu}{h}$, $\mu = \pm 1$, where $h > 0$ is small, and write $P_. = \frac{1}{h^2} h^2 P_.$. Then,

$$\widehat{\chi u}(\frac{\mu}{h}) = \text{"tr}\,[\widehat{\chi}(\frac{1}{h}(\mu - \sqrt{h^2 P_.})) + \widehat{\chi}(\frac{1}{h}(\mu + \sqrt{h^2 P_.}))]_0^1\text{"}.$$

As already noticed, the function

$$f_h(z) = \widehat{\chi}(\frac{1}{h}(\mu - \sqrt{z})) + \widehat{\chi}(\frac{1}{h}(\mu + \sqrt{z})) \tag{9.6}$$

is entire since the odd powers of the square root of z dissappear from the series expansions.

For $z < 0$, write $z = -t$. Then $f_h(z) = \widehat{\chi}(\frac{1}{h}(\mu - i\sqrt{t})) + \widehat{\chi}(\frac{1}{h}(\mu + i\sqrt{t}))$, so even though $\widehat{\chi} \in \mathcal{S}(\mathbf{R})$, we may have exponential growth when $t >> h^2$. Fortunately, $h^2 P_. \geq -h^2 C$, so we are only concerned with the region $0 \leq t \leq Ch^2$, where $f_h(z)$ and all its derivatives are $\mathcal{O}(h^\infty)$.

We next look at the case of positive z. For $0 \leq z \leq 1 - \frac{1}{\mathcal{O}(1)}$, we have $|\mu - \sqrt{z}|, |\mu + \sqrt{z}| \sim 1$, so

$$\partial_z^k f_h(z) = \mathcal{O}_k(h^\infty), \tag{9.7}$$

also in this region.

For $z \geq 1 + \frac{1}{\mathcal{O}(1)}$, we have $|\mu - \sqrt{z}|$, $|\mu + \sqrt{z}| \sim 1 + \sqrt{z}$, so

$$\partial_z^k f_h(z) = \mathcal{O}_{N,k}(h^N(1+|z|)^{-N}), \forall k, N \in \mathbf{N}. \tag{9.8}$$

Noticing now that $h^2 P_\cdot$ satisfy all the assumptions of Theorem 2.2 with variable h, we can combine the preceding remarks about f_h and Proposition 2.1 to conclude that the contribution from the region $|z-1| \geq \frac{1}{\mathcal{O}(1)}$ to "$\mathrm{tr}\,[f_h(h^2 P_\cdot)]_0^1$" is $\mathcal{O}(h^\infty)$ (in the sense that we change the "trace" only by $\mathcal{O}(h^\infty)$, if we replace $f_h(z)$ by $\psi(z) f_h(z)$, where $\psi \in C_0^\infty(\mathbf{R})$ is equal to 1 near $z = 1$).

It remains to consider the contributions from the inteval $\alpha \leq z \leq \beta$, where $0 < \alpha < 1 < \beta$, and we shall study $f_h(z)$ for complex z satisfying,

$$\alpha \leq \mathrm{Re}\, z \leq \beta, \ |\mathrm{Im}\, z| \leq \gamma. \tag{9.9}$$

For such a z we have $\mathrm{Re}\sqrt{z} = \sqrt{\frac{|z|+\mathrm{Re}\, z}{2}}$, $\mathrm{Im}\sqrt{z} = (\mathrm{sgn}\,\mathrm{Im}\, z)\sqrt{\frac{|z|-\mathrm{Re}\, z}{2}}$,

$$\mathrm{Im}\sqrt{z} = \frac{\mathrm{Im}\, z}{\sqrt{2(|z|+\mathrm{Re}\, z)}}, \tag{9.10}$$

so

$$\frac{|\mathrm{Im}\, z|}{\sqrt{2(|\beta+i\gamma|+\beta)}} \leq |\mathrm{Im}\sqrt{z}| \leq \frac{|\mathrm{Im}\, z|}{2\sqrt{\alpha}}. \tag{9.11}$$

Since $\chi \in C_0^\infty(]a,b[)$, we have according to the Paley-Wiener theorem:

$$|\widehat{\chi}(\tau)| \leq \mathcal{O}_N(1)(1+|\tau|)^{-N} \times \begin{cases} e^{b\mathrm{Im}\,\tau}, & \mathrm{Im}\,\tau \geq 0, \\ e^{a\mathrm{Im}\,\tau}, & \mathrm{Im}\,\tau \leq 0. \end{cases} \tag{9.12}$$

If $\mu = 1$, we see from this, that the second term in the RHS of (9.6) is $\mathcal{O}(h^\infty)$ with all its z-derivatives for $\alpha \leq z \leq \beta$, while the first term of the RHS extends holomorphically to the region (9.9) and is

$$\mathcal{O}_N(1)\langle\frac{\mu-\sqrt{z}}{h}\rangle^{-N} e^{-\frac{a}{\sqrt{2(|\beta+i\gamma|+\beta)}}\mathrm{Im}\, z} \text{ for } \mathrm{Im}\, z \geq 0$$

and

$$\mathcal{O}_N(1)\langle\frac{\mu-\sqrt{z}}{h}\rangle^{-N} e^{-\frac{b}{2\sqrt{\alpha}}\mathrm{Im}\, z} \text{ for } \mathrm{Im}\, z \leq 0.$$

Similarly for $\mu = -1$, the first term of the RHS of (9.5) is $\mathcal{O}(h^\infty)$ with all its z-derivatives on the interval $[\alpha, \beta]$ while the second term extends holomorphically to the region (9.9) and is $\mathcal{O}_N(1)\langle\frac{\mu+\sqrt{z}}{h}\rangle^{-N} e^{\frac{a}{\sqrt{2(|\beta+i\gamma|+\beta)}}\mathrm{Im}\, z}$ for $\mathrm{Im}\, z \leq 0$ and $\mathcal{O}_N(1)\langle\frac{\mu+\sqrt{z}}{h}\rangle^{-N} e^{\frac{b}{2\sqrt{\alpha}}\mathrm{Im}\, z}$ for $\mathrm{Im}\, z \geq 0$.

We can then apply the variant of Theorem 2.2 given after Theorem 2.2 up to Proposition 2.3, and obtain the following: Let $h^2\lambda_\nu(P.)$ denote the resonances of $h^2P.$ in the intersection Ω_- of the region (9.9) with the closed lower half-plane. Then when $\lambda \to -\infty$, we get:

$$\widehat{\chi u}(\lambda) = [\sum_{h^2\lambda_\nu(P.)\in\Omega_-} \hat{\chi}(\frac{1}{h}(-1+\sqrt{h^2\lambda_\nu(P.)})) + \mathcal{O}(h^\infty)]_0^1,$$

or equivalently:

$$\widehat{\chi u}(\lambda) = [\sum_{\lambda_\nu(P.)\in\lambda^2\Omega_-} \hat{\chi}(\lambda+\sqrt{\lambda_\nu(P.)})]_0^1 + \mathcal{O}(|\lambda|^{-\infty}) \tag{9.13}$$

When $\lambda \to +\infty$, we change the sign of the angle of complex scaling in section 5, and define the resonances $h^2\lambda_\nu(P.)$ in Ω_+, where Ω_+ is the intersection of the region (9.9) with the closed upper halfplane. Theorem 2.2 remains valid with the obvious modifications, and we get

$$\widehat{\chi u}(\lambda) = [\sum_{\lambda_\nu(P.)\in\lambda^2\Omega_+} \hat{\chi}(\lambda-\sqrt{\lambda_\nu(P.)})]_0^1 + \mathcal{O}(|\lambda|^{-\infty}) \tag{9.14}$$

10. Lower bounds near the real axis

In this section we review the work [35] and extend the results to even dimensions. Let $u(t)$ be the distribution defined in the preceding section. To start with, we make the general assumptions of the preceding section, but we also assume that P_0 has no resonances in a set of the form $\{\mu \in \mathbf{C}; |z| \geq C, 0 \leq \arg z \leq 1/C\}$. Here and in the remainder of the section, we work with the convention that the resonances are defined in the upper half-plane (by switching the sign of the angle of scaling in section 5). In many cases, it turns out that $u(t)$ has conormal singularities on the positive half- axis and we can then apply a Tauberian argument to get lower bounds on the density of resonances in certain logarithmic neighborhoods of the positive half-axis. We start by discussing the Tauberian argument (of [35]) before discussing some applications.

Let $\Lambda = \{\mu_j\}$ be a discrete subset of $\{\mu \in \mathbf{C}; |z| \geq C, 0 \leq \arg z \leq 1/C\}$ for some fixed $C \geq 1$, where the μ_j may be repeated with some multiplicity, and assume that for some $n_1 > 0$, we have:

$$N(r) =_{\text{def}} \#\{\mu_j \in \Lambda; |\mu_j| \leq r\} = \mathcal{O}(r^{n_1}). \tag{10.1}$$

For $\rho > 0$, we define

$$\Lambda_\rho = \{\mu_j \in \Lambda; \operatorname{Im}\mu_j \leq \rho\log|\mu_j|\}.$$

$$N_\rho(r) = \#\{\mu_j \in \Lambda_\rho; \operatorname{Re}\mu_j \le r\}.$$

Let $u \in \mathcal{D}'(\mathbf{R})$, and assume that for every $\chi \in C_0^\infty(]0,\infty[)$, we have:

$$\widehat{\chi u}(\lambda) = \sum_{\mu_j \in \lambda\sqrt{\Omega_+}} \hat{\chi}(\lambda - \mu_j) + \mathcal{O}(\lambda^{-\infty}),\ \lambda \to +\infty, \tag{10.2}$$

where Ω_+ was defined in the preceding section. We then have the following minor modification of a result of [35]:

Theorem 10.1 *Let $k \in \mathbf{R}$, $d, b > 0$ and suppose that for all $\phi \in C_0^\infty(]0,\infty[)$ with support in a sufficiently small neighborhood of d and with $\phi(d) = 1$, we have:*

$$|\widehat{\phi u}(\lambda)| \ge (b - o(1))\lambda^k,\ \lambda \to +\infty.$$

Then for every $\epsilon > 0$ and $\rho > \frac{n_1 - k}{d - \epsilon^2}$, we have
a) If $k \ge 0$, then $N_\rho(r) \ge (B - o(1))r^{k+1}$, $B = \frac{b}{2\pi(k+1)}$, and

$$\sum_{\mu_j \in \Lambda_\rho, |\operatorname{Re}\mu_j| \le r} e^{-(d-\epsilon)\operatorname{Im}\mu_j} \ge (B - o_\epsilon(1))r^{k+1}.$$

b) If $k < 0$, then for all $\delta > 0$, there exists $r(\delta) > 0$, such that $N_\rho(r) \ge r^{1-\delta}$, for $r > r(\delta)$.

Proof. We will here only prove the part a). Let $\phi \in C_0^\infty(]-1,1[)$ with $\phi(0) = \frac{1}{2\pi}$, $\hat{\phi} \ge 0$. For $\gamma > 0$, we put $\phi_{\gamma,d}(t) = \phi(\frac{1}{\gamma}(t - d))$, so that $\widehat{\phi_{\gamma,d}}(\tau) = \gamma\hat{\phi}(\gamma\tau)e^{-id\tau}$. If γ is small enough, then according to (10.2) and the assumption in the theorem:

$$|\sum_{\mu_j \in \Lambda} \widehat{\phi_{\gamma,d}}(\lambda - \mu_j)| \ge (\frac{b}{2\pi} - o_\gamma(1))\lambda^k,\ \lambda \to +\infty. \tag{10.3}$$

Here we also used that

$$\sum_{\mu_j \in \Lambda \setminus \lambda\sqrt{\Omega_+}} \widehat{\phi_{\gamma,d}}(\lambda - \mu_j) = \mathcal{O}(\lambda^{-\infty}),$$

which is an easy consequence of the Paley-Wiener theorem:

$$|\widehat{\phi_{\gamma,d}}(\zeta)| \le \gamma C_M e^{(d\pm\gamma)\operatorname{Im}\zeta}(1 + |\gamma\zeta|)^{-M},\ \pm\operatorname{Im}\zeta \ge 0,\ M \in \mathbf{N}.$$

Using this estimate, we shall also estimate away the contribution to the sum in (10.3) from the μ_j's in $\Lambda \setminus \Lambda_\rho$. Using the bound on $N(r)$ we get for

$\lambda \geq 2$ with constants depending on M, γ, $d-\gamma>0$:

$$\sum_{\mu_j\in\Lambda\setminus\Lambda_\rho} |\widehat{\phi_{\gamma,d}}(\lambda-\mu_j)| \leq C \sum_{\mu_j\in\Lambda\setminus\Lambda_\rho} e^{-(d-\gamma)\mathrm{Im}\,\mu_j}(1+|\lambda-\mu_j|)^{-M}$$

$$\leq C \sum_{\mu_j\in\Lambda\setminus\Lambda_\rho} |\mu_j|^{-(d-\gamma)\rho}(1+|\lambda-|\mu_j||)^{-M} \leq C\int_1^\infty t^{-(d-\gamma)\rho}(1+|\lambda-t|)^{-M}dN(t)$$

$$\leq C(-(1+|\lambda-1|)^{-M}N(1) - \int_1^\infty \frac{d}{dt}(t^{-(d-\gamma)\rho}(1+|\lambda-t|)^{-M})N(t)dt)$$

$$\leq \tilde{C}\int_1^\infty |\frac{d}{dt}(t^{-(d-\gamma)\rho}(1+|\lambda-t|)^{-M})|t^{n_1}dt = \mathcal{O}(\lambda^{n_1-(d-\gamma)\rho}) = o(\lambda^k),$$

where we chose M large enough and used that that $n_1-(d-\gamma)\rho<k$ if γ is small enough.

Combining this estimate with (10.3), we get

$$|\sum_{\mu_j\in\Lambda_\rho} \widehat{\phi_{\gamma,d}}(\lambda-\mu_j)| \geq (\frac{b}{2\pi} - o_\gamma(1))\lambda^k,\ \lambda\to+\infty. \tag{10.4}$$

For $\epsilon>\gamma>0$, we introduce

$$\psi(\tau) = \sup_{\sigma\leq 0} |e^{-(d-\epsilon)\sigma}\widehat{\phi_{\gamma,d}}(\tau+i\sigma)|,\ \tau\in\mathbf{R}.$$

We claim that

$$\psi(\tau) = \gamma\hat{\phi}(\gamma\tau) + \frac{\gamma^2}{\epsilon-\gamma}\mathcal{O}_M(\langle\gamma\tau\rangle^{-M}). \tag{10.5}$$

First, it is immediate that $\psi(\tau)\geq\gamma\hat{\phi}(\gamma\tau)$. In order to have the upper bound, we write for $\sigma\leq 0$:

$$e^{\epsilon\sigma}|\hat{\phi}(\gamma\tau+i\gamma\sigma)| \leq e^{\epsilon\sigma}|\hat{\phi}(\gamma\tau+i\gamma\sigma)-\hat{\phi}(\gamma\tau)| + \hat{\phi}(\gamma\tau),$$

where the first term of the RHS is

$$f \leq e^{\epsilon\sigma}\gamma|\sigma|\int_0^1 |\hat{\phi}'(\gamma\tau+is\gamma\sigma)|ds \leq C_M e^{\epsilon\sigma}\gamma|\sigma|\int_0^1 e^{s|\gamma||\sigma|}(1+|\gamma\tau|+|s\gamma\sigma|)^{-M}ds$$

$$\leq C_M'\gamma e^{-(\epsilon-\gamma)|\sigma|}|\sigma|(1+|\gamma\tau|)^{-M} \leq \tilde{C}_M\frac{\gamma}{\epsilon-\sigma}\langle\gamma\tau\rangle^{-M},$$

which gives the required upper bound in (10.5).

We next show that for $\gamma<\epsilon^2$,

$$\int_{-\infty}^r \psi(\tau)d\tau = (1+\mathcal{O}(\gamma^{\frac{1}{2}}))H(r) + \mathcal{O}_M(\langle\gamma r\rangle^{-M}), \tag{10.6}$$

where $H(r) = 1_{[0,+\infty[}(r)$ is the Heaviside function: Since

$$\int_{-\infty}^{+\infty} \widehat{\phi}(\tau)d\tau = 2\pi\phi(0) = 1,$$

we have

$$\int_{-\infty}^{r} \gamma\widehat{\phi}(\gamma\tau)d\tau = \int_{-\infty}^{\gamma r} \widehat{\phi}(\tau)d\tau = \begin{cases} \mathcal{O}_M(\langle r\gamma\rangle^{-M}),\ r \leq 0, \\ 1 + \mathcal{O}_M(\langle r\gamma\rangle^{-M}),\ \gamma \geq 0. \end{cases}$$

This shows that the integral of the first term of the RHS of (10.5) gives the expected contribution to (10.6). It remains to treat the contribution from the remainder in (10.5). Since $\epsilon \geq \sqrt{\gamma}$, we have

$$\frac{\gamma^2}{\epsilon - \gamma}\mathcal{O}_M(\langle\gamma\tau\rangle^{-M}) = \gamma^{1/2}\mathcal{O}_M(\gamma\langle\gamma\tau\rangle^{-M})$$

so the integral of this can be treated the same way, and we have verified (10.6).

Introduce the positive measure

$$\mu(dr) = \sum_{\mu_j \in \Lambda_\rho} e^{-(d-\epsilon)\mathrm{Im}\,\mu_j}\delta(r - \mathrm{Re}\,\mu_j)dr.$$

Using (10.4) and the subsequent definition of ψ, we get:

$$(\frac{b}{2\pi} - o_\gamma(1))\lambda^k \leq \sum_{\mu_j\in\Lambda_\rho} |\widehat{\phi_{\gamma,d}}(\lambda - \mu_j)| \leq$$

$$\sum_{\mu_j\in\Lambda_\rho} \psi(\lambda - \mathrm{Re}\,\mu_j)e^{-(d-\epsilon)\mathrm{Im}\,\mu_j} = \int \psi(\lambda - \tau)\mu(d\tau) = \psi * \mu(\lambda),\ \lambda \geq 1 \tag{10.7}$$

Notice that $M(r) =_{\mathrm{def}} \mu([0,r]) \leq N_\rho(r)$ and that $\int \psi(\lambda - \tau)\mu(d\tau) \leq \mathcal{O}_{\gamma,M}(|\lambda|^{-M})$ for $\lambda \leq -1$. Then integrating (10.7) from 1 to $r \geq 1$, we get:

$$\begin{aligned}(\frac{b}{2\pi(k+1)} - o_\gamma(1))r^{k+1} &\leq \int_{-\infty}^{r} (\psi * \mu)(\lambda)d\lambda \\ = \int \psi(t)(\int_{-\infty}^{r-t} \mu(d\lambda))dt &= \int \psi(t)M(r-t)dt \\ &= \int \psi(r-\tau)M(\tau)d\rho. \end{aligned}\tag{10.8}$$

Since $M(r) = \mathcal{O}(r^{n_1})$, we deduce from (10.5) that for $\delta \in]0,1[$, when $r \geq 1$:

$$\int_{\mathbf{R}\setminus[r-r^\delta, r+r^\delta]} \psi(r-\tau)M(\tau)d\tau = \mathcal{O}_{\gamma,M}(r^{n_1-\delta M}),\ \forall M. \tag{10.9}$$

Then

$$(\frac{b}{2\pi(k+1)} - o_{\gamma,\delta}(1))r^{k+1} \leq \int_{r-r^\delta}^{r+r^\delta} \psi(r-\tau)M(\tau)d\tau$$

$$\leq (\int_{r-r^\delta}^{r+r^\delta} \psi(r-\tau)d\tau)M(r+r^\delta) = \int_{-r^\delta}^{r^\delta} \psi(t)d\tau M(r+r^\delta)$$

$$= M(r+r\delta)(1+\mathcal{O}(\gamma^{1/2}) + \mathcal{O}_{\gamma,M}(r^{-\delta M})).$$

Since we can choose γ arbitrarily small, part a) of the theorem follows easily by a substitution in r. □

Let $\mathcal{O} \subset\subset \mathbf{R}^n$, $n \geq 2$ be an open set with C^∞ boundary. Let P_1 be the self-adjoint realization of $-\Delta$ on $\mathbf{R}^n \setminus \mathcal{O}$ with Dirichlet, Neumann or Robin conditions on the boundary. Let P_0 be $-\Delta$ on $\mathbf{R}^n$. Then we can define u as in the preceding section and (9.14) holds for $\chi \in C_0^\infty(]0,+\infty[)$, where the only contributions come from the resonances λ_j of P_1 (here with the convention that $\operatorname{Im} \lambda_j \geq 0$). With $\mu_j = \sqrt{\lambda_j}$, we define $N_\rho(r)$ as above.

In $\mathbf{R}^n \setminus \mathcal{O}$ or rather in a certain topological space sitting over $\mathbf{R}^n \setminus \mathcal{O}$, we can define generalized bicharacteristics, also called C^∞-rays, in the sense of [24]. For a more complete discussion of this and for some other points below, we refer to the book [27] by Petkov-Stoyanov, and further references, given there. Roughly, a generalized bicharacteristics is a curve $I \ni t \mapsto \gamma(t) \in \mathbf{R}^n \setminus \mathcal{O}$, where I is some interval, such that $\gamma'(t)$ exists and is constant of norm 1 on every open subinterval on which $\gamma(t) \in \mathbf{R}^n \setminus \overline{\mathcal{O}}$. If $\gamma(t) \in \mathbf{R}^n \setminus \overline{\mathcal{O}}$ for $t_0 - \epsilon < t < t_0$ and $\gamma(t_0) \in \partial\mathcal{O}$ and $\gamma'(t_0 - 0) \notin T\partial\mathcal{O}$ (the tangent space of $\partial\Omega$), then it is required that γ reflects at $\gamma(t_0)$ according to the rules of geometrical optics, so that $\gamma(t) \in \mathbf{R}^n \setminus \overline{\mathcal{O}}$ for small $t - t_0 > 0$, and $\gamma'(t_0+0) \neq \gamma'(t_0-0)$ and $\gamma'(t_0+0) - \gamma'(t_0-0)$ is a multiple of the normal of $\partial\mathcal{O}$ at $\gamma(t_0)$. Such a point t_0 is called a transversal reflection point. When γ hits the boundary tangentially, the description is a little more complicated, and we refer to [24], [27] for a more complete decription of C^∞-rays near such diffractive points.

A C^∞-ray: $\mathbf{R} \ni t \mapsto \gamma(t) \in \mathbf{R}^n \setminus \mathcal{O}$ is called periodic with period $T > 0$ (or T-periodic) if $\gamma(t+T) = \gamma(t)$ for all $t \in \mathbf{R}$. If γ is T-periodic, we let $T_\gamma^\#$ denote the primitive period, i.e. the smallest period > 0 of γ. Clearly, $T = kT_\gamma^\#$, for some $1 \leq k \in \mathbf{N}$. We say that γ is transversally reflected if all boundary points of γ are points of transversal reflection.

If γ is a T-periodic transversally reflected C^∞-ray then one can define a corresponding linearized Poincaré map P_γ (also depending on the choice of the period T), which can be viewed, up to symplectic conjugation, as a symplectic map $P_\gamma : \mathbf{R}^{2(n-1)} \to \mathbf{R}^{2(n-1)}$. We say that γ is non-degenerate if $\det(I - P_\gamma) \neq 0$.

Let γ be a C^∞-ray and assume:

$$\gamma \text{ is a transversally reflected periodic non-degenerate } C^\infty\text{-ray of period } T_\gamma > 0 \text{ and of primitive period } T_\gamma^{\#}. \tag{10.10}$$

$$\text{There are no other } T_\gamma\text{-periodic } C^\infty\text{-rays, up to time translations and time reversals.} \tag{10.11}$$

Then in a neighborhood of $t = T_\gamma$, we have

$$u(t) \equiv \frac{T_\gamma^{\#}}{\pi} \mathrm{Re}\,(e^{i\frac{\pi}{2}\beta_\gamma}(t - T_\gamma - i0)^{-1}) |\det(P_\gamma - I)|^{-1/2} \bmod L^1_{\mathrm{loc}}. \tag{10.12}$$

Here β_γ is a real number which depends not only on γ, but also on the type of boundary condition, and whose actual value is unimportant in the following. The corresponding result for $\mathrm{tr}\cos t\sqrt{P_\Omega}$, when $P = -\Delta$ on Ω with Dirichlet, Neumann or Robin condition and Ω is bounded with smooth boundary was obtained in [10] (see also [27]) very much as a consequence of the general result on propagation of C^∞-singularities in [24]. Then (10.12) follows, if we notice that the property of finite propagation speed for supports of solutions to the wave equation implies that for every $T > 0$, we have

$$u(t) = 2(\mathrm{tr}\cos t\sqrt{P_{\Omega\setminus\mathcal{O}}} - \mathrm{tr}\cos t\sqrt{P_\Omega}), \text{ for } -T < t < T,$$

if $\Omega = B(0, R)$ and $R > 0$ is large enough depending on T. Here $P_{\Omega\setminus\mathcal{O}}$, P_Ω denote the realizations of $-\Delta$ on Ω and $\Omega \setminus \mathcal{O}$ respectively, with the appropriate boundary conditions.

If $\phi \in C_0^\infty(]0, \infty[)$ has its support close to T_γ and $\phi(T_\gamma) = 1$, then it follows from (10.12) that

$$|\widehat{\phi u}(\lambda)| = \frac{T_\gamma^{\#}}{|\det(P_\gamma - I)|^{1/2}} + o(1),\ \lambda \to +\infty, \tag{10.13}$$

so the assumptions of the previous theorem are satisfied with $d = T_\gamma$, $b = T_\gamma^{\#}|\det(P_\gamma - I)|^{-1/2}$, $k = 0$. Consequently for $\rho > n/T_\gamma$, we have

$$N_\rho(r) \geq (\frac{b}{2\pi} - o(1))r, \tag{10.14}$$

where $N_\rho(r)$ is previously defined with $\mu_j = \sqrt{\lambda_j}$, and λ_j being the resonances in a an angle attached from above to the real axis. For odd dimenions ≥ 3, this result is due to [35], (even if not stated there in the same generality).

The next application is also due to [35] in the case of odd dimensions ≥ 3. Let $S(R) = \{x \in \mathbf{R}^{n+1}; |x| = R\}$ be the sphere of radius $R \geq 1$, equipped with the induced metric $g_{S(R)}$. We will assume that R is sufficiently large. Consider

$$X_R = (S(R) \setminus B_{S(R)}(x_0, 1)) \cup (S^{n-1} \times]0,1[) \cup (\mathbf{R}^n \setminus B(0,1)). \quad (10.15)$$

Here x_0 is some fixed point on $S(R)$ and $B_{S(R)}(x_0, 1))$ denotes the open ball in $S(R)$ of center x_0 and radius 1. We give X_R the structure of a smooth manifold by introducing a parametrization of X_R near $S^{n-1} \times]0,1[$ of the form (t, ω), $-1 < t < 2$, $\omega \in S^{n-1}$ with the following rules of assigning a corresponding point in X_R: First we identify the tangent space of $S(R)$ at x_0 with $\mathbf{R}^n$ in a way to get a linear isometry between the two spaces. Then by using geodesic coordinates, we can identify points in $B_{S(R)}(x_0, 2) \setminus \{x_0\}$ with corresponding points (r, ω) with $\omega \in S^{n-1}$, $0 < r < 2$. If $1 \leq t < 2$, then we let (t, ω) correspond to a point in $B_{S(R)}(x_0, 2) \setminus B_{S(R)}(x_0, 1)$ in the way just described. If $0 < t < 1$, then (t, ω) designs a point in $]0,1[\times S^{n-1}$ in the obvious way. If $-1 < t \leq 0$, then the corresponding point should be $(1-t)\omega \in \mathbf{R}^n$.

Using this parametrization, we can easily construct a metric g_R on X_R which coincides with that of $S(R)$ in $S(R) \setminus B_{S(R)}(x_0, 3/2)$ and with that of, $\mathbf{R}^n$ in $\mathbf{R}^n \setminus B(0, 3/2)$ and which has a perfectly uniform behaviour in the "coordinates" (t, ω), when $R \to \infty$.

Let $-P_1$ be the corresponding Laplace operator Δ_{g_R} on X_R. As before we choose $P_0 = -\Delta_{\mathbf{R}^n}$.

The geodesics on $S(R)$ are all periodic of period $2\pi R$ and most of these geodesics are also geodesics on X_R, we therefore expect $u(t)$ to have strong singularities at the points $2\pi Rk$, $k \in \mathbf{Z}$. In [35] it was indeed proved that with $\phi_{d,\gamma}$ as in the proof of Theorem 10.1, we have

$$|\widehat{\phi_{2\pi Rk,\epsilon} u}|(\lambda) \geq (2\pi R^n + \mathcal{O}(R))\lambda^{n-1} + \mathcal{O}_{R,k,\epsilon}(\lambda^{n-2}), \quad (10.16)$$

for $\epsilon > 0$ small enough. We conclude that for every $\rho > 0$, we have $N_\rho(r) \geq \frac{1}{C(\rho)} r^n$, for $r \geq r(\rho)$.

This estimate was obtained in [35] in the case of odd dimensions and is new in the case of even dimensions.

In the 3-dimensional case, Fahry and Tsanov [6] have recently obtained the same lower bounds in much thinner neighborhoods of the real axis. I was unable to see how they obtained from [35] some crucial uniformity w.r.t. k, in their estimate (3.2).

It has not been possible to include here any discussion of the Lax-Phillips conjecture (see [19], [26]) or of the consequences that can been drawn from the singular behaviour of $u(t)$ in the Poisson formula when $t \to$

0 (see [36], [40], [22], [29]). One might expect that many questions around the Lax-Phillips conjecture can now be studied also in even dimensions and perhaps in long-range situations.

11. Upper bounds near the real axis

In this setion, we review some upper bounds on the density of resonances near the real axis.

We start with a result from [30]. Let

$$P = -h^2\Delta + V(x) \tag{11.1}$$

be a semi-classical Schrödinger operator on $\mathbf{R}^n$, $n \geq 2$, where V is a real-valued analytic potential with a holomorphic extenstion $\widetilde{V}$ to a set of the form $\{x \in \mathbf{C}^n; |\mathrm{Im}\, x| \leq \langle \mathrm{Re}\, x\rangle/C\}$ with $\widetilde{V(x)} \to 0$, $x \to \infty$. Let $E_0 > 0$. Then by the method of complex scaling, we can define the resonances λ_j of P in a h-independent neighborhood of E_0, and we have $\mathrm{Im}\, \lambda_j \leq 0$.

Let $p(x,\xi) = \xi^2 + V(x)$ be the semi-classical symbol of P, and let $H_p = 2\xi \cdot \frac{\partial}{\partial x} - \nabla V(x) \cdot \frac{\partial}{\partial \xi}$ be the corresponding Hamilton field. Following [8], we introduce $\Gamma_\pm = \{(x,\xi) \in p^{-1}([E_0 - \epsilon_0, E_0 + \epsilon_0]); |\exp(tH_p)(x,\xi)| \not\to \infty, t \to \mp\infty\}$, $K = \Gamma_+ \cap \Gamma_-$. Here $\epsilon_0 > 0$ is small. Then $\Gamma_\pm$ are closed and K is compact. A basic result implicit in [14] says that if $K = \emptyset$, then there are no resonances in some h-independent neighborhood of $[E_0 - \epsilon_0, E_0 + \epsilon_0]$, when h is small enough.

We assume that H_p generates a hyperbolic dynamical system near K in the following sense: Define $\widehat{K}$, $\widehat{\Gamma_\pm}$ as above with ϵ_0 replaced by $2\epsilon_0$. Assume:

In a neighborhood Ω_{ρ_0} of every point $\rho_0 \in K$, we can represent $\widehat{\Gamma}_\pm$ as a union of closed disjoint C^1 manifolds of dimension $n+1$ such that if $\rho \in \Omega_{\rho_0} \cap \widehat{\Gamma}_+$ and if $E_\rho^+ = T_\rho(\widehat{\Gamma}_{+,\rho})$ (tangent space of $\widehat{\Gamma}_{+,\rho}$ at ρ), where $\widehat{\Gamma}_{+,\rho}$ is the corresponding leaf, then E_ρ^+ depends continuously on $\rho \in \Omega_{\rho_0} \cap \Gamma_+$ and contains $H_p(\rho)$. Same assumption with "+" replaced by "−". (11.2)

$$E_\rho^+ \text{ and } E_\rho^- \text{ intersect transversally for every } \rho \in K. \tag{11.3}$$

We also assume that there exists a constant $C > 0$, such that

$$\|d(\exp tH_p)(v)\| \leq Ce^{-t/C}\|v\|,\ v \in T_\rho(\mathbf{R}^{2n})/E_\rho^+,\ \rho \in K,\ t \geq 0, \tag{11.4}$$

$$\|d(\exp(-tH_p))(v)\| \leq Ce^{-t/C}\|v\|,\ v \in T_\rho(\mathbf{R}^{2n})/E_\rho^-,\ \rho \in K,\ t \geq 0. \tag{11.5}$$

Here $d(\exp tH_p)$ is considered as a map $T(\mathbf{R}^{2n})/E_\rho^\pm \to T(\mathbf{R}^{2n})/E^\pm_{\exp tH_p(\rho)}$, and we equip the various spaces with their natural (induced) Euclidean norms.

We say that $d \geq 0$ is a Minkowski codimension of $\widehat{K}$ if

$$\limsup_{\epsilon\to 0} \epsilon^{-d}\mathrm{Vol}\,\{(x,\xi)\in \mathbf{R}^{2n};\ \mathrm{dist}\,((x,\xi),\widehat{K}) \leq \epsilon\} < +\infty. \tag{11.6}$$

We then have

Theorem 11.1 *Under the above assumptions, let d be a Minkowski codimension of $\widehat{K}$. Then there is a constant $C_0 > 0$ such that for $0 < h \leq 1/C_0$, $C_0 h \leq \delta \leq 1/C_0$, the number of resonances in the rectangle $]-\epsilon_0/2, \epsilon_0/2[-i[0,\delta[$ is $\leq C_0\delta^d h^{-n}$.*

The proof of this result is based on the theory in [14], with additional work in finding and using escape functions of limited regularity. The use of the Weyl-inequality without any considerations of determinants is also introduced here, and is also used in obtaining the other results that we review below. Unfortunately it would lead us too far to outline the proof of Theorem 11.1, but we wanted to recall the result, because it goes quite far in linking resonances to properties of dynamical systems, the analogous results should be obtained in other settings, and also, the trace formula in Theorem 2.2, is perhaps an encouragement to pursue the efforts towards even finer results, involving also lower bounds.

We next review an upper bound from [37] in a general abstract setting. Let $n \geq 2$ and let $\mathcal{H}$ be a complex separable Hilbert space with an orthogonal decomposition,

$$\mathcal{H} = \mathcal{H}_K \oplus L^2(\mathbf{R}^n \setminus K),$$

where $K \subset \mathbf{R}^n$ is a bounded convex set. As in section 2, we use the notion of restrictions or characteristic functions to denote the corresponding projection operators. Let $P : \mathcal{H} \to \mathcal{H}$ be an unbounded self-adjoint operator with $\mathcal{D} \subset \mathcal{H}$. Assume that

$$\mathcal{D}_{|\mathbf{R}^n\setminus K} \subset H^2(\mathbf{R}^n \setminus K),$$

and conversely that if $u \in H^2(\mathbf{R}^2\setminus K)$ vanishes near K, then $u \in \mathcal{D}$. Assume

$$(Pu)_{|\mathbf{R}^n\setminus K} = -\Delta(u_{|\mathbf{R}^n\setminus K}),\ u \in \mathcal{D},$$

$$1_K(P+i)^{-1} \text{ is compact.}$$

Let $K_\epsilon = \{x \in \mathbf{R}^n;\ \mathrm{dist}\,(x,K) \leq \epsilon\}$ for small $\epsilon > 0$. Put $\mathcal{H}_{K_\epsilon} = \mathcal{H}_K \oplus L^2(K_\epsilon \setminus K)$. Define $P_\epsilon^\#$ on this space with domain

$$\mathcal{D}_\epsilon^\# = \{u \in \mathcal{H}_{K_\epsilon};\ \chi u \in \mathcal{D},\ (1-\chi)u \in H^2(K_\epsilon) \cap H_0^1(\mathrm{int}\,(K_\epsilon))\},$$

where $\chi \in C_0^\infty(\mathrm{int}\,(K_\epsilon))$, $\chi = 1$ near K. For $u \in \mathcal{D}_\epsilon^\#$, we put

$$P_\epsilon^\# u = P(\chi u) + (-\Delta)((1-\chi)u) \in \mathcal{H}_{K_\epsilon}^\#.$$

Then $P_\epsilon^\#$ is self-adjoint with discrete spectrum. Introduce the counting function for the positive eigenvalues:

$$\Phi_\epsilon(r) = \#(\sigma(P_\epsilon^\#) \cap]0, r^2[).$$

Assume that in the limit $r \to +\infty$:

$$\Phi_\epsilon(r) = (1 + o_\epsilon(1))\Phi_\epsilon^\#(r),\ \Phi_\epsilon^\#(r) = \int_0^r \phi_\epsilon(\zeta)d\zeta,$$

where ϕ_ϵ satisfies:

$$1/\tilde{C}(C) \leq \phi_\epsilon(r_1)/\phi_\epsilon(r_2) \leq \tilde{C}(C),\ \text{for } 1/C \leq r_1/r_2 \leq C,\ r_1, r_2 \geq C(\epsilon),$$

with C, $\tilde{C}(C)$, $C(\epsilon)$ positive constants.

Let $\mu_j = \sqrt{\lambda_j}$, where λ_j are the resonances of P in a small sector attached to the positive half-axis from below, as in section 5 and define

$$N_\theta(r) = \#\{\mu_j; 1 \leq |\mu_j| \leq r, -\theta \leq \arg \mu_j \leq 0\}.$$

The main result in [37] is:

Theorem 11.2 *Under the assumptions above, we have for small $\theta > 0$:*

$$N_\theta(r) \leq (1 + C\epsilon(\theta))\Phi_{\epsilon(\theta)}(r) + C\epsilon(\theta)r^n,\ \textit{for } r \geq r(\theta),$$

where $\epsilon(\theta) = \theta^{2/7}$ in general, and $\epsilon(\theta) = \theta^{2/5}$, when K is strictly convex with smooth boundary.

Using the Weyl asymptotics for the eigenvalues of second order operators in bounded domains, we get the following consequence ([37]):

Theorem 11.3 *Let $\mathcal{O} \subset\subset \mathbf{R}^n$ be open with smooth boundary such that $\mathbf{R}^n \setminus \overline{\mathcal{O}}$ is connected. Let P be an elliptic second order operator with smooth coefficients on $\mathbf{R}^n \setminus \mathcal{O}$, equal to $-\Delta$ near infinity, and equipped with Dirichlet boundary conditions. Let $p(x,\xi) \geq 0$ be the principal symbol. If "ch" denotes "convex hull of" and $\Omega = \mathrm{ch}\,(\mathrm{supp}\,(P+\Delta) \cup \mathcal{O}) \setminus \mathcal{O}$, where $\mathrm{supp}\,(P+\Delta)$ denotes the closure of the set of points in $\mathbf{R}^n \setminus \mathcal{O}$, where $P \neq -\Delta$, then*

$$N_\theta(r) \leq \frac{(1 + \mathcal{O}(\theta^{2/7}))r^n}{(2\pi)^n} \int_{\Omega \times \mathbf{R}^n \cap p^{-1}([0,1])} dx d\xi + \mathcal{O}(\theta^{2/7})r^n,\ r > r(\theta).$$

We shall explain some ideas in the proof of Theorem 11.2 and concentrate on the case of a general convex and compact K. One uses complex scaling which is adapted to K and constructs m.t.r. submanifolds of $\mathbf{C}^n$ of the form

$$\Gamma_\epsilon = \{z = x + if_\epsilon'(x);\ x \in \mathbf{R}^n\},$$

where $\epsilon \in]0, \epsilon_0]$, $\epsilon_0 > 0$, so that the following holds:
i) $K \subset \Gamma_\epsilon$,
ii) $x + if_\epsilon'(x) = (1 + i\epsilon)x$ for $|x| \geq C$,
iii) $\widetilde{P}_\epsilon = -\Delta_{|\Gamma_\epsilon}$ is uniformly elliptic also w.r.t. ϵ.
iv) For $x \in K$: $\widetilde{p}_\epsilon(x,\xi) = \xi^2$. Here $\widetilde{p}_\epsilon$ denotes the principal symbol of $\widetilde{P}_\epsilon$.
v) For $0 \leq \operatorname{dist}(x, K) \leq C_2\epsilon$, we have $-\theta_0 \leq \arg \widetilde{p}_\epsilon(x,\xi) \leq 0$, where $\tan\theta_0 = 2\sqrt{2}$.
vi) For $C_2\epsilon \leq \operatorname{dist}(x, K) \leq C$, we have $-\theta_0 \leq \arg \widetilde{p}_\epsilon(x,\xi) \leq -\epsilon^2/C$.
vii) For $\operatorname{dist}(x, K) \geq C$, we have $\arg \widetilde{p}_\epsilon(x,\xi) \leq -\theta_1$, $\tan\theta_1 = 2\epsilon/(1-\epsilon^2)$.

In order to obtain this, one uses geometric considerations, to find a function $\phi = \phi_\epsilon \in C^\infty(\mathbf{R}^n;\mathbf{R})$, such that

$$\phi_{|K} \leq 0, \tag{11.7}$$

$$\phi(x) \geq \frac{1}{C}\operatorname{dist}(x, K) - C\epsilon,\ \ x \in \mathbf{R}^n \setminus K, \tag{11.8}$$

$$\frac{\epsilon}{C} \leq \phi''(x) \leq C\epsilon, \tag{11.9}$$

$$\phi(x) = x^2/2 - a,\ \ |x| \geq C, \tag{11.10}$$

$$|\phi'(x)| \leq C, |x| \leq C. \tag{11.11}$$

One then takes

$$f(x) = C_2\epsilon g(\phi(x)/C_1\epsilon),$$

with $C_1 >> C_2 >> C$, where $g \in C^\infty(\mathbf{R})$ is a convex function with $g(t) = 0$, for $t \leq 0$, $g(t) = t -$ Const., for $t \geq 1$. We do not go into the details of this and simply give the calculation of $\widetilde{p}_\epsilon$: With $z = x + if'(x)$, $f = f_\epsilon$, we get the relation between tangent vectors, $\delta z = (1 + if''(x))\delta x$, and the corresponding relation for cotangent vectors, $\zeta = (1 + if''(x))^{-1}\xi$. Hence $\widetilde{p}_\epsilon(x,\xi) = ((1+if''(x))^{-1}\xi)^2$. Using that $(1+if''(x))^{-1} = (1 - if''(x))(1 + f''(x)^2)^{-1}$, we get

$$\widetilde{p}_\epsilon(x,\xi) = ((1 - if''(x))\widetilde{\xi})^2 = \langle (1 - f''(x)^2)\widetilde{\xi}, \widetilde{\xi}\rangle - 2i\langle f''(x)\widetilde{\xi}, \widetilde{\xi}\rangle,$$

with $\widetilde{\xi} = (1 + f''(x)^2)^{-1}\xi$. The construction of f also gives $\|f''\| \leq 1/\sqrt{2}$, so $|\xi| \sim |\widetilde{\xi}|$.

Let P_ϵ be the realization of P on Γ_ϵ, defined as P_θ is section 5. The resonances of P in $\{z \in \mathbf{C}; |z| \geq 1, 0 \leq -\arg z \leq \epsilon/C\}$ can then be identified with the square roots of eigenvalues of P_ϵ in the same set.

Introduce the auxiliary function

$$f(z) = z + \sqrt{z^2 - 1},$$

defined for all complex z which do not belong to the vertical half-lines $\pm 1 + i[0, +\infty[$, and with the branch of the square root chosen so that $f(0) = -i$. Then $f(\pm 1) = \pm 1$ and f is bijective: $\{z \in \mathbf{C}; \operatorname{Re} z < 0\} \to \{z \in \mathbf{C}; \operatorname{Re} z < 0, |z| > 1\}$. The inverse is given by $f^{-1}(w) = \frac{1}{2}(w + w^{-1})$.

For $g(\epsilon) > 0$ small, put $F_\epsilon(z) = F(z - ig(\epsilon))$, $F(z) = f(2z - 3)$. Using some pseudodiffererential and functional calculus from [34], we can then define $F_\epsilon(h^2 P_\epsilon)$, and verify that the eigenvalues of this operator in $1 \leq |w| \leq 1 + \delta$, $\operatorname{Im} w \leq 0$, for $\delta \leq \epsilon/C$ are precisely the images in this set under F_ϵ of the resonances of $h^2 P_\epsilon$. One can also show:

For $0 < h < h(\epsilon) > 0$, we have $B_\epsilon \geq 1$, where $B_\epsilon = F_\epsilon(P_\epsilon)^* F_\epsilon(P_\epsilon)$.

If $\mu_1 \leq \mu_2 \leq ..$ are the eigenvalues of $\sqrt{B_\epsilon}$ and $\delta \leq \epsilon^2/C$, and $M_{\delta,\epsilon}(h) = \#\{\mu_j; \mu_j \leq 1+\delta\}$, then $M_{\epsilon,\delta} \leq \Phi_\epsilon(h^{-1}(2 + \mathcal{O}(\delta^2))) - \Phi_\epsilon(h^{-1}(1 - \mathcal{O}(\delta^2))) + \mathcal{O}(\epsilon)h^{-n}$.

If $\widetilde{\lambda}_1, \widetilde{\lambda}_2, ..,$ are the eigenvalues of $F_\epsilon(h^2 P_\epsilon)$, with $|\widetilde{\lambda}_1| \leq |\widetilde{\lambda}_2| \leq .. \leq 1 + \delta$, then we have the Weyl inequalities,

$$\mu_1 \cdot .. \cdot \mu_k \leq |\widetilde{\lambda}_1| \cdot .. \cdot |\widetilde{\lambda}_k|.$$

Let $N_{\epsilon,\delta}(h)$ be the the number of $\widetilde{\lambda}_j$ with $|\widetilde{\lambda}_j| \leq 1 + \delta$. For $\delta_1 < \delta_2/2$, we then get with $M = M_{\epsilon,\delta_2}$, $N = N_{\epsilon,\delta_1}$, if $N \geq M$:

$$1^M (1 + \delta_2)^{N-M} \leq (1 + \delta_1)^N,$$

so

$$N \leq M \frac{\log(1 + \delta_2)}{\log \frac{1+\delta_2}{1+\delta_1}} = (1 + \mathcal{O}(\frac{\delta_1}{\delta_2}))M.$$

The remainder of the proof is then book-keeping. We choose $\delta_1/\delta_2 << 1$. Applying F_ϵ^{-1}, we get a bound on the number of resonances inside an ellipse with focal points $1 + ig(\epsilon), 2 + ig(\epsilon)$ of diameter $1 + \delta_1 + 1/(1 + \delta_1) = 2 + \mathcal{O}(\delta_1^2)$ and width $1 + \delta_1 - 1/(1 + \delta_1) = 2\delta_1 + \mathcal{O}(\delta_1^2)$. The required smallness of h depends on ϵ and on the choice of $g(\epsilon)$ but not on the other constants in the estimates above. To finish the proof one finally takes a geometric progression of h's and tries to cover the required sector as economically possible with dilated ellipses.

The result in Theorem 11.3 is far from perfect. It is quite possible that the method of scaling can be further improved even in the general convex case, so that we can get smaller powers of ϵ in the errors. Also, if we restrict the attention to the case $\mathcal{O} = \emptyset$ (in order to fix the ideas) it is quite possible (and indicated by preliminary results by M.Zerzeri [42]) that we can replace $\mathrm{vol}\,\mathrm{ch}\,\mathrm{supp}(P+\Delta)$ by the generally smaller phase space volume of the union of certain trapped trajectories. Further one may sometimes ask for estimates in certain parabolic neighboroods of the real axis (in analogy with those of Theorem 11.1.

In [38] we looked at the exterior Dirichlet problem in $\mathbf{R}^n \setminus \mathcal{O}$ when $\mathcal{O}$ is strictly convex with smooth boundary and showed, by using scaling up to the boundary, that $N_\theta(r) = \mathcal{O}(\theta^{3/2})r^n$, $r \geq r(\theta)$. The example of the ball, shows that the exponent 3/2 is sharp. Later Hargé-Lebeau [12] obtained results about the Gevrey regularity in the time direction for the wave-equation with an obstacle as above. They used the same type of complex scaling up to the boundary as in [38] and made the additional and important observation that the angle of scaling $\pi/3$ is very convenient. As a consequence they showed that there are only finitely many resonances in an inverse cubic neighborhood of the positive real axis, a result wich was previously known in dimensions 2 and 3 ([2], [7]) and in general dimension for analytic boundaries ([3]). The work [12] promted us to improve the estimates on the resonance density for C^∞ boundaries in [39] and in [32],[43] we continued with the analytic case. In the following we shall mainly concentrate on [32], since the estimates in that case depend on dynamical properties, but the proof relies a lot on [39], and the latter work contains some fine estimates that (so far) were not possible to carry over to the case of analytic boundaries.

To be more precise, we know from the works [2], [7], [12], [3], that if $\mathcal{O} \subset\subset$ is open strictly convex with C^∞ boundary, then there exists a constant $C_0 > 0$ such that for every $\epsilon > 0$, there are at most finitely many resonances ($\mu_j = \sqrt{\lambda_j}$, with λ_j as defined in section 5) in

$$\operatorname{Re} z \geq 1,\ \operatorname{Im} \geq -(C_0 - \epsilon)(\operatorname{Re} z)^{1/3}. \tag{11.12}$$

An explicit value of C_0 can apparently be obtained from the works cited above and was also obtained in [39], [32]. In the general C^∞-case, we can take $C_0 = C_\infty$, where

$$C_\infty = 2^{-1/3} \cos \frac{\pi}{6} \zeta_1 \inf_{\nu \in S(\partial\Omega)} Q(\nu)^{2/3} \tag{11.13}$$

In the case of analytic boundary $\partial\mathcal{O}$, we can take $C_0 = C_a$, where

$$C_a = 2^{-1/3} \cos \frac{\pi}{6} \zeta_1 \sup_{T>0} \inf_{\gamma\,\mathrm{boundary geodesic}} \frac{1}{T} \int_0^T Q(\gamma'(t))^{2/3} dt. \tag{11.14}$$

Here $S(\partial\mathcal{O})$ denotes the tangent sphere bundle, i.e. the bundle of normalized tangent vectors, (for the induced Riemannian metric on the boundary). $Q(\nu)$ denotes the curvature of the boundary in the direction ν, or in other words the second fundamental form defined by $Q(\nu) = \langle \phi''(x)\nu, \nu\rangle$, $\nu \in S_x(\partial\mathcal{O})$, where ϕ is a convex smooth function such that $\phi = 0$, $\|\nabla\phi\| = 1$ on $\partial\mathcal{O}$. In (11.14) we only consider boundary geodesics γ such that the derivative γ' is normalized.

In order to state the result, we shall work on $S(\partial\mathcal{O})$. Put

$$\zeta_1(\nu) = (2Q(\nu))^{2/3}\zeta_1, \tag{11.15}$$

where $0 > -\zeta_1 > -\zeta_2 > ..$ are the zeros of the Airy function, which we define up to a non-vanishing factor to be the solution of $(D_t^2 + t)\mathrm{Ai}(t) = 0$ for $t \in \mathbf{R}$, which is rapidly decaying when $t \to \infty$. Let $\Phi_t : S(\partial\mathcal{O}) \to S(\partial\mathcal{O})$ be the geodesic flow, and put

$$\zeta_1^T(\nu) = \frac{1}{T}\int_0^T \zeta_1(\Phi_t(\nu))dt, \tag{11.16}$$

$$\zeta_{1,\min}^T = \inf \zeta_1^T. \tag{11.17}$$

We also introduce the almost everywhere limit whose existence is assured by the Birkhoff ergodic theorem:

$$\zeta_1^\infty = \lim_{T\to\infty} \zeta_1^T \tag{11.18}$$

General arguments give:

$$\sup_T \zeta_{1,\min}^T = \lim_{T\to\infty} \zeta_{1,\min}^T \le \operatorname{ess\,inf} \zeta_1^\infty. \tag{11.19}$$

Put

$$W_\infty(\mu) = \int_{S\partial\mathcal{O}} (\mu - \cos\frac{\pi}{6}\zeta_1^\infty)_+^{1/2} dS, \tag{11.20}$$

where dS is the natural Euclidean volume element on $S(\partial\mathcal{O})$. The main result in [32] is:

Theorem 11.4 *There exists a constant $C > 0$ depending on the obstacle $\mathcal{O}$, such that if $k > 0$, $\mu \le \cos\frac{\pi}{6}\lim_{T\to\infty}\zeta_{1,\min}^T + 1/C$ and $N(k,\mu)$ denotes the number of resonances ($\mu_j = \sqrt{\lambda_j}$) situated in the closed lower half plane, above the parabola $\operatorname{Im} z = a(\operatorname{Re} z - k)^2 - \mu k^{1/3}$, which crosses the real axis at the two points $k \pm \sqrt{\frac{\mu}{2} r_0} k^{2/3}$, with $r_0 > 0$ some fixed constant > 0, is*

$$\le \frac{\sqrt{2r_0}}{(2\pi)^{n-1}} k^{n-1-\frac{1}{3}} (W_\infty(\cos\frac{\pi}{6}\lim_{T\to\infty}\zeta_{1,\min}^T + 3(\mu - \cos\frac{\pi}{6}\lim_{T\to\infty}\zeta_{1,\min}^T)) + o(1)), \tag{11.21}$$

when $k \to +\infty$.

As one can see from the the proof, the W_∞-term is a disguised phase space volume. If $\mu \leq \cos\frac{\pi}{6}\lim_{T\to\infty}\zeta_{1,\min}^T$, then the leading term in (11.21) is zero, in agreement with the pole-free region result. What may be more remarkable is that if $\lim_{T\to\infty}\zeta_{1,\min}^T < \operatorname{ess\,inf}\zeta_{1,\min}$ then we have the same conclusion if $0 \leq \mu - \cos\frac{\pi}{6}\lim\zeta_{1,\min}^T$ is small. (Here we run into delicate questions about the boundary dynamics, and we have no example showing that this interesting phenomenon may occur.) Zworski [43] has obtained more precise results for surfaces of revolution. When the boundary is only C^∞, the theorem remains valid provided that we refrain from taking time averages, or in other words that we have to replace $\lim_{T\to\infty}\zeta_{1,\min}^T$, by $\inf\zeta_1(\nu)$ and ζ_1^∞ by ζ_1. See [39] and appendix b (jointly with Zworski) in [32].

One would get a much more natural estimate if one were able to replace the factor 3 in the estimate by 1. The loss appears in the application of the Weyl inequalities, and we have not seen any trick similar to the one in the proof of Theorem 11.2, which would help us.

In order to give some ideas about the proof, we start by discussing the corresponding result for C^∞ boundaries, where there can be no averaging involved. We shall do scaling up to the boundary, and it turns out that the interesting things happen near the boundary, so we will only discuss that region. Near the boundary, we choose geodesic coordinates $(x', x_n) \in \partial\mathcal{O}\times[0,+\infty[$, so that x_n becomes the distance from the point described by x to the boundary, and x' is the corresponding boundary point. In these coordinates, we get the well known description of $-h^2\Delta$:

$$\begin{aligned}-h^2\Delta = (hD_{x_n})^2 - 2x_nQ(x', hD_{x'}) + R(x', hD_{x'})\\ +\mathcal{O}(x_n^2(hD_{x'})^2) + \mathcal{O}(h^2D_x) + \mathcal{O}(h^2).\end{aligned} \tag{11.22}$$

Here we take $h = k^{-1}$ and $Q = h^2Q(x', D_{x'})$, $R = h^2R(x', D_{x'})$ are positive elliptic operators. The principal symbol of Q can be identified with the second fundamental form and to leading order, $-R = \Delta_{\partial\mathcal{O}}$, the Laplace-Beltrami operator on $\partial\mathcal{O}$. We now use complex scaling which near the boundary and in geodesic coordinates takes the form: $(x', x_n) \mapsto (x', e^{i\pi/3}x_n)$. Let Γ be the image contour. Then we get the scaled operator,

$$P_\Gamma = e^{-2\pi i/3}((hD_{x_n})^2+2x_nQ(x', hD_{x'}))+R(x', hD_{x'})+\mathcal{O}(..)+\ldots \tag{11.23}$$

This is a degenerate elliptic operator, so we have a degenerate elliptic boundary value problem, and following the general philosophy for such problems, we view P_Γ as a vector valued h-differential operator in the tangential variables with operator valued symbol:

$$\begin{aligned}P_\Gamma(x', \xi') &= P_\Gamma(x, \xi', hD_{x_n}))\\ &= e^{-2\pi i/3}((hD_{x_n})^2 + 2x_nQ(x', \xi')) + R(x', \xi') + \ldots\end{aligned} \tag{11.24}$$

The eigenvalues for the Dirichlet problem on the positive half-axis for the operator $(hD_{x_n})^2 + 2x_nQ(x', \xi')$ are of the form,

$$h^{2/3}\zeta_j(x', \xi') = h^{2/3}(2Q(x', \xi'))^{2/3}\zeta_j, \tag{11.25}$$

so the eigenvalues of $P_\Gamma(x', \xi')$ become, if we ignore the $\mathcal{O}$-terms: $R(x', \xi') + e^{-2\pi i/3}h^{2/3}\zeta_j$.

From this we can conclude with some work, that if $\omega_0 = \Re\omega_0 + ir_0$, belongs to the open first quadrant, then $P_\Gamma(x, hD_x; h)$ has no eigenvalues in the disc of center ω_0 and radius

$$r_0 + \cos\frac{\pi}{6}h^{2/3} \inf_{R(x',\xi')=\mathrm{Re}\,\omega_0} \zeta_1(x', \xi') - \mathcal{O}(h),$$

which is essentially the result on absence of resonances near the real axis in the case of C^∞ boundary.

Also with quite a lot of work, it is possible to estimate the accumulation of small eigenvalues of $\sqrt{(P_\Gamma - \omega_0)^*(P_\Gamma - \omega_0)}$ in terms of $W(\mu) = \int_{S(\partial\mathcal{O})}(\mu - \cos\frac{\pi}{6}\zeta_1)_+^{1/2}dS$, and combining this with the Weyl inequalities in a rather precise way, we can get the C^∞ analogue of Theorem 11.4 essentially due to [39].

In the case of analytic boundaries, it is possible to use exponential weights microlocally in the bounary variables, and global FBI-transforms provide a convenient frame-work. Let X be a compact analytic Riemannian manifold. Let $\phi(\alpha, y)$ be an analytic function on $\{(\alpha, y) \in T^*X \times X; \mathrm{dist}\,(\alpha_x, y) \leq 1/C\}$, such that:

(A) ϕ is holomorphic and $= \mathcal{O}(|\langle\alpha_\xi\rangle|)$ on $\{(\alpha, y); |\mathrm{Im}\,\alpha_x|, |\mathrm{Im}\,y| \leq 1/C, |\mathrm{Im}\,\alpha_\xi| \leq \frac{1}{C}|\langle\alpha_\xi\rangle|\}$.

(B) $\phi(\alpha, \alpha_x) = 0$, $(\partial_\eta\phi)(\alpha, \alpha_x) = -\alpha_\xi$, $\mathrm{Im}\,(\partial_y^2\phi)(\alpha, \alpha_x) \sim |\langle\alpha_\xi\rangle| \cdot I$. Here we expressed the conditions in terms of canonical coordinates induced by some system of local coordinate charts, and used the notation: $\alpha = (\alpha_x, \alpha_\xi)$. By Taylor expansion, we get $\phi(\alpha, y) = \alpha_\xi \cdot (\alpha_x - y) + \mathcal{O}(1)\langle\alpha_\xi\rangle|\alpha_x - y|^2$.

If a is a suitable elliptic analytic symbol and $\chi \in C^\infty(X \times X)$ is equal to one near the diagonal and has its support in a small neighborhood of the diagonal, then we introduce the global FBI-transform:

$$Tu(\alpha; h) = \int e^{\frac{i}{h}\phi(\alpha,y)}a(\alpha, y; h)\chi(\alpha_x, y)u(y)dy, \tag{11.26}$$

taking distributions on X into holomorphic functions in α defined on some neighborhood of T^*X in a suitable complexification $\widetilde{T^*X}$ of this manifold. It is possible to construct an approximate left inverse of T which works up to certain exponentially small errors, but we will not go into the details of that essentially well-known fact here, and simply notice that the theory is very close to the one developed on $\mathbf{R}^n$ in [14].

Let $\Lambda \subset \widetilde{T^*X}$ be an I-Lagrangian manifold, i.e. a manifold which is Lagrangian for the real symplectic form which is equal to the imaginary part; $\operatorname{Im}\sigma$, of the complexification σ of the standard symplectic form on T^*X. We assume that Λ is close to T^*X in the C^∞-sense and coincides with T^*X outside a bounded set (in the fiber directions). Locally we can then find a smooth function H on Λ, such that

$$dH = -\operatorname{Im}(\alpha_\xi \cdot d\alpha_x)_{|\Lambda}. \tag{11.27}$$

We now assume that

(C) (11.27) has a *global* solution $H \in C^\infty(\Lambda; \mathbf{R})$.

Notice that we can normalize the choice of H by requiring $H(\alpha)$ to be 0 for large α_ξ.

An example of such a manifold can be produced from a function $G \in C_0^\infty(\widetilde{T^*X}; \mathbf{R})$, that we can view as a weight. For $t \in \mathbf{R}$, $|t|$ small, put $\Lambda_t = \exp(tH_G^{\operatorname{Im}\sigma})(T^*X)$. Then the assumption (C) is fulfilled.

Definition. Let Λ, H be as above so that (C) holds. For $m \in \mathbf{R}$ and for $|t|$ small, we put

$$H(\Lambda; \langle\alpha_\xi\rangle^m) = \{u \in \mathcal{D}'(X); T_\Lambda u \in L^2(\Lambda; e^{-2H/h}|\langle\alpha_\xi\rangle|^{2m} d\alpha)\}, \tag{11.28}$$

equipped with the natural norm. We observe that the norm also depends on the choice of H, and actually these spaces coincide with the Sobolev spaces, because of the condition that Λ should coincide with T^*X far away. It is rather the exponentially weighted norms introduced here which are interesting.

If $P(x, hD_x; h) = \sum_{|\alpha|\leq m} a_k(x;h)(hD_x)^k$ is an h-differential operator with analytic coefficients, uniformly bounded in a complex neighborhood of X, then for $|t|$ small enough, P is $\mathcal{O}(1) : H(\Lambda_t; \langle\alpha_\xi\rangle^m) \to H(\Lambda_t; 1)$, and can be viewed as an h-pseudor with leading symbol $P_{|\Lambda} + \mathcal{O}(h\langle\alpha_\xi\rangle^m)$.

In the proof of Theorem 11.4 we apply the above theory with $X = \partial\mathcal{O}$, $t = \mathcal{O}(h^{2/3})$ depending on h. Let $G(x', \xi')$ be a smooth realvalued function on $T^*\partial\mathcal{O}$ with compact support (in ξ). We then consider the scaled operator P_Γ on $H(\Lambda_{h^{2/3}}) \otimes L^2_{x_n}$ (with suitable modifications when x_n becomes large). According to the previous remark we can then view P_Γ as a vector valued h-pseudor operator with leading symbol

$$\begin{aligned}
&\widetilde{P}(x', \xi') = \\
&= e^{-2\pi i/3}((hD_{x_n})^2 + 2x_n Q(x', \xi')) + R(x', \xi') - ih^{2/3}(H_R G)(x', \xi') + \mathcal{O}... \\
&= e^{-2\pi i/3}((hD_{x_n})^2 + 2x_n Q(x', \xi') + \tfrac{h^{2/3}}{\cos\frac{\pi}{6}} H_R G(x', \xi')) + R + \mathcal{O}...
\end{aligned} \tag{11.29}$$

and this means that the discussion we gave in the C^∞ case can be applied with $\zeta_j(x',\xi')$ replaced by

$$\widetilde{\zeta}_j(x',\xi') = \zeta_j(x',\xi') + \frac{1}{\cos\frac{\pi}{6}} H_R G(x',\xi'). \tag{11.30}$$

It is natural to try to choose G so that that the infimum of $\widetilde{\zeta}_1$ over the cosphere bundle Σ: $R(x',\xi') = 1$ becomes as large as possible, and the natural way of doing this is to average.

The vectorfield $\nu = H_R$ conserves the Liouville measure on Σ. Let k be a smooth (at least Lipschitz) functions on $\mathbf{R}$ except for a jump $+1$ at 0 and assume also that k has compact support. Put $k_T(t) = k(t/T)$. Then if $v \in L^\infty(\mathbf{R})$, the convolution $u = k_T * v$ satisfies $\frac{du}{dt} = v - \frac{1}{T}\ell_T * v$, where $\ell_T = \ell(t/T)$, and $-\ell$ is the derivative of $k_{\mathbf{R}\setminus\{0\}} \in L^\infty(\mathbf{R})$, so that $\int \ell dt = 1$.

Let q be a real C^∞-function on Σ. Put $G_T = -\int k_T(-s) q \circ \exp(s\nu) ds$. Then,

$$q + \nu(G_T) = \frac{1}{T}\int \ell_T(-s) q \circ \exp(s\nu) ds.$$

With a suitable k, we have $\ell = 1_{[-1,0]}$. Then

$$q + \nu(G_T) = \frac{1}{T}\int_0^T q \circ \exp(s\nu) ds =_{\text{def}} q^T.$$

Applying the discussion to ζ_1, we may find G, so that on Σ:

$$\widetilde{\zeta}_1 = \zeta_1^T = \frac{1}{T}\int_0^T \zeta_1 \circ \exp(sH_R) ds.$$

Applying the arguments outlined for the C^∞-case, we get the estimates in the theorem with W_∞ replaced by a function W_T, which is defined the same way, but with ζ_1^∞ replaced by ζ_1^T, and with $\lim_{T\to\infty} \zeta_{1,\min}^T$ replaced by $\zeta_{1,\min}^T$. The last step in the proof is then to use the ergodic theorem to check that we can pass to the limit $T \to \infty$.

An extension of the results concerning absence of resonances in inverese cubic neighborhoods of the reals, for stricly convex obstacles with Gevrey boundary, has recently been obtained by B. and R. Lascar.

References

1. J.Aguilar, J.M.Combes, *A class of analytic perturbations for one-body Schrödinger Hamiltonians,* Comm. Math. Phys. 22(1971), 269-279.
2. V.M.Babich, N.S.Grigoreva, *The analytic continuation of the resolvent of the exterior three dimensional problem for the Laplace operator to second sheet,* Funcional Analysis i Prilozen. 8(1)(1974), 71-74.

3. C.Bardos, G.Lebeau, J.Rauch, *Scattering frequencies and Gevrey 3 singularities,* Inv. Mat. 90(1987), 77-114.
4. C.Bardos, J.C.Guillot, J.Ralston, *La relation de Poisson pour l'équation des ondes dans un ouvert non borné,* Comm. PDE 7(1982), 905-958.
5. M.Dimassi, J.Sjöstrand, Lecture notes in preparation.
6. L.S.Fahry, V.V.Tsanov, *Scattering poles for connected sums of Euclidean space and Zoll manifolds,* Ann. I.H.P. (phys. th.) 65(2)(1996), 163-174.
7. V.B.Filippov, A.B.Zayev, *Rigorous justification of the asymptotic solutions of sliding wave type,* Journal of Soviet Math. 30(2)(1985), 2395-2406.
8. C.Gérard, J.Sjöstrand, *Semiclassical resonances generated by a closed trajectory of hyperbolic type,* Comm. Math.Phys.,108(1987), 391-421.
9. I.Gohberg, M.Krein, *Introduction to the theory of linear non-self-adjoint operators,* Amer. Math. Soc., Providence, RI, (1969).
10. V.Guillemin, R.Melrose, *The Poisson summation formula for manifolds with boundary,* Adv. Math. 32(1979), 128-148.
11. L.Guillopé, M.Zworski, *Scattering asymptotics for Riemannian surfaces,* Preprint(1996).
12. T.Hargé, G.Lebeau, *Diffraction par un convexe,* Inv. Mat. 118(1)(1994).
13. B.Helffer, A.Martinez, *Comparaison entre les diverses notions de résonances,* Helv. Phys. Acta 60(1987), 992-1003.
14. B.Helffer, J.Sjöstrand, *Résonances en limite semiclassique.* Bull. de la SMF 114(3), Mémoire 24/25(1986).
15. B.Helffer, J.Sjöstrand, *Equation de Schrödinger avec champ magnétique et équation de Harper,* Springer Lecture Notes in Physics 345(1989), 118-197.
16. L.Hörmander, *On the singularities of solutions of partial differential equations,* in "Proc. internat. conf. on functional analysis and related topics, (Tokyo, 1969)", Univ. of Tokyo Press, Tokyo (1970).
17. L.Hörmander, J.Wermer, *Uniform approximation on compact sets in* $\mathbf{C}^n$, Math. Scand. 23(1968), 5-21.
18. W.Hunziker, *Distorsion analyticity and molecular resonance curves,* Ann. Inst. Poincaré, Phys. Théor. 45(1986), 339-358.
19. M.Ikawa, *Singular perturbation of symbolic flows and poles of the zeta functions,* Osaka J. Math. 27(1990), 281-300 and 29(1992), 161-174.
20. P.Lax, R.Phillips, *Scattering theory,* Academic Press, New York(1967).
21. R.Melrose, *Scattering theory and the trace of the wave group,* J. Funct. Anal. 45(1982), 29-40.
22. R.Melrose, *Geometric scattering theory,* Cambridge University Press (1995).
23. R.Melrose, *Polynomial bounds on the distribution of poles in scattering by an obstacle,* Journées Equations aux dérivées partielles, St Jean de Monts (1984).
24. R.Melrose, J.Sjöstrand, *Singularities in boundary value problems, I and II,* Comm. Pure Appl. Math. 31(1978), 593-617, and 35(1982), 129-168.
25. L.Nedelec, Work in preparation.
26. V.Petkov, *Sur la conjecture de Lax et Phillips pour un nombre fini d'obstacles strictement convexes,* Séminaire équations aux dérivées partielles, exposé no 11 (1995-96).
27. V.Petkov, L.Stoyanov, *Geometry of reflecting rays and inverse spectral problems,* J.Wiley & Sons, (1992).
28. D.Robert, *Autour de l'approximation semi-classique,* Progress in Mathematics, 68, Birkhäuser (1987).
29. A.Sa Bareto, M.Zworski, *Existence of poles in potential scattering,* preprint (1995).
30. J.Sjöstrand, *Geometric bounds on the density of resonances for semiclassical problems,* Duke Mathematical Journal, 60(1)(1990), 1-57.
31. J.Sjöstrand, *Microlocal analysis for the periodic magnetic Schrödinger equation and related questions,* CIME-lectures, Montecatini, (Juillet 1989), Springer Lecture Notes in Math. 1495 (1991), 237-332.

32. J.Sjöstrand, *Density of resonances for strictly convex analytic obstacles,* Can. J. Math., 48(2)(1996), 397-447
33. J.Sjöstrand, G.Vodev, *Asymptotics of the number of Rayleigh resonances,* preprint, (Dec. 1995, revised Sept. 1996),
34. J.Sjöstrand, M.Zworski, *Complex scaling and the distribution of scattering poles,* Journal of the AMS, 4(4)(1991), 729-769.
35. J.Sjöstrand, M.Zworski, *Lower bounds on the number of scattering poles,* Comm. in P.D.E., 18(5-6)(1993), 847-857.
36. J.Sjöstrand, M.Zworski, *Lower bounds on the number of scattering poles, II,* J. Funct. An., 123(2)(1994), 336-367.
37. J.Sjöstrand, M.Zworski, *Distribution of scattering poles near the real axis,* Comm.P.D.E., 17(5-6)(1992), 1021-1035.
38. J.Sjöstrand, M.Zworski, *Estimates on the number of scattering poles near the real axis for strictly convex obstacles,* Ann. Inst. Fourier, 43(3)(1993), 769-790.
39. J.Sjöstrand, M.Zworski, *The complex scaling method for scattering by strictly convex obstacles,* Ark. f. Matematik, 33(1)(1995), 135-172.
40. G.Vodev, *Asymptotics on the number of scattering poles for degenerate perturbations of the Laplacian,* J. Funct. An. 138(2)(1996), 295-310.
41. G.Vodev, Sharp bounds on the number of scattering poles for perturbations of the Laplacian, Comm. Math. Phys., 146(1992), 205-216.
42. M.Zerzeri, Thesis, in preparation
43. M.Zworski, Appendix C in [32].
44. M.Zworski, *Sharp polynomial bounds on the number of scattering poles,* Duke Math. J. 59(1989), 311-323.

Index

The manufacturer's authorised representative in the EU is Springer Nature Customer Service Centre GmbH, Europaplatz 3, 69115 Heidelberg, Germany. If you have any concerns regarding our products, please contact ProductSafety@springernature.com

Printed and bound by CPI Group (UK) Ltd, Croydon, CR0 4YY
23/04/2026
02095593-0007